Springer Proceedings in Mathematics & Statistics

Volume 144

Springer Proceedings in Mathematics & Statistics

This book series features volumes composed of selected contributions from workshops and conferences in all areas of current research in mathematics and statistics, including operation research and optimization. In addition to an overall evaluation of the interest, scientific quality, and timeliness of each proposal at the hands of the publisher, individual contributions are all refereed to the high quality standards of leading journals in the field. Thus, this series provides the research community with well-edited, authoritative reports on developments in the most exciting areas of mathematical and statistical research today.

More information about this series at http://www.springer.com/series/10533

Filippo Bracci · Jisoo Byun
Hervé Gaussier · Kengo Hirachi
Kang-Tae Kim · Nikolay Shcherbina
Editors

Complex Analysis and Geometry

KSCV10, Gyeongju, Korea, August 2014

Springer

Editors
Filippo Bracci
Department of Mathematics
Università di Roma "Tor Vergata"
Rome
Italy

Jisoo Byun
Department of Mathematics Education
Kyungnam University
Changwon
Korea

Hervé Gaussier
Institut Fourier
Université Joseph Fourier Grenoble
Grenoble
France

Kengo Hirachi
Graduate School of Mathematical Sciences
The University of Tokyo
Tokyo
Japan

Kang-Tae Kim
Department of Mathematics
Pohang University of Science and Technology (POSTECH)
Pohang
Korea

Nikolay Shcherbina
Department of Mathematics
Bergische Universität Wuppertal
Wuppertal
Germany

ISSN 2194-1009 ISSN 2194-1017 (electronic)
Springer Proceedings in Mathematics & Statistics
ISBN 978-4-431-55743-2 ISBN 978-4-431-55744-9 (eBook)
DOI 10.1007/978-4-431-55744-9

Library of Congress Control Number: 2015944727

Mathematics Subject Classification (2010): 14, 30, 32

Springer Tokyo Heidelberg New York Dordrecht London

Printed on acid-free paper

Springer Japan KK is part of Springer Science+Business Media (www.springer.com)

Preface

The KSCV Symposium, the Korean conference on Several Complex Variables, started in 1997 at Pohang University of Science and Technology (POSTECH) in an effort to promote the study of complex analysis and geometry in all dimensions. Since then, the conference met semi-regularly for about 10 years and then settled as a biannual conference. The sixth conference was held in 2002 as a satellite conference to the Beijing ICM. The symposia have been successful in the sense that many leading scholars in the field have participated from all over the world, and more importantly, many new researchers in this field, especially from Korea, have been brought up along with this effort.

The KSCV10 (the 10th) Symposium was held during 7–11 July 2014, as a satellite conference to the ICM again; this time the ICM was held in Seoul, Korea. It was clearly noticed by many that not only has the research level of the Korean SCV community but also that of the conference improved so much that the contents of the lectures will be useful to mathematicians in the field of complex analysis and geometry of the world. Therefore, the organizers as well as the participants of the conference agreed to have them organized into a book form; therefore, this proceedings volume was composed.

I would like to express deep thanks to all those who contributed their articles to this volume. We, the committee, of course, wish their study to flourish greatly.

April 2015

Filippo Bracci
Jisoo Byun
Hervé Gaussier
Kengo Hirachi
Kang-Tae Kim
Nikolay Shcherbina

Contents

Fatou Flowers and Parabolic Curves

Marco Abate

Abstract In this survey we collect the main results known up to now (July 2015) regarding possible generalizations to several complex variables of the classical Leau-Fatou flower theorem about holomorphic parabolic dynamics.

Keywords Local holomorphic dynamical systems · Parabolic points · Fatou flowers · Parabolic curves · Germs tangent to the identity

1 The Original Leau-Fatou Flower Theorem

In this survey we shall present the known generalizations of the classical Leau-Fatou theorem describing the local holomorphic dynamics about a parabolic point. But let us start with a number of standard definitions.

Definition 1.1 A *local n-dimensional discrete holomorphic dynamical system* (in short, a *local dynamical system*) is a holomorphic germ f of self-map of a complex n-dimensional manifold M at a point $p \in M$ such that $f(p) = p$; we shall denote by $\mathrm{End}(M, p)$ the set of such germs.

If f, g belongs to $\mathrm{End}(M, p)$ their composition $g \circ f$ is defined as germ in $\mathrm{End}(M, p)$; in particular, we can consider the sequence $\{f^k\} \subset \mathrm{End}(M, p)$ of *iterates* of $f \in \mathrm{End}(M, p)$, inductively defined by $f^0 = \mathrm{id}_M$ and $f^k = f \circ f^{k-1}$ for $k \geq 1$. The aim of local discrete dynamics is exactly the study of the behavior of the sequence of iterates.

Remark 1.1 In practice, we shall work with representatives, that is with holomorphic maps $f : U \to M$, where $U \subseteq M$ is an open neighborhood of $p \in U$, such that $f(p) = p$. The fact we are working with germs will be reflected in the freedom we have in taking U as small as needed. We shall also mostly (but not always) take $M = \mathbb{C}^n$ and $p = O$; indeed a choice of local coordinates φ for M centered at p

M. Abate (✉)
Dipartimento di Matematica, Università di Pisa, Largo Pontecorvo 5, 56127 Pisa, Italy
e-mail: abate@dm.unipi.it

F. Bracci et al. (eds.), *Complex Analysis and Geometry*, Springer Proceedings in Mathematics & Statistics 144, DOI 10.1007/978-4-431-55744-9_1

yields an isomorphism $\varphi_*\colon \mathrm{End}(M, p) \to \mathrm{End}(\mathbb{C}^n, O)$ preserving the composition by setting $\varphi_*(f) = \varphi \circ f \circ \varphi^{-1}$.

Definition 1.2 Let $f\colon U \to M$ be a representative of a germ in $\mathrm{End}(M, p)$. The *stable set* $K_f \subseteq U$ of f is the set of points $z \in U$ such that $f^k(z)$ is defined for all $k \in \mathbb{N}$; clearly, $p \in K_f$. If $z \in K_f$, the set $\{f^k(z)\}$ is the *orbit* of z; if $z \in U \setminus K_f$ we shall say that z *escapes*. The stable set depends on the chosen representative, but its germ at p does not; so we shall freely talk about the stable set of an element of $\mathrm{End}(M, p)$. An f*-invariant* set is a subset $P \subseteq U$ such that $f(P) \subseteq P$; clearly, the stable set is f-invariant.

Definition 1.3 A local dynamical system $f \in \mathrm{End}(M, p)$ is *parabolic* (and sometimes we shall say that p is a *parabolic fixed point* of f) if df_p is diagonalizable and all its eigenvalues are roots of unity; is *tangent to the identity* if $df_p = \mathrm{id}$. We shall denote by $\mathrm{End}_1(M, p)$ the set of local dynamical systems tangent to the identity in p.

Remark 1.2 If $f \in \mathrm{End}(M, p)$ is parabolic then a suitable iterate f^q is tangent to the identity; for this reason we shall mostly concentrate on germs tangent to the identity. Furthermore, if $f \in \mathrm{End}(M, p)$ is tangent to the identity then f^{-1} is a well-defined germ in $\mathrm{End}(M, p)$ still tangent to the identity.

Definition 1.4 The *order* $\mathrm{ord}_p(f)$ of a holomorphic function $f\colon M \to \mathbb{C}$ at $p \in M$ is the order of vanishing at p, that is the degree of the first non-vanishing term in the Taylor expansion of f at p (computed in any set of local coordinates centered at p). The *order* $\mathrm{ord}_p(F)$ of a holomorphic map $F\colon M \to \mathbb{C}^n$ at $p \in M$ is the minimum order of its components.

A germ $f \in \mathrm{End}(\mathbb{C}^n, O)$ can be represented by a n-tuple of convergent power series in n variables; collecting terms of the same degree we obtain the homogeneous expansion.

Definition 1.5 A *homogeneous map* of degree $d \geq 1$ is a map $P\colon \mathbb{C}^n \to \mathbb{C}^n$ where P is a n-tuple of homogeneous polynomials of degree d in n variables. The *homogeneous expansion* of a germ tangent to the identity $f \in \mathrm{End}_1(\mathbb{C}^n, O)$, $f \not\equiv \mathrm{id}_{\mathbb{C}^n}$, is the (unique) series expansion

$$f(z) = z + P_{\nu+1}(z) + P_{\nu+2}(z) + \cdots \tag{1.1}$$

where P_k is a homogeneous map of degree k, and $P_{\nu+1} \not\equiv O$. The number $\nu \geq 1$ is the *order* (or, sometimes, *multiplicity*) $\nu(f)$ of f at O, and $P_{\nu+1}$ is the *leading term* of f. It is easy to check that the order is invariant under change of coordinates, and thus it can be defined for any germ tangent to the identity $f \in \mathrm{End}_1(M, p)$; we shall denote by $\mathrm{End}_\nu(M, p)$ the set of germs tangent to the identity with order at least ν.

In the rest of this section we shall discuss the 1-dimensional case, where the homogeneous expansion reduces to the usual Taylor expansion

$$f(z) = z + a_{\nu+1} z^{\nu+1} + O(z^{\nu+2}) \tag{1.2}$$

with $a_{\nu+1} \neq 0$.

Definition 1.6 Let $f \in \mathrm{End}_1(\mathbb{C}, 0)$ be tangent to the identity given by (1.2). A unit vector $v \in S^1$ is an *attracting* (respectively, *repelling*) *direction* for f at 0 if $a_{\nu+1} v^\nu$ is real and negative (respectively, positive). Clearly, there are ν equally spaced attracting directions, separated by ν equally spaced repelling directions.

Example 1.1 To understand this definition, let us consider the particular case $f(z) = z + az^{\nu+1}$. If $v \in S^1$ is such that $av^\nu > 0$ then for every $z \in \mathbb{R}^+ v$ we have $f(z) \in \mathbb{R}^+ v$ and $|f(z)| > |z|$; in other words, the half-line $\mathbb{R}^+$ is f-invariant and repelled from the origin. Conversely, if $v \in S^1$ is such that $av^\nu < 0$ then $\mathbb{R}^+ v$ is again f-invariant but now $|f(z)| < |z|$ if $z \in \mathbb{R}^+ v$ is small enough; so there is a segment of $\mathbb{R}^+ v$ attracted by the origin.

Remark 1.3 If $f \in \mathrm{End}_1(\mathbb{C}, 0)$ is given by (1.2) then

$$f^{-1}(z) = z - a_{\nu+1} z^{\nu+1} + O(z^{\nu+2}).$$

In particular, if $v \in S^1$ is attracting (respectively, repelling) for f then it is repelling (respectively, attracting) for f^{-1}, and conversely.

To describe the dynamics of a tangent to the identity germ two more definitions are needed.

Definition 1.7 Let $v \in S^1$ be an attracting direction for a $f \in \mathrm{End}_1(\mathbb{C}, 0)$ tangent to the identity. The *basin* centered at v is the set of points $z \in K_f \setminus \{0\}$ such that $f^k(z) \to 0$ and $f^k(z)/|f^k(z)| \to v$ (notice that, up to shrinking the domain of f, we can assume that $f(z) \neq 0$ for all $z \in K_f \setminus \{0\}$). If z belongs to the basin centered at v, we shall say that the orbit of z *tends to* 0 *tangent to* v.

A slightly more specialized (but more useful) object is the following:

Definition 1.8 Let $f \in \mathrm{End}_1(\mathbb{C}, 0)$ be tangent to the identity. An *attracting petal* with attracting *central direction* $v \in S^1$ for f is an open simply connected f-invariant set $P \subseteq K_f \setminus \{0\}$ with $0 \in \partial P$ such that a point $z \in K_f \setminus \{0\}$ belongs to the basin centered at v if and only if its orbit intersects P. In other words, the orbit of a point tends to 0 tangent to v if and only if it is eventually contained in P. A *repelling petal* (with repelling central direction) is an attracting petal for the inverse of f.

We can now state the original *Leau-Fatou flower theorem*, describing the dynamics of a one-dimensional tangent to the identity germ in a full neighborhood of the origin (see, e.g., [M] for a modern proof):

Theorem 1.1 ([Le, F1, F2, F3]) *Let $f \in \mathrm{End}_1(\mathbb{C}, 0)$ be tangent to the identity of order $\nu \geq 1$. Let $v_1^+, \ldots, v_\nu^+ \in S^1$ be the ν attracting directions of f at the origin, and $v_1^-, \ldots, v_r^- \in S^1$ the ν repelling directions. Then:*

(i) *for each attracting (repelling) direction* v_j^+ (v_j^-) *we can find an attracting (repelling) petal* P_j^+ (P_j^-) *such that the union of these* 2ν *petals together with the origin forms a neighborhood of the origin. Furthermore, the* 2ν *petals are arranged cyclically so that two petals intersect if and only if the angle between their central directions is* π/ν.

(ii) $K_f \setminus \{0\}$ *is the (disjoint) union of the basins centered at the* ν *attracting directions.*

(iii) *If* B *is a basin centered at one of the attracting directions, then there is a function* $\chi: B \to \mathbb{C}$ *such that* $\chi \circ f(z) = \chi(z) + 1$ *for all* $z \in B$. *Furthermore, if* P *is the corresponding petal constructed in part* (i), *then* $\chi|_P$ *is a biholomorphism with an open subset of the complex plane containing a right half-plane — and so* $f|_P$ *is holomorphically conjugated to the translation* $z \mapsto z + 1$.

Definition 1.9 The function $\chi: B \to \mathbb{C}$ constructed in Theorem 1.1. (iii) is a *Fatou coordinate* on the basin B.

Remark 1.4 Up to a linear change of variable, we can assume that $a_{\nu+1} = -1$ in (1.2), so that the attracting directions are the ν-th roots of unity. Given $\delta > 0$, the set

$$D_{\nu,\delta} = \{z \in \mathbb{C} \mid |z^\nu - \delta| < \delta\} \tag{1.3}$$

has exactly ν connected components (each one symmetric with respect to a different ν-th root of unity), and it turns out that when $\delta > 0$ is small enough these components can be taken as attracting petals for f—even though to cover a neighborhood of the origin one needs slightly larger petals. The components of $D_{\nu,\delta}$ are distributed as petals in a flower; this is the reason why Theorem 1.1 is called "flower theorem".

So the union of attracting and repelling petals gives a pointed neighborhood of the origin, and the dynamics of f on each petal is conjugated to a translation via a Fatou coordinate. The relationships between different Fatou coordinates is the key to Écalle-Voronin holomorphic classification of parabolic germs (see, e.g., [A4] and references therein for a concise introduction to Écalle-Voronin invariants), which is however outside of the scope of this survey. We end this section with the statement of the Leau-Fatou flower theorem for general parabolic germs:

Theorem 1.2 ([Le, F1, F2, F3]) *Let* $f \in \mathrm{End}_1(\mathbb{C}, 0)$ *be of the form* $f(z) = \lambda z + O(z^2)$, *where* $\lambda \in S^1$ *is a primitive root of the unity of order* q. *Assume that* $f^q \not\equiv \mathrm{id}$. *Then there exists* $\mu \geq 1$ *such that* f^q *has order* $q\mu$, *and* f *acts on the attracting (respectively, repelling) petals of* f^q *as a permutation composed by* μ *disjoint cycles. Finally,* $K_f = K_{f^q}$.

In the subsequent sections we shall discuss known generalizations of Theorem 1.1 to several variables.

2 Écalle-Hakim Theory

From now on we shall work in dimension $n \geq 2$. So let $f \in \mathrm{End}_1(\mathbb{C}^n, O)$ be tangent to the identity; we would like to find a multidimensional version of the petals of Theorem 1.1.

If f had a non-trivial one-dimensional f-invariant curve passing through the origin, that is an injective holomorphic map $\psi\colon \Delta \to \mathbb{C}^n$, where $\Delta \subset \mathbb{C}$ is a neighborhood of the origin, such that $\psi(0) = O$, $\psi'(0) \neq O$ and $f\big(\psi(\Delta)\big) \subseteq \psi(\Delta)$ with $f|_{\psi(\Delta)} \not\equiv \mathrm{id}$, we could apply Leau-Fatou flower theorem to $f|_{\psi(\Delta)}$ obtaining a one-dimensional Fatou flower for f inside the invariant curve. In particular, if $z^o \in \psi(\Delta)$ belongs to an attractive petal, we would have $f^k(z^o) \to O$ and $[f^k(z^o)] \to [\psi'(0)]$, where $[\cdot]\colon \mathbb{C}^n \setminus \{O\} \to \mathbb{P}^{n-1}(\mathbb{C})$ is the canonical projection. The first observation we can make is that then $[\psi'(0)]$ cannot be any direction in $\mathbb{P}^{n-1}(\mathbb{C})$. Indeed:

Proposition 2.1 ([H2]) *Let $f(z) = z + P_{\nu+1}(z) + \cdots \in \mathrm{End}_1(\mathbb{C}^n, O)$ be tangent to the identity of order $\nu \geq 1$. Assume there is $z^o \in K_f$ such that $f^k(z^o) \to O$ and $[f^k(z^o)] \to [v] \in \mathbb{P}^{n-1}(\mathbb{C})$. Then $P_{\nu+1}(v) = \lambda v$ for some $\lambda \in \mathbb{C}$.*

Definition 2.1 Let $P\colon \mathbb{C}^n \to \mathbb{C}^n$ be a homogeneous map. A direction $[v] \in \mathbb{P}^{n-1}(\mathbb{C})$ is *characteristic* for P if $P(v) = \lambda v$ for some $\lambda \in \mathbb{C}$. Furthermore, we shall say that $[v]$ is *degenerate* if $P(v) = O$, and *non-degenerate* otherwise.

Remark 2.1 From now on, given $f \in \mathrm{End}_1(\mathbb{C}^n, O)$ tangent to the identity of order $\nu \geq 1$, every notion/object/concept introduced for its leading term $P_{\nu+1}$ will be introduced also for f; for instance, a (degenerate/non-degenerate) characteristic direction for $P_{\nu+1}$ will also be a (degenerate/non-degenerate) characteristic direction for f.

Remark 2.2 If $f \in \mathrm{End}_1(\mathbb{C}^n, O)$ is given by (1.1), then $f^{-1} \in \mathrm{End}_1(\mathbb{C}^n, O)$ is given by

$$f^{-1}(z) = z - P_{\nu+1}(z) + \cdots .$$

In particular, f and f^{-1} have the same (degenerate/non-degenerate) characteristic directions.

Remark 2.3 If $\psi\colon \Delta \to \mathbb{C}^n$ is a one-dimensional curve with $\psi(0) = O$ and $\psi'(0) \neq O$ such that $f|_{\psi(\Delta)} \equiv \mathrm{id}$, it is easy to see that $[\psi'(0)]$ must be a degenerate characteristic direction for f.

So if we have an f-invariant one-dimensional curve ψ through the origin then $[\psi'(0)]$ must be a characteristic direction. However, in general the converse is false: there are non-degenerate characteristic directions which are not tangent to any f-invariant curve passing through the origin.

Example 2.1 ([H2]) Let $f \in \mathrm{End}(\mathbb{C}^2, O)$ be given by

$$f(z,w)=\left(\frac{z}{1+z},\,w+z^2\right),$$

so that f is tangent to the identity of order 1, and $P_2(z,w)=(-z^2,z^2)$. In particular, f has a degenerate characteristic direction $[0:1]$ and a non-degenerate characteristic direction $[v]=[1:-1]$. The degenerate characteristic direction is tangent to the curve $\{z=0\}$, which is pointwise fixed by f, in accord with Remark 2.3. We claim that no f-invariant curve can be tangent to $[v]$.

Assume, by contradiction, that we have an f-invariant curve $\psi\colon \Delta\to\mathbb{C}^2$ with $\psi(0)=O$ and $[\psi'(0)]=[v]$. Without loss of generality, we can assume that $\psi(\zeta)=\bigl(\zeta,u(\zeta)\bigr)$ with $u\in\operatorname{End}(\mathbb{C},0)$. Then the condition of f-invariance becomes $f_2\bigl(\zeta,u(\zeta)\bigr)=u\bigl(f_1(\zeta,u(\zeta))\bigr)$, that is

$$u(\zeta)+\zeta^2=u\left(\frac{\zeta}{1+\zeta}\right). \tag{2.1}$$

Put $g(\zeta)=\zeta/(1+\zeta)$, so that $g^k(\zeta)=\zeta/(1+k\zeta)$; in particular, $g^k(\zeta)\to 0$ for all $\zeta\in\mathbb{C}\setminus\{-\frac{1}{n}\mid n\in\mathbb{N}^*\}$. This means that by using (2.1) we can extend u to $\mathbb{C}\setminus\{-\frac{1}{n}\mid n\in\mathbb{N}^*\}$ by setting

$$u(\zeta)=u\bigl(g^k(\zeta)\bigr)-\sum_{j=0}^{k-1}[g^j(\zeta)]^2$$

where $k\in\mathbb{N}$ is chosen so that $g^k(\zeta)\in\Delta$. Analogously, (2.1) implies that for $|\zeta|$ small enough one has

$$u\bigl(g^{-1}(\zeta)\bigr)+\bigl(g^{-1}(\zeta)\bigr)^2=u(\zeta);$$

so we can use this relation to extend u to all of $\mathbb{C}$, and then to $\mathbb{P}^1(\mathbb{C})$, because $g^{-1}(\infty)=-1$. So u is a holomorphic function defined on $\mathbb{P}^1(\mathbb{C})$, that is a constant; but no constant can satisfy (2.1), contradiction.

Remark 2.4 Ribón [R] has given examples of germs having no holomorphic invariant curves at all. For instance, this is the case for germs of the form $f(z,w)=(z+w^2,w+z^2+\lambda z^5)$ for all $\lambda\in\mathbb{C}$ outside a polar Borel set.

The first important theorem we would like to quote is due to Écalle [E] and Hakim [H2], and it says that we do always have a Fatou flower tangent to a non-degenerate characteristic direction, even when there are no invariant complex curves containing the origin in their relative interior. To state it, we need to define what is the correct multidimensional notion of petal.

Definition 2.2 A *parabolic curve* for $f\in\operatorname{End}_1(\mathbb{C}^n,O)$ tangent to the identity is an injective holomorphic map $\varphi\colon D\to\mathbb{C}^n\setminus\{O\}$ satisfying the following properties:

(a) D is a simply connected domain in $\mathbb{C}$ with $0 \in \partial D$;
(b) φ is continuous at the origin, and $\varphi(0) = O$;
(c) $\varphi(D)$ is f-invariant, and $(f|_{\varphi(D)})^k \to O$ uniformly on compact subsets as $k \to +\infty$.

Furthermore, if $[\varphi(\zeta)] \to [v]$ in $\mathbb{P}^{n-1}(\mathbb{C})$ as $\zeta \to 0$ in D, we shall say that the parabolic curve φ is *tangent* to the direction $[v] \in \mathbb{P}^{n-1}(\mathbb{C})$. Finally, a *Fatou flower* with ν *petals* tangent to a direction $[v]$ is a holomorphic map $\Phi\colon D_{\nu,\delta} \to \mathbb{C}$, where $D_{\nu,\delta}$ is given by (1.3), such that Φ restricted to any connected component of $D_{\nu,\delta}$ is a parabolic curve tangent to $[v]$, a *petal* of the Fatou flower. If ν is the order of f then we shall talk of a Fatou flower for f without mentioning the number of petals.

Then Écalle, using his resurgence theory (see, e.g., [S] for an introduction to Écalle's resurgence theory in one dimension), and Hakim, using more classical methods, have proved the following result (see also [W]):

Theorem 2.2 ([E, H2, H3]) *Let $f \in \mathrm{End}_1(\mathbb{C}^n, O)$ be tangent to the identity, and $[v] \in \mathbb{P}^{n-1}(\mathbb{C})$ a non-degenerate characteristic direction for f. Then there exists (at least) one Fatou flower tangent to $[v]$. Furthermore, for every petal $\varphi\colon \Delta \to \mathbb{C}^n$ of the Fatou flower there exists a injective holomorphic map $\chi\colon \varphi(\Delta) \to \mathbb{C}$ such that $\chi\big(f(z)\big) = \chi(z) + 1$ for all $z \in \varphi(\Delta)$.*

Definition 2.3 The function χ constructed in the previous theorem is a *Hakim-Fatou coordinate.*

Remark 2.5 A characteristic direction is a *complex* direction, not a real one; so it should not be confused with the attracting/repelling directions of Theorem 1.1. All petals of a Fatou flower are tangent to the same characteristic direction, but each petal is tangent to a different real direction inside the same complex (characteristic) direction. In particular, Fatou flowers of f and f^{-1} are tangent to the same characteristic directions (see Remark 2.2) but the corresponding petals are tangent to different real directions, as in Theorem 1.1.

In particular there exist parabolic curves tangent to $[1 : -1]$ for the system of Example 2.1 even though there are no invariant curves passing through the origin tangent to that direction.

Parabolic curves are one-dimensional objects in an n-dimensional space; it is natural to wonder about the existence of higher dimensional invariant subsets. A sufficient condition for their existence has been given by Hakim; to state it we need to introduce another definition.

Definition 2.4 Let $[v] \in \mathbb{P}^{n-1}(\mathbb{C})$ be a non-degenerate characteristic direction for a homogeneous map $P\colon \mathbb{C}^n \to \mathbb{C}^n$ of degree $\nu + 1 \geq 2$; in particular, $[v]$ is a fixed point for the meromorphic self-map $[P]$ of $\mathbb{P}^{n-1}(\mathbb{C})$ induced by P. The *directors* of P in $[v]$ are the eigenvalues $\alpha_1, \ldots, \alpha_{n-1} \in \mathbb{C}$ of the linear operator

$$\frac{1}{\nu}\left(d[P]_{[v]} - \mathrm{id}\right) : T_{[v]}\mathbb{P}^{n-1}(\mathbb{C}) \to T_{[v]}\mathbb{P}^{n-1}(\mathbb{C}).$$

As usual, if $f \in \mathrm{End}_1(\mathbb{C}^n, O)$ is of the form (1.2), then the *directors* of f in a non-degenerate characteristic direction $[v]$ are the directors of $P_{\nu+1}$ in $[v]$.

Remark 2.6 Definition 2.4 is equivalent to the original definition used by Hakim (see, e.g., [ArR]). Furthermore, in dimension 2 if $[v] = [1 : 0]$ is a non-degenerate characteristic direction of $P = (P_1, P_2)$ we have $P_1(1, 0) \neq 0$, $P_2(1, 0) = 0$ and the director is given by

$$\frac{1}{\nu} \frac{d}{d\zeta} \frac{P_2(1,\zeta) - \zeta P_1(1,\zeta)}{P_1(1,\zeta)}\bigg|_{\zeta=0} = \frac{1}{\nu}\left[\frac{\frac{\partial P_2}{\partial z_2}(1,0)}{P_1(1,0)} - 1\right].$$

Remark 2.7 Recalling Remark 2.2 one sees that a germ $f \in \mathrm{End}_1(\mathbb{C}^n, O)$ tangent to the identity and its inverse f^{-1} have the same directors at their non-degenerate characteristic directions.

Remark 2.8 The proof of Theorem 2.2 becomes simpler when no director is of the form $\frac{k}{\nu}$ with $k \in \mathbb{N}^*$; furthermore, in this case the parabolic curves enjoy additional properties (in the terminology of [AT1] they are *robust;* see also [Ro3]).

Definition 2.5 A *parabolic manifold* for a germ $f \in \mathrm{End}_1(\mathbb{C}^n, O)$ tangent to the identity is an f-invariant complex submanifold $M \subset \mathbb{C}^n \setminus \{O\}$ with $O \in \partial M$ such that $f^k(z) \to O$ for all $z \in M$. A *parabolic domain* is a parabolic manifold of dimension n. We shall say that M is *attached* to the characteristic direction $[v] \in \mathbb{P}^{n-1}(\mathbb{C})$ if furthermore $[f^k(z)] \to [v]$ for all $z \in M$.

Then Hakim has proved (see also [ArR] for the details of the proof) the following theorem:

Theorem 2.3 ([H3]) *Let* $f \in \mathrm{End}_1(\mathbb{C}^n, O)$ *be tangent to the identity of order* $\nu \geq 1$. *Let* $[v] \in \mathbb{P}^{n-1}(\mathbb{C})$ *be a non-degenerate characteristic direction, with directors* $\alpha_1, \ldots, \alpha_{n-1} \in \mathbb{C}$. *Furthermore, assume that* $\mathrm{Re}\,\alpha_1, \ldots, \mathrm{Re}\,\alpha_d > 0$ *and* $\mathrm{Re}\,\alpha_{d+1}, \ldots, \mathrm{Re}\,\alpha_{n-1} \leq 0$ *for a suitable* $d \geq 0$. *Then:*

(i) There exist (at least) ν *parabolic* $(d+1)$*-manifolds* $M_1, \ldots, M_\nu$ *of* $\mathbb{C}^n$ *attached to* $[v]$;

(ii) $f|_{M_j}$ *is holomorphically conjugated to the translation* $\tau(w_0, w_1, \ldots, w_d) = (w_0 + 1, w_1, \ldots, w_d)$ *defined on a suitable right half-space in* $\mathbb{C}^{d+1}$.

Remark 2.9 In particular, if all the directors of $[v]$ have positive real part, there is at least one parabolic domain. However, the condition given by Theorem 2.3 is not necessary for the existence of parabolic domains; see [Ri1, Us, AT3] for examples, and [Ro8] for conditions ensuring the existence of a parabolic domain when some directors have positive real part and all the others are equal to zero. Moreover, Lapan [L1] has proved that if $n = 2$ and f has a unique characteristic direction $[v]$ which is non degenerate then there exists a parabolic domain attached to $[v]$ even though the director is necessarily 0.

Two natural questions now are: how many characteristic directions are there? Does there always exist a non-degenerate characteristic direction? To answer the first question, we need to introduce the notion of multiplicity of a characteristic direction. To do so, notice that $[v] = [v_1 : \cdots : v_n] \in \mathbb{P}^{n-1}(\mathbb{C})$ is a characteristic direction for the homogeneous map $P = (P_1, \ldots, P_n)$ if and only if $v_h P_k(v) - v_k P_h(v) = 0$ for all $h, k = 1, \ldots, n$. In particular, the set of characteristic directions of P is an algebraic subvariety of $\mathbb{P}^{n-1}(\mathbb{C})$.

Definition 2.6 If the maximal dimension of the irreducible components of the subvariety of characteristic directions of a homogeneous map $P\colon \mathbb{C}^n \to \mathbb{C}^n$ is k, we shall say that P is *k-dicritical*; if $k = n$ we shall say that P is *dicritical;* if $k = 0$ we shall say that P is *non-dicritical.*

Remark 2.10 A homogeneous map $P\colon \mathbb{C}^n \to \mathbb{C}^n$ of degree d is dicritical if and only if $P(z) = p(z)z$ for some homogenous polynomial $p\colon \mathbb{C}^n \to C$ of degree $d - 1$. In particular, the degenerate characteristic directions are the zeroes of the polynomial p.

In the non-dicritical case we can count the number of characteristic directions, using a suitable multiplicity.

Definition 2.7 Let $[v] = [v_1 : \cdots : v_n] \in \mathbb{P}^{n-1}(\mathbb{C})$ be a characteristic direction of a homogeneous map $P = (P_1, \ldots, P_n)\colon \mathbb{C}^n \to \mathbb{C}^n$. Choose $1 \leq j_0 \leq n$ so that $v_{j_0} \neq 0$. The *multiplicity* $\mu_P([v])$ of $[v]$ is the local intersection multiplicity at $[v]$ in $\mathbb{P}^{n-1}(\mathbb{C})$ of the polynomials $z_{j_0} P_j - z_j P_{j_0}$ with $j \neq j_0$ if $[v]$ is an isolated characteristic direction; it is $+\infty$ if $[v]$ is not isolated.

Remark 2.11 The local intersection multiplicity $I(p_1, \ldots, p_k; z^o)$ of a set $\{p_1, \ldots, p_k\}$ of holomorphic functions at a point $z^o \in \mathbb{C}^n$ can be defined (see, e.g., [GH]) as

$$I(p_1, \ldots, p_k; z^o) = \dim \mathcal{O}_{n,z^o}/(p_1, \ldots, p_k),$$

where $\mathcal{O}_{n,z^o}$ is the local ring of germs of holomorphic functions at z^o, and the dimension is as vector space. It is easy to check that the definition of multiplicity of a characteristic direction does not depend on the index j_0 chosen. Furthermore, since the local intersection multiplicity is invariant under change of coordinates, we can use local charts to compute the local intersection multiplicity on complex manifolds.

Remark 2.12 When $n = 2$, the multiplicity of $[v] = [1 : v_2]$ as characteristic direction of $P = (P_1, P_2)$ is the order of vanishing at $t = v_2$ of $P_2(1, t) - tP_1(1, t)$; analogously, the multiplicity of $[0 : 1]$ is the order of vanishing at $t = 0$ of $P_1(t, 1) - tP_2(t, 1)$.

Then we have the following result (see, e.g., [AT1]):

Proposition 2.4 *Let $P\colon \mathbb{C}^n \to \mathbb{C}^n$ be a non-dicritical homogeneous map of degree $\nu + 1 \geq 2$. Then P has exactly*

$$\frac{1}{\nu}\left((\nu+1)^n-1\right)=\sum_{j=0}^{n-1}\binom{n}{j+1}\nu^j$$

characteristic directions, counted according to their multiplicity.

In particular, when $n = 2$ then a homogeneous map of degree $\nu + 1$ either is dicritical (and all directions are characteristic) or has exactly $\nu + 2$ characteristic directions. But all of them can be degenerate; an example is the following (but it is easy to build infinitely many others).

Example 2.2 Let $P(z, w) = (z^2w + zw^2, zw^2)$. Then the characteristic directions of P are $[1 : 0]$ and $[0 : 1]$, both degenerate. Using Remark 2.12, we see that $\mu_P([1 : 0]) = 3$ and $\mu_P([0 : 1]) = 1$.

So we cannot apply Theorem 2.2 to any germ of the form $f(z) = z + P(z) + \cdots$ when P is given by Example 2.2. However, as soon as the higher order terms are chosen so that the origin is an isolated fixed point then f does have parabolic curves:

Theorem 2.5 ([A2]) *Let $f \in End_1(\mathbb{C}^2, O)$ be tangent to the identity such that O is an isolated fixed point. Then f admits at least one Fatou flower tangent to some characteristic direction.*

In the next section we shall explain why this theorem holds, we shall give more general statements, and we shall give an example (Example 3.1) showing the necessity of the hypothesis that the origin is an isolated fixed point.

3 Blow-Ups, Indices and Fatou Flowers

In the previous section we saw that for studying the dynamics of a germ tangent to the identity it is useful to consider the tangent directions at the fixed point. A useful way for dealing with tangent directions consists, roughly speaking, in replacing the fixed point by the projective space of the tangent directions, in such a way that the new space is still a complex manifolds, where the tangent directions at the original fixed point are now points. We refer to, e.g., [GH] or [A1] for a precise description of this construction; here we shall limit ourselves to explain how to work with it.

Definition 3.1 Let M be a complex n-dimensional manifold, and $p \in M$. The *blow-up* of M of *center* p is a complex n-dimensional manifold $\tilde{M}$ equipped with a surjective holomorphic map $\pi\colon \tilde{M} \to M$ such that

(i) $E = \pi^{-1}(p)$ is a compact submanifold of $\tilde{M}$, the *exceptional divisor* of the blow-up, biholomorphic to $\mathbb{P}(T_pM)$;
(ii) $\pi|_{\tilde{M}\setminus E}\colon \tilde{M} \setminus E \to M \setminus \{p\}$ is a biholomorphism.

Let us describe the construction for $(M, p) = (\mathbb{C}^n, O)$; using local charts one can repeat the construction for any manifold. As a set, $\widetilde{\mathbb{C}^n}$ is the disjoint union of $\mathbb{C}^n \setminus \{O\}$ and $E = \mathbb{P}^{n-1}(\mathbb{C})$; we shall define a manifold structure using charts. For

$j = 1, \ldots, n$ let $U'_j = \{[v_1 : \cdots : v_n] \in \mathbb{P}^{n-1}(\mathbb{C}) \mid v_j \neq 0\}$, $U''_j = \{w \in \mathbb{C}^n \mid w_j \neq 0\}$ and $\tilde{U}_j = U'_j \cup U''_j \subset \widetilde{\mathbb{C}^n}$. Define $\chi_j \colon \tilde{U}_j \to \mathbb{C}^n$ by setting

$$\chi_j(q) = \begin{cases} \left(\frac{v_1}{v_j}, \ldots, \frac{v_{j-1}}{v_j}, 0, \frac{v_{j+1}}{v_j}, \ldots, \frac{v_n}{v_j}\right) & \text{if } q = [v_1 : \cdots : v_n] \in U'_j, \\ \left(\frac{w_1}{w_j}, \ldots, \frac{w_{j-1}}{w_j}, w_j, \frac{w_{j+1}}{w_j}, \ldots, \frac{w_n}{w_j}\right) & \text{if } q = (w_1, \ldots, w_n) \in U''_j. \end{cases}$$

We have

$$\chi_j^{-1}(w) = \begin{cases} [w_1 : \cdots : w_{j-1} : 1 : w_{j+1} : \cdots : w_n] & \text{if } w_j = 0, \\ (w_j w_, \ldots, w_j w_{j-1}, w_j, w_j w_{j+1}, \ldots, w_j w_n) & \text{if } w_j \neq 0, \end{cases}$$

and it is easy to check that $\{(\tilde{U}_1, \chi_1), \ldots, (\tilde{U}_n, \chi_n)\}$ is an atlas for $\widetilde{\mathbb{C}^n}$, with $\chi_j([0 : \cdots : 1 : \cdots : 0]) = O$ and $\chi_j\big(\tilde{U}_j \cap \mathbb{P}^{n-1}(\mathbb{C})\big) = \{w_j = 0\} \subset \mathbb{C}^n$. We can then define the projection $\pi \colon \widetilde{\mathbb{C}^n} \to \mathbb{C}^n$ in coordinates by setting

$$\pi \circ \chi_j^{-1}(w) = (w_1 w_j, \ldots, w_{j-1} w_j, w_j, w_{j+1} w_j, \ldots, w_n w_j);$$

it is easy to check that π is well-defined, that $\pi^{-1}(O) = \mathbb{P}^{n-1}(\mathbb{C})$ and that π induces a biholomorphism between $\widetilde{\mathbb{C}^n} \setminus \mathbb{P}^{n-1}(\mathbb{C})$ and $\mathbb{C}^n \setminus \{O\}$. Notice furthermore that $\widetilde{\mathbb{C}^n}$ has a canonical structure of line bundle over $\mathbb{P}^{n-1}(\mathbb{C})$ given by the projection $\tilde{\pi} \colon \widetilde{\mathbb{C}^n} \to \mathbb{P}^{n-1}(\mathbb{C})$ defined by

$$\tilde{\pi}(q) = \begin{cases} [v] & \text{if } q = [v] \in \mathbb{P}^{n-1}(\mathbb{C}), \\ [w_1 : \cdots : w_n] & \text{if } q = w \in \mathbb{C}^n \setminus \{O\}; \end{cases}$$

the fiber over $[v] \in \mathbb{P}^{n-1}(\mathbb{C})$ is given by the line $\mathbb{C}v \subset \mathbb{C}^n$.

Two more definitions we shall need later on:

Definition 3.2 Let $\pi \colon \tilde{M} \to M$ be the blow-up of a complex manifold M at $p \in M$. Given a subset $S \subset M$, the *full* (or *total*) *transform* of S is $\pi^{-1}(S)$, whereas the *strict transform* of S is the closure in $\tilde{M}$ of $\pi^{-1}(S \setminus \{O\})$.

Clearly, the full and the strict transform coincide if $p \notin S$; if $p \in S$ then the full transform is the union of the strict transform and the exceptional divisor. Furthermore, if S is a submanifold at p then its strict transform is $(S \setminus \{p\}) \cup \mathbb{P}(T_p S)$.

Definition 3.3 Let $f \in \mathrm{End}(M, p)$ be a germ such that df_p is invertible. Choose a representative (U, f) of the germ such that f is injective in U. Then the *blow-up* of f is the map $\tilde{f} \colon \pi^{-1}(U) \to \tilde{M}$ defined by

$$\tilde{f}(q) = \begin{cases} [df_p(v)] & \text{if } q = [v] \in E = \mathbb{P}(T_p M), \\ f(w) & \text{if } q = w \in U \setminus \{p\}. \end{cases}$$

In this way we get a germ about the exceptional divisor of a holomorphic self-map of the blow-up, given by the differential of f along the exceptional divisor and by f itself elsewhere, satisfying $\pi \circ \tilde{f} = f \circ \pi$. In particular, $K_{\tilde{f}} = \pi^{-1}(K_f) = \big(K_f \setminus \{O\}\big) \cup E$, and to study the dynamics of $\tilde{f}$ in a neighborhood of the exceptional divisor is equivalent to studying the dynamics of f in a neighborhood of p.

If $f \in \mathrm{End}_1(\mathbb{C}^n, O)$ is tangent to the identity, its blow-up $\tilde{f}$ in the chart (U_1, χ_1) is given by

$$\chi_1 \circ \tilde{f} \circ \chi_1^{-1}(w) = \begin{cases} w & \text{if } w_1 = 0, \\ \left(f_1(w_1, w_1w_2, \ldots, w_1w_n), \frac{f_2(w_1,w_1w_2,\ldots,w_1w_n)}{f_1(w_1,w_1w_2,\ldots,w_1w_n)}, \ldots, \frac{f_n(w_1,w_1w_2,\ldots,w_1w_n)}{f_1(w_1,w_1w_2,\ldots,w_1w_n)} \right) & \text{if } w_1 \neq 0; \end{cases}$$

similar formulas hold in the other charts. In particular, writing $w = (w_1, w')$, and $\chi_1 \circ \tilde{f} \circ \chi_1^{-1} = (\tilde{f}_1, \ldots, \tilde{f}_n)$, if f is tangent to the identity of order $\nu \geq 1$ and leading term $P_{\nu+1} = (P_{\nu+1,1}, \ldots, P_{\nu+1,n})$, we get

$$\begin{cases} \tilde{f}_1(w) = w_1 + w_1^{\nu+1} P_{\nu+1,1}(1, w') + O(w_1^{\nu+2}), \\ \tilde{f}_j(w) = w_j + w_1^{\nu}\big(P_{\nu+1,j}(1.w') - w_j P_{\nu+1,1}(1, w')\big) + O(w_1^{\nu+1}) \quad \text{if } j \neq 1. \end{cases} \tag{3.1}$$

It follows immediately that:

– if $\nu \geq 2$ then $\tilde{f}$ is tangent to the identity in all points of the exceptional divisor;
– if $\nu = 1$ then $\tilde{f}$ is tangent to the identity in all characteristic directions of f; in other points of the exceptional divisor the eigenvalues of the differential of $\tilde{f}$ are all equal to 1 but the differential is not diagonalizable.

This means that we can always repeat the previous construction blowing-up $\tilde{f}$ at a characteristic direction of f; this will be important in the sequel.

As a first application of the blow-up construction, let us use it for describing the dynamics of dicritical maps. If $f \in \mathrm{End}_1(\mathbb{C}^n, O)$ is dicritical, Theorem 2.2 yields a parabolic curve tangent to all directions outside a hypersurface of $\mathbb{P}^{n-1}(\mathbb{C})$ (notice that all directors are zero), and the same holds for f^{-1}. One can then summarize the situation as follows:

Proposition 3.1 ([Br1, Br2]) *Let $f \in \mathrm{End}_1(\mathbb{C}^n, O)$ be a dicritical germ tangent to the identity of order $\nu \geq 1$. Write $P_{\nu+1}(z) = p(z)z$, and let $D = \{[v] \in \mathbb{P}^{n-1}(\mathbb{C}) \mid p(v) = 0\}$. Then there are two open sets $U^+, U^- \subset \mathbb{C}^n \setminus \{O\}$ such that:*

(i) $\overline{U^+ \cup U^-}$ is a neighborhood of $\mathbb{P}^{n-1}(\mathbb{C}) \setminus D$ in the blow-up $\widetilde{\mathbb{C}^n}$ of O;
(ii) the orbit of any $z \in U^+$ converges to the origin tangent to a direction $[v] \in \mathbb{P}^{n-1}(\mathbb{C}) \setminus D$;
(iii) the inverse orbit (that is, the orbit under f^{-1}) of any $z \in U^-$ converges to the origin tangent to a direction $[v] \in \mathbb{P}^{n-1}(\mathbb{C}) \setminus D$.

Coming back to the general situation, when $f \in \mathrm{End}_1(\mathbb{C}^n, O)$ is tangent to the identity its blow-up $\tilde{f}$ fixes pointwise the exceptional divisor; more precisely, the fixed point set of $\tilde{f}$ is the full transform of the fixed point set of f, and in particular $\tilde{f}$ has a at least a hypersurface of fixed points. This is a situation important enough to deserve a special notation.

Definition 3.4 Let E be a connected (possibly singular) hypersurface in a complex manifold M. We shall denote by $\mathrm{End}(M, E)$ the set of germs about E of holomorphic self-maps of M fixing pointwise E.

If E is a hypersurface in a complex manifold M, we shall denote by $\mathscr{O}_M$ the sheaf of holomorphic functions on M, and by $\mathscr{I}_E$ the subsheaf of functions vanishing on E. Given $f \in \mathrm{End}(M, E)$, $f \not\equiv \mathrm{id}_M$, take $p \in E$. For every $h \in \mathscr{O}_{M,p}$, the germ $h \circ f$ is well-defined, and $h \circ f - h \in \mathscr{I}_{E,p}$. Following [ABT1] (see also [ABT2, ABT3]), we can then introduce a couple of important notions.

Definition 3.5 Let E be a connected hypersurface in a complex manifold M. Given $f \in \mathrm{End}(M, E)$, $p \in E$ and $h \in \mathscr{O}_{M,p}$, let $\nu_f(h) = \max\{\mu \in \mathbb{N} \mid h \circ f - h \in \mathscr{I}^{\mu}_{E,p}\}$. Then the *order of contact* ν_f of f with the identity along E is

$$\nu_f = \min\{\nu_f(h) \mid h \in \mathscr{O}_{M,p}\};$$

it can be shown ([ABT1]) that ν_f does not depend on $p \in E$. Furthermore, we say that f is *tangential* if $\min\{\nu_f(h) \mid h \in \mathscr{I}_{E,p}\} > \nu_f$ for some (and hence any; see again [ABT1]) $p \in E$.

Let $(z_1, \dots, z_n)$ be local coordinates in M centered at $p \in E$, and $\ell \in \mathscr{I}_{E,p}$ a reduced equation of E at p (that is, a generator of $\mathscr{I}_{E,p}$). If $(f_1, \dots, f_n)$ is the expression of f in local coordinates, it turns out [ABT1] that we can write

$$f_j(z) = z_j + \ell(z)^{\nu_f} g^o_j(z) \tag{3.2}$$

for $j = 1, \dots, n$, where there is a j_0 such that ℓ does not divide $g^o_{j_0}$; furthermore, f is tangential if and only if $\nu_f(\ell) > \nu_f$.

Remark 3.1 If E is smooth at p, we can choose local coordinates so that locally E is given by $\{z_1 = 0\}$, that is $\ell = z_1$. Then we can write

$$f_j(z) = z_j + z_1^{\nu_f} g^o_j(z)$$

with z_1 not dividing some g^o_j; and f is tangential if z_1 divides g^o_1, that is if $f_1(z) = z_1 + z_1^{\nu_f+1} h^o_1(z)$. More generally, if E has a normal crossing at p with $1 \leq r \leq n$ smooth branches, then we can choose local coordinates so that $\ell = z_1 \cdots z_r$, so that $f_j(z) = z_j + (z_1 \cdots z_r)^{\nu_f} g^o_j(z)$ with some $g^o_{j_0}$ not divisible by $z_1 \cdots z_r$; in this case f is tangential if and only if z_j divides g^o_j for $j = 1, \dots, r$. In particular, in the terminology of [A2] f is tangential if and only if it is nondegenerate and $b_f = 1$.

Definition 3.6 We say that $p \in E$ is a *singular point* for $f \in \mathrm{End}(M, E)$ (with respect to E) if $g_1^o(p) = \cdots = g_n^o(p)$ in (3.2); it turns out [ABT1] that this definition is independent of the local coordinates. Furthermore, the *pure order* (or *pure multiplicity*) $\nu_o(f, E)$ of f along E at p is

$$\nu_o(f, E) = \min\{\mathrm{ord}_O(g_1^o), \ldots, \mathrm{ord}_O(g_n^o)\} .$$

It is easy to see that the pure order does not depend on the local coordinates; in particular, p is singular for f with respect to E if and only if $\nu_o(f, E) \geq 1$. If E is the fixed point set of f at p then we shall talk of the *pure order* $\nu_o(f)$ of f at p.

Remark 3.2 When f is the blow-up of a germ $f_o \in \mathrm{End}_1(\mathbb{C}^n, O)$ tangent to the identity of order $\nu \geq 1$, then (3.1) implies that:

- f is tangential if and only if f_o is not dicritical; in particular, in this case being tangential is a generic condition;
- $\nu_f = \nu$ if f_o is not dicritical, and $\nu_f = \nu + 1$ if f_o is dicritical;
- if f_o is non dicritical, then $[v] \in \mathbb{P}^{n-1}(\mathbb{C})$ is singular for f if and only if it is a characteristic direction of f_o.

Using the notion of singular points we can generalize Proposition 2.1 as follows:

Proposition 3.2 ([ABT1]) *Let $E \subset M$ be a hypersurface in a complex manifold M, and $f \in \mathrm{End}(M, E)$, $f \not\equiv \mathrm{id}_M$, tangential to E. Let $p \in E$ be a smooth point of E which is non-singular for f. Then no infinite orbit of f can stay arbitrarily close to p, that is, there exists a neighborhood U of p such that for all $q \in U$ either the orbit of q lands on E or $f^{n_0}(q) \notin U$ for some $n_0 \in \mathbb{N}$. In particular, no infinite orbit is converging to p.*

More generally, we have:

Proposition 3.3 ([AT1]) *Let $f \in \mathrm{End}(\mathbb{C}^n, O)$ be of the form*

$$f_j(z) = \begin{cases} z_j + z_j \left(\prod_{h=1}^r z_h^{\nu_h}\right) g_j(z) & \text{for } 1 \leq j \leq r, \\ z_j + \left(\prod_{h=1}^r z_h^{\nu_h}\right) g_j(z) & \text{for } r+1 \leq j \leq n, \end{cases} \tag{3.3}$$

for suitable $1 \leq r < n$, with $\nu_1, \ldots, \nu_r \geq 1$ and $g_1, \ldots, g_n \in \mathcal{O}_{\mathbb{C}^n, O}$. Assume that $g_{j_0}(O) \neq 0$ for some $r + 1 \leq j_0 \leq n$. Then no infinite orbit can stay arbitrarily close to O.

A very easy example of this phenomenon, promised at the end of the previous section, is the following:

Example 3.1 Let $f(z, w) = (z, w + z^2)$. Then f is tangent to the identity at the origin; the fixed point set is $\{z = 0\}$, and thus O is not an isolated fixed point. We have $f^k(z, w) = (z, w + kz^2)$; therefore all orbits outside the fixed point set escape to infinity, and in particular no orbit converges to the origin. Notice that this germ has only one characteristic direction, which is degenerate (and tangent to the fixed point set). Moreover, f is tangential with order of contact 2 to its fixed point set, but the origin is not singular.

After these generalities, in the rest of this section we specialize to the case $n = 2$ and to tangential maps (because of Remark 3.2; see anyway [ABT1] for information on the dynamics of non-tangential maps). Take $f \in \mathrm{End}(M, E)$, where M is a complex surface and $E \subset M$ is a 1-dimensional curve smooth at $p \in E$, and assume that f is tangential to E with order of contact $\nu_f \geq 1$. Then we can choose local coordinates centered at p so that we can write

$$\begin{cases} f_1(z) = z_1 + z_1^{\nu_f+1} h_1^o(z), \\ f_2(z) = z_2 + z_1^{\nu_f} g_2^o(z), \end{cases} \tag{3.4}$$

where z_1 does not divide g_2^o; notice that $h_1^o(0, \cdot) = \frac{\partial g_1^o}{\partial z_1}(0, \cdot)$, where $g_1^o = z_1 h_1^o$. In particular, O is singular if and only if $g_2^o(O) = 0$. We then introduce the following definitions:

Definition 3.7 Let $f \in \mathrm{End}(M, E)$ be written in the form (3.4). Then:

- the *multiplicity* μ_p of f along E at p is $\mu_p = \mathrm{ord}_0\big(g_2^o(0, \cdot)\big)$, so that p is a singular point if and only if $\mu_p \geq 1$;
- the *transversal multiplicity* τ_p of f along E at p is $\tau_p = \mathrm{ord}_0\big(h_1^o(0, \cdot)\big)$;
- p is an *apparent singularity* if $1 \leq \mu_p \leq \tau_p$;
- p is a *Fuchsian singularity* if $\mu_p = \tau_p + 1$;
- p is an *irregular singularity* if $\mu_p > \tau_p + 1$;
- p is a *non-degenerate singularity* if $\mu_p \geq 1$ but $\tau_p = 0$;
- p is a *degenerate singularity* if $\mu_p, \tau_p \geq 1$;
- the *index* $\iota_p(f, E)$ of f at p along E is

$$\iota_p(f, E) = \nu_f \mathrm{Res}_0 \frac{h_1^o(0, \cdot)}{g_2^o(0, \cdot)} \ ;$$

- the *induced residue* $\mathrm{Res}_p^0(f)$ of f along E at p is

$$\mathrm{Res}_p^0(f) = -\iota_p(f, E) - \mu_p \ .$$

It is possible to prove (see [A2, ABT1, AT3]; notice that our index is ν_f times the residual index introduced in [A2]) that these definitions are independent of the local coordinates.

Remark 3.3 Recalling (3.1), we see that if f is obtained as the blow-up of a non-dicritical map f_o, and E is the exceptional divisor of the blow-up, then:

- the multiplicity of $[v]$ as characteristic direction of f_o is equal to the multiplicity of f along E at $[v]$;

– $[v]$ is a degenerate/non-degenerate characteristic direction of f_o if and only if it is a degenerate/non-degenerate singularity of f.

Furthermore, if we write $P_{\nu+1,1}(1,w) = \sum_{k=0}^{\nu+1} a_k w^k$ and $P_{\nu+1,2}(1,w) = \sum_{k=0}^{\nu+1} b_k w^k$ then $[1:0]$ is a characteristic direction if and only if $b_0 = 0$, non-degenerate if and only if moreover $a_0 \neq 0$, and (setting $b_{\nu+2} = 0$)

$$\frac{h_1^o(0,\zeta)}{g_2^o(0,\zeta)} = \frac{1}{\zeta} \frac{a_0 + \sum_{k=1}^{\nu+1} a_k \zeta^k}{(b_1 - a_0) + \sum_{k=1}^{\nu+1} (b_{k+1} - a_k)\zeta^k} .$$

So if $b_1 \neq a_0$ we have

$$\mu_O = 1, \quad \iota_{[1:0]}(f,E) = \frac{\nu a_0}{b_1 - a_0}, \quad \mathrm{Res}^0_{[1:0]}(f) = \frac{(\nu-1)a_0 + b_1}{a_0 - b_1} ;$$

moreover, if $a_0 \neq 0$ then $[1:0]$ is a non-degenerate characteristic direction with director $\alpha = (b_1 - a_0)/\nu a_0$. More generally, we have $\tau_{[1:0]} = \mathrm{ord}_0\big(P_{\nu+1,1}(1,w)\big)$, $\mu_{[1:0]} = \mathrm{ord}_0\big(P_{\nu+1,2}(1,w) - wP_{\nu+1,1}(1,w)\big)$ and

$$\iota_{[1:0]} = \frac{\nu a_{\mu-1}}{b_\mu - a_{\mu-1}}, \quad \mathrm{Res}^0_{[1:0]} = \frac{(\nu-\mu)a_{\mu-1} + \mu b_\mu}{a_{\mu-1} - b_\mu},$$

where $\mu = \mu_{[1:0]}$. In particular we obtain:

– if $[v]$ is non-degenerate characteristic direction of f_o with director $\alpha \neq 0$ then

$$\iota_{[v]}(f,E) = \frac{1}{\alpha};$$

– $[v]$ is a non-degenerate characteristic direction with non-zero director for f_o if and only if it is a Fuchsian singularity of multiplicity 1 for f.

Residues and indices are important for two reasons. First of all, we have the following *index theorem:*

Theorem 3.4 ([A2, ABT1]) *Let $E \subset M$ be a smooth compact Riemann surface in a complex surface M. Let $f \in \mathrm{End}(M,E)$, $f \not\equiv \mathrm{id}_M$, be tangential with order of contact ν; denote by $Sing(f) \subset E$ the finite set of singular points of f in E. Then*

$$\sum_{p \in Sing(f)} \iota_p(f) = \nu c_1(N_E), \qquad \sum_{p \in Sing(f)} Res^0_p(f) = -\chi(E),$$

where $c_1(N_E)$ is the first Chern class of the normal bundle N_E of E in M, and $\chi(E)$ is the Euler characteristic of E. In particular, when f is the blow-up of a nondicritical germ tangent to the identity and $E = \mathbb{P}^1(\mathbb{C})$ is the exceptional divisor we have

$$\sum_{p\in Sing(f)} \iota_p(f) = -\nu, \qquad \sum_{p\in Sing(f)} Res^0_p(f) = -2 .$$

Remark 3.4 Bracci and Tovena [BT] have defined a notion of index at non-smooth points of E allowing the generalization of Theorem 3.4 to non necessarily smooth compact Riemann surfaces, where in the statement $c_1(N_E)$ is replaced by the self-intersection $E \cdot E$.

The second reason is that the index can be used to detect the presence of parabolic curves. To state this precisely, we need a definition.

Definition 3.8 Let $f \in \mathrm{End}_1(\mathbb{C}^2, O)$ be tangent to the identity. We say that O is a *corner* if the germ of the fixed point set at the origin is locally reducible, that is has more than one irreducible component.

Then

Theorem 3.5 ([A2]) *Let $E \subset M$ be a smooth Riemann surface in a complex surface M, and take $f \in \mathrm{End}(M, E)$ tangential. Let $p \in E$ be a singular point, not a corner, such that $\iota_p(f) \notin \mathbb{Q}^+ \cup \{0\}$. Then there exists a Fatou flower with ν_f petals for f at p.*

Corollary 3.6 ([A2]) *Let $f \in \mathrm{End}_1(\mathbb{C}^2, O)$ be tangent to the identity, and assume that O is a nondicritical singular point. Let $[v] \in \mathbb{P}^1(\mathbb{C})$ be a characteristic direction, and $\tilde{f}$ the blow-up of f. If $[v]$ is not a corner for $\tilde{f}$ and $\iota_{[v]}(\tilde{f}) \notin \mathbb{Q}^+ \cup \{0\}$ then there exists a Fatou flower for f tangent to $[v]$.*

Theorem 2.5 is then a consequence of Theorem 3.4 and Corollary 3.6. Indeed, take $f \in \mathrm{End}_1(\mathbb{C}^2, O)$ tangent to the identity with an isolated fixed point at the origin. If O is dicritical, we can directly apply Theorem 2.2. Assume then O non-dicritical, and let $\tilde{f} \in \mathrm{End}(\widetilde{\mathbb{C}^2}, E)$ be the blow-up of f. Since O is non-dicritical, $\tilde{f}$ is tangential; Theorem 3.4 then implies that at least one characteristic direction $[v]$ has negative index. Since O is an isolated fixed point, the fixed point set of $\tilde{f}$ coincides with the exceptional divisor; therefore $[v]$ is not a corner, and Corollary 3.6 yields the Fatou flower we were looking for.

Remark 3.5 Bracci and Degli Innocenti (see [B, D]), using the definition of index introduced in [BT], have shown that Theorem 3.5 still holds when E is not smooth at p. Bracci and Suwa [BS] have also obtained a version of Theorem 3.5 when M has a (sufficiently tame) singularity at p.

Example 3.2 Let $f(z, w) = (z + z^2w + zw^2 + w^4, w + zw^2 + z^4)$. Then f is tangent to the identity at the origin of order 2, and the origin is an isolated fixed point. Furthermore, f is non-dicritical and it has (see Example 2.2) two characteristic directions, $[v_1] = [1 : 0]$ and $[v_2] = [0; 1]$, both degenerate. Working as in Remark 3.4 it is easy to see that $[v_1]$ is an irregular singularity of multiplicity 3 with index -2 and induced residue -1, and that $[v_2]$ is an apparent singularity of multiplicity 1, vanishing index, and induced residue -1. In particular, f admits a Fatou flower with 2 petals tangent to $[v_1]$.

Example 3.3 Let $f(z,w)=(z+w^2,w+z^3)$. Then f is tangent to the identity at the origin of order 1, and the origin is an isolated fixed point. Furthermore, f is nondicritical with only one characteristic direction $[v]=[1:0]$, which is degenerate of multiplicity 3, Fuchsian, with index -1 and induced residue -2. Therefore f admits a Fatou flower with one petal tangent to $[v]$; compare with Example 3.4.

There are still instances where Theorem 3.5 cannot be applied:

Example 3.4 Let $f(z,w)=(z+zw+w^3,w+2w^2+bz^3)$ with $b\neq 0$. This map is tangent to the identity, with order 1, and the origin is an isolated fixed point. Moreover, it has two characteristic directions: $[1:0]$, degenerate Fuchsian with multiplicity 2, index 1 and induced residue -3; and $[0:1]$, non-degenerate Fuchsian with multiplicity 1, index -2 and induced residue 1. Theorem 2.2 (as well as Corollary 3.6) yields a Fatou flower tangent to $[0:1]$; on the other hand, none of the results proven up to now say anything about direction $[1:0]$.

However, a deep result by Molino gives the existence of a Fatou flower in the latter case too:

Theorem 3.7 ([Mo]) *Let $E\subset M$ be a smooth Riemann surface in a complex surface M, and take $f\in\mathrm{End}(M,E)$ tangential with order of contact ν. Let $p\in E$ be a singular point, not a corner, such that $\nu_o(f)=1$ and $\iota_p(f)\neq 0$. Then there exists a Fatou flower for f at p. More precisely:*

(i) if p is an irregular singularity, or a Fuchsian singularity with $\iota_p(f)\neq\nu\mu_p$, then there exists a Fatou flower for f with $\nu+\tau_p(\nu+1)$ petals;
(ii) if p is a Fuchsian singularity with $\iota_p(f)=\nu\mu_p$ then there exists a Fatou flower for f with ν petals.

Remark 3.6 Even more precisely, when p is Fuchsian with $\mu_p\geq 2$ and $\iota_p(f)=\nu\mu_p$ then Molino constructs parabolic curves defined on the connected components of a set of the form

$$D_{\nu+1-\frac{1}{\mu_p},\delta}=\left\{\zeta\in\mathbb{C}\;\middle|\;|\zeta^{r+1-1/\mu_p}(\log\zeta)^{1-1/\mu_p}-\delta|<\delta\right\},$$

which has at least ν connected components with the origin in the boundary.

Corollary 3.8 ([Mo]) *Let $f\in\mathrm{End}_1(\mathbb{C}^2,O)$ be tangent to the identity of order $\nu\geq 1$, and assume that O is a nondicritical singular point. Let $[v]\in\mathbb{P}^1(\mathbb{C})$ be a characteristic direction, and $\tilde f$ the blow-up of f. If $[v]$ is not a corner for $\tilde f$, $\nu_o(\tilde f)=1$ and $\iota_{[v]}(\tilde f)\neq 0$ then there exists a Fatou flower for f with at least ν petals tangent to $[v]$.*

The assumption on the pure order in these statements seems to be purely technical; so it is natural to advance the following

Conjecture 3.9 Let $E\subset M$ be a smooth Riemann surface in a complex surface M, and take $f\in\mathrm{End}(M,E)$ tangential of order of contact ν. Let $p\in E$ be a singular point, not a corner, such that $\iota_p(f)\neq 0$. Then there exists a Fatou flower for f at p.

See Sect. 5, and in particular (5.3), for examples of systems having Fatou flowers at singular points with vanishing index.

Instrumental in the proofs of Theorems 3.5 and 3.7 is a reduction of singularities statement. We shall need a few definitions:

Definition 3.9 Let $f \in \mathrm{End}_1(\mathbb{C}^n, O)$ be tangent to the identity. A *modification* of f is a $\tilde{f} \in \mathrm{End}(M, E)$ obtained as the lifting of f to a finite sequence of blow-ups, where the first one is centered in O and the remaining ones are centered in singular points of the intermediate lifted maps contained in the exceptional divisor. A modification is *non-dicritical* if none of the centers of the blow-ups is dicritical. Associated to a modification $\tilde{f} \in \mathrm{End}(M, E)$ of f we have a holomorphic map $\pi \colon M \to \mathbb{C}^n$ such that $\pi^{-1}(O) = E$, $\pi|_{M \setminus E}$ is a biholomorphism between $M \setminus E$ and $\mathbb{C}^n \setminus \{O\}$, and $f \circ \pi = \pi \circ \tilde{f}$. The exceptional divisor E is the union of a finite number of copies of $\mathbb{P}^{n-1}(\mathbb{C})$, crossing transversally.

Definition 3.10 Let $f \in \mathrm{End}_1(M, p)$ be tangent to the identity, where M is a complex surface. In local coordinates centered at p, we can write $f(z) = z + \ell(z)g^o(z)$, where $\ell = \gcd(f_1 - z_1, f_2 - z_2)$ is defined up to units. We shall say that p is an *irreducible singularity* if:

(a) $\mathrm{ord}_p(\ell) \geq 1$ and $\nu_o(f) = 1$; and
(b) if λ_1, λ_2 are the eigenvalues of the linear part of g^o then either

($\star_1$) $\lambda_1, \lambda_2 \neq 0$ and $\lambda_1/\lambda_2, \lambda_2/\lambda_1 \notin \mathbb{N}$, or
($\star_2$) $\lambda_1 \neq 0, \lambda_2 = 0$.

It turns out that there always exists a modification with only dicritical or irreducible singularities:

Theorem 3.10 ([A2]) *Let $f \in \mathrm{End}_1(\mathbb{C}^2, O)$ be tangent to the identity, and assume that O is a singular point. Then there exists a non-dicritical modification $\tilde{f} \in \mathrm{End}(M, E)$ of f such that the singular points of $\tilde{f}$ on E are either irreducible or dicritical.*

Definition 3.11 Let $f \in \mathrm{End}_1(\mathbb{C}^2, O)$ be tangent to the identity. The modification of f satisfying the conclusion of Theorem 3.10 obtained with the minimum number of blow-ups is the *minimal resolution* of f.

It is easy to see that the techniques of the proof of Theorem 2.2 yield the existence of a Fatou flower at dicritical singularities, and at irreducible singularities of type ($\star_1$) which are not a corner; then the proof of Theorem 3.5 amounts to showing that if the index is not a non-negative rational number then the minimal resolution contains at least a singularity which is either dicritical or of type ($\star_1$) and not a corner. The proof of Theorem 3.7(i) consists in showing that, under those hypotheses, the minimal resolution must contain a non-degenerate singularity, which is not a corner and where one can apply Theorem 2.2; the proof of Theorem 3.7(ii) requires instead a technically hard extension of Theorem 2.2.

See also [Ro4, Ro7] and [LS] for other approaches to resolution of singularities for germs tangent to the identity in arbitrary dimension, and [AT2, AR] for the somewhat related problem of the identification of formal normal forms for germs tangent to the identity.

4 Parabolic Domains

Theorem 2.3 yields conditions ensuring the existence of parabolic domains attached to a non-degenerate characteristic direction. In dimension 2, Vivas has found conditions ensuring the existence of a parabolic domain attached to Fuchsian and irregular degenerate characteristic directions, and Rong has found conditions ensuring the existence of a parabolic domain attached to apparent degenerate characteristic directions. Very recently, Lapan [L2] has extended Rong's approach to cover more types of degenerate characteristic directions.

More precisely, Vivas has proved the following result:

Theorem 4.1 ([V1]) *Let $f \in \mathrm{End}_1(\mathbb{C}^2, O)$ be tangent to the identity of order $\nu \geq 1$, with O nondicritical. Let $[v] \in \mathbb{P}^1(\mathbb{C})$ be a degenerate characteristic direction, and $\tilde{f}$ the blow-up of f. Denote by $\mu \geq 1$ the multiplicity , by $\tau \geq 0$ the transversal multiplicity, by $\iota \in \mathbb{C}$ the index, and by $\nu_o \geq 1$ the pure order of $\tilde{f}$ at $[v]$. Assume that either*

(a) $[v]$ is Fuchsian (thus necessarily $\tau \geq 1$ because $[v]$ is degenerate) and

$$\mathrm{Re}\,\iota + \tau > 0\,, \qquad \left|\iota + \frac{\tau}{2} - \frac{\nu\mu}{2}\right| > \frac{\tau}{2} + \frac{\nu\mu}{2}\,;$$

or

(b) $[v]$ is Fuchsian, $\nu_o = 1$ and

$$\left|\iota - \frac{\mu\nu}{2}\right| < \frac{\mu\nu}{2}\,;$$

or

(c) $[v]$ is Fuchsian, $\nu_o > 1$ and

$$\mathrm{Re}\,\iota + \tau > 0, \qquad \left|\iota - \frac{(\nu+1)\tau}{2}\right| > \frac{(\nu+1)\tau}{2}\,;$$

or

(d) $[v]$ is irregular.

Then there is a parabolic domain attached to $[v]$.

See also Remark 6.4 for a comment about the conditions on ι and τ.

To state Rong's theorem, consider a germ $f \in \mathrm{End}_1(\mathbb{C}^2, O)$ tangent to the identity of order $\nu \geq 1$, and assume that $[1 : 0]$ is a characteristic direction of f. Then we can write

$$\begin{cases} f_1(z, w) = z + az^{r+1} + O(z^{r+2}) + w\alpha(z, w), \\ f_2(z, w) = w + bz^\nu w + dz^{s+1} + O(z^{s+2}) + O(wz^{\nu+1}) + w^2\beta(z, w), \end{cases} \tag{4.1}$$

with $\nu \leq r \leq +\infty$, $\nu + 1 \leq s \leq +\infty$, $\mathrm{ord}_O(\alpha) \geq \nu$, $\mathrm{ord}_O(\beta) \geq \nu - 1$, and $a \neq 0$ if $r < +\infty$ (respectively, $d \neq 0$ if $s < +\infty$). The characteristic direction $[1 : 0]$ is non-degenerate if and only if $r = \nu$; in this case the director is given by $\frac{1}{\nu}\left(\frac{b}{a} - 1\right)$. On the other hand, saying that $[1 : 0]$ is degenerate with $b \neq 0$ is equivalent to saying that $r > \nu$ and that $[1 : 0]$ has multiplicity 1 and transversal multiplicity at least 1; in particular, in this case it is an apparent singularity. Then Rong's theorem can be stated as follows:

Theorem 4.2 ([Ro9]) *Let $f \in \mathrm{End}_1(\mathbb{C}^2, O)$ be tangent to the identity of order $\nu \geq 1$ and written in the form* (4.1)*, so that* $[1 : 0]$ *is a characteristic direction. Assume that $r > \nu$ and $b \neq 0$, so that* $[1 : 0]$ *is an apparent degenerate characteristic direction. Suppose furthermore that $s > r$, and that $b^2/a \notin \mathbb{R}^+$ if $r = 2\nu$. Then there is a parabolic domain attached to* $[1 : 0]$.

To state Lapan's result we need to introduce a few definitions.

Definition 4.1 Let $f \in \mathrm{End}_1(\mathbb{C}^2, O)$ be tangent to the identity of order $\nu \geq 1$ with homogeneous expansion (1.1). We say that $[v] \in \mathbb{P}^1(\mathbb{C})$ is a *characteristic direction of degree* $s \geq \nu + 1$ if it is a characteristic direction of $P_{\nu+1}, \ldots, P_s$. We shall say that it is *non-degenerate in degree* $r + 1$, with $\nu < r < s$, if it is degenerate for $P_{\nu+1}, \ldots, P_r$ and non-degenerate for P_{r+1}.

For instance, if f is in the form (4.1) with $s < +\infty$, then $[1 : 0]$ is a characteristic direction of degree s. If furthermore $r+1 \leq s$ then it is non-degenerate in degree $r+1$.

Definition 4.2 Let $f \in \mathrm{End}_1(\mathbb{C}^2, O)$ be tangent to the identity of order $\nu \geq 1$ with homogeneous expansion (1.1). Assume that $[1 : 0] \in \mathbb{P}^1(\mathbb{C})$ is a characteristic direction of degree $s \geq \nu + 1$. Given $\nu + 1 \leq j \leq s$, the *j-order* of $[1 : 0]$ is the order of vanishing at 0 of $P_{j,2}(1, \cdot)$, where $P_j = (P_{j,1}, P_{j,2})$. We say that $[1; 0]$ is *of order one in degree* $t + 1$, with $\nu \leq t < s$, if the j-order of $[1 : 0]$ is larger than one for $\nu + 1 \leq j \leq t$ and of $(t + 1)$-order exactly equal to 1.

For instance, if $b \neq 0$ in (4.1) then $[1 : 0]$ is of order one in degree $\nu + 1$. More generally, if $b = 0$ and $[1 : 0]$ is of order one in degree $t + 1$ then we can replace the term $O(wz^{\nu+1})$ by $O(wz^{t+1})$.

Assume that $[1 : 0]$ is a characteristic direction of degree $s < +\infty$, non-degenerate in degree $r + 1 \leq s$ and of order one in degree $t + 1 \leq s$. Then we can write

$$\begin{cases} f_1(z, w) = z + az^{r+1} + O(z^{r+2}) + w\alpha(z, w), \\ f_2(z, w) = w + bz^{t+1}w + dz^{s+1} + O(z^{s+2}) + O(wz^{t+2}) + w^2\beta(z, w), \end{cases} \tag{4.2}$$

with $abd \neq 0$. Then Lapan's theorem can be stated as follows:

Theorem 4.3 ([L2]) *Let $f \in \mathrm{End}_1(\mathbb{C}^2, O)$ be tangent to the identity of order $\nu \geq 1$ and written in the form* (5.1), *so that* $[1:0]$ *is a characteristic direction of degree $s < +\infty$, non-degenerate in degree $\nu+1 \leq r+1 \leq s$, and of order one in degree $t+1 \leq s$. Assume that $t \leq r$ and $s > r+t-\nu$. Suppose furthermore that either*

(i) $r \neq t, 2t$, or
(ii) $r = t$ and $Re(b/a) > 0$, or
(iii) $r = 2t$ and $b^2/a \notin \mathbb{R}^+$.

Then there is a parabolic domain attached to $[1:0]$.

The assumptions of Theorem 4.2 imply that $[1:0]$ is a characteristic direction of degree s, non-degenerate in degree $\nu+1 < r+1 \leq s$, and of order one in degree $t+1 = \nu+1$; therefore Theorem 4.2 is a particular case of Theorem 4.3.

Parabolic domains are often used to build Fatou-Bieberbach domains, that is proper subsets of $\mathbb{C}^n$ biholomorphic to $\mathbb{C}^n$; see, e.g., [V2, SV] and references therein.

5 The Formal Infinitesimal Generator

A different approach to the study of parabolic curves in $\mathbb{C}^2$ has been suggested by Brochero-Martínez, Cano and López-Hernanz [BCL], and further developed by Câmara and Scárdua [CaS] and by Lopez-Hernanz and Sánchez [LS]. It consists in using the formal infinitesimal generator of a germ tangent to the identity. To describe their approach, we need to introduce several definitions.

Definition 5.1 We shall denote by $\widehat{\mathcal{O}}_n = \mathbb{C}[\![z_1, \ldots, z_n]\!]$ the space of formal power series in n variables. The *order* $\mathrm{ord}(\hat{\ell})$ of $\hat{\ell} \in \widehat{\mathcal{O}}_n$ is the lowest degree of a non-vanishing term in the Taylor expansion of $\hat{\ell}$. A *formal map* is a n-tuple of formal power series in n variables; the space of formal maps will be denoted by $\widehat{\mathcal{O}_n}^n$. We shall denote by $\widehat{\mathrm{End}}(\mathbb{C}^n, O)$ the set of formal maps with vanishing constant term; by $\widehat{\mathrm{End}_1}(\mathbb{C}^n, O)$ the subset of formal maps tangent to the identity, and by $\widehat{\mathrm{End}_\nu}(\mathbb{C}^n, O)$ the subset of formal maps tangent to the identity of order at least $\nu \geq 1$.

Definition 5.2 We shall denote by $\mathcal{X}_n$ the space of germs at the origin of holomorphic vector fields in $\mathbb{C}^n$. A *formal vector field* is an expression of the form $\hat{X} = \hat{X}_1 \frac{\partial}{\partial z_1} + \cdots + \hat{X}_n \frac{\partial}{\partial z_n}$ where $\hat{X}_1, \ldots, \hat{X}_n \in \widehat{\mathcal{O}}_n$ are the *components* of $\hat{X}$. The space of *formal vector fields* will be denoted by $\widehat{\mathcal{X}_n}$. The *order* $\mathrm{ord}(\hat{X})$ of $\hat{X} \in \widehat{\mathcal{X}_n}$ is the minimum among the orders of its components. We put $\widehat{\mathcal{X}_n^k} = \{\hat{X} \in \widehat{\mathcal{X}_n} \mid \mathrm{ord}(\hat{X}) \geq k\}$. If $\hat{X} \in \widehat{\mathcal{X}_n^k}$, the *principal part* of $\hat{X}$ will be the unique polynomial homogeneous vector field H_k of degree exactly k such that $\hat{X} - H_k \in \widehat{\mathcal{X}_n^{k+1}}$. A *characteristic direction* for $\hat{X}$ is an invariant line for H_k.

Remark 5.1 There is a clear bijection between $\widehat{\mathscr{X}_n}$ and $\widehat{\mathscr{O}_n}^n$ obtained by associating to a formal vector field the n-tuple of its components; so we shall sometimes identify formal vector fields and formal maps without comments. In particular, this bijection preserves characteristic directions.

If $X \in \mathscr{X}_n$ is a germ of holomorphic vector field vanishing at the origin (that is, of order at least 1), the associated time-1 map f_X will be a well defined germ in $\mathrm{End}(\mathbb{C}^n, O)$, that can be recovered as follows (see, e.g., [BCL]):

$$f_X = \sum_{k\geq 0} \frac{1}{k!} X^{(k)}(\mathrm{id}) , \qquad (5.1)$$

where $X^{(k)}$ is the k-th iteration of X thought of as derivation of $\mathrm{End}(\mathbb{C}^n, O)$. Now, not every germ in $\mathrm{End}(\mathbb{C}^n, O)$ can be obtained as a time-1 map of a convergent vector field (see, e.g., [IY, Theorem 21.31]). However, it turns out that the right-hand side of (5.1) is well-defined as a formal map for all $X \in \widehat{\mathscr{X}_n^1}$.

Definition 5.3 The *exponential map* $\exp\colon \widehat{\mathscr{X}_n^1} \to \widehat{\mathrm{End}(\mathbb{C}^n, O)}$ is defined by the right-hand side of (5.1).

When $k \geq 2$, if $\hat{X} \in \widehat{\mathscr{X}_n^k}$ has principal part H_k then it is easy to check that

$$\exp(\hat{X}) = \mathrm{id} + H_k + \mathrm{h.o.t.} ; \qquad (5.2)$$

in particular, the exponential of a formal vector field of order k is a formal map tangent to the identity of order $k - 1$. Takens (see, e.g., [IY, Theorem 3.17]) has shown that on the formal level the exponential map is bijective:

Proposition 5.1 *The exponential map* $\exp\colon \widehat{\mathscr{X}_n^{\nu+1}} \to \widehat{\mathrm{End}_\nu}(\mathbb{C}^n, O)$ *is bijective for all* $\nu \geq 1$.

Definition 5.4 If $\hat{f} \in \widehat{\mathrm{End}_\nu}(\mathbb{C}^n, O)$, the unique formal vector field $\hat{X} \in \widehat{\mathscr{X}_n^{\nu+1}}$ such that $\exp(\hat{X}) = \hat{f}$ is the *formal infinitesimal generator* of $\hat{f}$.

The idea now is to read properties of a holomorphic germ tangent to the identity from properties of its formal infinitesimal generator, using Theorem 2.2 as bridge for going back from the formal side to the holomorphic side.

The first observation is that if $\pi\colon (\widetilde{\mathbb{C}^2}, E) \to (\mathbb{C}^2, O)$ is the blow-up of the origin, $\hat{X} \in \widehat{\mathscr{X}_2^2}$ is a formal vector field and $[v] \in E$ is a characteristic direction of (the principal part of) $\hat{X}$ then we can find a formal vector field $\hat{X}_{[v]} \in \widehat{\mathscr{X}_2^2}$ such that $d\pi(\hat{X}_{[v]}) = \hat{X} \circ \pi$. This lifting is compatible with the exponential in the following sense:

Proposition 5.2 ([BCL]) *Let* $f \in \mathrm{End}_1(\mathbb{C}^2, O)$ *be tangent to the identity with formal infinitesimal generator* $\hat{X} \in \widehat{\mathscr{X}_2^2}$, *and let* $\tilde{f} \in End(\widetilde{\mathbb{C}^2}, E)$ *be the blow-up of* f. *Let* $[v] \in E$ *be a characteristic direction of* f, *and denote by* $\tilde{f}_{[v]}$ *the germ of* $\tilde{f}$ *at* $[v]$. *Then* $\tilde{f}_{[v]} = \exp(\hat{X}_{[v]})$.

In particular, Brochero-Martínez, Cano and López-Hernanz's proofs of Theorems 2.5 and 3.5 go as follows: let $\hat{X} \in \mathcal{X}_2^2$ be the formal infinitesimal generator of $f \in \mathrm{End}_1(\mathbb{C}^2, O)$ with an isolated fixed point (so that $\hat{X}$ has an isolated singular point at the origin). Then the formal version of Camacho-Sad's theorem [CS] (see also [Ca]) shows that we can find a finite composition $\pi: (M, E) \to (\mathbb{C}^2, O)$ of blow-ups at singular points and a smooth point $p \in E$ such that the lifting $\hat{X}_p$ of $\hat{X}$, in suitable coordinates centered at p adapted to E (in the sense that E is given by the equation $\{z = 0\}$ near p), has the expression

$$\hat{X}_p(z, w) = z^m \left(\left(\lambda_1 z + O(z^2)\right) \frac{\partial}{\partial z} + \left(\lambda_2 w + O(z)\right) \frac{\partial}{\partial w} \right)$$

with $\lambda_1 \neq 0$, $\lambda_2/\lambda_1 \notin \mathbb{Q}^+$ and $m \geq \mathrm{ord}(\hat{X}) - 1$. Then $\exp(\hat{X}_p)$ has the form

$$\exp(\hat{X}_p)(z, w) = \left(z + \lambda_1 z^{m+1} + O(z^{m+2}), w + \lambda_2 z^m w + O(z^{m+1})\right),$$

which has a non-degenerate characteristic direction transversal to E — and hence a Fatou flower outside the exceptional divisor. By Proposition 5.2, $\exp(\hat{X}_p)$ is the blow-up of $\exp(\hat{X}) = f$; therefore projecting this Fatou flower down by π we get a Fatou flower for f.

In [CaS] and [LS] this approach has been pushed further showing how to relate formal separatrices and parabolic curves.

Definition 5.5 A *formal curve* $\hat{C}$ in $(\mathbb{C}^2, 0)$ is a reduced principal ideal of $\widehat{\mathcal{O}}_2$. Any generator of the ideal is an *equation* of the curve; the equation is defined up to an unit in $\widehat{\mathcal{O}}_2$. The *tangent cone* of a formal curve $\hat{C}$ is the set of zeros of the homogeneous part of least degree of any equation of $\hat{C}$; the *tangent directions* to $\hat{C}$ are the points in $\mathbb{P}^1(\mathbb{C})$ determined by the tangent cone.

Remark 5.2 It is known that a formal curve $\hat{C}$ is irreducible if and only if it has a unique tangent direction.

Definition 5.6 Let $\hat{X} \in \widehat{\mathcal{X}_2^2}$. A *singular formal curve* for $\hat{X}$ is a formal curve $\hat{C} = (\hat{\ell})$ such that $\hat{X} = \hat{\ell}\hat{X}_1$ for some $\hat{X} \in \widehat{\mathcal{X}_2^1}$. A *formal separatrix* of $\hat{X}$ is a formal curve $\hat{C} = (\hat{\ell})$ such that $\hat{X}(\hat{\ell}) \in (\hat{\ell})$. Clearly singular formal curves are formal separatrices.

The corresponding notions for germs tangent to the identity are:

Definition 5.7 Let $f \in \mathrm{End}_1(\mathbb{C}^2, O)$. A formal curve $\hat{C} = (\hat{\ell})$ is a *formal separatrix* for f if $\hat{\ell} \circ f \in (\hat{\ell})$. In particular, this means that f acts by composition on $\widehat{\mathcal{O}}_2/(\hat{\ell})$; if the action is the identity, we say that $\hat{C}$ is *completely fixed* by f. Notice that $\hat{C}$ is completely fixed by f if and only if we can write $f = \mathrm{id} + \hat{\ell}\hat{g}$ for some $\hat{g} \in \widehat{\mathcal{O}}_2^2$.

Proposition 5.3 ([CaS]) *Let $\hat{X} \in \widehat{\mathcal{X}_2^2}$ be the formal infinitesimal generator of $f \in$ $End_1(\mathbb{C}^2, O)$. Then:*

(i) *a formal curve is a formal separatrix for f if and only if it is a formal separatrix for $\hat{X}$;*
(ii) *a formal curve is completely fixed for f if and only if it is a singular formal curve for $\hat{X}$;*
(iii) *a completely fixed curve for f always has a convergent equation;*
(iv) *the tangent directions to a formal separatrix are characteristic directions for f, and the tangent directions to a completely fixed curve are degenerate characteristic directions for f.*

Let $\hat{C} = (\hat{\ell})$ be a formal curve, and $[v] \in \mathbb{P}^1(\mathbb{C})$ a tangent direction to $\hat{C}$. If $\pi: (\widetilde{\mathbb{C}^2}, E) \to (\mathbb{C}^2, O)$ is the blow-up of the origin, we can find a formal curve $\pi^*\hat{C}_{[v]} = (\hat{\ell}_{[v]})$ at $[v]$ such that $\hat{\ell}_{[v]} = \hat{\ell} \circ \pi$; the tangent directions to $\pi^*\hat{C}_{[v]}$ are higher order tangent directions of $\hat{C}$. This construction can be iterated, and it gives a way of lifting formal curves along a finite sequence of blow-ups. Using this idea, and a generalization of Hakim's technique, López-Hernanz and Sánchez have been able to prove the following

Theorem 5.4 *(López-Hernanz and Sánchez,* [LS]*) Let $f \in \mathrm{End}_1(\mathbb{C}^2, O)$ be a germ tangent to the identity admitting a formal separatrix $\hat{C}$ not completely fixed. Then f or f^{-1} (or both) admit a parabolic curve tangent to (a tangent direction of) $\hat{C}$.*

Remark 5.3 Actually, [LS, Theorem 1] gives the more precise statement that the parabolic curve $\phi: D \to \mathbb{C}^2$ is *asymptotic* to the formal separatrix $\hat{C}$. This means that there exists a formal parametrization $\hat{\gamma} \in \widehat{\mathscr{O}_1}^2$ of $\hat{C}$ such that for every $N \in \mathbb{N}$ there exists $c_N > 0$ such that

$$|\phi(\zeta) - (J_N\hat{\gamma})(\zeta)| \le c_N|\zeta|^{N+1}$$

for all $\zeta \in D$, where $J_N\hat{\gamma}$ is the N-th jet of $\hat{\gamma}$. A formal parametrization of $\hat{C}$ is a formal map $\hat{\gamma} \in \widehat{\mathscr{O}_1}^2$ such that $g \in \hat{C}$ if and only if $g \circ \hat{\gamma} \equiv 0$.

Remark 5.4 In [CaS] Câmara and Scárdua claimed that under the hypotheses of Theorem 5.4 f must admit a parabolic curve tangent to $\hat{C}$. Unfortunately, the core of their argument was [CaS, Proposition 2.12], and in its proof they forgot to consider vector fields of the form $\hat{X}^o(z,w) = z(1+\lambda w^p)\frac{\partial}{\partial z} + w^{p+1}\frac{\partial}{\partial w}$, where their approach does not work.

The proof of Theorem 5.4 has three steps. First of all, the authors show that, assuming the existence of a formal separatrix not completely fixed, after a finite number of blow-ups the germ f can be brought in the following normal form:

$$\begin{cases} f_1(z,w) = z + z^{\nu+p+1}\big(\lambda + \psi(z,w)\big)\,, \\ f_2(z,w) = w + z^{\nu}\big(b(z) + a(z)w + O(w^2)\big)\,, \end{cases} \tag{5.3}$$

with $p \ge 0$, $\lambda \ne 0$, $\mathrm{ord}_O\,\psi \ge 1$, $a(0) \ne 0$ and $b(0) = 0$. For a germ in this form, a *real attracting direction* is a $\tau \in S^1$ such that $\tau^{\nu+p}\lambda = -1$. Then, generalizing

Hakim's proof of Theorem 2.2, the authors show that if f has a real attracting direction τ such that $\mathrm{Re}\left(\frac{a(0)}{\lambda\tau^p}\right) < 0$ then f admits a parabolic curve, that turns out to be asymptotic to $\hat{C}$. Finally, they show that at least one between f and f^{-1} have a real attracting direction satisfying the given condition.

Notice that a germ in the form (5.3) has pure order 1, but vanishing index if $p \geq 1$; so we cannot apply Theorem 3.7. On the other hand, there are germs tangent to the identity admitting parabolic curves thanks to Theorem 3.7 but without formal separatrices not completely fixed:

Example 5.1 Let $f = \exp(z^\nu X^o)$, with $\nu \geq 2$ and

$$\hat{X}^o(z, w) = z\big(\lambda + A(z, w)\big)\frac{\partial}{\partial z} + \big(z + \lambda w + B(z, w)\big)\frac{\partial}{\partial w} ,$$

with $\lambda \neq 0$, $\mathrm{ord}_O(A) \geq 1$ and $\mathrm{ord}_O(B) \geq 2$. Then

$$f(z, w) = \Big(z + z^{\nu+1}\big(\lambda + A(z, w)\big) + O(z^{2\nu+1}), w + z^\nu\big(z + \lambda w + B(z, w)\big) + O(z^{2\nu})\Big) .$$

The germ f has pure order 1 and the linear part of g^o is not diagonalizable; since the index of f at O along the fixed point set is 1, Theorem 3.7 yields a Fatou flower. Furthermore, f has a unique (degenerate) characteristic direction, $[0 : 1]$. Blowing up and looking in the coordinates centered at $[0 : 1]$ we get

$$\begin{aligned}\tilde{f}(u, w) &= \Big(u + u^{\nu+1}w^\nu\big(-u + w\hat{A}(u, w)\big), w + u^\nu w^{\nu+1}\big(\lambda + u + w\hat{B}(u, w)\big)\Big)\\ &= \exp(u^\nu w^\nu \tilde{X}^o) ,\end{aligned}$$

where $\tilde{X}^o = u\hat{C}(u, w)\frac{\partial}{\partial u} + w\big(\lambda + \hat{D}(u, w)\big)\frac{\partial}{\partial w}$ with $\mathrm{ord}_O(\hat{C}), \mathrm{ord}_O(\hat{D}) \geq 1$. Since the linear part of $\tilde{X}^o$ is diagonalizable, X^o has exactly two formal separatrices, necessarily given by the axes. It follows that all formal separatrices of $\tilde{f}$ are completely fixed; so to $\tilde{f}$ we cannot apply Theorem 5.4, but $\tilde{f}$ still has a Fatou flower because f does.

The paper [LS] also indicates a way to adapt these techniques to more than two variables. However, it should be kept in mind that [AT1] contains examples in $\mathbb{C}^3$ of germs tangent to the identity without parabolic curves asymptotic to formal separatrices.

6 Homogeneous Vector Fields and Geodesics

None of the results presented up to now (with the partial exception of Proposition 3.1) describe the dynamics in a full neighborhood of the fixed point, and so in this sense they are not a complete generalization of the Leau-Fatou flower theorem. As far as we know, up to now the only techniques able to give results in a full neighborhood are the ones introduced in [AT3], that we shall briefly describe now.

We have seen that every germ tangent to the identity can be realized as the time-1 map of a formal vector field of order at least 2; and that a lot of information can be deduced from the principal part of this vector field, principal part which is a *homogeneous* vector field. Furthermore, in dimension 1 Camacho-Shcherbakov theorem (see [C, Sh]) says that every germ tangent to the identity is locally topologically conjugated to the time-1 map of a homogeneous vector field. So time-1 maps of homogeneous vector fields clearly are an important class of examples; and the insights we obtain from their study (and, more generally, from the study of the real dynamics of homogeneous vector fields) can shed light on the dynamics of more general germs tangent to the identity.

The work described in [AT3] had exactly the aim of studying the real dynamics of homogenous vector fields in $\mathbb{C}^n$; for the sake of clarity, here we shall summarize only the results in $\mathbb{C}^2$ only, referring to [AT3] for more general statements.

Let $H \in \mathscr{X}_2^{\nu+1}$ be a homogeneous vector field in $\mathbb{C}^2$ of degree $\nu + 1 \geq 2$. It clearly determines a homogeneous self-map of $\mathbb{C}^2$ of the same degree; in particular, we can adapt to H all the definitions we introduced for homogeneous self-maps (degenerate/non-degenerate characteristic directions, multiplicities, index, induced residue, being dicritical).

Definition 6.1 Let $H \in \mathscr{X}_2$ be a homogeneous vector field in $\mathbb{C}^2$. A *characteristic line* for H is a line $L_v = \mathbb{C}v$ which is H-invariant, that is such that $[v] \in \mathbb{P}^1(\mathbb{C})$ is a characteristic direction.

If $L_v = \mathbb{C}v$ is a characteristic line then integral lines of H issuing from points in L_v stay inside L_v. If $[v]$ is degenerate, H vanishes identically along L_v, and so the dynamics there is trivial. If $[v]$ is non-degenerate, then the dynamics inside L_v is one-dimensional, and can be summarized as follows:

Lemma 6.1 *Let $[v] \in \mathbb{P}^1(\mathbb{C})$ be a non-degenerate characteristic direction of a homogeneous vector field $H = H_1\frac{\partial}{\partial z_1} + H_2\frac{\partial}{\partial z_2} \in \mathscr{X}_2^{\nu+1}$ of degree $\nu + 1 \geq 2$. Choose a representative $v \in \mathbb{C}^2$ so that $H(v) = v$. Then the real integral curve of H issuing from $\zeta_0 v \in L_v$ is given by*

$$\gamma_{\zeta_0 v}(t) = \frac{\zeta_0 v}{(1 - \zeta_0^\nu \nu t)^{1/\nu}} .$$

In particular no (non-constant) integral curve is recurrent, and we have:

(a) if $\zeta_0^\nu \notin \mathbb{R}^+$ then $\lim_{t\to+\infty} \gamma_{\zeta_0 v}(t) = O$;
(b) if $\zeta_0^\nu \in \mathbb{R}^+$ then $\lim_{t\to(\zeta_0^\nu \nu)^{-1}} \|\gamma_{\zeta_0 v}(t)\| = +\infty$.

Lemma 6.1 completely describes the dynamics of dicritical homogeneous vector fields. More precisely, we see that every non-degenerate characteristic line contains a Fatou flower; thus in this case Theorem 2.2 becomes trivial, and we can shift our interest to the understanding of the dynamics outside the characteristic lines. To do so we need to introduce a new ingredient:

Definition 6.2 Let ∇^o be a meromorphic connection on $\mathbb{P}^1(\mathbb{C})$ (see [IY] for an introduction to meromorphic connections), and denote by $\mathrm{Sing}(\nabla^o)$ the set of its poles. A *geodesic* for ∇^o is a smooth real curve $\sigma\colon I \to \mathbb{P}^1(\mathbb{C}) \setminus \mathrm{Sing}(\nabla^o)$ such that

$$\nabla^o_{\dot\sigma}\dot\sigma \equiv O.$$

The main result allowing the understanding of the real dynamics of homogeneous vector fields is the following:

Theorem 6.2 ([AT3]) *Let $H \in \mathscr{X}_2^{\nu+1}$ be a non-dicritical homogeneous vector field of degree $\nu + 1 \geq 2$ in $\mathbb{C}^2$, and denote by V_H the complement in $\mathbb{C}^2$ of the characteristic lines of H. Then there exists a meromorphic connection ∇^o on $\mathbb{P}^1(\mathbb{C})$, whose poles are a (possibly proper) subset of the characteristic directions of H, such that:*

(i) if $\gamma\colon I \to V_H$ is a real integral curve of H then its direction $[\gamma]\colon I \to \mathbb{P}^1(\mathbb{C})$ is a geodesic for ∇^o; conversely,
(ii) if $\sigma\colon I \to \mathbb{P}^1(\mathbb{C})$ is a geodesic for ∇^o then there exists exactly ν real integral curves $\gamma_1, \ldots, \gamma_\nu\colon I \to V_H$ of H, differing only by the multiplication by a ν-th root of unity, whose direction is given by σ, that is such that $\sigma = [\gamma_j]$.

Remark 6.1 If $H = H_1\frac{\partial}{\partial z_1} + H_2\frac{\partial}{\partial z_2} \in \mathscr{X}_2^{\nu+1}$ is a homogeneous vector field of degree $\nu + 1 \geq 2$, the meromorphic 1-form representing ∇^o in the standard chart centered at $0 \in \mathbb{P}^1(\mathbb{C})$ is given by ([AT3])

$$\eta^o = -\left[\nu\frac{H_1(1,\zeta)}{R(\zeta)} + \frac{R'(\zeta)}{R(\zeta)}\right] d\zeta,$$

where $R(\zeta) = H_2(1,\zeta) - \zeta H_1(1,\zeta)$; a similar formula, exchanging the rôle of z_1 and z_2, holds in the standard chart centered at $\infty \in \mathbb{P}^1(\mathbb{C})$. In particular recalling (5.2), (3.1) and Definition 3.7 we see that the poles of ∇^o are singular points for the blow-up $\tilde f$ of the time-1 map of H, and that the residue of ∇^o at a pole $p \in \mathbb{P}^1(\mathbb{C})$ coincides with the induced residue of $\tilde f$ at p.

Furthermore, in [AT3] we introduced another meromorphic connection ∇ defined on the ν-th tensor power $N_E^{\otimes\nu}$ of the normal bundle N_E of the exceptional divisor $E = \mathbb{P}^1(\mathbb{C})$ in the blow-up of the origin in $\mathbb{C}^2$. The meromorphic 1-form representing ∇ in the standard chart centered at $0 \in \mathbb{P}^1(\mathbb{C})$ is given by

$$\eta = -\nu\frac{H_1(1,\zeta)}{R(\zeta)}\, d\zeta.$$

Therefore the poles of ∇ are exactly the Fuchsian and irregular characteristic directions of $\tilde f$, and the residue of ∇ at a pole $p \in \mathbb{P}^1(\mathbb{C})$ coincides with the opposite of the index of $\tilde f$ at p.

So the study of the real integral curves of H is reduced to the study of the geodesics of a meromorphic connection on $\mathbb{P}^1(\mathbb{C})$. This study is subdivided in two parts: the study of the global behavior of geodesics, and the study of the local behavior nearby the poles. It turns out that the global behavior is related to the induced residues, while the local behavior is mainly related to the index. To state our results we need a couple of definitions.

Definition 6.3 A geodesic $\sigma\colon [0,\ell] \to \mathbb{P}^1(\mathbb{C})$ for a meromorphic connection ∇^o is *closed* if $\sigma(\ell) = \sigma(0)$ and $\sigma'(\ell)$ is a positive multiple of $\sigma'(0)$; it is *periodic* if $\sigma(\ell) = \sigma(0)$ and $\sigma'(\ell) = \sigma'(0)$.

Contrarily to the case of Riemannian geodesics, closed geodesics are not necessarily periodic; see [AT3]. The (induced) residue allows to recognize closed and periodic geodesics:

Proposition 6.3 ([AT3]) *Let ∇^o be a meromorphic connection on $\mathbb{P}^1(\mathbb{C})$, with poles $\{p_0, p_1, \dots, p_r\}$, and set $S = \mathbb{P}^1(\mathbb{C}) \setminus \{p_0, \dots, p_r\} \subseteq \mathbb{C}$. Let $\sigma\colon [0,\ell] \to S$ be a geodesic with $\sigma(0) = \sigma(\ell)$ and no other self-intersections; in particular, σ is an oriented Jordan curve. Let $\{p_1, \dots, p_g\}$ be the poles of ∇^o contained in the interior of σ. Then σ is a closed geodesic if and only if*

$$\sum_{j=1}^{g} \operatorname{Re}\operatorname{Res}_{p_j}(\nabla^o) = -1 ,$$

and it is a periodic geodesic if and only if

$$\sum_{j=1}^{g} \operatorname{Res}_{p_j}(\nabla^o) = -1 .$$

If σ is closed, it can be extended to an infinite length geodesic $\sigma\colon J \to S$, where J is a half-line (possibly $J = \mathbb{R}$). Moreover,

(i) if $\sum_{j=1}^{g} \operatorname{Im}\operatorname{Res}_{p_j}(\nabla) < 0$ then $\sigma'(t) \to O$ as $t \to +\infty$ and $|\sigma'(t)| \to +\infty$ as t tends to the other end of J;

(ii) if $\sum_{j=1}^{g} \operatorname{Im}\operatorname{Res}_{p_j}(\nabla) > 0$ then $\sigma'(t) \to O$ as $t \to -\infty$ and $|\sigma'(t)| \to +\infty$ as t tends to the other end of J.

Corollary 6.4 *Let $\gamma\colon \mathbb{R} \to \mathbb{C}^2 \setminus \{O\}$ be a non-constant periodic integral curve of a homogeneous vector field H of degree $\nu + 1 \geq 2$. Then the characteristic directions $[v_1], \dots, [v_g] \in \mathbb{P}^1(\mathbb{C})$ surrounded by $[\gamma]$ satisfy*

$$\sum_{j=1}^{g} \operatorname{Res}^0_{[v_j]}(H) = -1 ,$$

where $\mathrm{Res}^0_{[v_j]}(H)$ *denotes the induced residue at* $[v_j]$ *of the blow-up of the time-*1 *map of* H.

Closed but not periodic geodesics correspond to integral curves converging to the origin on one side and escaping to infinity on the other side; the convergence to the origin occurs along a spiral, and thus the time-1 map has orbits converging to the origin without being tangent to any direction. This can actually happen; see [AT3] for an example.

Definition 6.4 Let $\sigma\colon I \to S$ be a curve in $S = \mathbb{P}^1(\mathbb{C}) \setminus \{p_0, \ldots, p_r\}$. A *simple loop in* σ is the restriction of σ to a closed interval $[t_0, t_1] \subseteq I$ such that $\sigma|_{[t_0,t_1]}$ is a simple loop τ. If $p_1, \ldots, p_g$ are the poles of ∇ contained in the interior of τ, we shall say that τ *surrounds* $p_1, \ldots, p_g$.

Definition 6.5 A *saddle connection* for a meromorphic connection ∇^o on $\mathbb{P}^1(\mathbb{C})$ is a maximal geodesic $\sigma\colon (\varepsilon_-, \varepsilon_+) \to \mathbb{P}^1(\mathbb{C})$ (with $\varepsilon_- \in [-\infty, 0)$ and $\varepsilon_+ \in (0, +\infty]$) such that $\sigma(t)$ tends to a pole of ∇^o both when $t \uparrow \varepsilon_+$ and when $t \downarrow \varepsilon_-$. A *graph of saddle connections* is a connected graph in $\mathbb{P}^1(\mathbb{C})$ made up of saddle connections.

Then we have a Poincaré-Bendixson type theorem, describing the asymptotic behavior of geodesics:

Theorem 6.5 ([AT3, AB]) *Let* $\sigma\colon [0, \varepsilon_0) \to S$ *be a maximal geodesic for a meromorphic connection* ∇^o *on* $\mathbb{P}^1(\mathbb{C})$, *where* $S = \mathbb{P}^1(\mathbb{C}) \setminus \{p_0, \ldots, p_r\}$ *and* $p_0, \ldots, p_r$ *are the poles of* ∇^o. *Then either*

(i) $\sigma(t)$ *tends to a pole of* ∇^o *as* $t \to \varepsilon_0$*; or*
(ii) σ *is closed, and then surrounds poles* $p_1, \ldots, p_g$ *with* $\sum_{j=1}^{g} \mathrm{Re}\,\mathrm{Res}_{p_j}(\nabla^o) = -1$*; or*
(iii) the ω*-limit set of* σ *in* $\mathbb{P}^1(\mathbb{C})$ *is given by the support of a closed geodesic surrounding poles* $p_1, \ldots, p_g$ *with* $\sum_{j=1}^{g} \mathrm{Re}\,\mathrm{Res}_{p_j}(\nabla^o) = -1$*; or*
(iv) the ω*-limit set of* σ *in* $\mathbb{P}^1(\mathbb{C})$ *is a graph of saddle connections whose complement in* $\mathbb{P}^1(\mathbb{C})$ *has a connected component containing* $p_1, \ldots, p_g$ *with* $\sum_{j=1}^{g} \mathrm{Re}\,\mathrm{Res}_{p_j}(\nabla^o) = -1$*; or*
(v) σ *intersects itself infinitely many times, and in this case every simple loop of* σ *surrounds a set of poles whose sum of residues has real part belonging to* $(-3/2, -1) \cup (-1, -1/2)$.

In particular, a recurrent geodesic either intersects itself infinitely many times or is closed.

Corollary 6.6 *Let* H *be a homogeneous holomorphic vector field on* $\mathbb{C}^2$ *of degree* $\nu + 1 \geq 2$, *and let* $\gamma\colon [0, \varepsilon_0) \to \mathbb{C}^2$ *be a recurrent maximal integral curve of* Q. *Then* γ *is periodic or* $[\gamma]\colon [0, \varepsilon_0) \to \mathbb{P}^1(\mathbb{C})$ *intersects itself infinitely many times.*

Remark 6.2 We have examples (see [AT3]) of cases (i), (ii), (iii) and (v), but not yet of case (iv).

It is worthwhile to notice that the maximal geodesics of generic meromorphic connections will behave as in case (i), because the other cases require that a particular relationships between the (induced) residues should hold. In particular, this means that the direction of a maximal real integral curve of a generic homogeneous vector field will go from a characteristic line to a characteristic line (possibly the same); the next step then consists in understanding what happens nearby characteristic lines. It turns out that we can find holomorphic normal forms for apparent and Fuchsian singularities, and formal normal forms for irregular singularities; and we shall see that the local behavior is mostly related to the index.

To key behind this local study is the following

Theorem 6.7 ([AT3]) *Let N_E be the normal bundle of the exceptional divisor of the blow-up (M, E) of the origin in $\mathbb{C}^2$. Then for every $\nu \geq 1$ there exists a holomorphic ν-to-1 covering $\chi_\nu \colon \mathbb{C}^2 \setminus \{O\} \to N_E^{\otimes \nu} \setminus E$ satisfying $\pi \circ \chi_\nu(v) = [v]$, where $\pi \colon N_E^{\otimes \nu} \to E = \mathbb{P}^1(\mathbb{C})$ is the canonical projection, such that for every homogeneous vector field $H \in \mathcal{X}_2^{\nu+1}$ of degree $\nu + 1$ the push-forward $d\chi_\nu(H)$ defines a global holomorphic vector field G on the total space of $N_E^{\otimes \nu}$. In particular, a real curve $\gamma \colon I \to \mathbb{C}^2 \setminus \{O\}$ is an integral curve for H if and only if $\chi_\nu \circ \gamma$ is an integral curve for G.*

Definition 6.6 The field G is the *geodesic field* associated to the homogeneous vector field H. The reason of the name is that the projections on $\mathbb{P}^1(\mathbb{C})$ of the integral curves of G are geodesics for the connection ∇^o associated to H.

The point is that the field G has a form well suited to reduction to normal form. Indeed, if we denote by ζ the usual coordinate on $\mathbb{C} \subset \mathbb{P}^1(\mathbb{C})$ centered at the origin, and by v the corresponding coordinate on the fibers of $N_E^{\otimes \nu}$, which over $\mathbb{C}$ is canonically trivialized, we have

$$G(\zeta, v) = R(\zeta) v \frac{\partial}{\partial \zeta} + \nu H_1(1, \zeta) v^2 \frac{\partial}{\partial v} ,$$

where $R(\zeta) = H_2(1, \zeta) - \zeta H_1(1, \zeta)$ as before; and a similar formula holds in the usual coordinates centered at $\infty \in \mathbb{P}^1(\mathbb{C})$. In particular, we can read the multiplicity and the transversal multiplicity (and hence the type of singularity) of a characteristic direction of H in the order of vanishing of the components of the geodesic field.

Since G is a vector field on the total space of a line bundle, it is natural to consider only changes of coordinates preserving the bundle structure, that is changes of coordinates of the form

$$(\zeta, v) \mapsto \big(\psi(\zeta), \xi(\zeta) v\big) ,$$

where ψ a germ of biholomorphism and ξ is a non-vanishing holomorphic function. It turns out that these changes of coordinates are enough to obtain normal forms around apparent and Fuchsian singularities.

For apparent singularities we have the following theorem:

Theorem 6.8 ([AT3]) *Let $[v] \in \mathbb{P}^1(\mathbb{C})$ be an apparent characteristic direction of multiplicity $\mu \geq 1$ of a homogeneous vector field $H \in \mathscr{X}_2^{\nu+1}$. Then we can find local coordinates centered at $[v]$ such that in these coordinates the geodesic field G associated to H is given by*

$$G = \begin{cases} \zeta v \dfrac{\partial}{\partial \zeta} & \text{if } \mu = 1, \\ \zeta^\mu (1 + a\zeta^{\mu-1}) v \dfrac{\partial}{\partial \zeta} & \text{for some } a \in \mathbb{C} \text{ if } \mu > 1. \end{cases}$$

Furthermore, if $\mu > 1$ then $a \in \mathbb{C}$ is a holomorphic invariant, the apparent index.

In particular, around an apparent singularity the geodesic field G is explicitly integrable. Studying the integral lines of G and rephrasing the results in terms of the integral curves of H we obtain the following corollary:

Corollary 6.9 ([AT3]) *Let $H \in \mathscr{X}_2^{\nu+1}$ be a homogeneous vector field on $\mathbb{C}^2$ of degree $\nu+1 \geq 2$. Let $[v] \in \mathbb{P}^1(\mathbb{C})$ be an apparent singularity of H of multiplicity $\mu \geq 1$ (and apparent index $a \in \mathbb{C}$ if $\mu > 1$). Then:*

(i) if the direction $[\gamma(t)] \in \mathbb{P}^1(\mathbb{C})$ of an integral curve $\gamma\colon [0, \varepsilon) \to \mathbb{C}^2 \setminus \{O\}$ of H tends to $[v]$ as $t \to \varepsilon$ then $\gamma(t)$ tends to a non-zero point of the characteristic leaf $L_v \subset \mathbb{C}^2$;

(ii) no integral curve of H tends to the origin tangent to $[v]$;

(iii) there is an open set of initial conditions whose integral curves tend to a non-zero point of L_v;

(iv) if $\mu = 1$ or $\mu > 1$ and $a \neq 0$ then H admits periodic orbits of arbitrarily long periods accumulating at the origin.

In particular, in case (iv) the time-1 map of H has both periodic orbits accumulating at the origin (small cycles), when the period of the integral curve is rational, and orbits whose closure a is a closed Jordan curve, when the period of the integral curve is irrational; both phenomena cannot happen in one variable.

The holomorphic classification of Fuchsian singularities is the following:

Theorem 6.10 ([AT3]) *Let $[v] \in \mathbb{P}^1(\mathbb{C})$ be a Fuchsian characteristic direction of multiplicity $\mu \geq 1$, transversal multiplicity $\tau = \mu - 1 \geq 0$ and index $\iota \in \mathbb{C}^*$ of a homogeneous vector field $H \in \mathscr{X}_2^{\nu+1}$. Then we can find local coordinates centered at $[v]$ such that in these coordinates the geodesic field G associated to H is given:*

(i) if $\tau + \iota \notin \mathbb{N}^*$ by

$$\zeta^{\mu-1}\left(\zeta v \frac{\partial}{\partial \zeta} + \iota v^2 \frac{\partial}{\partial v}\right);$$

(ii) if $n = \tau + \iota \in \mathbb{N}^*$ *by*

$$\zeta^{\mu-1}\left(\zeta v\frac{\partial}{\partial \zeta} + \iota v^2(1 + a\zeta^n)\frac{\partial}{\partial v}\right)$$

for a suitable $a \in \mathbb{C}$ *which is a holomorphic invariant, the* resonant index.

When the resonant index is zero the integral curves of the geodesic field can be expressed in terms of elementary functions and easily studied. This is not the case when the resonant index is different from zero; however we are able to obtain the following description of the integral curves of H nearby Fuchsian characteristic directions:

Corollary 6.11 ([AT3]) *Let* $H \in \mathscr{X}_2^{\nu+1}$ *be a homogeneous vector field on* $\mathbb{C}^2$ *of degree* $\nu+1 \geq 2$. *Let* $[v] \in \mathbb{P}^1(\mathbb{C})$ *be a Fuchsian singularity of* H *of multiplicity* $\mu \geq 1$, *transversal multiplicity* $\tau = \mu - 1 \geq 0$ *and index* $\iota \in \mathbb{C}^*$ *(and resonant index* $a \in \mathbb{C}$ *if* $\tau + \iota \in \mathbb{N}^*$*). Then:*

(i) if the direction $[\gamma(t)] \in \mathbb{P}^1(\mathbb{C})$ *of an integral curve* $\gamma\colon [0, \varepsilon) \to \mathbb{C}^2 \setminus \{O\}$ *of* H *tends to* $[v]$ *as* $t \to \varepsilon$ *and* γ *is not contained in the characteristic leaf* L_v *then*

- *(a) if* $\tau + \mathrm{Re}\,\iota > 0$ *and* $\left|\iota + \frac{\tau}{2}\right| > \frac{\tau}{2}$ *then* $\gamma(t)$ *tends to the origin;*
- *(b) if* $\tau + \iota = 0$, *or* $\tau + \mathrm{Re}\,\iota < 0$, *or* $\tau + \mathrm{Re}\,\iota > 0$ *and* $\left|\iota + \frac{\tau}{2}\right| < \frac{\tau}{2}$, *then* $\|\gamma(t)\|$ *tends to* $+\infty$*;*
- *(c) if* $\tau + \mathrm{Re}\,\iota > 0$ *and* $\left|\iota + \frac{\tau}{2}\right| = \frac{\tau}{2}$ *then* $\gamma(t)$ *accumulates a circumference in* L_v.

Furthermore there is a neighbourhood $U \subset \mathbb{P}^1(\mathbb{C})$ *of* $[v]$ *such that an integral curve* γ *issuing from a point* $z_0 \in \mathbb{C}^2 \setminus L_v$ *with* $[z_0] \in U \setminus \{[v]\}$ *must have one of the following behaviors, where* $\hat{U} = \{z \in \mathbb{C}^2 \setminus \{O\} \mid [z] \in U\}$*:*

(ii) if $\tau + \mathrm{Re}\,\iota < 0$ *then*

- *(a) either* $\gamma(t)$ *escapes* $\hat{U}$, *and this happens for a Zariski open dense set of initial conditions in* $\hat{U}$*; or*
- *(b)* $[\gamma(t)] \to [v]$ *but* $\|\gamma(t)\| \to +\infty$*;*

in particular, no integral curve outside L_v *converge to the origin tangent to* $[v]$*;*

(iiii) if $\tau + \mathrm{Re}\,\iota = 0$ *but* $\tau + \iota \neq 0$ *then*

- *(a) either* $\gamma(t)$ *escapes* $\hat{U}$*; or*
- *(b)* $\gamma(t) \to O$ *without being tangent to any direction, and* $[\gamma(t)]$ *is a closed curve or accumulates a closed curve in* $\mathbb{P}^1(\mathbb{C})$ *surrounding* $[v]$*; or*
- *(c)* $\|\gamma(t)\| \to +\infty$ *without being tangent to any direction, and* $[\gamma(t)]$ *is a closed curve in* $\mathbb{P}^1(\mathbb{C})$ *surrounding* $[v]$*;*

in particular, no integral curve outside L_v *converge to the origin tangent to* $[v]$*;*

(iv) if $\tau + \iota = 0$ *then*

(a) *either $\gamma(t)$ escapes $\hat{U}$, and this happens for an open set $\hat{U}_1 \subset \hat{U}$ of initial conditions; or*
(b) *$[\gamma(t)] \to [v]$ with $\|\gamma(t)\| \to +\infty$, and this happens for an open set $\hat{U}_2 \subset \hat{U}$ of initial conditions such that $\hat{U}_1 \cup \hat{U}_2$ is dense in $\hat{U}$; or*
(c) *γ is a periodic integral curve with $[\gamma]$ surrounding $[v]$;*

in particular, no integral curve outside L_v converge to the origin tangent to $[v]$, but we have periodic integral curves of arbitrarily long period accumulating the origin;

(v) *if $\tau + \mathrm{Re}\,\iota > 0$ and $a = 0$ then $[\gamma(t)] \to [v]$ for an open dense set $\hat{U}_0$ of initial conditions in $\hat{U}$, and γ escapes $\hat{U}$ for $z \in \hat{U} \setminus \hat{U}_0$; moreover,*

(a) *if $\left|\iota + \frac{\tau}{2}\right| > \frac{\tau}{2}$ then $\gamma(t) \to O$ tangent to $[v]$ for all $z \in \hat{U}_0$;*
(b) *if $\left|\iota + \frac{\tau}{2}\right| < \frac{\tau}{2}$ then $\|\gamma(t)\| \to +\infty$ tangent to $[v]$ for all $z \in \hat{U}_0$;*
(c) *if $\left|\iota + \frac{\tau}{2}\right| = \frac{\tau}{2}$ then $\gamma(t)$ accumulates a circumference in L_v.*

Remark 6.3 We conjecture that Corollary 6.11.(v) should hold also when $a \neq 0$.

Remark 6.4 This result must be compared with Theorems 2.3 and 4.1. We already noticed that a non-degenerate characteristic direction $[v]$ with non-zero director δ is a Fuchsian singularity of multiplicity 1, and hence transversal multiplicity 0. Then Corollary 6.11 says that if $\mathrm{Re}\,\iota < 0$ (that is $\mathrm{Re}\,\delta < 0$) then no orbit of the time-1 map of H outside of L_v converges to the origin tangent to $[v]$, whereas if $\mathrm{Re}\,\iota > 0$ (that is $\mathrm{Re}\,\delta > 0$) and ι is not a positive integer (or $a = 0$ if $\iota \in \mathbb{N}^*$) then the orbits under the time 1-map of H converge to the origin tangent to $[v]$ for an open (and dense in a conical neighbourhood of $[v]$) set of initial conditions, providing the existence of a parabolic domain in accord with Theorem 2.3.

If instead $\tau > 0$ and $\tau + \iota \notin \mathbb{N}^*$ (or $a = 0$ if $\tau + \iota \in \mathbb{N}^*$) then Corollary 6.11 yields a parabolic basin when $\tau + \mathrm{Re}\,\iota > 0$ and $\left|\iota + \frac{\tau}{2}\right| > \frac{\tau}{2}$, which is a condition strictly weaker than the condition found in Theorem 4.1(a); this suggests that there might be room for improvement in the statement of the latter theorem.

Putting all of this together we can finally have a completely description of the dynamics for a substantial class of examples. For instance, we get the following:

Corollary 6.12 ([AT3]) *Let $H \in \mathcal{X}_2^{\nu+1}$ be a homogeneous vector field on $\mathbb{C}^2$ of degree $\nu+1 \geq 2$. Assume that H is non-dicritical and all its characteristic directions are Fuchsian of multiplicity 1. Assume moreover that for no set of characteristic directions the real part of the sum of the induced residues belongs to the interval $(-3/2, -1/2)$. Let $\gamma\colon [0, \varepsilon_0) \to \mathbb{C}^2$ be a maximal integral curve of H. Then:*

(a) *If $\gamma(0)$ belongs to a characteristic leaf L_{v_0}, then the image of γ is contained in L_{v_0}. Moreover, either $\gamma(t) \to O$ (and this happens for a Zariski open dense set of initial conditions in L_{v_0}), or $\|\gamma(t)\| \to +\infty$.*
(b) *If $\gamma(0)$ does not belong to a characteristic leaf then either*

(i) *γ converges to the origin tangentially to a characteristic direction $[v_0]$ whose index has positive real part; or* *(ii)* $\|\gamma(t)\| \to +\infty$ *tangentially to a characteristic direction* $[v_0]$ *whose index has negative real part.*

Furthermore, case (i) *happens for a Zariski open set of initial conditions.*

Remark 6.5 The conditions in Corollary 6.12 imply that there must be at least one index with positive real part. Indeed, if the multiplicity is 1 then the induced residue is one less the opposite of the index. So assuming that no sum of $1 \le g \le \nu + 2$ induced residues has real part belonging to the interval $(-3/2, -1/2)$ is equivalent to saying that no sum of $1 \le g \le \nu + 2$ indices has real part belonging to the interval $(\frac{1}{2} - g, \frac{3}{2} - g)$. Assume, by contradiction, that no index has positive real part; than the real part of all of them should be less than $-1/2$. So the real part of the sum of two indices must be less than $-1 < -1/2$; so it should be less than $-3/2$. Arguing by induction on g one then shows that the sum of the real part of all indices should be less than $\frac{1}{2} - (\nu + 2) = -\nu - 3/2 < -\nu$, against Theorem 3.4, contradiction.

Example 6.1 Corollary 6.12 describes completely the dynamics of most vector fields of the form

$$H(z, w) = \left(\rho z^2 + (1 + \tau)zw\right)\frac{\partial}{\partial z} + \left((1 + \rho)zw + \tau w^2\right)\frac{\partial}{\partial w} .$$

Indeed such a vector field has exactly three Fuchsian characteristic directions with multiplicity 1 and indices respectively ρ, τ and $-1 - \rho - \tau$; so the conditions required by Corollary 6.12 are satisfied as soon as $\mathrm{Re}\,\rho$, $\mathrm{Re}\,\tau \notin (-1/2, 1/2)$ and $\mathrm{Re}(\rho + \tau) \notin (-3/2, -1/2)$.

7 Other Systems with Parabolic Behavior

Another situation where Fatou flowers can exist is when the eigenvalues of the differential are all equal to 1 but the differential is not necessarily diagonalizable. The reason is that we can reduce to the tangent to the identity case by using a suitable sequence of blow-ups:

Theorem 7.1 ([A1]) *Let* $f \in \mathrm{End}(\mathbb{C}^n, O)$ *be such that all eigenvalues of* df_O *are equal to* 1. *Then there exist a complex n-dimensional manifold* M, *a holomorphic projection* $\pi : M \to \mathbb{C}^n$, *a canonical point* $\mathbf{e} \in M$ *and a germ around* $\pi^{-1}(O)$ *of holomorphic self-map* $\tilde{f} : M \to M$ *such that:*

(i) π *restricted to* $M \setminus \pi^{-1}(O)$ *is a biholomorphism between* $M \setminus \pi^{-1}(O)$ *and* $\mathbb{C}^n \setminus \{O\}$;
(ii) $\pi \circ \tilde{f} = f \circ \pi$;
(iii) $\mathbf{e}$ *is a fixed point of* $\tilde{f}$ *where* $\tilde{f}$ *is tangent to the identity.*

It should be remarked that the projection $\pi : M \to \mathbb{C}^n$ is obtained as a sequence of blow-ups whose centers are not necessarily reduced to points, and depend on the

Jordan structure of df_O. Furthermore π is chosen in such a way that the interesting part of the dynamics of $\tilde{f}$ is outside the exceptional divisor E (which is not in general pointwise fixed by $\tilde{f}$), allowing the study of the dynamics of f by means of the dynamics of $\tilde{f}$. For instance, we can get the following

Proposition 7.2 ([A1]) *Let* $f = (f_1, \ldots, f_n) \in \mathrm{End}(\mathbb{C}^n, O)$ *be such that* df_O *is not diagonalizable with all eigenvalues equal to* 1. *Without loss of generality we can suppose that* df_O *is in Jordan form with* ρ *blocks of order respectively* $\mu_1 \geq \mu_2 \geq \cdots \geq \mu_\rho \geq 1$, *where* $\mu_1 + \cdots + \mu_\rho = n$. *Assume that* $\mu_1 > \mu_2$ *and that the coefficient of* $(z_1)^2$ *in* f_{μ_1} *is not zero. Then* f *admits a parabolic curve tangent to* $[1 : 0 : \cdots : 0]$.

In dimension 2, using the tools introduced in [A2], one can get a cleaner result:

Corollary 7.3 ([A2]) *Let* $f \in \mathrm{End}(\mathbb{C}^2, O)$ *be such that* df_O *is a Jordan block with eigenvalue* 1, *and assume that the origin is an isolated fixed point. Then* f *admits a parabolic curve tangent to* $[1 : 0]$.

See [Ro6] (and [A3] for a particular example) for a detailed study of the existence of parabolic domains for germs in $\mathrm{End}(\mathbb{C}^2, O)$ with non-diagonalizable differential.

Finally, I would like to mention that parabolic curves, parabolic domains and Fatou flowers appear also in non-parabolic dynamical systems. This is not surprising in semi-parabolic systems, that is when the eigenvalues of the differential are either equal to 1 or in modulus strictly less than 1 (see, e.g., [N, H1, Ri2, U1, U2, Ro5]), or in quasi-parabolic systems, where the eigenvalues of the differential are either equal to 1 or have modulus equal to 1 (see, e.g., [BM, Ro1, Ro2]). On the other hand, a recent surprising discovery is that they also appear in *multi-resonant systems*, whose differential is not parabolic at all but whose eigenvalues satisfies some resonance relation; see, e.g., [BZ, BRZ, RV, BR] for the main results of this very interesting theory.

Acknowledgments Partially supported by FIRB 2012 project "Geometria Differenziale e Teoria Geometrica delle Funzioni".

References

[A1] Abate, M.: Diagonalization of non-diagonalizable discrete holomorphic dynamical systems. Am. J. Math. **122**, 757–781 (2000)

[A2] Abate, M.: The residual index and the dynamics of holomorphic maps tangent to the identity. Duke Math. J. **107**, 173–206 (2001)

[A3] Abate, M.: Basins of attraction in quadratic dynamical systems with a Jordan fixed point. Nonlinear Anal. **51**, 271–282 (2002)

[A4] Abate, M.: Discrete holomorphic local dynamical systems. In: Gentili, G., Guénot, J., Patrizio, G. (eds.) Holomorphic Dynamical Systems. Lecture Notes in Mathematics, 1998, pp. 1–55. Springer, Berlin (2010)

[AB] Abate, M., Bianchi, F.: A Poincaré-Bendixson theorem for meromorphic connections on compact Riemann surfaces. Preprint, arxiv:1406.6944 (2014)

[ABT1] Abate, M., Bracci, F., Tovena, F.: Index theorems for holomorphic self-maps. Ann. Math. **159**, 819–864 (2004)

[ABT2] Abate, M., Bracci, F., Tovena, F.: Index theorems for holomorphic maps and foliations. Indiana Univ. Math. J. **57**, 2999–3048 (2008)

[ABT3] Abate, M., Bracci, F., Tovena, F.: Embeddings of submanifolds and normal bundles. Adv. Math. **220**, 620–656 (2009)

[AR] Abate, M., Raissy, J.: Formal Poincaré-Dulac renormalization for holomorphic germs. Disc. Cont. Dyn. Syst. **33**, 1773–1807 (2013)

[AT1] Abate, M., Tovena, F.: Parabolic curves in $\mathbb{C}^3$. Abstr. Appl. Anal. **2003**, 275–294 (2003)

[AT2] Abate, M., Tovena, F.: Formal classification of holomorphic maps tangent to the identity. Disc. Cont. Dyn. Sys. 1–10 (2005)

[AT3] Abate, M., Tovena, F.: Poincaré-Bendixson theorems for meromorphic connections and holomorphic homogeneous vector fields. J. Diff. Equ. **251**, 2612–2684 (2011)

[ArR] Arizzi, M., Raissy, J.: On Écalle-Hakim's theorems in holomorphic dynamics. In: A. Bonifant, et al. (eds.) Progress in complex dynamics (pp. 387–449). Princeton University Press, Princeton (2014)

[B] Bracci, F.: The dynamics of holomorphic maps near curves of fixed points. Ann. Scuola Norm. Sup. Pisa **2**, 493–520 (2003)

[BM] Bracci, F., Molino, L.: The dynamics near quasi-parabolic fixed points of holomorphic diffeomorphisms in $\mathbb{C}^2$. Am. J. Math. **126**, 671–686 (2004)

[BRZ] Bracci, F., Raissy, J., Zaitsev, D.: Dynamics of multi-resonant biholomorphisms. Int. Math. Res. Not. 4772–4797 (2013)

[BR] Bracci, F., Rong, F.: Dynamics of quasi-parabolic one-resonant biholomorphisms. J. Geom. Anal. **24**, 1497–1508 (2014)

[BS] Bracci, F., Suwa, T.: Residues for singular pairs and dynamics of biholomorphic maps of singular surfaces. Int. J. Math. **15**, 443–466 (2004)

[BT] Bracci, F., Tovena, F.: Residual indices of holomorphic maps relative to singular curves of fixed points on surfaces. Math. Z. **242**, 481–490 (2002)

[BZ] Bracci, F., Zaitsev, D.: Dynamics of one-resonant biholomorphisms. J. Eur. Math. Soc. **15**, 179–200 (2013)

[Br1] Brochero-Martínez, F.E.: Groups of germs of analytic diffeomorphisms in $(\mathbb{C}^2, 0)$. J. Dyn. Control Sys. **9**, 1–32 (2003)

[Br2] Brochero-Martínez, F.E.: Dinámica de difeomorfismos dicríticos en $(\mathbb{C}^n, 0)$. Rev. Semin. Iberoam. Mat. **3**, 33–40 (2008)

[BCL] Brochero, F.E., Martínez, F., Cano, L.: López-Hernanz: Parabolic curves for diffeomorphisms in $\mathbb{C}^2$. Publ. Mat. **52**, 189–194 (2008)

[C] Camacho, C.: On the local structure of conformal mappings and holomorphic vector fields. Astérisque **59–60**, 83–94 (1978)

[CS] Camacho, C., Sad, P.: Invariant varieties through singularities of holomorphic vector fields. Ann. Math. **115**, 579–595 (1982)

[CaS] Câmara, L., Scárdua, B.: A Fatou type theorem for complex map germs. Conf. Geom. Dyn. **16**, 256–268 (2012)

[Ca] Cano, J.: Construction of invariant curves for singular holomorphic vector fields. Proc. Am. Math. Soc. **125**, 2649–2650 (1997)

[D] Degli Innocenti, F.: Holomorphic dynamics near germs of singular curves. Math. Z. **251**, 943–958 (2005)

[E] Écalle, J.: Les fonctions résurgentes. Tome III: L'équation du pont et la classification analytique des objects locaux. Prépublications Math. Orsay 85-05, Université de Paris-Sud, Orsay (1985)

[F1] Fatou, P.: Sur les équations fonctionnelles. I. Bull. Soc. Math. France **47**, 161–271 (1919)

[F2] Fatou, P.: Sur les équations fonctionnelles. II. Bull. Soc. Math. France **48**, 33–94 (1920)

[F3] Fatou, P.: Sur les équations fonctionnelles. III. Bull. Soc. Math. France **48**, 208–314 (1920)

[GH] Griffiths, P., Harris, J.: Principles of algebraic geometry. Wiley, New York (1978)
[H1] Hakim, M.: Attracting domains for semi-attractive transformations of $\mathbb{C}^p$. Publ. Mat. **38**, 479–499 (1994)
[H2] Hakim, M.: Analytic transformations of $(\mathbb{C}^p, 0)$ tangent to the identity. Duke Math. J. **92**, 403–428 (1998)
[H3] Hakim, M.: Transformations Tangent to the Identity. Stable Pieces of Manifolds, Preprint (1997)
[IY] Ilyashenko, Y., Yakovenko, S.: Lectures on Analytic Differential Equations. American Mathematical Society, Providence (2008)
[L1] Lapan, S.: Attracting domains of maps tangent to the identity whose only characteristic direction is non-degenerate. Int. J. Math. **24**(1350083), 1–27 (2013)
[L2] Lapan, S.: Attracting domains of maps tangent to the identity in $\mathbb{C}^2$ with characteristic direction of multiple degrees. Preprint, arxiv:1501.00244 (2015)
[Le] Leau, L.: Étude sur les équations fonctionelles à une ou plusieurs variables. Ann. Fac. Sci. Toulouse **11**, E1–E110 (1897)
[LS] López-Hernanz, L., Sánchez, F.S.: Parabolic curves of diffeomorphisms asymptotic to formal invariant curves. Preprint, arxiv:1411.2945 (2014)
[M] Milnor, J.: Dynamics in One Complex Variable. Annals of Mathematical Studies, vol. 160. Princeton University Press, Princeton (2006)
[Mo] Molino, L.: The dynamics of maps tangent to the identity and with non vanishing index. Trans. Am. Math. Soc. **361**, 1597–1623 (2009)
[N] Nishimura, Y.: Automorphismes analytiques admettant des sous-variétés de points fixés attractives dans la direction transversale. J. Math. Kyoto Univ. **23**, 289–299 (1983)
[RV] Raissy, J., Vivas, L.: Dynamics of two-resonant biholomorphisms. Math. Res. Lett. **20**, 757–771 (2013)
[R] Ribón, J.: Families of diffeomorphisms without periodic curves. Mich. Math. J. **53**, 243–256 (2005)
[Ri1] Rivi, M.: Local behaviour of discrete dynamical systems. Ph.D. Thesis, Università di Firenze (1999)
[Ri2] Rivi, M.: Parabolic manifolds for semi-attractive holomorphic germs. Mich. Math. J. **49**, 211–241 (2001)
[Ro1] Rong, F.: Quasi-parabolic analytic transformations of $\mathbb{C}^n$. J. Math. Anal. Appl. **343**, 99–109 (2008)
[Ro2] Rong, F.: Quasi-parabolic analytic transformations of $\mathbb{C}^n$. Parabolic manifolds. Ark. Mat. **48**, 361–370 (2010)
[Ro3] Rong, F.: Robust parabolic curves in $\mathbb{C}^m$ $(m \geq 3)$. Houst. J. Math. **36**, 147–155 (2010)
[Ro4] Rong, F.: Absolutely isolated singularities of holomorphic maps of $\mathbb{C}^n$ tangent to the identity. Pac. J. Math. **246**, 421–433 (2010)
[Ro5] Rong, F.: Parabolic manifolds for semi-attractive analytic transformations of $\mathbb{C}^n$. Trans. Am. Math. Soc. **363**, 5207–5222 (2011)
[Ro6] Rong, F.: Local dynamics of holomorphic maps in $\mathbb{C}^2$ with a Jordan fixed point. Mich. Math. J. **62**, 843–856 (2013)
[Ro7] Rong, F.: Reduction of singularities of holomorphic maps of $\mathbb{C}^2$ tangent to the identity. Publ. Math. Debrecen **83**(4), 537–546 (2013)
[Ro8] Rong, F.: The non-dicritical order and attracting domains of holomorphic maps tangent to the identity. Int. J. Math. **25**(1450003), 1–10 (2014)
[Ro9] Rong, F.: New invariants and attracting domains for holomorphic maps in $\mathbb{C}^2$ tangent to the identity. Preprint, to appear in Publ. Mat. (2014)
[S] Sauzin, D.: Introduction to 1-summability and resurgence. Preprint, arxiv:1405.0356 (2014)
[Sh] Shcherbakov, A.A.: Topological classification of germs of conformal mappings with identity linear part. Moscow Univ. Math. Bull. **37**, 60–65 (1982)
[SV] Stensones, B., Vivas, L.: Basins of attraction of automorphisms in $\mathbb{C}^3$. Ergodic Theory Dyn. Syst **34**, 689–692 (2014)

[U1] Ueda, T.: Local structure of analytic transformations of two complex variables, I. J. Math. Kyoto Univ. **26**, 233–261 (1986)
[U2] Ueda, T.: Local structure of analytic transformations of two complex variables. II. J. Math. Kyoto Univ. **31**, 695–711 (1991)
[Us] Ushiki, S.: Parabolic fixed points of two-dimensional complex dynamical systems. Sūrikaisekikenkyūsho Kōkyūroku **959**, 168–180 (1996)
[V1] Vivas, L.: Degenerate characteristic directions for maps tangent to the identity. Indiana Univ. Math. J. **61**, 2019–2040 (2012)
[V2] Vivas, L.R.: Fatou-Bieberbach domains as basins of attraction of automorphisms tangent to the identity. J. Geom. Anal. **22**, 352–382 (2012)
[W] Weickert, B.: Attracting basins for automorphisms of $\mathbb{C}^2$. Invent. Math. **132**, 581–605 (1998)

A CR Proof for a Global Estimate of the Diederich–Fornaess Index of Levi-Flat Real Hypersurfaces

Masanori Adachi

Abstract Yet another proof is given for a global estimate of the Diederich–Fornaess index of relatively compact domains with Levi-flat boundary, namely, the index must be smaller than or equal to the reciprocal of the dimension of the ambient space. This proof reveals that this kind of estimate makes sense and holds also for abstract compact Levi-flat CR manifolds.

Keywords Diederich–Fornaess index · CR geometry · Levi-flat real hypersurface

1 Introduction

The *Diederich–Fornaess index* $\eta(\Omega)$ of a $\mathscr{C}^\infty$-smoothly bounded domain Ω in a complex manifold X is a numerical index on the strength of certain pseudoconvexity of its boundary $\partial\Omega$. In this paper, we consider the index in the sense that $\eta(\Omega)$ is defined to be the supremum of the exponents $\eta \in (0, 1]$ admitting a $\mathscr{C}^\infty$-smooth defining function of $\partial\Omega$, say $\rho : (\partial\Omega \subset) U \to \mathbb{R}$, so that $-|\rho|^\eta$ is strictly plurisubharmonic in $U \cap \Omega$; if no such η is allowed, we let $\eta(\Omega) = 0$.

For instance, if a defining function attains $\eta = 1$, it gives a strictly plurisubharmonic defining function of $\partial\Omega$ and the boundary is strictly pseudoconvex. The pseudoconvexity of $\partial\Omega$ is clearly necessary for $\eta(\Omega)$ to be positive; a much stronger condition is actually necessary and sufficient, the existence of a defining function ρ such that the complex hessian of $-\log|\rho|$ is bounded from below by a hermitian metric of X near the boundary $\partial\Omega$ as observed by Ohsawa and Sibony [OS].

M. Adachi (✉)
Center for Geometry and its Applications, Pohang University of Science and Technology, Pohang 790-784, Republic of Korea

M. Adachi
Graduate School of Mathematics, Nagoya University, Nagoya 464-8602, Japan

Present Address:
Department of Mathematics, Faculty of Science and Technology, Tokyo University of Science, Noda 278-8510, Japan
e-mail: adachi_masanori@ma.noda.tus.ac.jp

F. Bracci et al. (eds.), *Complex Analysis and Geometry*, Springer Proceedings in Mathematics & Statistics 144, DOI 10.1007/978-4-431-55744-9_2

The Diederich–Fornaess index $\eta(\Omega)$ being positive means that the boundary $\partial\Omega$ is well approximated by a family of strictly pseudoconvex real hypersurfaces from inside. The original motivation of the study of Diederich and Fornaess [DF] was to construct such an approximation on pseudoconvex domains in Stein manifolds, and the index is considered to measure certain strength of the approximation. Since then, the meaning of the index has been intensively studied in relation to the global regularity in the $\overline{\partial}$-Neumann problem, in particular, pathologies occurring on the worm domain. See for example [FS, §1], [Be] and references therein.

Under such circumstances, Fu and Shaw [FS] and Brinkschulte and the author [AB1] reached a general estimate for the Diederich–Fornaess index of weakly pseudoconvex domains. Here we state the result in a restricted form, for domains with $\mathscr{C}^\infty$-smooth Levi-flat boundary:

Theorem 1 *([FS],* see also *[AB1] and [A2]) Let Ω be a relatively compact domain with $\mathscr{C}^\infty$-smooth Levi-flat boundary M in a complex manifold of dimension $(n+1) \geq 2$. Then the Diederich–Fornaess index $\eta(\Omega)$ must be $\leq 1/(n+1)$.*

The purpose of this paper is to give yet another proof of Theorem via an estimate on the Levi-flat boundary M without looking inside Ω directly. The idea is to identify the usual Diederich–Fornaess index $\eta(\Omega)$ with its counterpart $\eta(M)$ on the Levi-flat boundary based on the author's previous work [A1].

Definition 1 Let M be an oriented $\mathscr{C}^\infty$-smooth Levi-flat CR manifold. The *Diederich–Fornaess index* $\eta(M)$ of M is defined to be the supremum of $\eta \in (0, 1)$ admitting a $\mathscr{C}^\infty$-smooth hermitian metric h^2 of the holomorphic normal bundle $N_M^{1,0}$ of M so that

$$i\Theta_h - \frac{\eta}{1-\eta} i\alpha_h \wedge \overline{\alpha_h} > 0$$

holds on M as quadratic forms on the holomorphic tangent bundle $T_M^{1,0}$ of M; if no such η is allowed, we let $\eta(M) = 0$. Here the forms α_h and Θ_h denote the leafwise Chern connection form and its curvature form of $N_M^{1,0}$ with respect to h^2 respectively. (See Sect. 2 for their precise definitions.)

In our setting, $\eta(\Omega)$ agrees with $\eta(M)$ as we will see in Lemma 3, and Theorem follows from the following main lemma.

Lemma 1 *Let M be a compact $\mathscr{C}^\infty$-smooth Levi-flat CR manifold of dimension $(2n+1) \geq 3$. Then the Diederich–Fornaess index $\eta(M)$ must be $\leq 1/(n+1)$.*

The organization of this paper is as follows. In Sect. 2, we provide preliminaries on CR geometry and confirm that the two notions of Diederich–Fornaess index, $\eta(\Omega)$ and $\eta(M)$, actually coincide for Levi-flat real hypersurfaces based on previous works. In Sect. 3, after proving Lemma 1, we give a remark that the substantial content of Lemma 1 has been already pointed out by Bejancu and Deshmukh [BD] in manner of differential geometry, and conclude this paper with an open question.

2 Preliminaries

2.1 Levi-Flat CR Manifold

Let us recall the notion of Levi-flat CR manifold briefly. In the sequel, "smooth" means infinitely differentiable.

Let M be a smooth manifold of dimension $(2n+1) \geq 3$. A *CR structure* (of hypersurface type) of M is given by a subbundle $T_M^{0,1} \subset \mathbb{C} \otimes_{\mathbb{R}} TM$ satisfying the following conditions:

- $T_M^{0,1}$ is a smooth $\mathbb{C}$-subbundle $T_M^{0,1} \subset \mathbb{C} \otimes_{\mathbb{R}} TM$ of $\mathrm{rank}_{\mathbb{C}}\, n$;
- $T_M^{1,0} \cap T_M^{0,1} =$ (the zero section) where $T_M^{1,0} := \{v \in \mathbb{C} \otimes_{\mathbb{R}} TM \mid \overline{v} \in T_M^{0,1}\}$;
- $[\Gamma(T_M^{0,1}), \Gamma(T_M^{0,1})] \subset \Gamma(T_M^{0,1})$

where $\Gamma(\cdot)$ denotes the space of smooth sections of the bundle, and the bracket means the Lie bracket of complexified vector fields. The pair $(M, T_M^{0,1})$ is called a *CR manifold*, which is regarded as an abstraction of real hypersurfaces in complex manifolds associated with their (anti-)holomorphic tangent bundles.

We say that a CR manifold $(M, T_M^{0,1})$ is *Levi-flat* if it satisfies further integrability condition

$$[\Gamma(T_M^{1,0} \oplus T_M^{0,1}), \Gamma(T_M^{1,0} \oplus T_M^{0,1})] \subset \Gamma(T_M^{1,0} \oplus T_M^{0,1}). \tag{1}$$

This is equivalent to say that the real distribution $H_M := \mathrm{Re}(T_M^{1,0} \oplus T_M^{0,1}) \subset TM$ of $\mathrm{rank}_{\mathbb{R}}\, 2n$ is integrable in the sense of Frobenius. It follows from Frobenius' theorem and Newlander–Nirenberg's theorem that the distribution H_M defines a smooth foliation $\mathscr{F}$ by complex manifolds on M, namely, we have an atlas consisting of foliated charts. We call $\mathscr{F}$ the *Levi foliation*.

For a Levi-flat CR manifold $(M, T_M^{0,1})$, we shall refer to $T_M^{1,0}$ as the *holomorphic tangent bundle* of M and call the quotient $\mathbb{C}$-line bundle $N_M^{1,0}$,

$$0 \to T_M^{1,0} \oplus T_M^{0,1} \to \mathbb{C} \otimes_{\mathbb{R}} TM \xrightarrow{\pi} N_M^{1,0} \to 0,$$

the *holomorphic normal bundle*. This is because $T_M^{1,0}$ agrees with the holomorphic tangent bundle of the leaves of the Levi foliation $\mathscr{F}$. Note that our holomorphic tangent bundle is distinct from $(\mathbb{C} \otimes_{\mathbb{R}} TM)/T_M^{0,1}$ and our *(p, q)-form* on M means a section of $\bigwedge^p (T_M^{1,0})^* \otimes \bigwedge^q (T_M^{0,1})^* \subset \bigwedge^{p+q} (T_M^{1,0} \oplus T_M^{0,1})^*$.

Now let us consider a Levi-flat CR manifold, simply denoted by M, and define the form α_h mentioned in Sect. 1. Fix a smooth hermitian metric h^2 of $N_M^{1,0}$; in our convention, we denote by $h : N_M^{1,0} \to \mathbb{R}$ the map given by the norm induced from h^2 on $(N_M^{1,0})_p$ for each $p \in M$. Pick a local smooth section ξ of $N_M^{1,0}$ around $p \in M$ so that it is both normalized by h^2 and real, i.e., $\overline{\xi} = \xi$, which is determined up to its sign. Using such a ξ, we define the $(1, 0)$-form $\alpha_h : T_M^{1,0} \to \mathbb{C}$ so as to satisfy

$$\pi([v, \widetilde{\xi}]_p) = -\alpha_h(v_p)\xi_p \qquad (2)$$

for $v_p \in (T_M^{1,0})_p$ where $\widetilde{\xi}$ and v are any lift and extension of ξ and v_p to local sections of $\mathbb{C} \otimes_{\mathbb{R}} TM$ respectively. Here we used the Levi-flatness (1) to assure that α_h is independent of the choice of ξ, $\widetilde{\xi}$ and v. We define $\overline{\alpha}_h(\overline{v_p}) := \overline{\alpha_h(v_p)}$, the complex-conjugate $(0, 1)$-form of α_h.

Remark 2 The left hand side of (2) is the covariant derivative of ξ along v_p with respect to a complex Bott connection of the Levi foliation $\mathscr{F}$ and the form α is considered to measure the size of infinitesimal holonomy of $\mathscr{F}$ with respect to h^2.

We give the $(1, 1)$-form $\Theta_h : T_M^{1,0} \otimes T_M^{0,1} \to \mathbb{C}$ by

$$\begin{aligned}\Theta_h(v_p \otimes \overline{w}_p) &:= v_p\alpha_h(\overline{w}) - \overline{w}_p\alpha_h(v) - \alpha_h([v, \overline{w}]_p) \\ &= -\overline{w}_p\alpha_h(v) - \alpha_h([v, \overline{w}]_p)\end{aligned}$$

where v and $\overline{w}$ are arbitrary extensions of v_p and $\overline{w}_p$ to local sections of $T_M^{1,0}$ and $T_M^{0,1}$ respectively. We again used the Levi-flatness (1) for the last term to be defined.

2.2 Description on Foliated Charts

Although we have defined the forms α_h and Θ_h in a coordinate-free manner, their descriptions on foliated charts are convenient in actual computations. Here we briefly introduce them.

Take a *foliated chart* $(U, (z_U, t_U))$ of the Levi-flat CR manifold M, a chart $(z_U, t_U) : U \to \mathbb{C}^n \times \mathbb{R}$ so that $T_M^{1,0}|U$ agrees with the pull-back bundle of $T^{1,0}\mathbb{C}^n \subset \mathbb{C} \otimes_{\mathbb{R}} T(\mathbb{C}^n \times \mathbb{R})$. Any coordinate change between intersecting foliated charts, say $(U, (z_U, t_U))$ and $(V, (z_V, t_V))$, are of the form

$$z_U = z_U(z_V, t_V), \quad t_U = t_U(t_V)$$

where z_U is holomorphic in z_V. A *leaf* N of $\mathscr{F}$ is a connected complex manifold injectively immersed in M such that z_U is holomorphic and t_U is locally constant on $U \cap N$ for any foliated chart $(U, (z_U, t_U))$. Our manifold M is decomposed into the direct sum of the leaves of $\mathscr{F}$. A *CR function* on M, a $\mathbb{C}$-valued function which is annihilated by vectors in $T_M^{0,1}$ by its definition, agrees with a function which is *leafwise* holomorphic, namely, holomorphic in z_U on any foliated chart $(U, (z_U, t_U))$.

On a foliated chart $(U, (z_U = (z_U^1, z_U^2, \cdots, z_U^n), t_U))$, we may trivialize $T_M^{1,0}$ and $N_M^{1,0}$ by using

$$\left\{\frac{\partial}{\partial z_U^1}, \frac{\partial}{\partial z_U^2}, \cdots, \frac{\partial}{\partial z_U^n}\right\} \quad \text{and} \quad \frac{\partial}{\partial t_U}$$

respectively. This description illustrates that $T_M^{1,0}$ and $N_M^{1,0}$ are *locally trivial CR vector bundles*, smooth vectors bundles with local trivialization covers whose transition functions are CR. The transition functions of $N_M^{1,0}$ are much better; They are leafwise constant.

Some computations show that on a foliated chart $(U, (z_U, t_U))$, the forms α_h and Θ_h for a given hermitian metric h^2 of $N_M^{1,0}$ are described as

$$\alpha_h = \sum_{j=1}^{n} \frac{\partial \log h_U}{\partial z_U^j} dz_U^j,$$

$$\Theta_h = \sum_{j,k=1}^{n} \frac{\partial^2(-\log h_U)}{\partial z_U^j \partial \overline{z}_U^k} dz_U^j \wedge d\overline{z}_U^k$$

where $h_U := h(\frac{\partial}{\partial t_U})$. We can see that α_h and Θ_h agree with the leafwise Chern connection and curvature form of $N_M^{1,0}$ with respect to h^2 respectively up to a positive multiplicative constant.

2.3 The Diederich–Fornaess Index

In this section, we confirm that the two notions of Diederich–Fornaess index given in Sect. 1 coincide for Levi-flat real hypersurfaces.

Let Ω be a relatively compact domain with smooth Levi-flat boundary M in a complex manifold of dimension ≥ 2. We introduce here terms for intermediate notions that appeared in the definition of the Diederich–Fornaess indices. The *Diederich–Fornaess exponent* η_ρ of a fixed defining function $\rho : (\partial\Omega \subset)U \to \mathbb{R}$ of $\partial\Omega$ is the supremum of the exponents $\eta \in (0, 1]$ such that $-|\rho|^\eta$ is strictly plurisubharmonic in $U \cap \Omega$; if no such η is allowed, we let $\eta_\rho = 0$. We also define the *Diederich–Fornaess exponent* η_h of a fixed hermitian metric h^2 of $N_M^{1,0}$ in the same manner. The Diederich–Fornaess indices are clearly the supremum of the corresponding Diederich–Fornaess exponents.

Lemma 3 *We have* $\eta(\Omega) = \eta(M)$.

Proof It is proved in [A1, Theorem 1.1] that one can construct a smooth hermitian metric h_ρ^2 of $N_M^{1,0}$ from a given smooth defining function ρ of M with $\eta_\rho > 0$ so that $\eta_\rho = \eta_{h_\rho}$. Hence, $\eta(\Omega) \leq \eta(M)$.

To derive the other inequality, it suffices to show that any hermitian metric h^2 of $N^{1,0}$ with $\eta_h > 0$, which condition is equivalent to $i\Theta_h > 0$ as quadratic forms on $T_M^{1,0}$, can be obtained by the construction above from a defining function of M. This inverse construction originates from the work of Brunella [Br] where he proved that this is possible if the Levi foliation of M extends to a holomorphic foliation on a neighborhood of M. Although the extended holomorphic foliation may not exist in our setting, we are able to apply refined constructions explained in [O, §1], [BI, Proposition 1], or [A2, Proposition 3.1] and finish the proof. □

Remark 4 We have restricted ourselves not to formulate the results for Levi-flat real hypersurfaces with finite differentiability because we have a technical problem at this point. The construction from defining functions to hermitian metrics in [A1] loses one order in differentiability since taking its normal derivative, although the inverse constructions in [**?**] or [A2] do not give us a gain in differentiability. So we cannot simply state that any $\mathscr{C}^k$-smooth hermitian metric can be obtained from a $\mathscr{C}^k$ or $\mathscr{C}^{k+1}$-smooth defining function for $2 \le k < \infty$ unlike in the case $k = \infty$.

3 The Proof of Lemma 1 and a Remark

3.1 Proof of Lemma 1

Now we shall give the proof of Lemma 1.

Proof of Lemma 1 Suppose the contrary: $\eta(M) > 1/(n+1)$. By definition, there exists a smooth hermitian metric of $N_M^{1,0}$, say h^2, such that

$$i\Theta_h - \frac{1}{n} i\alpha_h \wedge \overline{\alpha}_h > 0$$

as quadratic forms on $T_M^{1,0}$.

By taking a double covering of M if necessary, we may assume that M is oriented. We let $\eta := h_U dt_U$ where t_U is the transverse coordinate of a positively-oriented foliated chart $(U, (z_U, t_U))$ and $h_U := h(\frac{\partial}{\partial t_U})$. Then we see that η is a well-defined 1-form on M, and that $\Theta_h \wedge \eta$, $\alpha_h \wedge \eta$ and $\overline{\alpha_h} \wedge \eta$ make sense as differential forms on M regardless of the choice of extensions of α_h or Θ_h to tensors on $\mathbb{C} \otimes_{\mathbb{R}} TM$. Among these forms, we can show the equalities $(d\alpha_h) \wedge \eta = \Theta_h \wedge \eta$ and $d\eta = (\alpha_h + \overline{\alpha_h}) \wedge \eta$ from straightforward computation on the foliated chart.

Now we obtain by direct computation that

$$\begin{aligned}
&d\left((i\Theta_h - \frac{1}{n} i\alpha_h \wedge \overline{\alpha_h})^{n-1} \wedge i\alpha_h \wedge \eta\right) \\
&= (n-1)(i\Theta_h - \frac{1}{n} i\alpha_h \wedge \overline{\alpha_h})^{n-2} \wedge \frac{1}{n} i\Theta_h \wedge i\alpha_h \wedge \overline{\alpha_h} \wedge \eta \\
&\quad + (i\Theta_h - \frac{1}{n} i\alpha_h \wedge \overline{\alpha_h})^{n-1} \wedge (i\Theta_h - i\alpha_h \wedge \overline{\alpha_h}) \wedge \eta \\
&= (i\Theta_h - \frac{1}{n} i\alpha_h \wedge \overline{\alpha_h})^{n} \wedge \eta,
\end{aligned}$$

and Stokes' theorem yields a contradiction:

$$
\begin{aligned}
0 &< \int_M (i\Theta_h - \frac{1}{n} i\alpha_h \wedge \overline{\alpha_h})^n \wedge \eta \\
&= \int_M d\left((i\Theta_h - \frac{1}{n} i\alpha_h \wedge \overline{\alpha_h})^{n-1} \wedge i\alpha_h \wedge \eta \right) \\
&= 0.
\end{aligned}
$$

□

Remark 5 The proof shows in particular that $\int_M i\Theta_h \wedge \eta = \int_M i\alpha_h \wedge \overline{\alpha}_h \wedge \eta$ always holds when $\dim_{\mathbb{R}} M = 3$. This equality well explains the behavior of the Diederich–Fornaess exponent of an explicit example described in [A2, §5].

3.2 The Approach of Bejancu and Deshmukh

We give a remark that the substantial content of Lemma 1 has been already observed by Bejancu and Deshmukh [BD] in the context of differential geometry.

Remark 6 When $\dim_{\mathbb{R}} M = 3$, the integrand $(i\Theta_h - i\alpha_h \wedge \overline{\alpha_h}) \wedge \eta$ was used in [BD] to show that the totally real Ricci curvature of compact Levi-flat real hypersurfaces in Kähler surfaces cannot be everywhere positive.

Let us explain this coincidence. Suppose that we have an oriented smooth Levi-flat real hypersurface M in a Kähler surface (X, ω). We restrict on M the Kähler metric ω as a Riemannian metric and consider its Levi-Civita connection ∇^M and Ricci curvature Ric^M. We also consider the Gauss–Kronecker curvature $G_{\mathscr{F}/M}$ of the leaves of the Levi foliation $\mathscr{F}$ in M. Take the signed distance function to M with respect to the given Kähler metric ω and induce a hermitian metric h^2 of $N_M^{1,0}$ from it. Then, we can observe by direct computation that

$$
\begin{aligned}
4(i\Theta_h - i\alpha_h \wedge \overline{\alpha_h}) &= (\mathrm{Ric}^M(\xi,\xi) - 2G_{\mathscr{F}/M})\, \omega|T_M^{1,0} \otimes T_M^{0,1} \\
&= (\mathrm{Ric}^M(\xi,\xi) - \frac{1}{2}\|d\eta\|^2 + \|\nabla^M \xi\|^2)\, \omega|T_M^{1,0} \otimes T_M^{0,1}
\end{aligned}
$$

where ξ is the Reeb vector field of M chosen so that it is normalized and orthogonal to H_M with respect to ω and positively directed, and η is the metric dual of ξ. The last line is exactly the integrand used in [BD]. We leave the details of this computation to the reader, who can find the techniques needed in [AB2, BD].

3.3 Open Question

We conclude this paper with stating an open question explicitly.

Question Can we formulate the Diederich–Fornaess index for any CR manifold of hypersurface type? Needless to say, it should agree with the Diederich–Fornaess index of its complemental domain when it is realized as the boundary real hypersurface of a domain in a complex manifold. Can we prove the global estimate of Fu and Shaw, and Brinkschulte and the author for this index in its full generality?

Acknowledgments The author is partially supported by an NRF grant 2011-0030044 (SRC-GAIA) of the Ministry of Education, the Republic of Korea, and a JSPS Grant-in-Aid for Young Scientists (B) 26800057. The author gratefully acknowledges an enlightening discussion with J. Brinkschulte. He is also grateful to T. Inaba for his useful remarks.

References

[A1] Adachi, M.: A local expression of the Diederich–Fornaess exponent and the exponent of conformal harmonic measures. Bull. Braz. Math. Soc. (N.S.) **46**, 65–79 (2015)

[A2] Adachi, M.: On a global estimate of the Diederich–Fornaess index of Levi-flat real hypersurfaces. Adv. Stud. Pure. Math., to appear. arXiv:1410.2693

[AB1] Adachi, M., Brinkschulte, J.: A global estimate for the Diederich–Fornaess index of weakly pseudoconvex domains. Nagoya Math. J., to appear. arXiv:1401.2264

[AB2] Adachi, M., Brinkschulte, J.: Curvature restrictions for Levi-flat real hypersurfaces in complex projective planes. Ann. Inst. Fourier (Grenoble), to appear. arXiv:1410.2695

[BD] Bejancu, A., Deshmukh, S.: Real hypersurfaces of $\mathbf{CP}^n$ with non-negative Ricci curvature. Proc. Am. Math. Soc. **124**, 269–274 (1996)

[Be] Berndtsson, B., Charpentier, P.: A Sobolev mapping property of the Bergman kernel. Math. Z. **235**, 1–10 (2000)

[BI] Biard, S., Iordan, A.: Non existence of Levi flat hypersurfaces with positive normal bundle in compact Kähler manifolds of dimension ≥ 3. Preprint. arXiv:1406.5712

[Br] Brunella, M.: On the dynamics of codimension one holomorphic foliations with ample normal bundle. Indiana Univ. Math. J. **57**, 3101–3113 (2008)

[DF] Diederich, K., Fornaess, J.E.: Pseudoconvex domains: bounded strictly plurisubharmonic exhaustion functions. Invent. Math. **39**, 129–141 (1977)

[FS] Fu, S., Shaw, M.-C.: The Diederich–Fornæss exponent and non-existence of Stein domains with Levi-flat boundaries. J. Geom. Anal. (2014). doi:10.1007/s12220-014-9546-6

[OS] Ohsawa, T., Sibony, N.: Bounded p.s.h. functions and pseudoconvexity in Kähler manifold. Nagoya Math. J. **149**, 1–8 (1998)

[O] Ohsawa, T.: Nonexistence of certain Levi flat hypersurfaces in Kähler manifolds from the viewpoint of positive normal bundles. Publ. Res. Inst. Math. Sci. **49**, 229–239 (2013)

Unbounded Pseudoconvex Domains in $\mathbb{C}^n$ and Their Invariant Metrics

Taeyong Ahn, Hervé Gaussier and Kang-Tae Kim

Abstract In this article, we introduce a method to study the positivity and the completeness of the Bergman metric for a broad collection of unbounded domains.

Keywords Bergman metric · Positivity · Completeness · Hahn-Lu comparison · Unbounded domain

1 Some Problems

Whether a complex manifold admits a positive-definite and complete Bergman metric (as well as other invariant metrics) has attracted much attention for quite some time. While the Bergman metric of any bounded domain in $\mathbb{C}^n$ is positive-definite, the completeness was extensively studied and has satisfactory conclusion (cf. [Ohs, Diede] et al.). For manifolds, there are not many theorems in this direction; one of the well-known theorems is Theorem H of [GW]. In between, there are unbounded domains of $\mathbb{C}^n$. As pointed out in [HTS], even some of the most basic-looking features of bounded domains have turned out nontrivial when asked upon the general unbounded domains in $\mathbb{C}^n$.

T. Ahn
Center for Geometry and Its Applications, POSTECH, Pohang 790-784, The Republic of Korea
e-mail: triumph@postech.ac.kr

H. Gaussier
Univ. Grenoble Alpes, IF, 38000 Grenoble, France

H. Gaussier
CNRS, IF, 38000 Grenoble, France
e-mail: herve.gaussier@ujf.grenoble.fr

K. Kim (✉)
Center for Geometry and Its Applications and Department of Mathematics, POSTECH, Pohang 790-784, The Republic of Korea
e-mail: kimkt@postech.ac.kr

F. Bracci et al. (eds.), *Complex Analysis and Geometry*, Springer Proceedings in Mathematics & Statistics 144, DOI 10.1007/978-4-431-55744-9_3

Naturally, we wished to introduce a method for studying the positivity and the completeness of the Bergman metric for unbounded domains. But before focussing more narrowly, we feel that it may be appropriate to point out some more related problems whose answers are still in order.

Question 1.1 Which unbounded domains possess the property that their Bergman metric is positive-definite and complete?

One quick example is the domain

$$\Omega := \{(z, w) \in \mathbb{C}^2 \colon |w| < \exp(-|z|^2)\}.$$

This domain contains the complex line $\{(z, w) \in \mathbb{C}^2 \colon w = 0\}$ and hence is not Kobayashi-hyperbolic and not biholomorphic to any bounded domain. But its volume is finite, as one can check by a direct computation. Therefore any constant function is square integrable. More generally, all holomorphic polynomial functions are of L^2. Thus the Bergman kernel exists. Moreover, in an unpublished note, the 3rd named author in a communication with S. Shimizu of Tohoku University (Japan) found explicit formulae of the Bergman kernel and metric; the result, showing the positivity and the completeness of the Bergman metric, is presented in [AGK].

Now, consider the famous examples (for $\varepsilon = 0, 1$)

$$\Omega_{\mathrm{KN}}^{\varepsilon} := \{(z, w) \in \mathbb{C}^2 \colon \operatorname{Re} w + \varepsilon |zw|^2 + |z|^8 + \frac{15}{7} \operatorname{Re} |z|^2 z^6 < 0\}$$

called the Kohn-Nirenberg domain. The case $\varepsilon = 0$ was studied by Herbort [Herb]; its Bergman metric is positive and complete. The other case is not so much different, but requires some care. See [AGK] for details; the main theme of [AGK] is to set up a method that can be used in this direction of study, and we shall present a brief survey on this in the later part of this article.

A more general question is the following:

Question 1.2 Is the Kohn-Nirenberg domain biholomorphic to a bounded domain?

No progress has been made toward the solution of this problem as far as the authors are aware of, at least at the time of this writing. However, this may be interesting to see: according to Bedford and Fornaess [BF], the domain $\Omega_{\mathrm{KN}}^{\varepsilon}$ has a holomorphic peak function at the origin, say h, that is continuous up to the boundary and enjoys exponential decay. Now if one considers the map $f \colon \Omega_{\mathrm{KN}}^{\varepsilon} \to \mathbb{C}^3$ defined by

$$f(z, w) = (h(z, w), zh(z, w), wh(z, w)), \forall (z, w) \in \Omega_{\mathrm{KN}}^{\varepsilon}$$

then $f(\Omega_{\mathrm{KN}}^{\varepsilon})$ is a bounded subset of $\mathbb{C}^3$. Notice that f is also 1-1 and holomorphic. This domain is, therefore, Caratheodory (and hence also Kobayashi) hyperbolic.

Question 1.3 Which unbounded convex domains are biholomorphic to a bounded convex domain?

For a complex manifold to be biholomorphic to a bounded domain, it must be Kobayashi hyperbolic. Then any Kobayashi hyperbolic convex domain is known to be biholomorphic to a bounded domain. However it is still unclear when the bounded realization of the unbounded convex hyperbolic domain should be convex.

This problem is more attractive when one recalls the theorem of Vinberg, Piatetskii-Shapiro and Gindikin which says that every bounded homogenous domain (hence in particular Kobayashi hyperbolic) is biholomorphic to a Siegel domain (of the second kind). Since every Siegel domain is an unbouned convex domain, the above question transforms into: "*Which homogeneous domains are biholomorphic to a bounded convex domain?*" It is an open conjecture that all homogeneous domains biholomorphic to a bounded convex domain is a bounded symmetric domain. [Known as Gindikin's problem.]

The main focus of this article, however, is upon the Bergman metric, its positivity and completeness.

2 A Remark on Hahn-Lu Comparision Theorem

Toward Question 1.1, the following (slight) modification of the statement turns out to be useful:

Theorem 2.1 (Hahn-Lu Comparison Theorem, II) *If the Bergman kernel K_M of a complex manifold M satisfies the condition $K_M(p, p) \neq 0$ at $p \in M$ then, for the Caratheodory pseudometric c_M and the Bergman (pseudo) metric β_M, it holds that*

$$(c_M(p, v))^2 \leq \beta_M|_p(v, v), \forall v \in T_pM.$$

Here, the Caratheodory pseudometric is the classic concept defined upon the family $\mathscr{H}(M, D)$ of the holomorphic functions of M with image contained in the unit open disc D in $\mathbb{C}$:

$$c_M(p, v) = \sup\{|d\psi_p(v)| : \psi \in \mathscr{H}(M, D), \psi(p) = 0\}.$$

The Bergman (pseudo) metric at p is defined by the matrix representation (with respect to a holomorphic local coordinate system) with its jk-th entry

$$\beta_{jk}(p) = \left.\frac{\partial^2}{\partial z_j \partial \bar{z}_k}\right|_p \log K_*(z, z),$$

where K_* is the complex-valued function defined by

$$K_M(z, w) = K_*(z, w) dz_1 \wedge \ldots \wedge dz_n \wedge d\bar{w}_1 \wedge \ldots \wedge d\bar{w}_n.$$

We recall that K_M denotes the Bergman kernel of M. Notice that, if $K_M(p,p) \neq 0$, then the Bergman metric is nonnegative, but it may not in general be positive (or, positive-definite in this case). This is why we used the term "pseudo" in parentheses.

The theorem was proved in 1958 by Q.-K. Lu (the transliteration of his Chinese name then was written as K.-H. Look) for the bounded domains and was proved independently by K. T. Hahn around 1976 for complex manifolds with the assumption that both Caratheodory and Bergman metrics are positive [Lu, Hahn76]. Their proofs were essentially the same.

Then we observed that the "short" proof presented by Hahn in [Hahn78] actually demonstrated more than what was written in the statement of the theorem. Indeed, the above statement is just the result of the arguments there.

On the other hand, the significance of this observation is that this current version can be useful for showing the positivity of the Bergman metric; notice that *the Bergman metric is positive whenever the kernel is nonzero and the Caratheodory metric is positive.*

3 A Technique for Positivity and Completeness of Bergman Metric

From the preceding section, it became obvious that one should study the class of square integrable functions. The following theorem provides a technique which has turned out to be useful.

First we set the notation: we put $\delta_U(z) := \min\{1, \text{ dist }(z, \mathbb{C}^n \setminus U)\}$, where U is an open subset in $\mathbb{C}^n$ and "dist" means the Euclidean distance. Moreover, with a holomorphic function g, define $Z(g) = \{z \colon g(z) = 1\}$.

Theorem 3.1 (Ahn et al. [AGK]) *Let Ω be a domain in $\mathbb{C}^n$. If $p \in \partial\Omega$ satisfies the following two properties:*

1. *There exists an open neighborhood V of p and a function $g \in \mathscr{O}(V)$ supporting $V \cap \Omega$ at p.*
2. *There are constants r_1, r_2, r_3 with $0 < r_1 < r_2 < r_3 < 1$ and $B^n(p, r_3) \subset V$, and there exists a Stein neighborhood U of $\overline{\Omega}$ and a function $h \in \mathscr{O}(\Omega \cup V) \cap \mathscr{O}^*(V)$ satisfying*

$$Z_g \cap U \cap \overline{B^n(p, r_2)} \setminus B^n(p, r_1) = \emptyset \qquad (\dagger)$$

and

$$|h(z)|^2 \leq C_0 \frac{\delta_U(z)^{2n}}{(1 + \|z\|^2)^2}, \forall z \in \Omega \qquad (\ddagger)$$

for some positive constant C_0, then Ω admits a holomorphic peak function at p.

The reader would agree that construction of globally bounded holomorphic functions with peaking property at boundary points should be useful in the light of the exposition of this article, but would ask a question, incidentally, whether the conditions in the hypothesis of this theorem can be met in a broad collection of domains. We shall discuss this point in the next section with examples.

4 Examples

4.1 The Kohn-Nirenberg and Fornaess Domains

The first is the Kohn-Nirenberg domain mentioned earlier:

$$\Omega^0_{\mathrm{KN}} := \{(z, w) \in \mathbb{C}^2 : \rho(z, w) = \mathrm{Re}\, w + |z|^8 + \frac{15}{7}\mathrm{Re}\, |z|^2 z^6 < 0\}$$

For this domain we use the holomorphic peak-function, say f, constructed in [BF]. It enjoys the following properties:

- f vanishes nowhere,
- $|f|$ decays exponentially at infinity. In particular it is square-integrable on $\Omega_\varepsilon = \{\rho < \varepsilon(|z|^8 + |w|)\}$,
- f peaks at $(0, 0)$, i.e., $|f(z)| < 1$ for every $z \in \Omega_{\mathrm{KN}}$, and $f(z) \to 1$ as $z \in \Omega_{\mathrm{KN}}$ approaches $(0, 0)$.

Thus one may let this f take the role of h. Then this constructs, via Theorem 3.1, global peak functions at every strongly pseudoconvex boundary point. Then the other properties (such as the facts that all weakly pseudoconvex boundary points are in the orbit of $(0, 0)$ via the action by translation, and that the domain is weighted-homogeneous) combined shall imply that the domain is complete Caratheodory hyperbolic. Then the comparison theorem of Hahn-Lu applies here and one obtains that the Bergman metric is positive and complete.

The other domain

$$\Omega^1_{\mathrm{KN}} := \{(z, w) \in \mathbb{C}^2 : \rho(z, w) = \mathrm{Re}\, w + |zw|^2 + |z|^8 + \frac{15}{7}\mathrm{Re}\, |z|^2 z^6 < 0\}$$

as well as similar domains constructed by Fornaess [Fspso] can be shown, after some minor adjustments, to satisfy the same conclusion that their respective Bergman metric is positive and complete.

4.2 Other Domains

Herbort presented in [Herb] the concept of domains with diagonal type, and showed that their Bergman metrics are positive and complete. Our method is more general in its nature and shows that the same conclusion holds for all domains defined by a weighted-homogeneous pluri-subharmonic polynomial defining function regardless the dimension.

Notice that Theorem 3.1 concerns only the technique of finding a global peak function starting with a local holomorphic support function. This procedure is independent of the particular features of the defining function such as weighted-homogeneity. In fact this theorem works for the domain defined by $|w| < \exp(-|z|^2)$ for instance.

5 Remarks and More Questions

Earlier, the following theorem was discovered:

Theorem 5.1 (Chen et al. [CKO]) *If* $\rho\colon \mathbb{C}^n \to \mathbb{R}$ *is a pluri-subharmonic function with* $\lim_{\|z\|\to\infty} \rho(z) = +\infty$, *then the Bergman metric of the domain* $\Omega = \{(z_0, \ldots, z_n) \in \mathbb{C}^{n+1} : \operatorname{Re} z_0 > \rho(z_1, \ldots, z_n)\}$ *is positive and complete.*

The proof-arguments appear quite different from ours. It is of interest to compare the two methods. For instance one can observe from the assumptions in Theorem 5.1 that Ω admits a global holomorphic peak function at infinity. Indeed, $(z \in \Omega,\ ||z|| \to \infty) \Leftrightarrow (z \in \Omega,\ |z_0| \to +\infty)$. Denote by Φ that peak holomorphic function :

$$\forall z \in \Omega,\ |\Phi(z)| < 1 \text{ and } \lim_{|z|\to\infty} \Phi(z) = 1.$$

Since the Caratheodory (pseudo)metric decreases under the action of holomorphic maps, then $\lim_{z\in\Omega,\ |z|\to\infty} d^C_\Omega(z, z^0) = +\infty$ for every $z^0 \in \Omega$, where d^C_Ω denotes the Caratheodory pseudodistance on Ω. The same condtion is satisfied for the Bergman metric according to the Hahn-Lu comparison Theorem. It seems particularly interesting to investigate the existence of holomorphic peak functions at boundary points of Ω.

Finally, we would like to suggest another problem. Fornaess constructed in [F04] a manifold exhausted by a sequence of biholomorphic images of the open unit ball, which admits a nonconstant bounded plurisubharmonic function, and yet its Kobayashi metric is identically zero. This is called a *Short-*$\mathbb{C}^k$ for every $k \geq 2$. Not much has been studied on this manifold. Although the problem seems to be beyond our techniques, we would like to close this article with the following question.

Question 5.1 Does a Short-$\mathbb{C}^k$ admit Bergman metric?

For further details, the reader is invited to read [AGK].

Acknowledgments Research of the first and the third named authors is supported in part by the grant 2011-0030044 (The SRC-GAIA) of the NRF of Korea. Part of the contents of this article was presented by the first named author in The KSCV10 Symposium, GyeongJu, Korea, in August 2014.

References

[AGK] Ahn, T., Gaussier, H., Kim, K.-T.: Positivity and completeness of invariant metrics (2014) (Preprint)

[BF] Bedford, E., Fornaess, J.E.: A construction of peak functions on weakly pseudoconvex domains. Ann. Math. **107**, 555–568 (1978)

[CKO] Chen, B.-Y., Kamimoto, J., Ohsawa, T.: Behavior of the Bergman kernel at infinity. Math. Z. **248**, 695–798 (2004)

[Diede] Diederich, K.: Uber die 1. und 2. Ableitungen der Bergmanschen Kernfunktion und ihr Randverhalten. Math. Ann. **203**, 129–170 (1973)

[Fspso] Fornaess, J.E.: Peak points on weakly pseudoconvex domains. Math. Ann. **227**, 173–175 (1977)

[F04] Fornaess, J.E.: Short $\mathbb{C}^k$, Complex analysis in several variables memorial conference of Kiyoshi Oka's Centennial Birthday, 95108. Adv. Stud. Pure Math. **42**, 141–146 (2004) (Math. Soc. Tokyo, Japan)

[GW] Greene, R.E., Wu, H.: Function theory on manifolds which possess a pole. Lecture Notes in Mathematics, vol. 699. Springer, Berlin (1979)

[Hahn76] Hahn, K.T.: On completeness of the Bergman metric and its subordinate metrics. Proc. Nat. Acad. Sci. U.S.A. **73**(12), 4294 (1976)

[Hahn78] Hahn, K.T.: Inequality between the Bergman metric and Caratheodory differential metric. Proc. Am. Math. Soc. **68–2**, 193–194 (1978)

[HTS] Harz, T., Shcherbina, N., Tomassini, G.: On defining functions for unbounded pseudoconvex domains (2014). arxiv:1405.2250

[Herb] Herbort, G.: Invariant metric and peak functions on pseudoconvex domains of homogeneous diagonal type. Math. Z. **209**, 223–243 (1992)

[Lu] Look, K.H. (= Q.-K. Lu): Schwarz lemma and analytic invariants. Sci. Sin. **7**, 453–504 (1958)

[Ohs] Ohsawa, T.: A remark on the completeness of the Bergman metric. Proc. Japan Acad. Ser. A Math. Sci. **57–4**, 238–240 (1981)

Abstract Basins of Attraction

Leandro Arosio

Abstract Abstract basins appear naturally in different areas of several complex variables. In this survey we want to describe three different topics in which they play an important role, leading to interesting open problems.

Keywords Canonical models · Bedford's conjecture · Loewner theory in several variables

1 The Construction of Abstract Basins

In recent years strong links among three different areas of research in several complex variables were discovered. It was indeed shown that the Bedford conjecture, the Loewner theory and the theory of models for holomorphic self-maps revolve around a common concept introduced by Fornæss–Stensønes [FS04] in 2004: the abstract basin of attraction. Let $\mathbb{B}^q$ denote the open unit ball in $\mathbb{C}^q$, and let $(\varphi_n : \mathbb{B}^q \to \mathbb{B}^q)_{n\geq 0}$ be a family of univalent (holomorphic injective) self-maps. We can think of this family as a dynamical system whose evolution law may change in time, and we call it a *non-autonomous univalent dynamical system.* If $\varphi_n = \varphi$ for all $n \geq 0$, then we call it *autonomous.* If $0 \leq n \leq m$ we denote by $\varphi_{n,m}$ the composition $\varphi_{m-1} \circ \varphi_{m-2} \circ \cdots \circ \varphi_n$. To construct the abstract basin we take the direct limit of (φ_n), that is, we consider the following equivalence relation on the product $\mathbb{B}^q \times \mathbb{N}$: let $0 \leq n \leq m$, then $(x, n) \simeq (y, m)$ if and only if $\varphi_{n,m}(x) = y$. The set $\Omega = \mathbb{B}^q \times \mathbb{N}/_\sim$ is the *abstract basin of attraction (or the tail space) of* (φ_n). It comes naturally endowed with a family of mappings $(f_n : \mathbb{B}^q \to \Omega)_{n\geq 0}$ which satisfy

$$f_n = f_m \circ \varphi_{n,m}, \quad 0 \leq n \leq m,$$

L. Arosio (✉)
Dipartimento Di Matematica, Università di Roma "Tor Vergata",
Via Della Ricerca Scientifica 1, 00133 Rome, Italy
e-mail: arosio@mat.uniroma2.it

F. Bracci et al. (eds.), *Complex Analysis and Geometry*, Springer Proceedings in Mathematics & Statistics 144, DOI 10.1007/978-4-431-55744-9_4

and which give a complex structure to Ω. The abstract basin Ω satisfies the following universal property: if Λ is a complex manifold, and $(g_n : \mathbb{B}^q \to \Lambda)_{n\geq 0}$ is a family of holomorphic mappings which satisfy

$$g_n = g_m \circ \varphi_{n,m}, \quad 0 \leq n \leq m, \tag{1.1}$$

then there exists a holomorphic mapping $\Psi : \Omega \to \Lambda$ such that $f_n = \Psi \circ g_n$ for all $n \geq 0$. The mapping Ψ is univalent if and only if the mappings g_n are univalent for all $n \geq 0$, and its image is the domain $\bigcup_{n\geq 0} g_n(\mathbb{B}^q) \subset \Lambda$.

Remark 1.1 The same construction works for a non-autonomous univalent dynamical system with $\mathbb{R}^+$ as an index set, that is, a family of univalent mappings $(\varphi_{s,t} : \mathbb{B}^q \to \mathbb{B}^q)_{0\leq s\leq t}$ satisfying the *evolution equation*

$$\varphi_{u,t} \circ \varphi_{s,u} = \varphi_{s,t}, \quad 0 \leq s \leq u \leq t. \tag{1.2}$$

If $t \mapsto \varphi_{s,t}(z)$ is locally lipschitz, uniformly on compacta in z, then $(\varphi_{s,t})$ is called an *evolution family*.

We know by construction that the basin Ω is the growing union of the subdomains $f_n(\mathbb{B}^q)$ which are biholomorphic to $\mathbb{B}^q$, however, the complex structure of Ω may be very complicated. Using a striking example due to Fornæss [FO76] it is easy to see that there exists an abstract basin which is not Stein and which is not biholomorphic to a domain of $\mathbb{C}^q$.

The name "abstract basin of attraction" deserves an explanation. Assume that there exists a non-autonomus dynamical system given by a family of automorphisms $(\Phi_n : X \to X)$, where X is a q-dimensional complex manifold, and a univalent mapping $h : \mathbb{B}^q \to X$ such that for all $n \geq 0$ we have

$$h \circ \varphi_n = \Phi_n \circ h.$$

For all $n \geq 0$, let $\Phi_{0,n} := \Phi_{n-1} \circ \Phi_{n-2} \circ \cdots \circ \Phi_0$. Then the family $g_n := \Phi_{0,n}^{-1} \circ h$ satisfies equation (1.1), and thus, by the universal property, the abstract basin of attraction Ω of (φ_n) is biholomorphic to the domain of X given by

$$\bigcup_{n\geq 0} g_n(\mathbb{B}^q) = \{x \in X : \Phi_{0,n}(x) \in h(\mathbb{B}^q) \text{ eventually}\}.$$

Assume now that the origin is attracting for the dynamical system (φ_n), that is $\varphi_{0,n}(z) \to 0$ for all $z \in \mathbb{B}^q$. Then $\Phi_{0,n}(x) \to h(0)$ for all $x \in h(\mathbb{B}^q)$, which implies that the abstract basin of attraction Ω of (φ_n) is biholomorphic to the "actual" basin of attraction of (Φ_n) at the point $h(0)$, that is the domain of X defined by

$$\{x \in X : \Phi_{0,n}(x) \to h(0)\}.$$

2 Bedford's Conjecture

Let $\varphi\colon \mathbb{B}^q \to \mathbb{B}^q$ be a univalent mapping fixing the origin such that the spectrum of the differential at the origin $d_0\varphi$ is contained in the punctured disc $\Delta \setminus \{0\}$. Then the autonomous univalent dynamical system associated with φ is attracting at the origin, and its abstract basin of attraction is biholomorphic to $\mathbb{C}^q$ by the Poincaré–Dulac theory (see e.g. [R88]).

It is natural to ask whether this holds true if we consider non-autonomous dynamical systems $(\varphi_n\colon \mathbb{B}^q \to \mathbb{B}^q)$ whose contraction rate at the origin is uniformly bounded from above and from below.

Conjecture 2.1 ([FS04]) Let $(\varphi_n\colon \mathbb{B}^q \to \mathbb{B}^q)$ be a non-autonomous univalent dynamical system such that $\varphi_n(0) = 0$ for all $n \geq 0$ and that

$$a\|z\| \leq \|\varphi_n(z)\| \leq b\|z\|, \quad n \geq 0,\ 0 < a \leq b < 1.$$

Then the abstract basin Ω of (φ_n) is biholomorphic to $\mathbb{C}^q$.

Remark 2.1 Condition $b < 1$ ensures that $\varphi_{0,n}(z) \to 0$ for all $z \in \mathbb{B}^q$. If we drop the condition $a > 0$ there is the following counterexample by Fornæss [FOR03]. Set

$$\varphi_n\colon (z, w) \mapsto (z^2 + a_n w, a_n z),$$

where $|a_0| < 1$ and $|a_{n+1}| \leq a_n^2$. Then Ω admits a non-constant bounded plurisubharmonic function, and thus is not biholomorphic to $\mathbb{C}^2$. Such a domain is called a *short* $\mathbb{C}^2$, since it shares several invariants with $\mathbb{C}^2$ without being biholomorphic to $\mathbb{C}^2$.

Conjecture 2.1 is still open, and is deeply studied since Fornæss–Stensønes [FS04] proved that it is stronger than the well-known *Bedford conjecture*:

Conjecture 2.2 ([BED]) Let X be a complex manifold endowed with a Riemannian metric, and let $f\colon X \to X$ be an automorphism which acts hyperbolically on some invariant compact subset $K \subset X$. If $p \in K$, then the stable manifold $\Sigma(p)$ is biholomorphic to $\mathbb{C}^k$, where k is the stable dimension.

Jonsson–Varolin [JV02] showed that the Bedford conjecture is true for every p in a subset of K which is of full-measure with respect to any invariant probability measure on K, and Abate–Abbondandolo–Majer [A14] showed that $\Sigma(p)$ is biholomorphic to $\mathbb{C}^k$ when the negative Lyapounov exponents of f at p are well defined.

Fornæss–Stensønes [FS04] proved that Ω is always biholomorphic to a domain of $\mathbb{C}^q$. This implies in particular that Ω has to be Stein, and it is also easy to see that the Kobayashi pseudometric and pseudodistance of Ω are zero everywhere. The main approach to Conjecture 2.1 consists in adapting the Poincaré–Dulac method to the non-autonomous setting, and this is why arithmetic relations between a and b play a capital role. Wold [WO05] showed that if $b^2 < a$, then Ω is biholomorphic to $\mathbb{C}^q$. Abbondandolo–Majer [A014] showed that in $\mathbb{B}^2$ this condition can be loosen to $b^{29/14} < a$. Very recently Peters–Smit [PT98] obtained $b^{\frac{11}{5}} < a$, and, assuming all $d_0\varphi_n$ diagonal, $b^3 < a$.

Other interesting "weak monotonicity" relations among the eigenvalues of $d_0\varphi_n$ are considered in [PE07, A14]. See [E35] for a survey on non-autonomous basins and the Bedford conjecture.

3 Loewner's Theory

The Loewner PDE in the unit disc was introduced by Loewner [LOE3] in 1923 while he was working on the Bieberbach conjecture. It was later developed by Kufarev [KUF43] and Pommerenke [PM65] and is today one of the principal tools in geometric function theory. Recently Bracci–Contreras–Díaz-Madrigal [BC12, BD09] introduced a framework for a very general Loewner theory which works in complete Kobayashi hyperbolic manifolds (see also [AB11]). We consider the case of $\mathbb{B}^q$. They give the following definition, which is a generalization to the setting of continuous time of the concept of non-autonomous univalent dynamical system. A *Herglotz vector field* on $\mathbb{B}^q$ is a function $h(z,t)\colon \mathbb{B}^q \times \mathbb{R}^+ \to \mathbb{C}^q$ such that

1. $z \mapsto h(z,t)$ is a semicomplete holomorphic vector field for a.e. $t \in \mathbb{R}^+$,
2. $t \mapsto h(z,t)$ is measurable and locally bounded, uniformly on compacta in z.

They prove that a Herglotz vector field h is semicomplete in the sense that the ODE

$$\frac{dz(t)}{dt} = h(z,t)$$

has a solution flow given by an evolution family $(\varphi_{s,t}\colon \mathbb{B}^q \to \mathbb{B}^q)_{0\le s\le t}$.

The Loewner PDE in several complex variables was studied by Pfaltzgraff, Graham, Duren, Kohr, Hamada, and others (see [PF74, P75, DG10, GK02, GH08]). In [AH13], the Loewner PDE was generalized to the setting of Herglotz vector fields in the following way:

$$\frac{\partial f_t(z)}{\partial t} = -d_z f_t\, h(z,t), \quad \text{a.e. } t \ge 0, z \in \mathbb{B}^q, \tag{3.1}$$

where $h(z,t)$ is a given Herglotz vector field and where the unknown $(f_t\colon \mathbb{B}^q \to \mathbb{C}^q)$ is a family of univalent mappings such that $t \mapsto f_t(z)$ is locally lipschitz, uniformly on compacta in z.

The following result shows that solutions exist and are essentially unique if we do not restrict ourselves to solutions with values in $\mathbb{C}^q$.

Theorem 3.1 ([AH13]) *The Loewner PDE (3.1) admits a univalent solution $(f_t\colon \mathbb{B}^q \to \Omega)$, where Ω is the abstract basin of attraction of the evolution family $(\varphi_{s,t})$ associated with the Herglotz vector field $h(z,t)$. If $(g_t\colon \mathbb{B}^q \to Q)$ is another solution,* *where Q is a q-dimensional complex manifold, then there exists a holomorphic mapping $\Psi : \Omega \to Q$ such that*

$$g_t = \Psi \circ f_t.$$

The abstract basin Ω is called the *Loewner range* of $h(z, t)$ (or of $(\varphi_{s,t})$).

Remark 3.1 This result transforms the analytic problem of finding a univalent solution $(f_t : \mathbb{B}^q \to \mathbb{C}^q)$ for the Loewner PDE (3.1) to the geometric problem of understanding whether the Loewner range Ω of $(\varphi_{s,t})$ is biholomorphic to a domain of $\mathbb{C}^q$. For example, in one variable, we know that Ω is non-compact and simply connected (since it is the growing union of discs). Thus, by the uniformization theorem, it has to be biholomorphic to $\mathbb{C}$ or to the disc Δ. In either case it is biholomorphic to a domain of $\mathbb{C}$, and hence we obtain that the Loewner PDE in one variable always admits a univalent solution, as proved by Contreras–Díaz-Madrigal–Gumenyuk [ON10] with a different method.

This embedding problem was solved in [AW13] using two major tools: a result of Docquier–Grauert [DO60] which implies that for all $0 \leq s \leq t$ the pair $(f_s(\mathbb{B}^q), f_t(\mathbb{B}^q))$ is Runge, and Andersén–Lempert theory.

Theorem 3.2 *The Loewner range Ω of $(\varphi_{s,t})$ is biholomorphic to a domain of $\mathbb{C}^q$. As a consequence, the Loewner PDE (3.1) always admits a univalent solution $(f_t : \mathbb{B}^q \to \mathbb{C}^q)$.*

An interesting open question is to find conditions for a Herglotz vector field $h(z, t)$ which ensure that its Loewner range Ω is biholomorphic to $\mathbb{C}^q$. Some conditions are given in [AR11, AR13, AR12]. We can also formulate a continuous-time analogue of the Bedford conjecture. Recall that if A is a linear endomorphism of $\mathbb{C}^q$, we denote $m(A) := \min\{\mathtt{Re}\langle Az, z\rangle : |z| = 1\}$ and $k(A) := \max\{\mathtt{Re}\langle Az, z\rangle : |z| = 1\}$.

Conjecture 3.1 Let $h(z, t)$ a Herglotz vector field on $\mathbb{B}^q$ of the form $h(z, t) = A(t)z + O(|z|^2)$. Assume that

1. $m(A(t)) > 0$ for all $t \geq 0$ and $\int_0^\infty m(A(t))d(t) = \infty$,
2. there exists $\ell \in \mathbb{R}^+$ such that $\ell m(A(t)) \geq k(A(t))$, for all $t \geq 0$.

Then the Loewner range of $h(z, t)$ is biholomorphic to $\mathbb{C}^q$.

Consider the evolution family $(\varphi_{s,t})_{0 \leq s \leq t}$ associated with $h(z, t)$. Notice that any discretization of the index set $\mathbb{R}^+$ gives as a result a (discrete) non-autonomous dynamical system $(\varphi_{n,m})_{0 \leq n \leq m}$. The abstract basins of the two families are easily seen to be biholomorphic. The assumptions in Conjecture 3.1 allow to discretize the index set $\mathbb{R}^+$ in such a way that $(\varphi_{n,m})_{0 \leq n \leq m}$ satisfies the assumption of the Bedford conjecture. Thus Conjecture 3.1 is weaker than the Bedford conjecture.

For a survey on Loewner theory, see [AB10].

4 Models

The idea of using representation models to understand the local dynamics of holomorphic self-maps goes back to the birth of complex dynamics itself, that is the introduction in 1870 of the Schröder equation [E70, SC70]. Let $f\colon \Delta \to \Delta$ be a holomorphic self-map of the unit disc fixing the origin. Assume that $0 < |f'(0)| < 1$. Then the origin is an attracting fixed point, and the Schröder equation is the following:

$$\sigma \circ f = f'(0) \circ \sigma, \tag{4.1}$$

where $\sigma\colon \Delta \to \mathbb{C}$ is an unknown holomorphic function. This equation was solved in 1884 by Königs [KO84], which showed that there exists a holomorphic solution σ, which is unique if we impose $\sigma(0) = 0,\ \sigma'(0) = 1$.

If f has no interior fixed points, then by the Denjoy–Wolff theorem there exists a point $a \in \partial\Delta$ called the *Denjoy–Wolff point* such that f^n converges to a uniformly on compacta. Moreover the *dilation of* f at a is defined as the following non-tangential limit:

$$\angle \lim_{z\to a} f'(z) = \lambda \in (0, 1].$$

The mapping f is called *hyperbolic* iff $\lambda \in (0, 1)$, and is called *parabolic* iff $\lambda = 1$.

Let $\mathbb{H}$ denote the upper half-plane. If f is hyperbolic, Valiron [VA31] proved in 1931 that there exists a holomorphic function $\sigma\colon \Delta \to \mathbb{H}$ such that

$$\sigma \circ f = \frac{1}{\lambda}\sigma,$$

and any other solution is a positive multiple of σ. Notice that the growing union $\bigcup_{m\in\mathbb{N}} \lambda^m\sigma(\Delta)$ fills the whole half-plane $\mathbb{H}$. If f is parabolic, Pommerenke–Baker [PO79, BA79] proved in 1979 that there exists a holomorphic function $\sigma\colon \Delta \to \mathbb{C}$ such that

$$\sigma \circ f = \sigma + 1.$$

In this case we have two cases for the complex structure of the growing union $\bigcup_{m\in\mathbb{N}} (\sigma(\Delta) - m)$. Recall that, if $z_m := f^m(z_0)$ is an orbit, its *step* $s(f, z_0)$ is defined as $\lim_{m\to\infty} k_\Delta(z_m, z_{m+1})$, where k_Δ denotes the Poincaré distance of the disc Δ. Such a limit exists thanks to the non-expansiveness of the Poincaré distance. We have the following dichotomy:

1. for any orbit (z_m) we have $s(f, z_0) = 0$ (*zero-step*),
2. for any orbit (z_m) we have $s(f, z_0) > 0$ (*nonzero-step*).

The union $\bigcup_{m\in\mathbb{N}} (\sigma(\Delta) - m)$ fills the whole $\mathbb{C}$ in the zero-step case, and is biholomorphic to Δ in the nonzero-step case.

Let now $f\colon \mathbb{B}^q \to \mathbb{B}^q$. If the origin is an attracting fixed point, then the Poincaré–Dulac theory applies and one can solve a generalized Schröder equation in several complex variables (see e.g. Rosay–Rudin [R88]). If there are no fixed points in $\mathbb{B}^q$, then as in one variable there exists a *Denjoy–Wolff point* $a \in \partial\mathbb{B}^q$, such that $f^n \to a$ uniformly on compact subsets. The *dilation of* f at a is defined as

$$\liminf_{z \to a} \frac{1 - \|f(z)\|}{1 - \|z\|} = \lambda \in (0, 1].$$

Again the mapping f is called *hyperbolic* iff $\lambda \in (0, 1)$, and is called *parabolic* iff $\lambda = 1$. Zero-step and nonzero-step are defined as in the disc using the Kobayshi distance instead of the Poincaré distance (notice however that it is not a dichotomy anymore).

There are several generalizations by Bracci, Poggi-Corradini, Gentili, Bayart, Jury (see [G05, BGP10, JU10, BA08]) of the Valiron and Abel equations in the unit ball which require additional regularity at the Denjoy–Wolff point $a \in \mathbb{B}^q$. All are obtained by scaling limit arguments. In 1981 Cowen [CO81] unified the Schröder, Valiron and Abel equations in a unique framework, introducing the concept of *model* (without naming it). Models in several complex variables were recently introduced in [AM73]. If $f : \mathbb{B}^q \to \mathbb{B}^q$ is a univalent mapping, a *semi-model for* f is given by a complex manifold Ω (*the base space*), an automorphism ψ of Ω and a holomorphic mapping $h : \mathbb{B}^q \to \Omega$ such that $h \circ f = \psi \circ h$. We also assume $\Omega = \cup_{m \in \mathbb{N}} \psi^{-m}(h(\mathbb{B}^q))$. If h is univalent, then we call (Ω, ψ, h) a *model for* f.

Theorem 4.1 ([AM73]) *Every univalent mapping* $f : \mathbb{B}^q \to \mathbb{B}^q$ *admits a model* (Ω, h, ψ)*, where* Ω *is the abstract basin of attraction of the autonomous univalent dynamical system associated with* f*. Moreover, if* (Λ, ℓ, ϕ) *is any semi-model for* f*, then there exists a surjective holomorphic map* $\Psi : \Omega \to \Lambda$ *such that the following diagram commutes:*

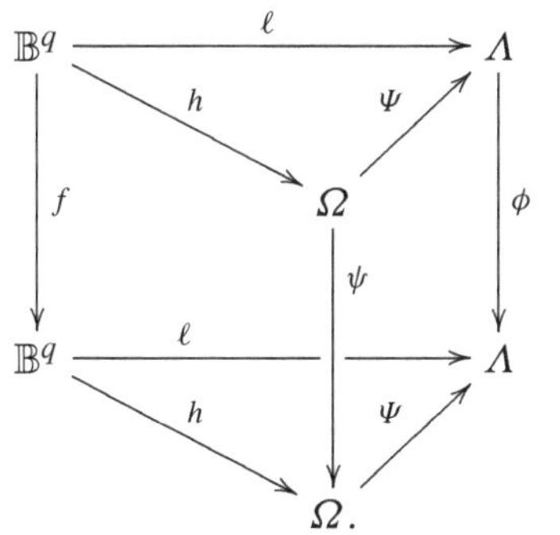

The complex structure of Ω is not known in general. However, we can single out a semi-model on a possibly lower dimensional ball $\mathbb{B}^k$ which contains all the Kobayashi pseudodistance information of the model. Since Ω is the growing union of domains which are biholomorphic to $\mathbb{B}^q$, there exists, by a result of Fornæss–Sibony [FS81], a surjective holomorphic submersion $r : \Omega \to \mathbb{B}^k$, where $0 \le k \le q$, on whose fibers the Kobayashi pseudo-distance of Ω vanishes. This implies that the automorphism

ψ preserves the fibers, inducing an automorphism τ of $\mathbb{B}^k$ such that the following diagram commutes:

$$\begin{array}{ccc}
\mathbb{B}^q & \xrightarrow{f} & \mathbb{B}^q \\
\downarrow h & & \downarrow h \\
\Omega & \xrightarrow{\psi} & \Omega \\
\downarrow r & & \downarrow r \\
\mathbb{B}^k & \xrightarrow{\tau} & \mathbb{B}^k.
\end{array}$$

Set $\ell := r \circ h$. The semi-model $(\mathbb{B}^k, \ell, \tau)$ is called the *canonical (Kobayashi hyperbolic) semi-model*: every other semi-model with Kobayashi hyperbolic base space is a factor of this one.

Let us discuss what happens in the unit disc. If f is hyperbolic with dilation λ, then by the Valiron equation it has a model with base space Δ, and the automorphism ψ is hyperbolic with dilation λ. The canonical semi-model coincides with the model. If f is parabolic nonzero-step, then by the Abel equation it has a model with base space Δ, and the automorphism ψ is parabolic. The canonical semi-model coincides with the model. If f is parabolic zero-step the base space of the model is $\mathbb{C}$ and thus the base space of the canonical semi-model is a single point. In [AM73], the first two cases are generalized to several complex variables.

Theorem 4.2 *Let $f : \mathbb{B}^q \to \mathbb{B}^q$ be a hyperbolic (resp. parabolic nonzero-step) univalent mapping with dilation λ. Let $(\mathbb{B}^k, \ell, \tau)$ be its canonical semi-model. Then $k \geq 1$ and τ is an hyperbolic (resp. parabolic) automorphism with the same dilation λ.*

Corollary 4.1 *Let $f : \mathbb{B}^q \to \mathbb{B}^q$ be a hyperbolic univalent mapping with dilation λ. Then there exists a holomorphic solution $\Theta : \mathbb{B}^q \to \mathbb{H}$ to the Valiron equation*

$$\Theta \circ f = \frac{1}{\lambda}\Theta.$$

The following questions are open.

Question 4.1 It would be natural to conjecture the following dichotomy (which is true in one variable and for linear fractional mappings of $\mathbb{B}^q$): if $f : \mathbb{B}^q \to \mathbb{B}^q$ is a parabolic univalent mapping and $(\mathbb{B}^k, \ell, \tau)$ is its canonical semi-model, then either τ is a parabolic automorphism, or $k = 0$ and the canonical semi-model is trivial.

It is known that the automorphism τ cannot be hyperbolic, however it is an open question whether a parabolic univalent mapping $f : \mathbb{B}^q \to \mathbb{B}^q$ can have a canonical semi-model $(\mathbb{B}^k, \ell, \tau)$ with $k > 0$ and τ with an interior fixed point.

Question 4.2 If $f : \mathbb{B}^q \to \mathbb{B}^q$ is univalent, are the fibers of the holomorphic submersion $\Omega \to \mathbb{B}^k$ biholomorphic to $\mathbb{C}^{q-k}$? Is it true that Ω is biholomorphic to $\mathbb{B}^k \times \mathbb{C}^{q-k}$?

Remark 4.1 By a result of Fornæss–Sibony [FS81], if $k = q - 1$, then the answer to Question 4.2 is affirmative.

Acknowledgments Supported by the ERC grant "HEVO - Holomorphic Evolution Equations" n. 277691.

References

[A14] Abate, M., Abbondandolo, A., Majer, P.: Stable manifolds for holomorphic automorphisms. J. Reine Angew. Math. **690**, 217–247 (2014)

[E35] Abbondandolo, A., Arosio, L., Fornæss, J.E., Majer, P., Peters, H., Raissy, J., Vivas, L.: A survey on non-autonomous basins in several complex variables. arXiv:1311.3835 (preprint)

[A014] Abbondandolo, A., Majer, P.: Global stable manifolds in holomorphic dynamics under bunching conditions. Int. Math. Res. Not. IMRN **14**, 4001–4048 (2014)

[AB10] Abate, M., Bracci, F., Contreras, M.D., Díaz-Madrigal, S.: The evolution of Loewner's differential equations. Newslett. Eur. Math. Soc. **78**, 31–38 (2010)

[AR11] Arosio, L.: Resonances in Loewner equations. Adv. Math. **227**, 1413–1435 (2011)

[AR12] Arosio, L.: Basins of attraction in Loewner equations. Ann. Acad. Sci. Fenn. Math. **37**(2), 563–570 (2012)

[AR13] Arosio, L.: Loewner equations on complete hyperbolic domains. J. Math. Anal. Appl. **398**(2), 609–621 (2013)

[AB11] Arosio, L., Bracci, F.: Infinitesimal generators and the Loewner equation on complete hyperbolic manifolds. Anal. Math. Phys. **1**(4), 337–350 (2011)

[AM73] Arosio, L., Bracci, F.: Canonical models in holomorphic iteration. Trans. Am. Math. Soc. doi:10.1090/tran/6593. arXiv:1401.6873

[AH13] Arosio, L., Bracci, F., Hamada, H., Kohr, G.: An abstract approach to Loewner chains. J. Anal. Math. **119**, 89–114 (2013)

[AW13] Arosio, L., Bracci, F., Wold, E.F.: Solving the Loewner PDE in complete hyperbolic starlike domains of $\mathbb{C}^N$. Adv. Math. **242**, 209–216 (2013)

[BA79] Baker, I.N., Pommerenke, C.: On the iteration of analytic functions in a half-plane II. J. London Math. Soc. (2) **20**(2), 255–258 (1979)

[BA08] Bayart, F.: The linear fractional model on the ball. Rev. Mat. Iberoam. **24**(3), 765–824 (2008)

[BED] Bedford, E.: Open problem session of the Biholomorphic Mappings Meeting at the American Institute of Mathematics. Palo Alto, California (2000)

[BC12] Bracci, F., Contreras, M.D., Díaz-Madrigal, S.: Evolution families and the Loewner Equation I: the unit disc. J. Reine Angew. Math. (Crelle's J.) **672**, 1–37 (2012)

[BD09] Bracci, F., Contreras, M.D., Díaz-Madrigal, S.: Evolution families and the Loewner Equation II: complex hyperbolic manifolds. Math. Ann. **344**, 947–962 (2009)

[G05] Bracci, F., Gentili, G.: Solving the Schröder equation at the boundary in several variables. Mich. Math. J. **53**(2), 337–356 (2005)

[BGP10] Bracci, F., Gentili, G., Poggi-Corradini, P.: Valiron's construction in higher dimensions. Rev. Mat. Iberoam. **26**(1), 57–76 (2010)

[ON10] Contreras, M.D., Díaz-Madrigal, S., Gumenyuk, P.: Loewner chains in the unit disc. Rev. Mat. Iberoamericana **26**, 975–1012 (2010)

[CO81] Cowen, C.C.: Iteration and the solution of functional equations for functions analytic in the unit disk. Trans. Am. Math. Soc. **265**(1), 69–95 (1981)

[DO60] Docquier, F., Grauert, H.: Levisches Problem und Rungescher Satz für Teilgebiete Steinscher Mannigfaltigkeiten. Math. Ann. **140**, 94–123 (1960)

[DG10] Duren, P., Graham, I., Hamada, H., Kohr, G.: Solutions for the generalized Loewner differential equation in several complex variables. Math. Ann. **347**(2), 411–435 (2010)
[FO76] Fornæss, J.E.: An increasing sequence of Stein Manifolds whose limit is not Stein. Math. Ann. **223**, 275–277 (1976)
[FOR03] Fornæss, J.E.: Short $\mathbb{C}^k$. Mathematical Society of Japan, Advanced Studies in Pure Mathematics (2003)
[FS81] Fornæss, J.E., Sibony, N.: Increasing sequences of complex manifolds. Math. Ann. **255**(3), 351–360 (1981)
[FS04] Fornæss, J.E., Stensønes, B.: Stable manifolds of holomorphic hyperbolic maps. Int. J. Math. **15**(8), 749–758 (2004)
[GK02] Graham, I., Hamada, H., Kohr, G.: Parametric representation of univalent mappings in several complex variables. Can. J. Math. **54**, 324–351 (2002)
[GH08] Graham, I., Hamada, H., Kohr, G., Kohr, M.: Asymptotically spirallike mappings in several complex variables. J. Anal. Math. **105**, 267–302 (2008)
[JV02] Jonsson, M., Varolin, D.: Stable manifolds of holomorphic diffeomorphisms. Invent. Math. **149**, 409–430 (2002)
[JU10] Jury, T.: Valiron's theorem in the unit ball and spectra of composition operators. J. Math. Anal. Appl. **368**(2), 482–490 (2010)
[KO84] Königs, G.: Recherches sur les intégrales de certaines équations fonctionnelles. Ann. Sci. École Norm. Sup. **1**(3), 3–41 (1884)
[KUF43] Kufarev, P.P.: On one-parameter families of analytic functions. Mat. Sb. **13**, 87–118 (1943). in Russian
[LOE3] Loewner, C.: Untersuchungen über schlichte konforme Abbildungen des Einheitskreises. Math. Ann. **89**, 103–121 (1923)
[PE07] Peters, H.: Perturbed basins. Math. Ann. **337**(1), 1–13 (2007)
[PT98] Peters, H., Smit, I.M.: Adaptive trains for attracting sequences of holomorphic automorphisms. arXiv:1408.0498 (preprint)
[PM65] Pommerenke, C.: Über die Subordination analytischer Funktionen. J. Reine Angew. Math. **218**, 159–173 (1965)
[PO79] Pommerenke, C.: On the iteration of analytic functions in a half plane. J. London Mat. Soc. **19**(2, 3), 439–447 (1979)
[PF74] Pfaltzgraff, J.A.: Subordination chains and univalence of holomorphic mappings in $\mathbb{C}^n$. Math. Ann. **210**, 55–68 (1974)
[P75] Pfaltzgraff, J.A.: Subordination chains and quasiconformal extension of holomorphic maps in $\mathbb{C}^n$. Ann. Acad. Sci. Fenn. Ser. A I Math. **1**, 13–25 (1975)
[R88] Rosay, J.P., Rudin, W.: Holomorphic maps from $\mathbb{C}^n$ to $\mathbb{C}^n$. Trans. Am. Math. Soc. **310**(1), 47–86 (1988)
[E70] Schröder, E.: Über unendliche viele Algorithmen zur Auflösung der Gleichungen. Math. Ann. **2**, 317–365 (1870)
[SC70] Schröder, E.: Über iterirte Functionen. Math. Ann. **3**, 296–322 (1870)
[VA31] Valiron, G.: Sur l'itération des fonctions holomorphes dans un demi-plan. Bull. Sci. Math. **47**, 105–128 (1931)
[WO05] Wold, E.F.: Fatou-Bieberbach domains. Int. J. Math. **16**, 1119–1130 (2005)

Invertible Dynamics on Blow-ups of $\mathbb{P}^k$

Eric Bedford

Abstract This is a survey of some recent results on the iteration of (pseudo) automorphisms of blowups of k-dimensional projective space.

Keywords Pseudo automorphism · Birational map · Dynamical system · Rational manifold

1 Introduction

Let X be a complex manifold, and let f be an automorphism, i.e. biholomorphic self map, of X. We discuss the project of finding compact complex manifolds X which carry automorphisms f which are dynamically interesting. If the dimension of X is 1, then X is a compact Riemann surface, and if the genus is at least 2, then $Aut(X)$ is finite. The other two cases are when X is a torus, in which case the automorphisms are essentially translations, or $X = \mathbb{P}^1$, in which case the automorphisms are linear (fractional). Thus, to find maps with interesting dynamics, we must start with dimension 2. In this case, a Theorem of Cantat (see [C1, C2]) restricts the set of possible surfaces X. This raises the question to know exactly which surfaces X might arise in these cases. Some basic ergodic properties hold for all automorphisms with dynamical degree >1, but little is known about the topological properties of such maps. Here we focus on the case where X is a blowup of $\mathbb{P}^2$ and we describe some of the results that are known.

Next we will consider the case of dimension 3 and higher. In this case, we find that it is natural to widen our search to manifolds X which carry pseudo-automorphisms. These are birational maps whose indeterminate and exceptional behaviors only influence subvarieties of codimension 2 and greater. It is the purpose of this paper to discuss pseudo-automorphisms that can be obtained from birational maps of $\mathbb{P}^k$ by blowing up, and in this discussion we formulate a number of open questions.

E. Bedford (✉)
Stony Brook University, Stony Brook, NY 11794, USA
e-mail: ebedford@math.sunysb.edu

F. Bracci et al. (eds.), *Complex Analysis and Geometry*, Springer Proceedings in Mathematics & Statistics 144, DOI 10.1007/978-4-431-55744-9_5

2 Rational Surfaces

The group of automorphisms (biholomorphic self-maps) of complex projective space $\mathbb{P}^k$ is $PGL(\mathbb{C}, k+1)$, which may be written as linear fractional transformations of $\mathbb{C}^k$. We can widen this class of manifolds by blowing up. In terms of complex analysis, all the global holomorphic functions are constant, and this is not changed by blowing up. However, the set of biholomorphic self-mappings might change.

Let us start with dimension 2 and the classical example of the Cremona Involution, which on $\mathbb{C}^2$ is given by $(x, y) \mapsto (1/x, 1/y)$, and on projective space we write it as a mapping of degree 2:

$$J : [x_0 : x_1 : x_2] \mapsto [1/x_0 : 1/x_1 : 1/x_2] = [x_1x_2 : x_0x_2 : x_0x_1]$$

A rational (or meromorphic) mapping is said to be *regular* at a point p if it is holomorphic in a neighborhood of p. The *indeterminacy locus* of a rational map $f : X \dashrightarrow Y$, written $\mathscr{I}(f)$ is defined as the set of all points where f is not regular. It may be shown that if f is indeterminate at a point p, then f blows up p to a variety $V \subset Y$, and the dimension of V is at least one. This "blown up" image V may be defined in more than one way. One of them is simply the cluster set: $V = \bigcap_{\varepsilon>0} \text{closure}\,(f(B(p, \varepsilon) - \mathscr{I}(f)))$, where $B(p, \varepsilon)$ denotes the ball about p with radius ε.

If $W \subset X$ is any subvariety, and if $\mathscr{I}(f)$ does not contain any irreducible component of W, then $W - \mathscr{I}(f)$ is dense in W. The closure of the image $f(W - \mathscr{I}(f))$ is a subvariety of Y, and we call it the *strict transform* of W. A general fact is that the indeterminacy locus has codimension ≥ 2. Thus if $H \subset X$ is a hypersurface, then we may take its strict transform $f(H) \subset Y$. We say that a hypersurface H is *exceptional* if the codimension of $f(H)$ is ≥ 2.

With these definitions, we see that the indeterminacy locus of J is $\mathscr{I}(J) = \{e_0, e_1, e_2\}$, and the lines $\Sigma_j := \{x_j = 0\}$, $j = 0, 1, 2$, are exceptional. Since J is an involution, (i.e. $J^2 =$ identity), we see that J blows up e_j to Σ_j. Geometrically, J acts as an inversion in the coordinate triangle, as shown in Fig. 1.

Now we blow up of $\mathbb{P}^2$ at the point e_0. This is a new manifold X with a holomorphic projection $\pi : X \to \mathbb{P}^2$ with the properties: (1) the exceptional fiber $E_0 := \pi^{-1}(e_0)$

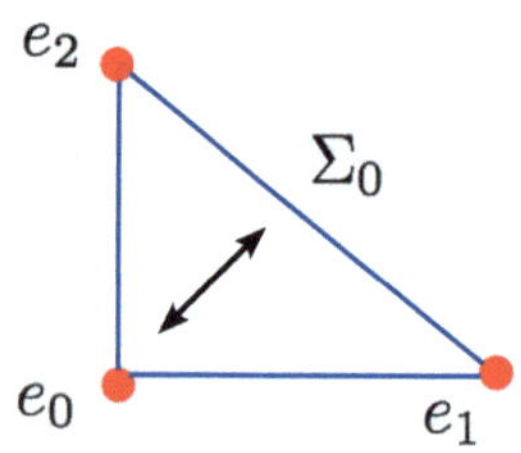

Fig. 1 Indeterminate point $e_0 = [1 : 0 : 0] \leftrightarrow$ exceptional curve $\Sigma_0 = \{x_0 = 0\}$

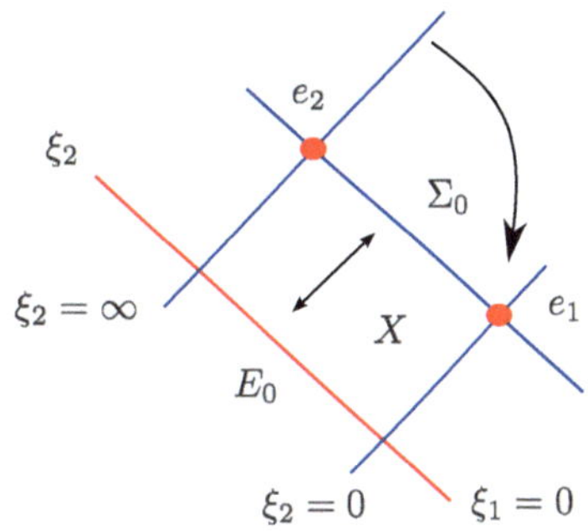

Fig. 2 The space X obtained by blowing up e_0; action of the induced map J_X

is isomorphic to $\mathbb{P}^1$, and (2) $\pi : X - E_0 \to \mathbb{P}^2 - \{e_0\}$ is biholomorphic. We may represent X in local coordinates (ξ_1, ξ_2) over $\mathbb{P}^2 - \Sigma_0$

$$\pi(\xi_1, \xi_2) = [1 : \xi_1 : \xi_1\xi_2] = [x_0 : x_1 : x_2],\ \pi^{-1}(x) = (\xi_1 = x_1/x_0, \xi_2 = x_2/x_1)$$

Figure 2 shows the new blowup space X in the (ξ_1, ξ_2) coordinate chart. The point e_0 has been replaced by a curve (called *exceptional* or *blowup divisor*) E_0. The projection π maps $X - E_0$ biholomorphically to $\mathbb{P}^2 - \{e_0\}$. Since $e_0 \notin \Sigma_0$, π^{-1} is holomorphic in a neighborhood of Σ_0, and via this biholomorphism, we have a new curve $\pi^{-1}\Sigma_0$ inside X. We write this again as Σ_0, although technically it is the *strict transform* of Σ_0 in X. We may also use π^{-1} to lift the curve $\Sigma_1 - \{e_0\}$ to $X - E_0$; inside the (ξ_1, ξ_2) coordinate chart, this corresponds $\xi_2 = \infty$. The closure of this set is a curve in X, which is the strict transform Σ_1' of Σ_1. Although the curves Σ_1 and Σ_1' are isomorphic, their normal bundles are not.

Now we describe the behavior of the induced map $J_X := \pi^{-1} \circ J \circ \pi$ in the part of the (ξ_1, ξ_2)-coordinate chart where $\xi_2 \neq 0, \infty$. J_X will map this set into $X - E_0$, which is mapped biholomorphically to $\mathbb{P}^2 - \{e_0\}$ by π. We represent its range in the homogeneous x-coordinates as

$$J_X(\xi_1, \xi_2) = J([1 : \xi_1 : \xi_1\xi_2]) = [1 : \xi_1^{-1} : \xi_1^{-1}\xi_2^{-1}] = [\xi_1\xi_2 : \xi_2 : 1]$$

which shows that J_X is regular in this coordinate chart.

To summarize: the induced map J_X has two points of indeterminacy e_1 and e_2. The exceptional locus of J_X consists of the strict transforms of Σ_1 and Σ_2. If we continue this process and blowup all 3 points e_0, e_1 and e_2, then we obtain a manifold Z, and the induced map J_Z is an automorphism of Z. This is pictured symbolically in Fig. 3, where all 3 blowup divisors are represented as red (thicker) segments, and the strict transforms of the Σ_j are blue (thinner) segments. Together, they form a hexagon, and the action of the induced map J_Z is to interchange opposite sides of the hexagon.

Thus we have started with a birational map and have obtained an automorphism on some blowup of $\mathbb{P}^2$. We will explore the question of how much more generally this might work: what are the birational maps of $\mathbb{P}^2$ that might lead to automorphisms after

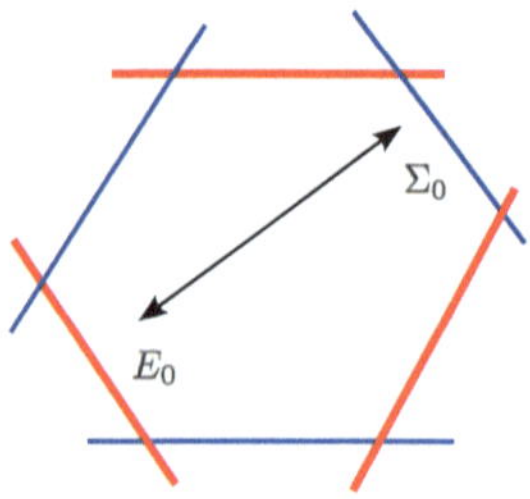

Fig. 3 The space Z obtained by blowing up e_0, e_1, e_2; the action of the induced map J_Z

some blowups? For instance, de Fernex and Ein [dFE] have shown: *If $f : X \dashrightarrow X$ is a birational map of finite order (i.e. if f^N is the identity for some N), then there is an iterated blowup $\pi : Z \to X$ such that the induced map f_Z is an automorphism of Z.*

3 Degree Complexity, or Dynamical Degree

Every holomorphic or rational map $f : \mathbb{P}^k \dashrightarrow \mathbb{P}^k$ is represented by polynomials $f = [f_0 : \cdots : f_k]$ of a common degree d. Dividing by the GCD, we may suppose that the degree d is minimal, and we set $\deg(f) := d$. The *dynamical degree* is the limit

$$\delta(f) := \lim_{n\to\infty} \left(\deg(f^n)\right)^{1/n}.$$

We may think of δ as the degree complexity, or as the exponential rate of degree growth of f^n as $n \to \infty$. We note that $1 \le \delta \le d$, and for "generic" f we have $\delta = d$ (see [FS1, FS2]). The cases we are interested in, however, are when f is an automorphism (or pseudoautomorphism) with $\delta > 1$, and in this case δ is algebraic but never rational (see [B]). In particular, it follows that $\delta < d$.

We may consider the lift $F = (F_0, \ldots, F_k)$ of f to a polynomial self-map of $\mathbb{C}^{k+1}$. In this case, $F^n = F \circ \cdots \circ F$ is the usual composition of polynomials, and the degree of F^n is d^n. Let us write $F^n = \psi_n F^{\langle n\rangle}$, where ψ_n is the GCD of F^n. In this case, we have $\deg(F^{\langle n\rangle}) \sim \delta^n$, and $\deg(\psi_n) \sim d^n - \delta^n$, so ψ_n carries almost all the degree of F^n.

The degree of f is closely related to how it pulls back hypersurfaces. Recall that $H^2(\mathbb{P}^k) = H^{1,1}(\mathbb{P}^k)$, and $H^2(\mathbb{P}^k; \mathbb{Z})$ is generated by the class of a hyperplane, which we will again write as H. If V is any hypersurface, then its class is given by $\deg(V) \cdot H \in H^2(\mathbb{P}^k; \mathbb{Z})$ corresponding to its degree. The pullback of the class of a hyperplane $H = \{\sum c_j x_j = 0\}$ is given by the class of $\{\sum c_j f_j = 0\}$. Thus $f^* H = \deg(f) H$.

Let β denote any Kähler form (for instance, the Fubini-Study form) with $\int_{\mathbb{P}^k} \beta^k = 1$. Then $\deg(V) = \int_V \beta^{k-1}$. It follows that for a generic hypersurface H, we have

$$\delta = \lim_{n\to\infty} \left(\mathrm{Vol}_{k-1}(f^{-n}(H))\right)^{1/n}$$
$$= \lim_{n\to\infty} \left(\int_{f^{-n}H} \beta^{k-1}\right)^{1/n} = \lim_{n\to\infty} \left(\int_{\mathbb{P}^k} \beta^{k-1} \wedge (f^n)^*\beta\right)^{1/n}$$

so the dynamical degree also measures the exponential rate of growth of $(k-1)$-dimensional volume under pullback.

It would be convenient if we could have $(f^n)^* = (f^*)^n$. In our case, that would mean that $\deg(f^n) = (\deg(f))^n$ (see [FS1, FS2]). In dimension $k = 2$, [DF] showed that there is an iterated blowup $\pi : X \to \mathbb{P}^2$ such that the induced map $f_X := \pi^{-1} \circ f \circ \pi$ does satisfy $(f_X^n)^* = (f_X^*)^n$. In this case, we let $\beta_X := \pi^*(\beta)$, so these integrals become:

$$\delta = \lim_{n\to\infty} \left(\int_{\mathbb{P}^2} \beta \wedge (f^n)^*\beta\right)^{1/n} = \lim_{n\to\infty} \left(\int_X \beta_X \wedge (f_X^n)^*\beta_X\right)^{1/n}$$
$$= \lim_{n\to\infty} \left(\int_X \beta_X \wedge (f_X^*)^n\beta_X\right)^{1/n} = \|f_X^*\|_{\mathrm{sp}}$$

where $\|\cdot\|_{\mathrm{sp}}$ denote the spectral radius, i.e. the modulus of the largest eigenvalue. The reason that the growth of β_X under $(f_X^*)^n$ gives the growth of $||(f_X^n)^*||$ is that since it is a Kähler form, it can lie in an eigenspace only if $\delta = 1$, and thus $(f_X^*)^n\beta_X$ must grow like the largest eigenvalue.

To give a simple example of the regularization of a map, we return to the map J from §1. In this case, we have $J^* = 2$, which means that J^* acts on the generator of $H^2(\mathbb{P}^2;\mathbb{Z})$ as multiplication by 2. On the other hand J^2 is the identity map, so $(J^2)^* = 1 \neq 4 = (J^*)^2$.

Now consider the space X, which was obtained by blowing up $\mathbb{P}^2$ at the points e_0, e_1, e_2. The cohomology group $H^2(X;\mathbb{Z})$ has the ordered basis $\langle H_X, E_0, E_1, E_2\rangle$, where H_X denotes the class of the strict transform of a generic hyperplane (line), and E_j denotes the class of the exceptional fiber over e_j. Let us see how to represent J_X^* with respect to this basis. Since J_X is an automorphism of X, we see that $J_X^* : E_j \leftrightarrow \Sigma_j'$. Next we need to represent Σ_j' with respect to this basis. Inside $\mathbb{P}^2$, we have that $\Sigma_j = H$ is the generator of $H^2(\mathbb{P}^2;\mathbb{Z})$. Now let us pull this back under π^*. It follows that $\pi^*H = H_X = \pi^*\Sigma_j$. On the other hand, suppose for instance that $j = 0$. Then Σ_0 contains e_1 and e_2, so when we pull back, we get the total preimage, and $\pi^*(\Sigma_0) = \Sigma_0' + E_1 + E_2$. This gives $J_X^* : E_j \mapsto \Sigma_j' = H_X - E_1 - E_2$. Finally, $J^{-1}\{\sum c_j x_j = 0\} = \{c_0x_1x_2 + c_1x_0x_2 + c_2x_0x_1 = 0\}$, so $J^{-1}\{\sum c_j x_j = 0\}$ contains $\{e_0, e_1, e_2\}$. This is a curve of degree 2, so its class in $H^2(\mathbb{P}^2)$ is $2H$. We apply π^* to $2H = J^{-1}\{\sum c_j x_j = 0\}$ and find that we have $2H_X = J^{-1}\{\sum c_j x_j = 0\}_X + \sum E_p$ inside $H^2(X)$. Since J_X is an automorphism, we obtain $J_X^* = J^{-1}\{\sum c_j x_j = 0\}_X = 2H_X - \sum E_p$. Writing this as a matrix with respect to our ordered basis, we find $J_X^* = \begin{pmatrix} 2 & 1 & 1 & 1 \\ -1 & 0 & -1 & -1 \\ -1 & -1 & 0 & -1 \\ -1 & -1 & -1 & 0 \end{pmatrix}$, which satisfies $(J_X^*)^2 = I$.

In the preceding heuristic argument we did not take into account the multiplicities of the E_p, which in fact turn out to be 1 in this case. In higher dimension, they are > 1.

4 Some Rational Surface Automorphisms with $\delta > 1$

We start with a particularly simple family of planar rational maps:

$$f_{a,b}(x,y) = \left(y, \frac{y+a}{x+b}\right) : \mathbb{C}^2 \dashrightarrow \mathbb{C}^2$$

We may write this in homogeneous coordinates $[x_0 : x_1 : x_2] = [1 : x : y]$ as

$$f_{a,b} : \quad [x_0 : x_1 : x_2] \mapsto [x_0(bx_0 + x_1) : x_2(bx_0 + x_1) : x_0(ax_0 + x_2)]$$

The points of indeterminacy are $\{e_1, e_2, p = [1 : -b : -a]\}$. The Jacobian determinant is $2x_0(bx_0 + x_1)(ax_0 + x_2)$, which vanishes on three lines, which are the exceptional curves of $f_{a,b}$. Choosing a linear map mapping this triangle to the standard coordinate triangle $\{x_0x_1x_2 = 0\}$, we may conjugate $f_{a,b}$ to a map of the form $L \circ J$, where L is linear, and J is the map from §1. The triangle of exceptional curves, shown in Fig. 4, is mapped as:

$$\Sigma_\beta = \{bx_0 + x_1 = 0\} \to e_2 \dashrightarrow \Sigma_0 \to e_1 \dashrightarrow \Sigma_\gamma = \{ax_0 + x_2 = 0\} \to q := [1 : -a : 0]$$

Now let us construct a new space Y by blowing up $\mathbb{P}^2$ at the points e_1 and e_2. We see that the induced map f_Y has one exceptional curve Σ_γ and one point of indeterminacy p. If the orbit of q lands on p, i.e. if $f^N q = p$, then we may blow up the orbit $q, fq, \ldots, f^N q = p$. If we start by blowing up q, then Σ_γ is no longer

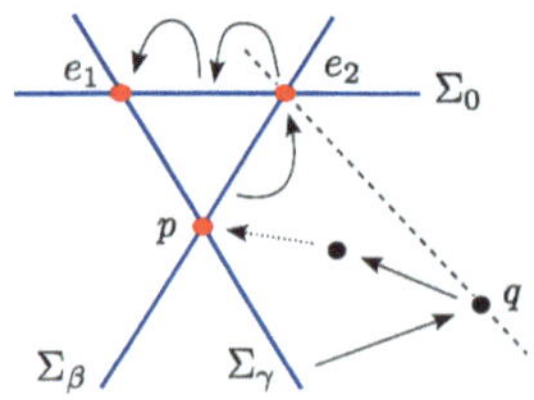

Fig. 4 The space Y obtained by blowing up e_1, e_2; action of the induced map f_Y

exceptional, but the blowup fiber Q over q is exceptional. Next, we blow up fq, so Q is now mapped to fQ (the fiber over q) and is no longer exceptional. We continue over the whole orbit and obtain an automorphism, since the fiber P over the indeterminate point p now maps in a regular way to the line $\overline{qe_2}$, which is drawn in (dashed) black in Fig. 4. (It is easy to see that this line is what we would expect because $p = \Sigma_\gamma \cap \Sigma_\beta$, and $\Sigma_\gamma \mapsto q$, $\Sigma_\beta \mapsto e_2$, so we expect p to be blown up to $\overline{qe_2}$.)

Let us define the set

$$\mathcal{V}_N := \{(a,b) \in \mathbb{C}^2 : f_{a,b}^N(q) = f_{a,b}^N(-a,0) = p = (-b,-a)\}$$

From ([BK2], Theorem 2) we have: *If $(a,b) \in \mathcal{V}_N$, then there is a blowup X such that the induced map f_X is an automorphism.*

Now let us suppose that $f_{a,b}$ is an automorphism and show how determine f_X^*. We may use the basis $\langle H_X, E_1, E_2, P = f^N Q, f^{N-1}Q, \ldots, Q\rangle$ as an ordered basis for $H^2(X;\mathbb{Z})$. Since f_X is an automorphism, we can read off the behavior of f_X^{-1} from Fig. 8:

$$f_X^* : \quad E_1 \mapsto \Sigma_0',\ E_2 \mapsto \Sigma_\beta',\ P = f^N Q \mapsto f^{N-1}Q \mapsto \cdots \mapsto Q \mapsto \Sigma_\gamma'$$

Further, as we saw in the case of J_X, we have

$$\Sigma_0' = H_X - E_1 - E_2, \quad \Sigma_\beta' = H_X - P - E_2, \quad \Sigma_\gamma' = H_X - E_1 - P$$

and

$$f_X^* : \quad H_X \mapsto 2H_X - E_1 - E_2 - P$$

This defines f_X^* on all the basis elements of $H^2(X)$, so we may now compute the characteristic polynomial of f_X^*, and we find that it is:

$$\chi_N(t) = t^{N+1}(t^3 - t - 1) + t^3 + t^2 - 1 \tag{4.1}$$

The dynamical degree of $f_{a,b}$ is the spectral radius of f_X^*, which is the largest root of χ_N. We conclude (cf [BK2]) that: *If $(a,b) \in \mathcal{V}_N$, and if $N \geq 7$, then $f_{a,b}$ is an automorphism of X with $\delta(f_{a,b}) > 1$.*

It is known that $\mathcal{V}_N \neq \emptyset$ for all N (see [M] and [BK3]), but: *It is not known whether $\mathcal{V}_N$ is discrete for $N \geq 7$.* For generic $(a,b) \in \mathbb{C}^2$, the dynamical degree of $f_{a,b}$ is $\delta_* \sim 1.324$, the largest root of $t^3 - t - 1$. *Does the cardinality of $\mathcal{V}_N$ grow like δ_*^N as $N \to \infty$?*

5 Connection Between Dynamical Degree and Length Growth: A Graphic Example

If we choose $(a, b) \in \mathscr{V}_7$, then the map $f_{a,b}$ in §3 will be an automorphism with $\delta \sim 1.17628$, which is the largest root of χ_7 from (4.1). By the discussion in §2, we know that the nth iterate of a (complex) line will have 2-dimensional area $\sim \delta^n$. This is closely related to the fact (see [C2]) that the entropy of $f_{a,b}$ is $\log \delta$. There is only one map $f_{a,b}$ (and its inverse) with $(a, b) \in \mathscr{V}_7 \cap \mathbb{R}^2$. The restriction of this map to $X_{\mathbb{R}}$, the real points of X was shown in [BK3] to have entropy $\log \delta$. Figure 5, taken from ([BK2], Figure A.1), shows the example of the real point $(a, b) \in \mathscr{V}_7$. This shows the image of a line L after n iterations, and inspection shows empirically that the length grows $\sim \delta^n$.

To represent the real projective plane, we have taken the usual polar coordinates (r, θ) and replaced them by modified polar coordinates (ρ, θ), where $\rho = \arctan(r)$. Thus lines appear to be circular in these coordinates. The exceptional curve Σ_0 is the outer circle bounding the picture, and the exceptional curves Σ_β and Σ_γ are labeled. The intersection points of any two of these exceptional curves are necessarily indeterminate. The image of Σ_γ is labeled "0", and the image of "0" is "1", etc. The cubic invariant curve is pictured, and the two fixed points are on this curve, close together, just above the number "7" (which is the label for the indeterminate point $\Sigma_\beta \cap \Sigma_\gamma$).

The forward image of the line L appears to be "bunched" at the points "0", "1", … These points are blown up in the construction of X, so the "bunching" is an artifact of the projection π, which takes all the points of a blowup fiber and collapses them to a point. It appears that $f^n L$ may converge to a lamination as $n \to \infty$, and it

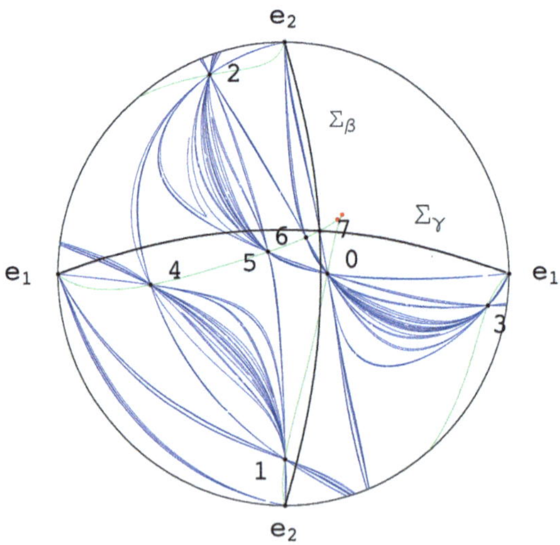

Fig. 5 $f(x, y) = \left(y, \dfrac{y+a}{x+b}\right)$, $a = -0.499497$, $b = -0.415761$

would be interesting to know whether this is true. (A related map was shown to have a lamination in [BD].)

We note that one of the fixed points is attracting, and the basin has full area (see [M, BK3]). Another graphical representation for this map is to choose a point of the basin and plot its orbit under f^{-1}, as was done in ([M], Fig. 1). It is striking to see the similarities between these figures, the main visual difference coming from the coordinate systems used: affine for ([M], Fig. 1) and compactified polar for Fig. 5 above.

6 Heuristic Picture: Dynamical Complexity Versus Degree Complexity

The main goal of this paper is to discuss maps with interesting dynamics, so let us give some examples of interesting dynamics. In dimension 1, we consider a rational map $f : \mathbb{P}^1 \to \mathbb{P}^1$ with degree$(f) = \delta(f) = \deg_{\text{top}}(f) > 1$. Such maps are not invertible. One of the basic results of the subject concerns the backward dynamics of f, i.e. the distribution of preimages of a point.

Theorem ([Br, FLM, L]). *For almost all points z_0 there is a limiting distribution of point masses over the preimages of z_0:*

$$\mu_f := \lim_{n\to\infty} \frac{1}{d^n} \sum_{\{a:f^n(a)=z_0\}} \delta_a .$$

This measure is *balanced*, which means that, locally, $f^*\mu_f = d \cdot \mu_f$. Thus, as we consider backward iteration, the different branches of f^{-1} give $\mu_f(S_j) = \mu_f(S)/d$. The effect is like Bernoulli trials, as illustrated in Fig. 6.

If $K \subset X$ is a compact set, and if $f : X \to X$, then we may define the stable set $W^s(K) := \{x \in X : \text{dist}(f^n(x), K) \to 0 \text{ as } n \to \infty\}$. The unstable set $W^u(x_0)$ is defined as above, with f replaced by f^{-1}. In dimension 1, if x_0 is a repelling periodic point, the stable set of a point $W^s(x_0)$ is just the set of all preimages of x_0. The Theorem above says that the asymptotic distribution of $W^s(x_0)$ is independent of x_0. In dimension 2, the stable set has the structure of a manifold (curve).

Stable Manifold Theorem. *Let $f : X \to X$ be an automorphism of a complex surface, let x_0 be a saddle fixed point, and let*

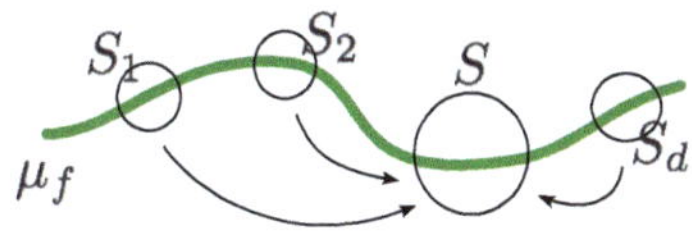

Fig. 6 Choosing preimages of a point is like flipping a d-sided coin

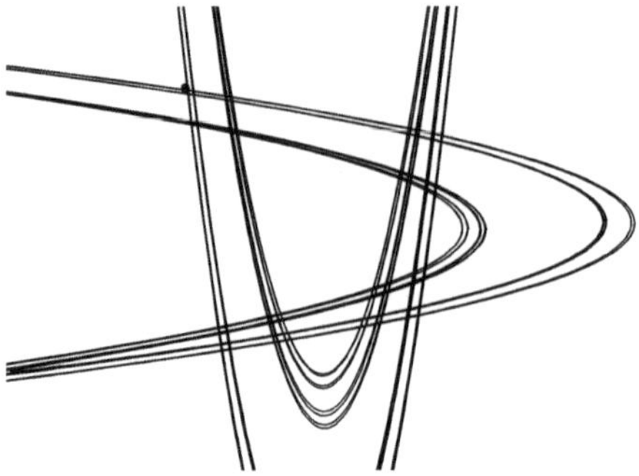

Fig. 7 Stable/unstable manifolds for the Horseshoe Map $f_{c,a}(x, y) = (c + ay - x^2, -x)$ with $c = 6.0, a = 0.8$

$$W^s(x_0) := \{x : \lim_{n\to\infty} \text{dist}(f^n x, x_0) = 0\}$$

be its stable set. Then there is an injective holomorphic immersion $\xi : \mathbb{C} \to X$ such that $\xi(\mathbb{C}) = W^s(x_0)$.

A classic example is the Horseshoe Map. Figure 7 is the $\mathbb{R}^2$ slice of a complex automorphism of $\mathbb{C}^2$. It shows the stable and unstable manifolds of a saddle fixed point x_0, which is marked in the upper left. The arcs which are essentially "left-right" (with bends) are contained in $W^u(x_0)$. They are cut off by the viewbox of the picture, but they are connected in $\mathbb{R}^2$. Similarly, the arcs which are oriented in the "up-down" sense are contained in the stable manifold $W^s(x_0)$. In the case of the horseshoe, the closure of $W^s(x_0)$ is a complicated set; a laminar structure is clearly visible.
We summarize some dynamical properties of the horseshoe map:

- All points outside $\overline{W^s(x_0)} \cup \overline{W^u(x_0)}$ escape to infinity in either forward or backward time.
- $\overline{W^s(x_0)} \cap \overline{W^u(x_0)} \cong$ Cantor set $\times$ Cantor set
- Saddle (periodic) points are dense in $\overline{W^s(x_0)} \cap \overline{W^u(x_0)}$
- Dynamics on Cantor set $\times$ Cantor set is conjugate to the shift on $\{0, 1\}^{\mathbb{Z}}$
- The closure $\overline{W^s(z_0)}$ is the same for all saddles z_0.

In the 1-dimensional case, the stable set is $W^s(x_0) = \bigcup_{n\geq 0} f^{-n}(x_0)$. This is an infinite set whose closure contains the Julia set. If we start with an element x_0 in the Julia set, then $f^{-n}(x_0)$ will also be contained in the Julia set, and it will fill it out as $n \to \infty$. By the Theorem at the beginning of this section, we may think of the invariant measure μ as describing how the sets $W^s(x_0)$ accumulate. We want to do something similar in the 2-dimensional case. Suppose that γ is an oriented curve in $\mathbb{R}^2$, or if Γ is a complex submanifold of $\mathbb{C}^2$. If γ has locally bounded length (or if Γ has locally bounded area), we can consider the current of integration $[\gamma]$, which acts on test forms by $\varphi \mapsto \int_\gamma \varphi$. In the case of the horseshoe, we cannot define the current of integration $[W^s(x_0)]$ directly, but we may construct a current by taking the average over arcs of stable manifolds in $\mathscr{W}^s$. Thus, instead of an invariant measure, we have a family of transversal measures, which assign mass to families of stable arcs. When we map $\mathscr{W}^s$ by f^{-1}, each individual stable manifolds is stretched, but

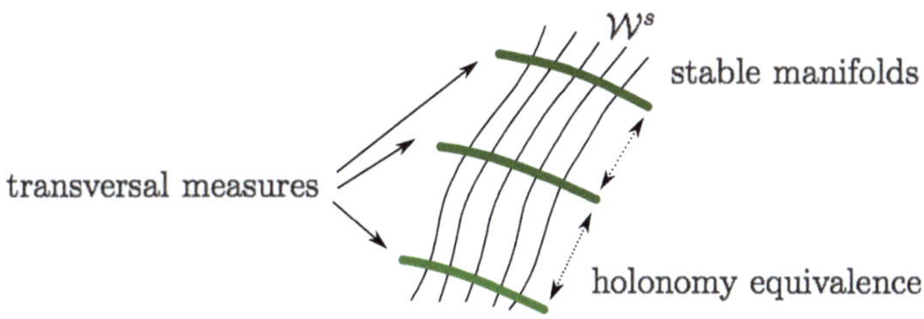

Fig. 8 Mapping by f^{-1} acts on the family of transversal measures

the action on the set of transversal measures is very much like the Bernoulli trials that we saw in the 1-dimensional case. Thus f^{-1} maps one stable manifold to another, mixing them in the same chaotic way we saw in dimension 1 (Fig. 8).

While this phenomenon is rather special for real mappings, such as the horseshoe, it happens quite generally in the complex case. If f is an automorphism of a complex surface with $\delta(f) > 1$, then there is an invariant current T^s which satisfies $f^*T^s = \delta T^s$. This current always exists, but it does not always have the same elegant geometric laminarity that we have seen in the case of the horseshoe. The current T^s can be sliced by a complex disk, and such slices $T^s|_D$ will serve as a family of "transversal measures". If we work in a more measure-theoretic sense, we can assign a laminar structure to T^s, and such currents have been useful in studying the dynamics of f. With Misha Lyubich and John Smillie, we have written a series of papers on the subject. See [D1, D2, D3] for a more recent treatment.

7 What Are the Compact Surfaces X Which Carry an $f \in Aut(X)$ with $\delta_1(f) > 1$?

Using the Kodaira classification of surfaces, Cantat gave the following answer:

Theorem [C1]. *Suppose that X is a compact complex manifold of (complex) dimension 2, and there is an automorphism f of X with $\delta(f) > 1$. Then X is a blowup of one of the following cases:*

- X is a torus, and f is a "standard" torus automorphism.
- X is a $K3$ surface, or a finite quotient of one of these.
- X is a rational surface

This leads immediately to the more precise question: *What are the surfaces and maps that actually occur for $K3$ or rational surfaces?* The $K3$ surfaces or rational surfaces which can carry nontrivial automorphisms are quite special and not easy to find.

The set of all $K3$ surfaces has dimension 20, and the set of the family of $K3$ surfaces that carry nontrivial automorphisms has smaller dimension. On the other hand, by [BK4], rational surfaces with automorphisms with $\delta > 1$ can occur in families of arbitrarily large dimension. This leads us to expect that a more interesting variety of dynamical behaviors will be found within the class of rational surfaces, so we concentrate on them, rather than on $K3$ surfaces.

Theorem (Nagata). *If a rational surface X carries an automorphism with $\delta > 1$, then X is an (iterated) blowup $\pi : X \to \mathbb{P}^2$ of the projective plane.*

Any rational surface X can be obtained from $\mathbb{P}^2$ by a series of blowings up and down. This Theorem says that X can be obtained by blowups alone.

Thus one possible approach is to look for birational maps $g : \mathbb{P}^2 \dashrightarrow \mathbb{P}^2$ which can be lifted to automorphisms, as was done in §2–4. Another approach which has been productive is to require that the automorphism have an invariant curve. McMullen gave a "synthetic" approach to this question, and Diller [Di] has shown all the possibilities for rational surface automorphisms with invariant curves, and which arise from quadratic maps of $\mathbb{P}^2$. Let us remark (see [BK3]) that while many maps have been constructed by starting from invariant curves, not all rational surface automorphisms have invariant curves.

If X is obtained from $\mathbb{P}^2$ by a sequence of N blowups, then $H^2(X;\mathbb{Z}) = \mathbb{Z}^{N+1}$, and the intersection product makes $H^2(X;\mathbb{Z})$ isometrically isomorphic to the Lorentz space $\mathbb{Z}^{1,N}$, with signature $(1, N)$. A Theorem of Nagata (see [Do1, M]) says that if f is an automorphism of X, then f^* must belong to the Weyl group W_N. Thus the question arises: *What are all the elements of W_N which can be realized by rational surface automorphisms?* That is, if $g \in W_N$, are there a surface X and an $f \in Aut(X)$ such that the map induced by f^* on $\mathbb{Z}^{1,N}$ coincides with g? If this is the case, then $\delta(f)$ is equal to the spectral radius of g. Uehara [Ueh] has shown that for any $g \in W_N$, there exists $f \in Aut(X)$ such that $\delta(f)$ is the spectral radius of g. In other words, the set of all dynamical degrees coincides with the spectral radii of elements of W_N. On the other hand, Diller [Di] and Uehara [Ueh] have shown that certain elements of W_N cannot be realized by automorphisms with invariant curves, leaving open the question: *What happens for maps without invariant curves?* The set of all dynamical degrees of surface automorphisms is well ordered and has other properties (see [BC]).

8 Pseudo-automorphisms

We were led to pseudo-automorphisms by the problem of finding all periodicities within the family $f_{a,b}$ of birational maps of $\mathbb{P}^3$:

$$f_{a,b}(x_1, x_2, x_3) = \left(x_2, x_3, \frac{a_0 + a_1x_1 + a_2x_2 + a_3x_3}{b_0 + b_1x_1 + b_2x_2 + b_3x_3}\right)$$

What are the parameters $a_j, b_j, 0 \le j \le 3$, such that $f^p =$ identity *for some p?* This question seems to have originated when the first of the period 8 maps below was found many decades ago by Lyness [Ly]. The second one was found by Csörnyei and

Laczkovich [CL]. The following is proved in [BK5], where the word "nontrivial" is also explained.

Theorem. *The only nontrivial periods that can appear are 8 and 12. The maps are*

$$(x, y, z) \to (y, z, (1 + y + z)/x) \text{ or } (y, z, (-1 - y + z)/x) \text{ period 8}$$

$$(x, y, z) \to \left(y, z, \left(\frac{\eta}{1-\eta} + \eta y + z\right) / \left(\eta^2 x\right)\right) \text{ period 12}$$

where η is a primitive cube root of -1.

The principle behind the method of proof is to consider the function $(a, b) \mapsto \delta(f_{a,b})$ on parameter space and identify the subvariety of parameters $\{(a, b) : \delta(f_{a,b}) = 1\}$, since any possible periodic map must lie here. The generic map in this family has dynamical degree equal to $\delta_* > 1$, and the scan of parameter space yielded a number of cases where $1 < \delta(f_{a,b}) < \delta_*$. In the most interesting cases, we blow up 2 points, then 2 lines, and then an orbit of curves (11 in one case and 19 in another) and arrive at a map without exceptional hypersurfaces. The first of these cases is from ([BK5], Theorem 1):

$$F_a : \quad (x, y, z) \mapsto (y, z, (a + \omega y + z)/x) \tag{8.1}$$

where $a \in \mathbb{C}$, $a \neq 0$, and ω is a primitive cube root of unity. After blowups, these maps are "regularized" to the point that they are almost automorphisms, except that there are a finite number of indeterminate curves, and points of these indeterminate curves are blown up to other curves. These maps are pseudo-automorphisms, and it is easy to deal with the map f^* on cohomology, but the existence of indeterminacy makes it tricky to analyze the pointwise dynamics.

In connection with the method described above, it would be interesting to know more generally for a family f_a of rational maps:

Is $a \mapsto \delta(f_a)$ lower semi-continuous?

and

Is $\{a : \delta(f_a) \leq t\}$ always a subvariety?

The answer is "yes" for birational surface maps (see [X], Theorem 1.6)).

The indeterminacy locus $\mathcal{I}(f)$ of any rational map $f : X \dashrightarrow Y$ has codimension at least 2. Thus if H is any hypersurface, we may define the image (strict transform) of H as the closure of $f(H - \mathcal{I}(f))$. We say that H is *exceptional* if the codimension of the strict transform of H is ≥ 2. A birational $f : X \dashrightarrow X$ is a *pseudo-automorphism* if neither f nor f^{-1} has an exceptional hypersurface.

Pseudo-automorphisms behave very much like automorphisms, and we expand our search to include this richer source of interesting maps. In dimension 2, all pseudo-automorphisms are in fact automorphisms. What happens in dimension > 2? Given that the blowups of $\mathbb{P}^2$ have yielded interesting automorphisms, it makes sense to ask: *Are there 3-folds X which are obtained as blowups of $\mathbb{P}^3$ and which carry automorphisms f with $\delta(f) > 1$?* Of course, we would expect such automorphisms to exist only in very special cases.

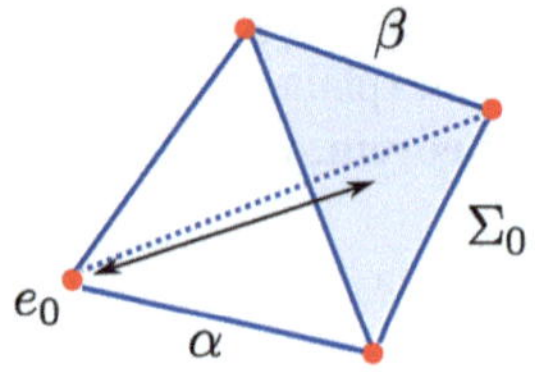

Fig. 9 Cremona involution blows up generic point of edge α to all of edge β

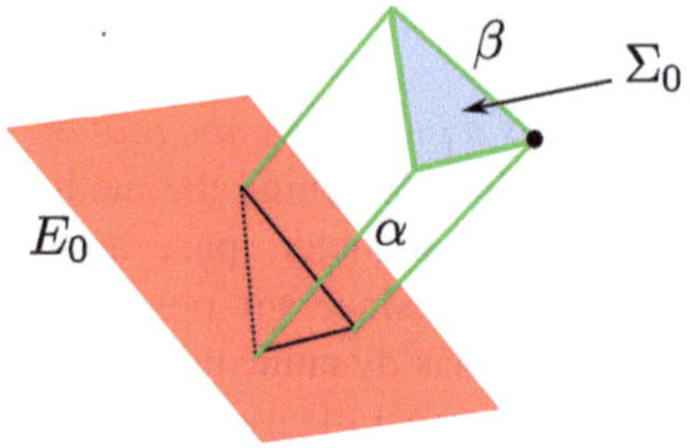

Fig. 10 After blowup of e_0: generic point of edge α still blows up to all of edge β

Theorem [T]. *If X is obtained from $\mathbb{P}^3$ by blowing up points and curves satisfying a certain condition, and if f is an automorphism of X, then $\delta_1(f) = \delta_2(f)$.*

Theorem [BaC]. *If X is obtained from $\mathbb{P}^k$ by blowing up points, then any automorphism f of X satisfies $\delta_\ell(f) = 1$ for all ℓ.*

The Cremona involution on $\mathbb{P}^3$ is the cubic map given by

$$J(x) = [1/x_0 : 1/x_1 : 1/x_2 : 1/x_3] = [x_1x_2x_3 : x_0x_2x_3 : x_0x_1x_3 : x_0x_1x_2]$$

which acts as an involution on the coordinate tetrahedron $e_j \leftrightarrow \Sigma_j$, $j = 0, 1, 2, 3$. We now see a new phenomenon: any non-vertex point of an edge of the tetrahedron is blown up by J to the skew edge. In Fig. 9, this means that any non-vertex point of α will be blown up to the whole edge β.

Let $\pi : X \to \mathbb{P}^3$ be the blowup of $\mathbb{P}^3$ at e_0. The 3 edges of the tetrahedron passing through e_0 are now separated as in Fig. 10. The induced map $J_X : X \dashrightarrow X$ maps $E_0 \leftrightarrow \Sigma_0$, so Σ_0 is no longer exceptional. The restricted map $J_X|_{E_0} : E_0 \to \Sigma_0$ "looks like" the 2D map J mapping $\mathbb{P}^2$ to itself: the black triangle inside E_0 is exceptional, and the dotted black line is mapped to the bold black dot in Σ_0 in Fig. 10. The edges of the tetrahedron (in green) are still indeterminate.

If we let $\pi : Y \to \mathbb{P}^3$ be the space obtained by blowing up all the vertices e_j, $j = 0, 1, 2, 3$, then the induced map J_Y will be a pseudo-automorphism of Y. In the space Y, the strict transforms of the edges of the tetrahedron are disjoint curves. Let $\pi : Z \to Y$ be the space obtained by blowing up the strict transforms of the edges of the tetrahedron. Thus J_Z is an automorphism of Z, a very simple instance of the Theorem of de Fernex and Ein mentioned above.

On the other hand, the example of F_a in (8.1) illustrates why this procedure may encounter difficulties if the map is not periodic; in [BK5] Theorem 4, it is shown that F_a is not birationally conjugate to an automorphism. Since blowing up or down is a birational operation, F_a cannot be turned into an automorphism by any sort of blowing up procedure. A heuristic explanation for this difficulty in the case of F_a is that we will need to blow up an orbit of curves, but the curves in the orbit are not pairwise disjoint.

9 Intermediate Dynamical Degrees

Let X be a manifold of dimension k, and let f be an automorphism of X. We measure the complexity of f in terms of dynamical degrees. We assume that X is Kähler, so the action f^* on cohomology respects the (p,q) bi-gradation $H^* = \oplus H^{p,q}$. If $1 \le \ell \le k$, then the ℓ-th dynamical degree is defined as the exponential rate of growth of the induced map f^* on $H^{\ell,\ell}(X)$:

$$\delta_\ell(f) := \limsup_{n\to\infty} ||f^{n*} : H^{\ell,\ell}(X) \to H^{\ell,\ell}(X)||^{1/n}$$

This is independent of the choice of norm on $H^{\ell,\ell}$. For an automorphism, we have $(f^n)^* = (f^*)^n$, so δ_ℓ is the same as the spectral radius of f^* acting on $H^{\ell,\ell}(X)$. It is also the same as the spectral radius of the restriction of f^* to $H^{2\ell}(X)$. As in §2, an equivalent definition is the exponential rate of growth of the integrals $\int \beta^{k-\ell} \wedge (f^n)^*(\beta^\ell)$ as $n \to \infty$.

In all cases, $\delta_\ell \ge 1$, $\delta_0 = 1$, and $\delta_k = \deg_{\mathrm{top}}$ is the topological or mapping degree of f. Since f is invertible, we have $\delta_k = 1$. By duality, $\delta_\ell(f) = \delta_{k-\ell}(f^{-1})$. For the intermediate degrees, $\ell \mapsto \log(\delta_\ell)$ is concave. Thus, if $\delta_1 = 1$, then $\delta_\ell = 1$ for all $0 \le \ell \le k$; and if $\delta_1 > 1$, then $\delta_\ell > 1$ for all $1 \le \ell \le k-1$.

Let us note that if $f : X \dashrightarrow Y$ is merely rational, then there is still a well-defined linear map $f^* : H^*(Y) \to H^*(X)$. In dealing with rational maps, birational conjugacy is a natural sense of equivalence. It was shown (see [DF] and [DS2]) that δ_ℓ is an invariant of birational conjugacy. However, the topological entropy is not a birational invariant (see [G1]), but [DS2] gives an inequality:

$$\mathrm{entropy}_{\mathrm{top}}(f) \le \max(\log(\delta_1), \dots, \log(\delta_k))$$

The only general class of non-holomorphic, rational maps for which the intermediate degrees has been computed is the case of monomial maps, and the following is nontrivial:

Theorem [Lin, FW]. *Let $A = (a_{p,q})$ be an integer matrix of size $k \times k$, and let*

$$f_A(x) = x^A = (x_1^{a_{1,1}} \cdots x_k^{a_{1,k}}, \dots, x_1^{a_{k,1}} \cdots x_k^{a_{k,k}}) = \left(\prod_q x_q^{a_{1,q}}, \dots, \prod_q x_q^{a_{k,q}}\right)$$

be the associated monomial map. Then for each $\ell \geq 1$, $\delta_\ell(f_A)$ *is the spectral radius of the* ℓ*-th exterior power* $\wedge^\ell A$ *of* A. *Equivalently, if* $|\lambda_1| \geq |\lambda_2| \geq \cdots \geq |\lambda_k|$ *are the eigenvalues of* A, *then* $\delta_\ell = |\lambda_1 \lambda_2 \cdots \lambda_\ell|$.

Intermediate degrees have also been determined by [A] and [KR], but despite their obvious importance, we have the problem:

Determine $\delta_\ell(f)$ *for* $1 < \ell < k$ *for other nontrivial* f.

10 Existence of Pseudo-automorphisms of Blowups of $\mathbb{P}^k$

We will try to find pseudo-automorphisms of the form $L \circ J$ for some L which is a nonsingular $(k+1) \times (k+1)$ matrix, and thus a linear automorphism of $\mathbb{P}^k$. The exceptional locus consists of the hyperplanes Σ_j, which are mapped:

$$f := L \circ J \; : \; \Sigma_j \to L_j$$

where L_j denotes the point of $\mathbb{P}^k$ defined by the jth column of the matrix L. We will have a pseudo-automorphism if

$$L_j \mapsto f(L_j) \mapsto \cdots \mapsto f^{m_j}(L_j) = e_{\sigma_j}, \quad f^\ell(L_j) \notin \bigcup_i \Sigma_i$$

where σ is a permutation of $\{0, \ldots, k\}$. Let $\pi : X \to \mathbb{P}^k$ be the blowup of the orbits of the L_j. The induced map $f_X := \pi^{-1} \circ f \circ \pi$ will be a pseudo-automorphism of X. In ([BK1], Theorem A.1) a formula is given for the characteristic polynomial of f_X:

Theorem. *The characteristic polynomial defining the dynamical degree* $\delta_1(L \circ J) = \delta_1(m_1, \ldots, m_k, \sigma)$ *is given by an explicit formula involving the orbit lengths* m_j *and the permutation* σ.

Thus the lengths m_j and the permutation σ specify the dynamical degree that will be produced. As a practical matter, however, the strategy given above for finding L is not feasible because it involves solving equations of very high degree in many variables. The relevant computations are possible, however, if we assume that all the centers of blowup lie in an invariant curve. The existence of automorphisms of blowups of $\mathbb{P}^2$ with invariant curves was studied by McMullen, Diller and Uehara. Perroni and Zhang brought the method of McMullen from dimension 2 to higher dimension and gave the abstract existence of a map with an invariant curve.

Theorem [PZ]. *For all* $k \geq 2$ *and* $d \geq 1$ *there exist infinitely many manifolds* X *obtained by blowing up points on* $(\mathbf{P}^k)^d$ *such that* X *carries a pseudoautomorphism with* $\delta > 1$.

With Diller and Kim, we were motivated by the desire to see the Perroni-Zhang maps more concretely, and for this we used the method of Diller [Di]. Let us consider a parametrized curve $\psi : \mathbb{C} \to \mathscr{C} \subset \mathbb{P}^k$. We say that $\mathscr{C}$ satisfies a group law if the following holds: For each hyperplane $H \subset \mathbb{P}^k$, the set of all solutions (with multiplicity) $t_1, \ldots, t_N$ of $\psi(t_i) \in H$ satisfies $\sum t_i = 0$. There are several cases of curves with group law; all the curves we work with have degree $k+1$. For instance, there is $\mathscr{C}_1 := \psi(\mathbb{C})$, which is the image of $t \mapsto \psi(t) = [1 : t : t^2 : \cdots : t^{k-1} : t^{k+1}]$ which is irreducible and has a cusp. There is also $\mathscr{C}_2 := \psi(\mathbb{C})$, which is the image of the set-valued map $t \mapsto \psi(t) = \{[1 : t : \cdots : t^k], [0 : \cdots : 0 : (-1)^{k-1} : t]\}$ and is the union of rational normal curve and a tangent line. Both of these curves are singular at $[0 : \cdots : 0 : 1] = \psi(\infty)$.

If there is an invariant curve, then the points to be blown up are of the form $\psi(t_j)$. The problem of finding the centers of blowup is thus transformed to a problem of determining the points $\{t_i\} \subset \mathbb{C}$. Using the group law on the curve, we obtain:

Theorem [BDK]. *For most choices of orbit lengths $(m_0, \ldots, m_k)$ and permutations σ, there is a matrix L such that the space $\pi : X \to \mathbb{P}^k$ obtained by blowing up the orbits yields an induced pseudo-automorphism $f_X : X \dashrightarrow X$.*

This method applies also to products $(\mathbb{P}^k)^d$. For this we use the following variant of J. We write a point of $(\mathbb{P}^k)^d$ as $(x, y^{(1)}, \ldots, y^{(d-1)})$ and set

$$x/y^{(j)} := (x_0/y_0^{(j)}, \ldots, x_k/y_k^{(j)}), \quad J(x, y^{(1)}, \ldots, y^{(d-1)}) = (1/x, x/y^{(1)}, \ldots, x/y^{(d-1)})$$

A linear map $L \in Aut((\mathbb{P}^k)^d$ has the form $L \circ \tau$ where $L := (L_1, \ldots, L_d)$, $L_j \in Aut(\mathbb{P}^k)$, and τ is a permutation of the factors $\mathbb{P}^k$.

This Cremona involution is discussed by Dolgachev [Do2, Do3] and Mukai [Muk] in connection with Weyl groups $W(p, q, r)$ which have T-shaped Coxeter diagrams. In this case, $p = k+1$, $q = d+1$, and $r-1$ is the number of blowups forming the space $\pi : X \to (\mathbb{P}^k)^d$, and $W(p, q, r) \subset GL(H^2(X; \mathbb{Z}), \mathbb{Z})$. It is shown in [BDK] that if a map of the form $f := L \circ \tau \circ J$ is a pseudo-automorphism of a space X obtained by blowing up $(\mathbb{P}^k)^d$, then the induced map f^* belongs to the Weyl group $W(p, q, r)$. The more general statement, however, is still open: *Let X be a blowup of $(\mathbb{P}^k)^d$, and let $\mathscr{G}$ denote the group generated by J and the linear maps $Aut((\mathbb{P}^k)^d)$. If $f \in \mathscr{G}$ induces a pseudo-automorphism of X, does if follow that $f^* \in W(p, q, r)$?*

A reflection group like $W(p, q, r)$ is generated by reflections $r_1, \ldots, r_N$. The Coxeter element of this group is given by the product $r_1 \cdots r_N$ of these reflections, where each of the r_j appears exactly once. This element is unique up to conjugation and represents the simplest element of reflection group with spectral radius > 1 (see [M] and [Do1]). The results of [PZ] and [BDK] also apply to the Coxeter elements of $W(p, q, r)$. For instance, let us consider the case with orbit lengths $(1, \ldots, 1, n)$, and the permutation $\sigma = (0\,1\,2\,\ldots k)$ is cyclic. In this case the action of f^* on $Pic(X)$ corresponds to the Coxeter element of the Weyl group $W(k+1, 2, n+k+1)$. The existence of such maps, representing the Coxeter element, was given by Perroni and Zhang. In [BDK] it is shown that we may write such a map as $f := L \circ J$, where L has the form:

$$L = \begin{pmatrix} 0 & 0 & 0 & \dots & 0 & 1 \\ \beta_1 & 0 & 0 & \dots & 0 & 1-\beta_1 \\ 0 & \beta_2 & 0 & \dots & 0 & 1-\beta_2 \\ \vdots & & & & & \vdots \\ 0 & 0 & 0 & \dots & \beta_k & 1-\beta_k \end{pmatrix}$$

with β_i being a rational function of δ, any root of the minimal polynomial χ_n which gives the dynamical degree of $L \circ J$. Different choices of invariant curve lead to different expressions for β_i as rational functions of δ.

In dimension 2, it appears that the majority of automorphisms $f = L \circ J$ of blowups of $\mathbb{P}^2$ do not have invariant curves. *Is it the case that "most" pseudo-automorphsms of blowups of* $(\mathbb{P}^k)^d$ *do not have invariant curves?*

11 Cohomological Hyperbolicity

A map $f : X \dashrightarrow X$ is said to be *cohomologically hyperbolic* if there is a unique $1 \le p \le k-1$ such that $\delta_p(f)$ is maximal. In this case, the maximal growth occurs uniquely in bidegree (p, p), which corresponds to codimension p. As was noted in [BDK], we have $\delta_1 = \delta_{k-1}$ for all maps $f = L \circ J$ which are pseudo-automorphisms of point blowups of $\mathbb{P}^k$. Thus f is not cohomologically hyperbolic when $k = 3$. We note the following open questions about maps $L \circ J$:

What are the intermediate dynamical degrees $\delta_\ell(L \circ J)$ *when* $k > 3$*?*

Can $L \circ J$ *be cohomologically hyperbolic for even* k *greater than 3?*

Theorem [DS1]. *If* f *is a cohomologically hyperbolic automorphism, then there are invariant currents* $T^{s/u}$*, and these may be used to form an invariant measure* μ *with interesting dynamical properties.*

Guedj [G2] has conjectured that in the presence of cohomological hyperbolicity, the basic ergodic properties of 2-dimensional maps should carry over to higher dimension. If f is not cohomologically hyperbolic, then it is not clear to what extent a result like this would remain valid, and it is not clear what approach will reveal the dynamics of such maps. It would be helpful if there could be an invariant fibration which would allow us to somehow study the dynamics with lower-dimensional objects and techniques. In dimension 2, cohomological hyperbolicity fails (for invertible maps) exactly when $\delta(f) = 1$. In this case, Diller-Favre have shown:

Theorem [DF]. *If* $f : X \dashrightarrow X$ *is a bimeromorphic surface map with* $\delta(f) = 1$*, then there is an invariant fibration.*

Let $f : X \dashrightarrow X$ be a meromorphic map. Suppose that there is a dominant, meromorphic map $\phi : X \dashrightarrow Y$ and a meromorphic map $g : Y \dashrightarrow Y$ such that $0 < \dim(Y) < \dim(X)$, and $g \circ \phi = \phi \circ f$. In other words, ϕ gives a meromorphic semiconjugacy from (f, X) to (g, Y). In this case, the sets $\{\phi = \text{const}\}$ form an *invariant fibration.*

In the presence of an invariant fibration, there is a *dynamical degree on the fiber*, written $\delta_j(f|\phi)$ (defined in [DN]), and it is related to the other dynamical degrees by:

Theorem [DN, DNT]. *Suppose that the map f has an invariant fibration as above. Then*

$$\delta_p(f) = \max_{\max\{0,p-k+\ell\}\le j\le\min\{p,\ell\}} \delta_j(g)\delta_{p-j}(f|\phi) .$$

As a consequence of this, one can show:

Theorem. *If X is a 3-fold, and $f : X \dashrightarrow X$ is a birational map with an invariant fibration, then $\delta_1 = \delta_2$. In this case, f is not cohomologically hyperbolic.*

The possibilities for invariant fibrations in the automorphisms of tori are discussed in [OT1, OT2]. Guedj also conjectured: *If f is not cohomologically hyperbolic, then f has an invariant fibration, or at least an invariant foliation.* In dimension 2, [KPR] have given a counter-example with $\delta_1 = \delta_2 = \deg_{top} = 2$, which is thus a non-invertible map.

If $\dim(X) = 3$, and if $f : X \dashrightarrow X$ is birational, then the condition that f is not cohomologically hyperbolic is equivalent to the condition that $\delta_1(f) = \delta_2(f)$. We now give a 3-dimensional counterexample which is invertible. Set

$$L_{a,c} = \begin{pmatrix} 0&0&0&1\\ 1&0&0&a\\ 0&1&0&0\\ 0&0&1&c \end{pmatrix}$$

with $a, c \in \mathbb{C}$ such that

$$na^2 + (n+1)ac + nc^2 = 0$$

for some $n \ge 2$, and let $J(x) = [1/x_0 : \cdots : 1/x_3]$ be the usual Cremona involution on $\mathbb{P}^3$.

Theorem [BCK]. *For $n \ge 2$, we set $f_{a,c} := L_{a,c} \circ J$. The dynamical degrees are $\delta_1(f) = \delta_2(f) > 1$. There is no (singular) foliation of dimension 1 or 2 which is invariant under $f_{a,c}$. In particular, there is no invariant (singular) fibration.*

Now let us sketch the structure of the map $f := f_{a,c}$ in the example.

$$f(\Sigma_0) = e_1 := [0:1:0:0], \quad f(\Sigma_1) = e_2, \quad f(\Sigma_2) = e_3$$

$$f(\Sigma_3) = p := [1:a:0:c]$$

Theorem. *Let Y denote $\mathbb{P}^3$ blown up at the points e_1 and e_3. Then the induced map f_Y is a dominant map of an invariant 4-cycle of surfaces:*

$$\Sigma_0 \to E_1 \to \Sigma_2 \to E_3 \to \Sigma_0$$

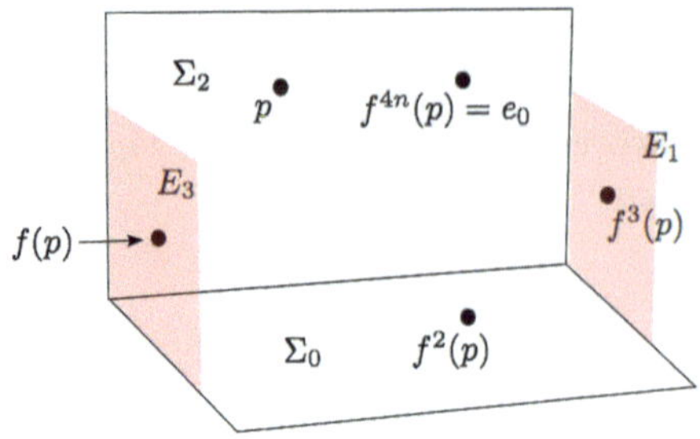

Fig. 11 Construction of Y

The orbit of the exceptional image point p is inside this invariant set (Fig. 11).

Theorem. *If $na^2 + (n+1)ac + nc^2 = 0$, then the f_Y-orbit of p lands on the point e_0. Let X denote the space obtained by blowing up the $4n+2$ points p, $f_Y(p)$, ..., $f_Y^{4n} p = e_0$, and e_3. Then the induced map f_X is a pseudo-automorphism.*

One difference between these maps and the [BDK] maps is that that there are two "levels" of blowup, and in fact none of the [BDK] maps is birationally conjugate to any of the [BCK] maps. The invariant 4-cycle of surfaces $\Gamma := \Sigma_0 \cup E_1 \cup \Sigma_2 \cup E_3$ plays an important role in understanding f.

Theorem. *For $g := f^4|_{\Sigma_0}$, the dynamical degree satisfies $\delta_1(g) > 1$, but it is not a Salem number. Thus g is not birationally conjugate to a surface automorphism.*

We note that the maps F_a of (8.1) were analyzed by means of an invariant 8-cycle of surfaces with similar properties. We may use [BK5], Theorem 1.5, to conclude that the maps in [BCK] are not birationally conjugate to an automorphism of a 3-dimensional manifold.

References

[A] Amerik, E.: A computation of invariants of a rational self-map. Ann. Fac. Sci. Toulouse Math. (6) **18**(3), 445–457 (2009)

[BaC] Bayraktar, T., Cantat, S.: Constraints on automorphism groups of higher dimensional manifolds. J. Math. Anal. Appl. **405**(1), 209–213 (2013)

[B] Bedford, E.: The dynamical degrees of a mapping. In: Proceedings of the Workshop Future Directions in Difference Equations, pp. 3–13, Colecc. Congr., vol. 69, Univ. Vigo, Serv. Publ., Vigo (2011)

[BCK] Bedford, E., Cantat, S., Kim, K.: Pseudo-automorphisms with no invariant foliation. J. Mod. Dyn. **8**, 221–250 (2014)

[BD] Bedford, E., Diller, J.: Real and complex dynamics of a family of birational maps of the plane: the golden mean subshift. Am. J. Math. **127**(3), 595–646 (2005)

[BDK] Bedford, E., Diller, J., Kim, K.: Pseudoautomorphisms with invariant curves. In: Proceedings of the Abel Symposium 2013, to appear. arXiv:1401.2386 (2013)

[BK1] Bedford, E., Kim, K.: On the degree growth of birational mappings in higher dimension. J. Geom. Anal. **14**(4), 567–596 (2004)

[BK2] Bedford, E., Kim, K.: Periodicities in linear fractional recurrences: degree growth of birational surface maps. Mich. Math. J. **54**(3), 647–670 (2006)

[BK3] Bedford, E., Kim, K.: Dynamics of rational surface automorphisms: linear fractional recurrences. J. Geom. Anal. **19**(3), 553–583 (2009)

[BK4] Bedford, E., Kim, K.: Continuous families of rational surface automorphisms with positive entropy. Math. Ann. **348**(3), 667–688 (2010)

[BK5] Bedford, E., Kim, K.: Dynamics of (pseudo) automorphisms of 3-space: periodicity versus positive entropy. Publ. Mat. **58**(1), 65–119 (2014)

[BC] Blanc, J., Cantat, S.: Dynamical degrees of birational transformations of projective surfaces. arXiv:1307.0361

[Br] Brolin, H.: Invariant sets under iteration of rational functions. Ark. Mat. **6**(1965), 103–144 (1965)

[C1] Cantat, S.: Dynamique des automorphismes des surfaces projectives complexes. C. R. Acad. Sci. Paris Sér. I Math. **328**(10), 901–906 (1999)

[C2] Cantat, S.: Quelques aspects des systèmes dynamiques polynomiaux: existence, exemples, rigidité. pp. 13–95, Panor. Synthèses, vol. 30, Soc. Math. France, Paris (2010)

[CL] Csörnyei, M., Laczkovich, M.: Some periodic and non-periodic recursions. Monatsh. Math. **132**(3), 215–236 (2001)

[dFE] de Fernex, T., Ein, L.: Resolution of indeterminacy of pairs. In: Algebraic Geometry, pp. 165–177. de Gruyter, Berlin (2002)

[Di] Diller, J.: Cremona transformations, surface automorphisms, and plane cubics (With an appendix by Dolgachev, I.). Mich. Math. J. **60**(2), 409–440 (2011)

[DF] Diller, J., Favre, C.: Dynamics of bimeromorphic maps of surfaces. Am. J. Math. **123**(6), 1135–1169 (2001)

[DN] Dinh, T.-C., Nguyên, V.-A.: Comparison of dynamical degrees for semi-conjugate meromorphic maps. Comment. Math. Helv. **86**(4), 817–840 (2011)

[DNT] Dinh, T-C., Nguyên, V.-A., Truong, T.T.: On the dynamical degrees of meromorphic maps preserving a fibration. Commun. Contemp. Math. **14**(6), 1250042, 18 pp (2012)

[DS1] Dinh, T.-C., Sibony, N.: Green currents for holomorphic automorphisms of compact Kähler manifolds. J. Am. Math. Soc. **18**(2), 291–312 (2005)

[DS2] Dinh, T.-C., Sibony, N.: Une borne supérieure pour l'entropie topologique d'une application rationnelle. Ann. Math. (2) **161**(3), 1637–1644 (2005)

[Do1] Dolgachev, I.: Reflection groups in algebraic geometry. Bull. Am. Math. Soc. (N.S.) **45**(1), 1–60 (2008)

[Do2] Dolgachev, I.: Infinite Coxeter groups and automorphisms of algebraic surfaces. In: The Lefschetz centennial conference, Part I (Mexico City, 1984), pp. 91–106, Contemp. Math., vol. 58, Am. Math. Soc., Providence, RI (1986)

[Do3] Dolgachev, I.: Weyl groups and Cremona transformations. In: Singularities, Part 1 (Arcata, Calif., 1981), pp. 283–294, Proc. Sympos. Pure Math., vol. 40, Am. Math. Soc., Providence, RI (1983)

[D1] Dujardin, R.: Structure properties of laminar currents on P^2. J. Geom. Anal. **15**(1), 25–47 (2005)

[D2] Dujardin, R.: Sur l'intersection des courants laminaires. Publ. Mat. **48**(1), 107–125 (2004)

[D3] Dujardin, R.: Laminar currents in P^2. Math. Ann. **325**(4), 745–765 (2003)

[FW] Favre, C., Wulcan, E.: Degree growth of monomial maps and McMullen's polytope algebra. Indiana Univ. Math. J. **61**(2), 493–524 (2012)

[FS1] Fornæss, J.E., Sibony, N.: Complex dynamics in higher dimensions. Notes partially written by Gavosto. E.A.: NATO Adv. Sci. Inst. Ser. C Math. Phys. Sci., vol. 439, Complex potential theory (Montreal, PQ, pp. 131–186. Kluwer Acad. Publ, Dordrecht (1993) (1994)

[FS2] Fornæss, J.E., Sibony, N.: Complex dynamics in higher dimension. II. In: Modern methods in complex analysis (Princeton, NJ, 1992), pp. 135–182, Ann. of Math. Stud., vol. 137, Princeton Univ. Press, Princeton, NJ (1995)

[FLM] Freire, A., Lopes, A., Mañé, R.: An invariant measure for rational maps. Bol. Soc. Brasil. Mat. **14**(1), 45–62 (1983)

[G1] Guedj, V.: Entropie topologique des applications méromorphes. Ergodic Theory Dyn. Syst. **25**(6), 1847–1855 (2005)

[G2] Guedj, V.: Propriétés ergodiques des applications rationelles. In: Quelques aspects des systèmes dynamiques polynômiaux, pp. 97–202, Panor. Synthèses, vol. 30, Soc. Math. France, Paris (2010)
[KPR] Kaschner, S.R., Pérez, R.A., Roeder, R.K.W.: Examples of rational maps of $\mathbb{C}P^2$ with equal dynamical degrees and no invariant foliation, arXiv:1309.4364
[KR] Koch, S., Roeder, R.: Computing dynamical degrees, arXiv:1403.5840
[Lin] Lin, J.-L.: Pulling back cohomology classes and dynamical degrees of monomial maps. Bull. Soc. Math. France **140**(4), 533–549 (2013) 2012
[Ly] Lyness, R.C.: Notes 1581,1847, and 2952, Math. Gazette **26** (1942), 62, **29** (1945), 231, and **45** (1961), 201
[L] Lyubich, M.: Entropy properties of rational endomorphisms of the Riemann sphere. Ergodic Theory Dyn. Syst. **3**(3), 351–385 (1983)
[M] McMullen, C.T.: Dynamics on blowups of the projective plane. Publ. Math. Inst. Hautes tudes Sci. No. 105, pp. 49–89 (2007)
[Muk] Mukai, S.: Geometric realization of T-shaped root systems and counterexamples to Hilbert's tenth problem. In: Algebraic transformation groups and algebraic varieties, vol. 132 of Encyclopedia Math. Sci., pp. 123–129. Springer, Berlin (2004)
[OT1] Oguiso, K., Truong, T.T.: Salem numbers in dynamics of Kähler threefolds and complex tori, arXiv:1309.4851
[OT2] Oguiso, K., Truong, T.T.: Explicit Examples of rational and Calabi-Yau threefolds with primitive automorphisms of positive entropy, arXiv:1306.1590
[PZ] Perroni, F., Zhang, D.-Q.: Pseudo-automorphisms of positive entropy on the blowups of products of projective spaces. Math. Ann. **359**(1–2), 189–209 (2014)
[T] Truong, T.T.: On automorphisms of blowups of $\mathbf{P}^3$, arXiv:1202.4224
[Ueh] Uehara, T.: Rational surface automorphisms with positive entropy, arXiv:1009.2143
[X] Xie, J.: Periodic points of birational maps on projective surfaces, arXiv:1106.1825

On Nazarov's Complex Analytic Approach to the Mahler Conjecture and the Bourgain-Milman Inequality

Zbigniew Błocki

Abstract We survey the several complex variables approach to the Mahler conjecture from convex analysis due to Nazarov. We also show, although only numerically, that his proof of the Bourgain-Milman inequality using estimates for the Bergman kernel for tube domains cannot be improved to obtain the Mahler conjecture which would be the optimal version of this inequality.

Keywords Mahler conjecture · Bergman kernel · Pluricomplex Green function

1 Introduction

Let K be a convex symmetric body in $\mathbb{R}^n$. This means that $K = -K$, K is convex, bounded, closed and has non-empty interior. The dual (or polar) body of K is given by

$$K' = \{y \in \mathbb{R}^n : x \cdot y \leq 1 \text{ for all } x \in K\},$$

where $x \cdot y = x_1 y_1 + \cdots + x_n y_n$. The Mahler volume of K is defined by

$$M(K) = \lambda_n(K)\lambda_n(K'),$$

where λ_n denotes the Lebesgue measure in $\mathbb{R}^n$. It is easy to see that it is independent of linear transformations and thus also on the inner product in $\mathbb{R}^n$. The Mahler volume is therefore an invariant of the Banach space $(\mathbb{R}^n, q_K)$, where q_K is the Minkowski functional of K:

$$q_K(x) = \inf\{t > 0 : t^{-1}x \in K\} = \sup\{x \cdot y : y \in K'\}.$$

Z. Błocki (✉)
Uniwersytet Jagielloński, Instytut Matematyki, Łojasiewicza 6, 30-348 Kraków, Poland
e-mail: Zbigniew.Blocki@im.uj.edu.pl; umblocki@cyf-kr.edu.pl

F. Bracci et al. (eds.), *Complex Analysis and Geometry*, Springer Proceedings in Mathematics & Statistics 144, DOI 10.1007/978-4-431-55744-9_6

The Blaschke-Santaló inequality says that the Mahler volume is maximal for balls:

$$\lambda_n(K)\lambda_n(K') \le (\lambda_n(\mathbb{B}_n^2))^2,$$

where for $p \ge 1$ we denote

$$\mathbb{B}_n^p = \{x \in \mathbb{R}^n : |x_1|^p + \cdots + |x_n|^p \le 1\}.$$

In fact, it holds without the assumption of symmetry but one has to assume that the interior of K contains the origin. Moreover, one has equality if and only if K is an ellipsoid, that is a linear image of $\mathbb{B}_n^2$. It was proved by Blaschke [B1, B2] for $n = 2$, $n = 3$, and by Santaló [S1] for arbitrary n (see also [SR]).

Mahler [M1] conjectured that $M(K)$ is minimized by cubes, that is

$$\lambda_n(K)\lambda_n(K') \ge \lambda_n(\mathbb{B}_n^1)\lambda_n(\mathbb{B}_n^\infty) = \frac{4^n}{n!},$$

where $\mathbb{B}_n^\infty = [-1, 1]^n$. It can be easily proved for $n = 2$: if K is a polygon with k vertices and $\widetilde{K}$ is the polygon with $k - 1$ vertices obtained from K by moving one vertex as in the following picture

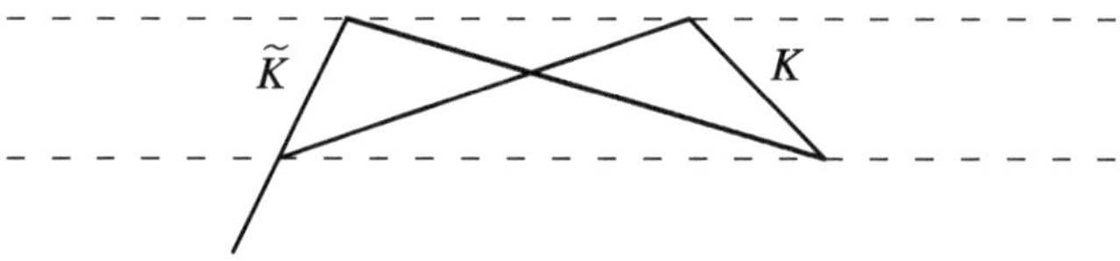

then $\lambda_2(\widetilde{K}) = \lambda_2(K)$ but one can show that $\lambda_2(\widetilde{K}') \ge \lambda_2(K')$.

Bourgain and Milman [BM] proved the following lower bound for the Mahler volume: there exists $c > 0$ such that

$$\lambda_n(K)\lambda_n(K') \ge c^n \frac{4^n}{n!}.$$

This is an important result in the theory of finitely-dimensional Banach spaces, it also has applications in number theory, see [BM]. We see that the Mahler conjecture is equivalent to this inequality with $c = 1$. The best known constant so far is $c = \pi/4$ and was obtained by Kuperberg [Ku].

One of possible difficulties with the Mahler conjecture is that if it is true then there would be more minimizers than cubes (and their linear images). We have $(\mathbb{B}_2^\infty)' = \mathbb{B}_2^1 \simeq \mathbb{B}_2^\infty$, where by $\simeq$ we denote the linear equivalence, and indeed for $n = 2$ the square is the only minimizer (up to linear transformations). However, for $n = 3$ the octahedron $\mathbb{B}_3^1 = (\mathbb{B}_3^\infty)'$ is not linearly equivalent to the cube $\mathbb{B}_3^\infty$. The conjecture for $n = 3$ is that the cube and octahedron are the only minimizers. For arbitrary n it should be so called Hansen-Lima bodies [HL]: these are intervals for $n = 1$ and in higher dimensions they are obtained by either taking products of lower-dimensional Hansen-Lima bodies or by taking their duals.

There is also a version of the Mahler conjecture for not necessarily symmetric bodies. Assuming that the origin is in the interior of K, it is expected that a centered simplex (that is the convex hull of affinely independent $v^1, \dots, v^{n+1} \in \mathbb{R}^n$ such that $v^1 + \dots + v^{n+1} = 0$) is the only minimizer, that is

$$\lambda_n(K)\lambda_n(K') \geq \frac{(n+1)^{n+1}}{(n!)^2}.$$

Recently Nazarov [N1] proposed a complex analytic approach to the Bourgain-Milman inequality and Mahler conjecture. Considering the Bergman kernel on the tube domain $\Omega = intK + i\mathbb{R}^n$ at the origin

$$K_\Omega(0,0) = \sup\{\frac{|f(0)|^2}{||f||^2_{L^2(\Omega)}} : f \in \mathcal{O}(\Omega) \cap L^2(\Omega),\ f \not\equiv 0\}$$

and using the formula for the Bergman kernel in tube domains of Rothaus [R1], see also [Hs], he proved the upper bound

$$K_\Omega(0,0) \leq \frac{n!}{\pi^n} \frac{\lambda_n(K')}{\lambda_n(K)}. \tag{1}$$

The main part of his paper was devoted to the proof of the lower bound

$$K_\Omega(0,0) \geq \left(\frac{\pi}{4}\right)^{2n} \frac{1}{(\lambda_n(K))^2}. \tag{2}$$

As is usually the case with lower bounds for the Bergman kernel, the main tool was Hörmander's estimate [H1]. Combining (1) with (2) we immediately obtain the Bourgain-Milman inequality with $c = (\pi/4)^3$.

In Sect. 2 we will present Nazarov's equivalent complex analytic formulation of the Mahler conjecture using the Paley-Wiener theorem. The upper bound (1) is explained in Sect. 3. We include the proof of Rothaus' [R1] integral formula for the Bergman kernel in tube domains, since it is not so well known. In Sect. 4 we discuss the lower bound using some simplifications from [Bln]. We also show that this approach cannot give the Mahler conjecture. We will see, although only numerically using *Mathematica*, that although the Bergman kernel for tube domains does behave well under taking products, it does not under taking duals.

The author is grateful for the invitation to the organizers of the 10th Korean Conference in Several Complex Variables held in August 2014 in Gyeong-Ju, especially to Kang-Tae Kim.

2 Equivalent SCV Formulation

Assume that K is a convex body in $\mathbb{R}^n$, not necessarily symmetric. For $u \in L^2(K')$ consider its Fourier transform

$$\widehat{u}(z) = \int_{K'} u(x)e^{-ix\cdot z}d\lambda(x), \quad z \in \mathbb{C}^n,$$

it is an entire holomorphic function. By the Schwarz inequality and the Parseval formula

$$|\widehat{u}(0)|^2 \leq \lambda_n(K') \int_{K'} |u|^2 d\lambda_n = \frac{\lambda_n(K')}{(2\pi)^n} \int_{\mathbb{R}^n} |\widehat{u}(x)|^2 d\lambda_n(x)$$

and we have equality for $u \equiv 1$ on K'. It is clear that $f = \widehat{u}$ satisfies

$$|f(z)| \leq Ce^{q_K(\mathrm{Im} z)}, \quad z \in \mathbb{C}^n, \tag{3}$$

for some $C > 0$. On the other hand, if $f \in \mathcal{O}(\mathbb{C}^n)$ satisfies (3) and is such that

$$\int_{\mathbb{R}^n} |f(x)|^2 d\lambda_n(x) < \infty \tag{4}$$

then by the Plancherel theorem $f = \widehat{u}$ for some $u \in L^2(\mathbb{R}^n)$ and by the Paley-Wiener theorem $\mathrm{supp}\, \mathrm{u} \subset \mathrm{K}'$. Therefore

$$\lambda_n(K') = (2\pi)^n \sup_{f \in \mathcal{P}, f \not\equiv 0} \frac{|f(0)|^2}{||f||^2_{L^2(\mathbb{R}^n)}},$$

where $\mathcal{P}$ denotes the family of entire holomorphic functions satisfying (3) and (4).

This way we have obtained a formula for the volume of the dual K' which is expressed only in terms of K, and not K'. It means that the Mahler conjecture is equivalent to finding $f \in \mathcal{O}(\mathbb{C}^n)$ with $f(0) = 1$, satisfying (3) and such that

$$\int_{\mathbb{R}^n} |f(x)|^2 d\lambda_n(x) \leq n! \left(\frac{\pi}{2}\right)^n \lambda_n(K)$$

in the symmetric case, and

$$\int_{\mathbb{R}^n} |f(x)|^2 d\lambda_n(x) \leq \frac{(n!)^2(2\pi)^n}{(n+1)^{n+1}} \lambda_n(K)$$

in the asymmetric one.

3 The Upper Bound

Nazarov [N1] showed that the upper bound (1) easily follows from the formula for the Bergman kernel in tube domains $\Omega = D + i\mathbb{R}^n$, where D is an arbitrary convex domain in $\mathbb{R}^n$:

$$K_\Omega(z, w) = \frac{1}{(2\pi)^n} \int_{\mathbb{R}^n} \frac{e^{(z+\bar{w})\cdot y}}{J_D(y)} d\lambda_n(y), \tag{5}$$

where

$$J_D(y) = \int_K e^{2x\cdot y} d\lambda_n(x)$$

(see [R1] and [Hs]). Indeed, for $y \in \mathbb{R}^n$ and $x_0 \in K$ using the fact that $(x_0 + K)/2 \subset K$ and that K is symmetric we get

$$J_K(y) \geq \frac{1}{2^n} \int_K e^{(x_0+x)\cdot y} d\lambda_n(x) \geq \frac{\lambda_n(K)}{2^n} e^{x_0 \cdot y}.$$

Therefore $J_K \geq 2^{-n} e^{q_{K'}}$ and to obtain (1) it is enough to observe that

$$\int_{\mathbb{R}^n} e^{-q_K} d\lambda_n = \int_0^\infty e^{-t} \lambda_n(\{q_K < t\}) dt = n! \lambda_n(K).$$

Proof (Proof of (5)*)* Take $\tilde{x} \in D$ and $r > 0$ such that $C_r := \tilde{x} + r(-1, 1)^n \subset D$. Then

$$J_D(y) \geq J_{C_r}(y) = e^{2\tilde{x}\cdot y} \frac{\sinh(2ry_1)}{y_1} \cdots \frac{\sinh(2ry_n)}{y_n}$$

and thus

$$\int_{\mathbb{R}^n} \frac{e^{2\tilde{x}\cdot y}}{J_D(y)} d\lambda(y) \leq \left(\frac{c}{r}\right)^{2n}, \tag{6}$$

where

$$c^2 = \frac{1}{2} \int_0^\infty \frac{t}{\sinh t} dt = \frac{\pi^2}{8}.$$

Since D is convex, we have $D + D = 2D$ and from (6) it follows in particular that the integral on the right-hand side of (5) is convergent.

For $u \in L^2(\mathbb{R}^n, J_D)$ and $z \in T_D$ set

$$\tilde{u}(z) = \int_{\mathbb{R}^n} u(y) e^{z\cdot y} d\lambda(y).$$

By (6) the integral is convergent and thus $\tilde{u}$ is holomorphic in T_D. It also follows that $h(y) := u(y)e^{\operatorname{Re} z\cdot y} \in L^2(\mathbb{R}^n)$ and we can write $\tilde{u}(z) = \widehat{h}(-\operatorname{Im} z)$. By the Parseval formula and the Fubini theorem

$$||\widetilde{u}||^2_{L^2(T_D)} = (2\pi)^n \int_K \int_{\mathbb{R}^n} |u(y)|^2 e^{2x\cdot y} d\lambda(y) d\lambda(x) = (2\pi)^n ||u||^2_{L^2(\mathbb{R}^n, J_D)}. \quad (7)$$

We claim that in fact the mapping

$$L^2(\mathbb{R}^n, J_D) \ni u \longmapsto \widetilde{u} \in A^2(T_D) \quad (8)$$

is onto. For $f \in A^2(T_D)$ approximating D by relatively compact subsets from inside and using the fact that $|f|^2$ is subharmonic we may assume that f is bounded in T_D. Multiplying f by functions of the form $e^{\varepsilon z\cdot z}$ we may even assume that it satisfies the estimate

$$|f(z)| \leq M e^{-\varepsilon |\mathrm{Im} z|^2} \quad (9)$$

for some positive constants M and ε. For a fixed $x \in D$ and $f_x(y) = f(x+iy)$ we have $f_x(y) = \widetilde{u}(x+iy)$ where $u(y) = (-2\pi)^{-n} \widehat{f_x}(y) e^{-x\cdot y}$. We have to prove that for a fixed y the definition of u is independent of x. From (9) it follows that we can differentiate under the sign of integration

$$\begin{aligned} &\frac{\partial}{\partial x_j} \int_{\mathbb{R}^n} f(x+ia) e^{-(x+ia)\cdot y} d\lambda(a) \\ &= \int_{\mathbb{R}^n} \left(\frac{\partial f}{\partial x_j}(x+ia) - y_j f(x+ia) \right) e^{-(x+ia)\cdot y} d\lambda(a). \end{aligned}$$

We have $\partial f/\partial x_j = -i \partial f/\partial a_j$ and by (9) we can also integrate by parts. Therefore

$$\begin{aligned} \int_{\mathbb{R}^n} \frac{\partial f}{\partial x_j}(x+ia) e^{-(x+ia)\cdot y} d\lambda(a) &= -i \int_{\mathbb{R}^n} \frac{\partial f}{\partial a_j}(x+ia) e^{-(x+ia)\cdot y} d\lambda(a) \\ &= \int_{\mathbb{R}^n} y_j f(x+ia) e^{-(x+ia)\cdot y} d\lambda(a) \end{aligned}$$

and therefore $u(y)$ is independent of x and the mapping (8) is onto.

By $K(z,w)$ denote the right-hand side of (5) and fix $w \in T_D$. Then $K(\cdot, w) = (2\pi)^{-n} \widetilde{v}$, where

$$v(y) = \frac{e^{\bar{w}\cdot y}}{J_D(y)} \in L^2(\mathbb{R}^n, J_D)$$

by (6). It follows from (7) that $K(\cdot, w) \in A^2(T_D)$ and to finish the proof we have to show that it has the reproducing property. For $f = \widetilde{u} \in A^2(T_D)$ where $u \in L^2(\mathbb{R}^n, J_D)$ by (7)

$$\langle f, K(\cdot, w)\rangle_{A^2(T_D)} = \frac{1}{(2\pi)^n} \langle \widetilde{u}, \widetilde{v}\rangle_{A^2(T_D)} = \langle u, v\rangle_{L^2(\mathbb{R}^n, J_D)} = \int_{\mathbb{R}^n} u(y) e^{w\cdot y} d\lambda(y) = f(w).$$

This finishes the proof of (5).

4 The Lower Bound

The lower bound (2) easily follows from a general lower bound for the Bergman kernel proved in [Bln]: if Ω is a pseudoconvex domain in $\mathbb{C}^n$ then for $w \in \Omega$ and $t \leq 0$

$$K_\Omega(w, w) \geq \frac{1}{e^{-2nt}\lambda_{2n}(\{G_\Omega(\cdot, w) < t\})}, \tag{10}$$

where

$$G_\Omega(z, w) = \sup\{u(z) \colon u \in PSH^-(\Omega), \ \limsup_{z \to w}(u(z) - \log|z - w|) < \infty\}$$

is the pluricomplex Green function of Ω. It was proved in [Bln] using the Donnelly-Fefferman [DF] estimate for $\bar\partial$ (which can be easily deduced from Hörmander's estimate, see [Ber]) and the tensor-power trick. A simpler proof using subharmonicity of sections of the Bergman kernel from [Ber2] was later given by Lempert [L2] (see [Bms]).

The estimate (10) has various consequences when we let $t \to -\infty$. For example for $n = 1$ it gives the Suita conjecture

$$c_\Omega(w)^2 \leq \pi K_\Omega(w, w),$$

where

$$c_\Omega(w) = \exp(\lim_{z \to w}(G_\Omega(z, w) - \log|z - w|))$$

is the logarithmic capacity of $\mathbb{C} \setminus \Omega$ with respect to w. It was originally proved in [Bin]. For arbitrary n if Ω is convex then using Lempert's theory [L1] one can obtain the estimate

$$K_\Omega(w, w) \geq \frac{1}{\lambda_{2n}(I_\Omega(w))}, \tag{11}$$

where

$$I_\Omega(w) = \{\varphi'(0) \colon \varphi \in \mathcal{O}(\Delta, \Omega), \ \varphi(0) = w\}$$

is the Kobayashi indicatrix (Δ is the unit disk in $\mathbb{C}$). This particular estimate for convex domains seems to be very accurate, see [BZ1, BZ2] for details.

Now let us come back to the case of the tube domain $\Omega = intK + i\mathbb{R}^n$ where K is a convex symmetric body in $\mathbb{R}^n$. Let $\varphi \in \mathcal{O}(\Delta, \Omega)$ be such that $\varphi(0) = 0$. By S denote the strip $\{|\mathrm{Re}\,\zeta| < 1\}$ in $\mathbb{C}$ and let $\Phi : S \to \Delta$ be biholomorphic with $\Phi(0) = 0$. By the Schwarz lemma for $u \in K'$

$$\left| \frac{\partial}{\partial\zeta}\bigg|_{\zeta=0} \Phi(\varphi(\zeta) \cdot u) \right| \leq 1$$

and since $|\Phi'(0)| = \pi/4$ we obtain

$$|\varphi'(0) \cdot u| \leq \frac{4}{\pi}.$$

It follows that

$$I_\Omega(0) \subset \frac{4}{\pi}(K'' + iK'') = \frac{4}{\pi}(K + iK)$$

and

$$\lambda_{2n}(I_\Omega(0)) \leq \left(\frac{4}{\pi}\right)^{2n} (\lambda_n(K))^2.$$

The estimate (11) now gives the lower bound (2).

It was conjectured in [Bln] that the following lower bound holds in tube domains

$$K_\Omega(0,0) \geq \left(\frac{\pi}{4}\right)^n \frac{1}{(\lambda_n(K))^2}. \tag{12}$$

It would be optimal because one can easily check using the product formula for the Bergman kernel that one has equality in (12) for the unit cube $K = [-1, 1]^n$.

We will show however that we do not have equality in (12) for all Hansen-Lima bodies. Take the octahedron

$$K = \mathbb{B}_3^1 = \{x \in \mathbb{R}^3 \colon |x_1| + |x_2| + |x_3| \leq 1\}.$$

One can then compute that

$$J_K(y) = \frac{y_1 \sinh(2y_1)}{(y_1^2 - y_2^2)(y_1^2 - y_3^2)} + \frac{y_2 \sinh(2y_2)}{(y_2^2 - y_1^2)(y_2^2 - y_3^2)} + \frac{y_3 \sinh(2y_3)}{(y_3^2 - y_1^2)(y_3^2 - y_2^2)}$$

when all coordinates y_j are different and that it extends to a positive smooth function in $\mathbb{R}^3$. One can then compute numerically using (5) that

$$K_\Omega(0,0) = \frac{1}{(2\pi)^3} \int_{\mathbb{R}^3} \frac{d\lambda_3}{J_K} = 0.2758\ldots \tag{13}$$

However, since $\lambda_n(\mathbb{B}_n^1) = 2^n/n!$, the right-hand side of (12) is equal to

$$\frac{9\pi^3}{1024} = 0.2725\ldots$$

This shows (although only numerically) that the Bergman kernel for tube domains does not behave well under taking duals. It is also clear that even proving optimal versions of the estimates (2) and (1) cannot give an optimal lower bound for the

Mahler volume and thus this Nazarov's approach to the Bourgain-Milman inequality cannot give its expected optimal form, that is the Mahler conjecture.

To make this argument precise and get rid of the numerical computation in (13), one could try to consider the n-dimensional octahedron

$$K_n = \mathbb{B}_n^1 = \{x \in \mathbb{R}^n : |x_1| + \cdots + |x_n| \le 1\}.$$

One can compute that

$$J_{K_n}(y) = \begin{cases} \displaystyle\sum_{j=1}^n \frac{y_j^{n-2}\cosh(2y_j)}{(y_j^2 - y_1^2)\dots(y_j^2 - y_{j-1}^2)(y_j^2 - y_{j+1}^2)\dots(y_j^2 - y_n^2)}, & n \text{ even} \\ \displaystyle\sum_{j=1}^n \frac{y_j^{n-2}\sinh(2y_j)}{(y_j^2 - y_1^2)\dots(y_j^2 - y_{j-1}^2)(y_j^2 - y_{j+1}^2)\dots(y_j^2 - y_n^2)}, & n \text{ odd} \end{cases}.$$

One could perhaps estimate J_{K_n} from above in such a way that it would imply that

$$\limsup_{n\to\infty} \left(\frac{1}{(n!)^2} \int_{\mathbb{R}^n} \frac{d\lambda_n}{J_{K_n}} \right)^{1/n} > \frac{\pi^2}{8}.$$

Another possibility would be to apply (11): it would be enough to show that there exists n such that if I_n is the Kobayashi indicatrix of the tube domain $int\, K_n + i\mathbb{R}^n$ at the origin then

$$\lambda_{2n}(I_n) < \frac{16^n}{(n!)^2\pi^n}.$$

This could perhaps be possible using Lempert's theory for tube domains developed by Zając [Z1].

Acknowledgments Partially supported by the Ideas Plus grant 0001/ID3/2014/63 of the Polish Ministry of Science and Higher Education.

References

[Ber] Berndtsson, B.: Weighted estimates for the $\overline{\partial}$-equation, Complex Analysis and Geometry, Columbus, Ohio, 1999, Ohio State Univ. Math. Res. Inst. Publ. 9, pp. 43–57, Walter de Gruyter (2001)

[Ber2] Berndtsson, B.: Subharmonicity properties of the Bergman kernel and some other functions associated to pseudoconvex domains. Ann. Inst. Fourier **56**, 1633–1662 (2006)

[B1] Blaschke, W.: Über affine Geometrie VII: Neue Extremeigenschaften von Ellipse und Ellipsoid, Ber. Verh. Sächs. Akad. Wiss. Leipzig Math.-Phys. Kl. **69**, 306–318 (1917)

[B2] Blaschke, W.: Affine Geometrie IX: Verschiedene Bemerkungen und Aufgaben, Ber. Verh. Sächs. Akad. Wiss. Leipzig Math.-Phys. Kl. **69**, 412–420 (1917)

[Bin] Błocki, Z.: Suita conjecture and the Ohsawa-Takegoshi extension theorem. Invent. Math. **193**, 149–158 (2013)

[Bln] Błocki, Z.: A lower bound for the Bergman kernel and the Bourgain-Milman inequality, Geometric Aspects of Functional Analysis, Israel Seminar (GAFA) 2011–2013. In: Klartag, B., Milman, E. (eds.) Lecture Notes in Mathematics, vol. 2116, pp. 53–63. Springer (2014)
[Bms] Błocki, Z.: Cauchy-Riemann meet Monge-Ampère. Bull. Math. Sci. **4**, 433–480 (2014)
[BZ1] Błocki, Z., Zwonek, W.: Estimates for the Bergman kernel and the multidimensional Suita conjecture. New York J. Math. **21**, 151–161 (2015)
[BZ2] Błocki, Z., Zwonek, W.: On the Suita conjecture for some convex ellipsoids in $\mathbb{C}^2$. arXiv:1409.5023, Experimental Math. (to appear)
[BM] Bourgain, J., Milman, V.: New volume ratio properties for convex symmetric bodies in $\mathbb{R}^n$. Invent. Math. **88**, 319–340 (1987)
[DF] Donnelly, H., Fefferman, C.: L^2-cohomology and index theorem for the Bergman metric. Ann. of Math. **118**, 593–618 (1983)
[HL] Hansen, A.B., Lima, Å.: The structure of finite-dimensional Banach spaces with the 3.2. intersection property. Acta Math. **146**, 1–23 (1981)
[H1] Hörmander, L.: L^2 estimates and existence theorems for the $\bar\partial$ operator. Acta Math. **113**, 89–152 (1965)
[Hs] Hsin, C.-I.: The Bergman kernel on tube domains. Rev. Un. Mat. Argentina **46**, 23–29 (2005)
[Ku] Kuperberg, G.: From the Mahler conjecture to Gauss linking integrals. Geom. Funct. Anal. **18**, 870–892 (2008)
[L1] Lempert, L.: La métrique de Kobayashi et la représentation des domaines sur la boule. Bull. Soc. Math. Fr. **109**, 427–474 (1981)
[L2] Lempert, L.: Private Communication, October (2013)
[M1] Mahler, K.: Ein Minimalproblem für konvexe Polygone. Mathematika B (Zutphen) **7**, 118–127 (1938)
[N1] Nazarov, F.: The Hörmander proof of the Bourgain-Milman theorem. In: Klartag, B., Mendelson, S., Milman, V.D. (eds.) Geometric Aspects of Functional Analysis, Israel Seminar 2006–2010. Lecture Notes in Mathematics, vol. 2050, pp. 335–343. Springer (2012)
[R1] Rothaus, O.S.: Some properties of Laplace transforms of measures. Trans. Amer. Math. Soc. **131**, 163–169 (1968)
[RZ] Ryabogin, D., Zvavitch, A.: Analytic methods in convex geometry. Lectures given at the Polish Academy of Sciences, November 2011, IMPAN Lecture Notes (to appear). http://www.impan.pl/Dokt/EN/SpLect/RZ2011.pdf
[SR] Saint-Raymond, J.: Sur le volume des corps convexes symétriques. In: Initiation Seminar on Analysis: G. Choquet, M. Rogalski, J. Saint-Raymond, 20th Year: 1980/1981, Exp. No. 11, pp. 25, Publ. Math. Univ. Pierre et Marie Curie, 46, Univ. Paris VI, Paris (1981)
[S1] Santaló, L.A.: An affine invariant for convex bodies of n-dimensional space. Portugaliae Math. **8**, 155–161 (1949). (in Spanish)
[Su] Suita, N.: Capacities and kernels on Riemann surfaces. Arch. Ration. Mech. Anal. **46**, 212–217 (1972)
[Z1] Zajac, S.: Complex geodesics in convex tube domains. arXiv:1303.0014, Ann. Scuola Norm. Sup. Pisa (to appear)

A Survey on Bergman Completeness

Bo-Yong Chen

Abstract We provide a survey of results on Bergman completeness of open complex manifolds

Keywords Bergman completeness

1 Introduction

Let M be a complex manifold of dimension n. Let $\mathscr{H}$ be the Hilbert space of holomorphic $n-$forms f on M satisfying

$$\left|\int_M f \wedge \bar{f}\right| < \infty.$$

Let $h_1, h_2, \ldots$ be a complete orthonormal basis for $\mathscr{H}$. We may define the Bergman kernel (form) K_M of M as

$$K_M(z, w) = \sum_j h_j(z) \wedge \overline{h_j(w)}.$$

Let $(z_1, z_2, \ldots, z_n)$ be a local coordinate system in M. Let

$$K_M(z) := K_M(z, z) = K^*(z) dz_1 \wedge \cdots \wedge dz_n \wedge d\bar{z}_1 \wedge \cdots \wedge d\bar{z}_n$$

where $K^*(z)$ is a locally defined function. If K^* is positive, then we may define the Bergman metric ds^2_M of M as

B.-Y. Chen (✉)
Department of Mathematics, Tongji University, Shanghai 200092, China
e-mail: boychen@tongji.edu.cn

B.-Y. Chen
School of Mathematical Sciences, Fudan University, Shanghai 200433, China

F. Bracci et al. (eds.), *Complex Analysis and Geometry*, Springer Proceedings in Mathematics & Statistics 144, DOI 10.1007/978-4-431-55744-9_7

$$ds_M^2 = \sum_{\alpha,\beta} \frac{\partial^2 \log K^*}{\partial z_\alpha \partial \bar{z}_\beta} dz_\alpha d\bar{z}_\beta.$$

We say that M possesses the Bergman metric if ds_M^2 is a Kähler metric on M. The Bergman distance d_B is the distance with respect to ds_M^2. A complex manifold is said to be Bergman complete if d_B is complete.

In contrast with compact complex manifolds, the *quantitative* complex analysis on *open* complex manifolds is far from well-developed, even in the case of Riemann surfaces! As the Bergman kernel and metric depend only on the complex structure, and they are invariant under biholomorphic transformations, thus they should occupy a central place in the study of open complex manifolds. This is essentially the theme of Kobayashi's ground-breaking paper [Kobayashi59], although it is titled *geometry of bounded domains*.

It is not difficult to see that M possesses the Bergman metric if and only if the holomorphic mapping

$$\tau : M \to \mathbb{P}(\mathscr{H}), \quad z \mapsto [h_1(z) : h_2(z) : \cdots]$$

is an immersion, where $\mathbb{P}(\mathscr{H})$ stands for the complex projective space of $\mathscr{H}$. Kobayashi's decisive observation is

$$ds_M^2 = \tau^*(ds_{FS}^2)$$

where ds_{FS}^2 is the Fubini-Study metric of $\mathbb{P}(\mathscr{H})$. It follows that for any given distinct points $z, w \in M$, the Bergman distance d_B and the Fubini-Study distance d_{FS} satisfy

$$d_B(z, w) \geq d_{FS}(\tau(z), \tau(w)).$$

Since

$$d_{FS}(\tau(z), \tau(w)) = \arccos \frac{|\sum_j h_j^*(z)\overline{h_j^*(w)}|}{\sqrt{\sum_j |h_j^*(z)|}\sqrt{\sum_j |h_j^*(w)|}}$$

where h_j^* is a local representation of h_j, we have

$$d_B(z, w) \geq \arccos \frac{|h_1^*(z)|}{\sqrt{\sum_j |h_j^*(z)|^2}} \geq \sqrt{1 - \frac{|h_1^*(z)|^2}{\sum_j |h_j^*(z)|^2}} = \sqrt{1 - \frac{|h_1^*(z)|^2}{K^*(z)}} \quad (1.1)$$

provided that we choose $\{h_j\}$ such that $h_j(w) = 0$ for all $j \geq 2$. From this Kobayashi reached the following

Kobayashi's criterion (cf. [Kobayashi59, Kobayashi61]). *Suppose there is a dense subset S of $\mathcal{H}$ such that for every $f \in S$ and for any infinite sequence $\{p_k\}$ of points in M which has no adherent point in M, there is a subsequence $\{p_{k_j}\}$ such that*

$$\frac{f(p_{k_j}) \wedge \overline{f(p_{k_j})}}{K_M(p_{k_j})} \to 0 \quad as \quad j \to \infty. \tag{1.2}$$

Then M is Bergman complete.

Let us give a short proof of Kobayashi's criterion. Suppose M is not Bergman complete, i.e. there is a d_B−Cauchy sequence $\{p_k\}$ which has no adherent point in M. Let $k_0 \in \mathbb{Z}^+$ satisfy

$$d_B(p_k, p_l) < 1/2 \qquad \forall k, l \geq k_0.$$

Since S is dense in $\mathcal{H}$, we may construct by using the Gram-Schmidt procedure on S a complete orthonormal basis $\{\tilde{h}_j\}$ of $\mathcal{H}$ such that every $\tilde{h}_j$ enjoys the same property as $f \in S$. Put $w = p_{k_0}$ in (1.1). We may write $h_1 = \sum_j a_j \tilde{h}_j$ with $\sum_j |a_j|^2 = 1$. Choose $j_0 \in \mathbb{Z}^+$ (depending only on p_{k_0}), such that $\sum_{j>j_0} |a_j|^2 \leq 1/4$. Put $h_{1,j_0} = \sum_{j=1}^{j_0} a_j \tilde{h}_j$. By the Cauchy-Schwarz inequality, we have

$$(h_1 - h_{1,j_0}) \wedge \overline{(h_1 - h_{1,j_0})} \leq \sum_{j>j_0} |a_j|^2 \sum_{j>j_0} \tilde{h}_j \wedge \overline{\tilde{h}_j} \leq \frac{1}{4} K_M,$$

so that

$$\frac{h_1 \wedge \overline{h_1}}{K_M} \leq \frac{2h_{1,j_0} \wedge \overline{h_{1,j_0}}}{K_M} + \frac{1}{2}.$$

Let $\{p_{k_j}\}$ be a subsequence of $\{p_k\}$ such that (1.2) is verified for h_{1,j_0}. Then $\frac{h_1 \wedge \overline{h_1}}{K_M}(p_{k_j}) < 3/4$ provided j sufficiently large. On the other hand, it follows from (1.1) that

$$\frac{h_1(p_{k_j}) \wedge \overline{h_1(p_{k_j})}}{K_M(p_{k_j})} > \frac{3}{4},$$

and we get a contradiction.

The goal of this article is to survey some results concerning Bergman completeness, built on Kobayashi's criterion. Due to my personal taste, I am not able to cover all interesting results in this direction. I must apologize to those authors whose papers are not mentioned here. One may consult the nice books of Jarnicki and Pflug [JarnickiPflug, JarnickiPflug2] for more references.

Nevertheless, Bergman completeness is only the first step to understand the geometry of the Bergman metric, much more works need to be done in future.

2 Bergman Completeness for Domains in $\mathbb{C}^n$

The first result concerning Bergman completeness was given by Bremermann:

Theorem 2.1 (cf. [Bremermann]) *Every bounded Bergman complete domain in $\mathbb{C}^n$ is pseudoconvex.*

Obviously, the converse is not true (e.g., the punctured disc). Thus it is natural to ask

Problem 2.1 (*cf.* [Kobayashi59]) Which bounded pseudoconvex domain in $\mathbb{C}^n$ is Bergman complete?

By using his criterion, Kobayashi showed that every bounded analytic polyhedron is Bergman complete. A useful consequence of Kobayashi's criterion is that

$$H^\infty(\Omega) \text{ lies dense in } \mathscr{H} \text{ and } \lim_{z\to\partial\Omega} K_\Omega(z) = \infty \tag{2.1}$$

implies Bergman completeness, where $H^\infty(\Omega)$ stands for the set of bounded holomorphic functions on Ω. For the sake of simplicity, we say that a bounded domain Ω is Bergman exhaustive if $\lim_{z\to\partial\Omega} K_\Omega(z) = \infty$.

The first general result toward Problem 2.1 is due to Ohsawa:

Theorem 2.2 (cf. [Ohsawa81]) *Every bounded pseudoconvex domain in $\mathbb{C}^n$ with a C^1 boundary is Bergman complete.*

The Bergman exhaustiveness follows from the following result of Pflug:

Theorem 2.3 (cf. [Pflug75]) *Let Ω be a bounded pseudoconvex domain in $\mathbb{C}^n$ and $p \in \partial\Omega$. Suppose there exist a sequence $\{z^\nu\} \subset \mathbb{C}^n\backslash\Omega$, and positive numbers $\beta \geq 1$, $r \leq 1$ such that $z^\nu \to p$ and*

$$B(z^\nu, r|z^\nu - p|^\beta) \cap \Omega = \emptyset.$$

Then Ω is Bergman exhaustive.

It is difficult to verify that $H^\infty(\Omega)$ lies dense in $\mathscr{H}$, yet it is easy to verify this property *locally*. Thus the following localization principle of the Bergman metric becomes important:

Proposition 2.1 (cf. [Ohsawa84]) *Let Ω be a bounded pseudoconvex domain in $\mathbb{C}^n$. Let $p \in \partial\Omega$ and let $V \subset\subset U$ be two bounded neighborhoods of p. Then there are constants $C_1, C_2 > 0$ such that*

$$C_1 ds^2_\Omega(z) \leq ds^2_{\Omega\cap U}(z) \leq C_2 ds^2_\Omega(z), \qquad \forall z \in \Omega \cap V.$$

This proposition may be proved by a standard application of Hörmander's L^2 estimates for the $\bar{\partial}$–operator (cf. [HormanderBook]). Without any regularity assumption on the boundary, Jarnicki and Pflug [JarnickiPflug89] proved that every bounded balanced domain is Bergman complete.

It follows from the work of Kerzman and Rosay [KerzmanRosay] that every bounded pseudoconvex domain with a C^1 boundary is *hyperconvex*, i.e., there exists a continuous plurisubharmonic (psh) function $\rho : \Omega \to [-1, 0)$ such that $\{\rho < -c\} \subset\subset \Omega$ for all $c > 0$. Another important class of hyperconvex domains are Teichmüller spaces of compact Riemann surfaces of genus ≥ 2 (cf. [Krushkal]). Blocki and Pflug [BlockiPflug] and Herbort [HerbortHyperconvex] proved independently the following result which has been a longstanding conjecture due to Kobayashi (see e.g., [KobayashiBook98]):

Theorem 2.4 *Every bounded hyperconvex domain in $\mathbb{C}^n$ is Bergman complete.*

Earlier, Ohsawa [OhsawaHyperconvex] has proved that every hyperconvex domain is Bergman exhaustive, which also initiates a program of studying asymptotic behavior of L^2 holomorphic objects through investigating the Green function (see also [Ohsawa95]).

Recall that the *pluricomplex Green function* $g_\Omega(z, w)$ of Ω is defined as

$$g_\Omega(z, w) = \sup\{u(z) : u < 0, u \in PSH(\Omega), u(z) \leq \log|z - w| + O(1) \text{ near } w\}$$

where $PSH(\Omega)$ stands for the set of psh functions on Ω.

The following result was discovered independently by Herbort and myself, and suggests that pluripotential theory would be essential for the study of Bergman completeness:

Proposition 2.2 (cf. [Chen99, HerbortHyperconvex]) *Let Ω be a bounded pseudoconvex domain in $\mathbb{C}^n$. Suppose there is a constant $c > 0$ such that*

$$\lim_{w \to \partial\Omega} |\{g_\Omega(\cdot, w) < -c\}| = 0 \tag{2.2}$$

where $|\cdot|$ stands for the (Euclidean) volume. Then Ω is Bergman complete.

Let me explain briefly the idea of proving the proposition. It suffices to verify Kobayashi's criterion. Given $f \in \mathscr{H}$ and $w \in \Omega$, we look for a new function $\tilde{f} \in \mathscr{H}$ (which actually depends on w) such that $\tilde{f}(w) = f(w)$ and $\|\tilde{f}\|_{L^2}$ tends to zero as $w \to \partial\Omega$. Since $K_\Omega(w) \geq |\tilde{f}(w)|^2/\|\tilde{f}\|^2_{L^2}$, it follows that

$$\frac{|f(w)|^2}{K_\Omega(w)} \leq \|\tilde{f}\|^2_{L^2} \to 0 \quad \text{as } w \to \partial\Omega.$$

The desired function $\tilde{f}$ is given by

$$\tilde{f} = \chi(\log(-g_\Omega(\cdot, w)))f - u$$

where χ is a standard cut-off function such that

$$\operatorname{supp}\chi(\log(-g_\Omega(\cdot, w))) \subset \{g_\Omega(\cdot, w) < -c\}.$$

Note that $\bar\partial \tilde f = 0$ if and only if

$$\bar\partial u = f\bar\partial\chi(\log(-g_\Omega(\cdot, w))).$$

Thanks to the L^2–estimates of Donnelly and Fefferman [DonnellyFefferman], we may find a solution u satisfying

$$\begin{aligned}
&\int_\Omega |u|^2 e^{-2ng_\Omega(\cdot,w)} \\
&\le \text{const.} \int_\Omega |f|^2|\chi'(\cdot)|^2|\bar\partial \log(-g_\Omega(\cdot, w))|^2_{i\partial\bar\partial\log(-g_\Omega(\cdot,w)+1)} e^{-2ng_\Omega(\cdot,w)} \\
&\le \text{const.} \int_{\{g_\Omega(\cdot,w)<-c\}} |f|^2.
\end{aligned}$$

Since u is holomorphic in a neighborhood of w, we see that $u(w) = 0$. Thus $\tilde f(w) = f(w)$ and

$$\begin{aligned}
\|\tilde f\|^2_{L^2} &\le 2\int_{\{g_\Omega(\cdot,w)<-c\}} |f|^2 + 2\int_\Omega |u|^2 \\
&\le \text{const.} \int_{\{g_\Omega(\cdot,w)<-c\}} |f|^2 \to 0 \quad \text{as } w \to \partial\Omega.
\end{aligned}$$

Thus we are done. To make the argument rigorous, we need to smooth $g_\Omega(\cdot, w)$ by a standard approximating procedure.

To prove Theorem 2.5, it suffices to verify (2.2) for bounded hyperconvex domains. Blocki and Pflug used the following results due to Blocki:

Proposition 2.3 (cf. [Blocki93]) *Let Ω be a bounded domain in $\mathbb{C}^n$. Assume that u, v are non-positive psh functions such that $u = 0$ on $\partial\Omega$. Then*

$$\int_\Omega |u|^n (dd^c v)^n \le n!\|v\|^{n-1}_\infty \int_\Omega |v|(dd^c u)^n. \tag{2.3}$$

Theorem 2.5 (cf. [Blocki96]) *Let Ω be a bounded hyperconvex domain in $\mathbb{C}^n$. Then there exists a solution ϕ of the following Monge-Ampere equation*

$$\det\left(\frac{\partial^2\phi}{\partial z_\alpha \partial\bar z_\beta}\right) = 1, \quad \phi \in C(\overline{\Omega}) \text{ and } \phi|_{\partial\Omega} = 0.$$

Put $u = g_\Omega(\cdot, w)$ and $v = \phi$ in (2.3), one gets

$$\begin{aligned} |\{g_\Omega(\cdot, w) < -1\}| &\le \int_\Omega |g_\Omega(\cdot, w)|^n (dd^c\phi)^n \\ &\le n!\|\phi\|_\infty^{n-1} \int_\Omega |\phi|(dd^c g_\Omega(\cdot, w))^n \\ &\le \text{const.}|\phi(w)| \to 0 \end{aligned}$$

as $w \to \partial\Omega$, for $(dd^c g_\Omega(\cdot, w))^n = \delta_w$ (cf. Demailly [Demailly82]).

Remark 2.1 Recently, the property of the function $\lambda(t) := |\{g_\Omega(\cdot, w) < -t\}|$ defined on $(0, \infty)$, in particular, the asymptotic behavior of $\lambda(t)$ as $t \to \infty$, has attracted much attention (see e.g., [BlockiSuita, BlockiBourgain, BerndtssonLempert]).

On the other side, there are many non-hyperconvex, Bergman complete domains (cf. [Chen99, HerbortHyperconvex, PflugZwonek03, PflugZwonek05]). For instance, one has the following

Proposition 2.4 (cf. [ChenEssay]) *Let D be a bounded pseudoconvex domain in $\mathbb{C}^n$ and let $\varphi > 0$ be a continuous psh function on D satisfying*

$$\liminf_{z\to\partial D} \frac{\varphi(z)}{\log 1/\delta_D(z)} = \infty.$$

Then (2.2) holds for the Hartogs domain $\Omega := \{(z, w) \in D \times \mathbb{C} : |w| < e^{-\varphi(z)}\}$, in particular, it is Bergman complete.

This result suggests that condition (2.2) is almost optimal for Bergman completeness, e.g., let D be a punctured disc and $\varphi(z)$ be psh on D satisfying $\varphi(z) \sim N \log 1/|z|$ as $z \to 0$, where N is a positive integer, then Ω would not be Bergman complete.

It is important to obtain *quantitative* lower estimates on the Bergman distance which implies completeness. Diederich and Ohsawa proved the following

Theorem 2.6 (cf. [DiederichOhsawa]) *Let $\Omega \subset \mathbb{C}^n$ be a bounded pseudoconvex domain with a C^2 boundary and let $z^0 \in \Omega$. Then the Bergman distance d_B satisfies*

$$d_B(z^0, z) \ge \text{const.} \log|\log \delta_\Omega(z)|$$

for all $z \in \Omega$ sufficiently close to $\partial\Omega$. Here δ_Ω stands for the (Euclidean) boundary distance.

The key idea of [DiederichOhsawa] is to use the following strengthening of Kobayashi' observation:

Proposition 2.5 (cf. [DiederichOhsawa]) *Let p_1, p_2 be distinct points in a bounded domain $\Omega \subset \mathbb{C}^n$. Suppose there exists a constant $C > 0$ such that for any $f \in \mathscr{H}$ with $\|f\|_{L^2} = 1$ there is another $\tilde{f} \in \mathscr{H}$ satisfying $\tilde{f}(p_1) = 0$, $\tilde{f}(p_2) = f(p_2)$, and $\|\tilde{f}\|_{L^2} \le C$, then $d_B(p_1, p_2) \ge C'$ where C' is a positive constant depending only on C.*

Proof Recall from (1.1) that

$$d_B(p_1, p_2) \geq \sqrt{1 - \frac{|h_1(p_2)|^2}{K_\Omega(p_2)}}$$

where $\{h_j\}$ is a complete orthonormal basis of $\mathscr{H}$ satisfying $h_j(p_1) = 0$ for all $j \geq 2$. If $|h_1(p_2)|^2 \leq \frac{1}{2} K_\Omega(p_2)$, then we have $d_B(p_1, p_2) \geq 1/\sqrt{2}$; otherwise, we may choose h_2 satisfying $|h_2(p_2)| \geq |h_1(p_2)|/C$, so that

$$\begin{aligned} d_B(p_1, p_2) \geq \sqrt{1 - \frac{|h_1(p_2)|^2}{K_\Omega(p_2)}} &= \sqrt{\frac{\sum_{j=2}^\infty |h_j(p_2)|^2}{K_\Omega(p_2)}} \\ &\geq \frac{|h_2(p_2)|}{\sqrt{K_\Omega(p_2)}} \geq \frac{|h_1(p_2)|}{C\sqrt{K_\Omega(p_2)}} \geq \frac{1}{\sqrt{2}C}. \end{aligned}$$

□

Built on the previous proposition, we may prove the following result through a similar argument as the proof of Proposition 2.6:

Proposition 2.6 (cf. [BlockiGreen], see also [ChenZhang]) *Let Ω be a bounded pseudoconvex domain in $\mathbb{C}^n$. Suppose that p_1, p_2 are distinct points in Ω satisfying*

$$\{g_\Omega(\cdot, p_1) < -1\} \cap \{g_\Omega(\cdot, p_2) < -1\} = \emptyset,$$

then $d_B(p_1, p_2) \geq \text{const}_n$.

Blocki improved substantially the result of Diederich-Ohsawa as follows

Theorem 2.7 (cf. [BlockiGreen]) *One has*

$$d_B(z^0, z) \geq \text{const.} |\log \delta_\Omega(z)| / \log|\log \delta_\Omega(z)|$$

for all $z \in \Omega$ sufficiently close to $\partial\Omega$.

The proof of Theorem 2.7 relies on Proposition 2.6 and the following quantitative estimate of g_Ω, which is also useful for other purposes (see e.g., [ChenFu11]):

Proposition 2.7 (cf. [BlockiGreen], see also [HerbortGreen] for a weaker result) *Let $\Omega \subset\subset \mathbb{C}^n$ be a pseudoconvex domain. Suppose there is a negative psh function ρ on Ω satisfying*

$$C_1 \delta_\Omega^a(z) \leq -\rho(z) \leq C_2 \delta_\Omega^b(z), \quad z \in \Omega$$

where $C_1, C_2 > 0$ and $a \geq b \geq 0$ are constants. Then there are positive numbers δ_0, C such that

$$\{g_\Omega(\cdot, w) < -1\} \subset \{C^{-1}\delta_\Omega(w)^{\frac{a}{b}} |\log \delta_\Omega(w)|^{-\frac{1}{b}} \leq \delta_\Omega \leq C\delta_\Omega(w)^{\frac{b}{a}} |\log \delta_\Omega(w)|^{\frac{n}{a}}\}$$

holds for any $w \in \Omega$ *with* $\delta_\Omega(w) \leq \delta_0$.

For planar domains, I showed the following

Theorem 2.8 (cf. [Chen00]) *Let* Ω *be a bounded domain in* $\mathbb{C}$. *If* Ω *is Bergman exhaustive, then it is Bergman complete.*

The converse does not hold. Zwonek [ZwonekExample] has constructed a Bergman complete Zalcman type domain, which is not Bergman exhaustive. By a Zalcman type domain we mean a planar domain defined by

$$\Delta \backslash \left(\bigcup_j \overline{\Delta}_j \cup \{0\} \right)$$

where $\{\Delta_j\}$ is a sequence of disjoint discs in the unit disc Δ. Zwonek's example also disproved an old conjecture due to Kobayashi [Kobayashi59] that Bergman completeness implies

$$\lim_{z \to \partial\Omega} |f(z)|^2 / K_\Omega(z) = 0$$

for *all* $f \in \mathscr{H}$. It is still unclear whether the converse of Kobayashi's criterion fails.

A characterization in terms of logarithmic capacity for Bergman exhaustive planar domains was given by Zownek:

Theorem 2.9 (cf. [ZwonekWiener]) *Let* Ω *be a bounded domain in* $\mathbb{C}$ *and* $p \in \partial\Omega$. *Then*

$$\lim_{z \to p} K_\Omega(z) = \infty$$

if and only if

$$\gamma_\Omega(z) := \int_{\delta_\Omega(z)}^{1/2} \frac{dt}{t^3 |\log(\text{cap}(\overline{\Delta_t(z)} \backslash D))|} \to \infty \quad \text{as } z \to p.$$

Here $\Delta_t(z)$ *stands for the disc with center* z *and radius* t.

Similar results on the Bergman metric were obtained in Pflug and Zwonek [PflugZwonek03]. By using these results, Wang [XuWang] was able to show that Bergman completeness is not a *quasiconformal* invariant for bounded planar domains. It is a classical result that (Green) hyperbolicity is a quasiconformal invariant for open Riemann surfaces.

It is well-known that every hyperbolic planar domain admits a canonical complete conformally invariant metric: the Poincaré metric of constant curvature -1. The following question is of classical interest:

Problem 2.2 What are relationships between the Bergman metric and the Poincaré metric?

I have not learnt any example that the Bergman metric is not dominated by the Poincaré metric. On the positive side, one has the following

Theorem 2.10 (cf. [ChenEssay]) *The Bergman metric and the Poincaré metric are equivalent on uniformly perfect domains. Both distances grow like* $|\log \delta_\Omega|$ *near* $\partial\Omega$.

A hyperbolic domain $\Omega \subset \mathbb{C}$ is said to be *uniformly perfect* if there exists a constant $c > 0$ such that for any boundary point $p \in \partial\Omega$ and $0 < r < \mathrm{diam}\partial\Omega$ there is a point $q \in \partial\Omega$ such that $cr \leq |q - p| \leq r$. For instance, the complement of the $\frac{1}{3}$−Cantor set in Δ is uniformly perfect. There are many equivalent definitions of uniform perfectness, as well as various interesting examples, among them of particular interest is the complement in $\mathbb{P}^1$ of the Julia set of a rational function of degree at least two (cf. [SugawaPerfect]).

We refer to [Wolpert, NikolovPflugZwonek] for various interesting results concerning the comparison of the Bergman metric with other invariant metrics on higher dimensional domains (usually with a highly complicated boundary).

The Bergman kernel and metric are deeply studied for some unbounded domains, e.g., Siegel domains of the second kind. Another interesting class of unbounded domains are model domains defined by

$$\Omega_\psi := \{(z', z_n) \in \mathbb{C}^n : \mathrm{Im}\, z_n > \psi(z')\}$$

where ψ is a psh function in $\mathbb{C}^{n-1}$.

Problem 2.3 When is Ω_ψ Bergman complete?

The answer is positive when ψ satisfies $\psi > 0$ and

$$\lim_{|z'| \to +\infty} \psi(z') = +\infty$$

(cf. [ChenKamimotoOhsawa], see also [PflugZwonek05] for related results). The case when ψ has singularities is more complicated and interesting. For instance, we have $K_{\Omega_\psi} = 0$ if $\psi(z') = \log|z'|$, whereas $K_\psi > 0$ if

$$\psi(z') \sim \log|z'| \text{ as } z' \to 0 \quad \text{and} \quad \psi(z') \sim |z| \text{ as } |z'| \to +\infty.$$

Recently, Ahn-Gaussier-Kim obtained a closely related result:

Theorem 2.11 (cf. [AhnKim]) *Let* Ω_{KN} *be the Kohn-Nirenberg domain defined by*

$$\Omega_{\mathrm{KN}} = \{(z_1, z_2) \in \mathbb{C}^2 : \mathrm{Im}\, z_2 > P_{2k}(z_1)\}$$

where P_{2k} *is a real-valued polynomial in* z_1 *and* $\bar{z}_1$ *satisfying (1)* $P_{2k}(tz) = t^{2k} P_{2k}(z)$ *for any* $t \in \mathbb{R}$ *and* $z \in \mathbb{C}$. *(2)* $\partial^2 P_{2k}/\partial z \partial \bar{z} > 0$ *on* $\mathbb{C}^* = \mathbb{C} - \{0\}$. *Then* Ω_{KN} *is complete with respect to the Carathéodory and Bergman metrics.*

3 Bergman Completeness for Open Complex Manifolds

For complex manifolds, one has to deal at first with the existence problem of the Bergman kernel or metric. The following is a classical one:

Theorem 3.1 (cf. [AhlforsSario]) *Every non-planar Riemann surface admits a nonzero square integrable holomorphic* $1-$*form, i.e., the Bergman kernel does not vanish.*

One of the most interesting class of open complex manifolds are universal coverings of a compact complex manifold with an infinite fundamental group. Suppose $\tilde{M}$ is a complex manifold and Γ is a free, properly discontinuous subgroup of the automorphism group $\mathrm{Aut}(\tilde{M})$ of $\tilde{M}$ such that $M := \tilde{M}/\Gamma$ is compact. The first Chern number c_1 of M is negative provided that $\tilde{M}$ possesses the Bergman metric. From the opposite direction, one may propose the following

Problem 3.1 Let M be a compact complex $n-$manifold with an infinite fundamental group and $c_1 < 0$. Is the Bergman kernel of the universal covering $\tilde{M}$ of M nonvanishing?

The answer is positive when $n \leq 2$. The case $n = 1$ is trivial. The proof for $n = 2$ is due to Claudon [Claudon]. It follows from Atiyah's L^2 index theorem and Miyaoka-Yau's inequality $c_2 \geq c_1^2/3$:

$$h^{2,0}_{(2)}(\tilde{M}) - h^{1,0}_{(2)}(\tilde{M}) + h^{0,0}_{(2)}(\tilde{M}) = \chi_{(2)}(\mathscr{O}_{\tilde{M}}) = \chi(\mathscr{O}_M) = \frac{c_1^2 + c_2}{12} \geq \frac{c_1^2}{9} > 0.$$

Every L^2 holomorphic function f on $\tilde{M}$ has to be constant in view of a L^p-Liouville theorem of Yau [Yau76]. Since $\tilde{M}$ is of infinite volume, f has to be zero, i.e., $h^{0,0}_{(2)}(\tilde{M}) = 0$, so that $h^{2,0}_{(2)}(\tilde{M}) > 0$, i.e., there exists a nonzero holomorphic $2-$form on $\tilde{M}$.

Conversely, I would like to ask

Problem 3.2 Let Ω be a bounded pseudoconvex domain in $\mathbb{C}^n$ with $n \geq 2$ (e.g. a bounded symmetric domain) and Γ a free, properly discontinuous subgroup of the automorphism group $\mathrm{Aut}(\Omega)$ of Ω. When does Ω/Γ possess a nonzero square integrable holomorphic $n-$form?

Kobayashi proposed the following criterion for the existence of the Bergman metric:

Proposition 3.1 (cf. [Kobayashi59]) *A complex manifold M possesses the Bergman metric provided the following two conditions are verified:*

(1) *For every* $w \in M$*, there exists a* n*−form* $f \in \mathscr{H}$ *such that* $f(w) \neq 0$*;*
(2) *For every* $w \in M$*, there are* n*−forms* $f_1, \ldots, f_n$ *in* $\mathscr{H}$ *satisfying* $f_\alpha(w) = 0$*,* $\frac{\partial f_\alpha^*}{\partial z_\beta}(w) = \delta_{\alpha\beta}$ *(Kronecker delta) for* $1 \leq \alpha, \beta \leq n$*. Here* f_α^**,* $1 \leq \alpha \leq n$*, are local representations of* f*.*

The concept of the pluricomplex Green function may be extended to complex manifolds, which depends only on the complex structure of the manifold. A criterion in terms of the pluricomplex Green function can be given as follows:

Proposition 3.2 (cf. [ChenZhang]) *Let* M *be a Stein manifold. Suppose for any* $w \in M$ *there is a positive number* $c > 0$ *such that* $\{g_M(\cdot, w) < -c\}$ *is relatively compact in* M*. Then* M *possesses the Bergman metric.*

From this we immediately know that every hyperbolic Riemann surface possesses the Bergman metric. Combining with a theorem of Carleson on removable singularities of L^2 holomorphic functions (see e.g., [Conway95]), we know that for any $\Omega \subset \mathbb{C}$

$$
\begin{aligned}
K_\Omega > 0 \text{ at one point } &\Longleftrightarrow K_\Omega > 0 \text{ everywhere} \\
&\Longleftrightarrow ds_\Omega^2 \text{ exists } \Longleftrightarrow \Omega \text{ is hyperbolic.}
\end{aligned}
$$

The situation is completely different for higher dimensional domains. Rosay and Rudin [RosayRudin] constructed a domain $\Omega \subset \mathbb{C}^2$ with finite volume, whereas there exists a surjective, locally biholomorphic map $F : \mathbb{C}^2 \to \Omega$. It follows that $K_\Omega(z) \geq 1/|\Omega| > 0$, i.e., $\partial\bar{\partial} \log K_\Omega$ is well-defined, whereas

$$
g_\Omega(z, w) \leq \inf \left\{ g_{\mathbb{C}^2}(\zeta, \eta) : \zeta \in F^{-1}(z),\ \eta \in F^{-1}(w) \right\} = -\infty
$$

for all $z, w \in \Omega$. It is unclear whether Ω can be made to be Bergman complete.

Let D be a parabolic domain and $U \subset\subset D$ a small disc. The domain $\Omega = D \backslash \overline{U}$ is hyperbolic so that it possesses the Bergman metric, which is not complete in view of Carleson's theorem. It is reasonable to ask

Problem 3.3[1] Let M be a parabolic Riemann surface and $U \subset\subset M$ a local coordinate disc. Is $M' := M \backslash \overline{U}$ always Bergman incomplete?

In their famous book [GreeneWuBook], Greene and Wu suggested to study the Bergman metric through Riemannian geometry. They proved the following

Theorem 3.2 (cf. [GreeneWuBook]) *Let* (M, g) *be a Kählerian Cartan-Hadamard manifold, let* o *be a fixed point in* M *and let* r *be the distance from* o*. Then*

[1]Recently, I got a counterexample.

(1) *If the inequality*

$$\text{sectional curvature} \le \frac{-A}{r^2(\log r)^{1-\varepsilon}}$$

holds outside a compact subset of M, where ε and A are positive constants, then M possesses the Bergman metric.

(2) *Suppose*

$$-\frac{B}{r^2} \le \text{sectional curvature} \le -\frac{A}{r^2}$$

holds outside a compact subset of M for some positive constants A and B, then the Bergman metric ds_M^2 satisfies $ds_M^2 \ge \text{const.}(1+r^2)^{-1}g$. In particular, M is Bergman complete.

(3) *Suppose*

$$-B \le \text{sectional curvature} \le -A$$

for some positive constants A and B, then $ds_M^2 \ge \text{const.}g$. In particular, M is Bergman complete.

Recall that a Cartan-Hadamard manifold is a complete, simply-connected Riemannian manifold of nonpositive sectional curvature. Greene-Wu conjectured that the hypothesized lower bound in (2) or (3) is unnecessary for the lower estimate of the Bergman metric, they even conjectured that M is Bergman complete under the assumptions in part (1).

In attempt to solve these conjectures, Zhang and I proved the following

Theorem 3.3 (cf. [ChenZhang]) *Let M be a Kählerian Cartan-Hadamard manifold, let o be a fixed point in M and let r be the distance from o. Then*

(1) *Suppose*

$$\text{sectional curvature} \le -\frac{A}{r^2}$$

outside a compact subset of M for suitable positive constant A, then the Bergman distance d_B satisfies

$$d_B(o,x) \ge \text{const.}\log r(x).$$

(2) *Suppose*

$$\text{sectional curvature} \le -A$$

for some positive constant A, then

$$d_B(o,x) \ge \text{const.}r(x).$$

Greene and Wu [GreeneWuBook] also showed that under the following weaker assumption

$$\text{sectional curvature} \leq \frac{-(1+\varepsilon)}{r^2 \log r}$$

outside a compact set, M has to be a *hyperconvex* manifold, i.e., there is a smooth strictly psh function $\rho : M \to [-1, 0)$ such that $\{\rho < -c\} \subset\subset M$ for all $c > 0$. Thus it is worthwhile to extend Theorem 2.5 as follows:

Theorem 3.4 (cf. [ChenHyperconvex]) *Every hyperconvex manifold is Bergman complete.*

Below we list some examples of hyperconvex manifolds beyond hyperconvex domains:

(a) Closed complex submanifolds of a hyperconvex domain Ω; these manifolds can be highly complicated even when Ω is the unit ball!
(b) Bounded pseudoconvex domains in $\mathbb{P}^n$ with a C^2 boundary (cf. [OhsawaSibony]).
(c) Sufficiently small neighborhoods of a totally real C^1 submanifold in a complex manifold (cf. [HarveyWells]).
(d) Regular coverings of a hyperconvex manifold (cf. [Vajaitu]).

Although the proof of the previous theorem is not basically different from [BlockiPflug], it still requires a few additional observations. Indeed, the following modified criterion for Bergman completeness was implicitly used:

Proposition 3.3 (cf. [ChenEssay]) *Let M be a Stein manifold which possesses the Bergman metric. Suppose that for any infinite sequence of points $\{p_k\}$ in M which has no adherent point in M, there are a subsequence $\{p_{k_j}\}$, a number $c > 0$ and a continuous volume form dV on M such that for any compact subset K of M, the related volume*

$$|K \cap \{g_M(\cdot, p_{k_j}) < -c\}|$$

tends to zero as $j \to \infty$, then M is Bergman complete.

Even for bounded hyperconvex domains, this criterion has the advantage of avoiding any use of the solution of the Monge-Ampere equation. Furthermore, it was used in [ChenEssay] to show that every Stein subvariety in a complex manifold admits a fundamental family of *Bergman complete* Stein neighborhoods, which improves a famous result of Siu [SiuNeighborhood].

As is well-known, every Stein manifold can be embedded holomorphically as a closed complex submanifold of some $\mathbb{C}^n$. It is natural to ask

Problem 3.4 Which closed complex submanifold of $\mathbb{C}^n$ is Bergman complete?

For instance, the preimage $\pi^{-1}(S) \subset \mathbb{C}^n$ of a smooth ample divisor S in an Abelian variety A is Bergman complete, where $\pi : \mathbb{C}^n \to A$ is the covering map (see e.g., [ChenEssay]). When $n > 2$, there is no nonconstant bounded holomorphic functions on $\pi^{-1}(S)$; I guess that the related pluricomplex Green function equals $-\infty$.

Even for a smooth analytic hypersurface M defined by $f = 0$ where f is an entire function in $\mathbb{C}^n$, it is still of great interest to find a criterion for Bergman completeness of M in terms of the function f.

Finally, let us look at Riemann surfaces from a different viewpoint. Consider at first an orientable surface M, i.e., a two-dimensional differentiable manifold. Let

$$ds^2 = E(x, y)dx^2 + 2F(x, y)dxdy + G(x, y)dy^2$$

where $EG - F^2 > 0$, $E > 0$, be a (smooth) Riemannian metric defined in local coordinates (x, y) of M. It is easy to see that every (paracompact) surface carries a (complete) Riemannian metric by means of patching up together local metrics through a partition of unity. By *isothermal parameters* we mean local coordinates (ξ, ζ) with $\xi = \xi(x, y)$, $\zeta = \zeta(x, y)$, such that

$$ds^2 = \lambda(\xi, \zeta)(d\xi^2 + d\zeta^2), \quad \lambda(\xi, \zeta) > 0.$$

Such isothermal parameters are known to exist by the famous Korn-Lichtenstein theorem, which goes back to Gauss. Thus M carries local complex coordinates $z = \xi + \zeta i$ so that it becomes a Riemann surface in classical sense. This observation is significant since the complex structure of a surface is often unknown, whereas the Riemannian metric can be analyzed through general theory of Riemannian geometry. From this viewpoint, assumptions relying on the complex structure are unnatural.

Now I formulate a basic problem:

Problem 3.5 Let M be an open Riemann surface with a complex structure induced by some complete Riemannian metric ds^2. Under which condition on ds^2 is the surface M Bergman complete?

As is well-known, popular conditions in Riemannian geometry are curvature, volume, etc. These are not strong enough for giving a criterion for Bergman completeness. Certain *global* condition is needed.

A nice global property of Riemannian manifolds is *isoperimetric inequalities*. Suppose M is a complete Riemannian n−manifold. Let $\mathscr{F}$ denote the set of precompact domains $\Omega \subset M$ with a smooth boundary. For $0 < \nu \leq \infty$, the ν−dimensional isoperimetric constant $I_\nu(M)$ of M is defined by

$$I_\nu(M) = \inf_{\Omega \in \mathscr{F}} |\partial\Omega|/|\Omega|^{1-1/\nu}.$$

Recently, I obtained the following

Theorem 3.5 (cf. [ChenRiemann]) *Let M be a complete Riemannian surface with the Gauss curvature bounded below by a constant. Let o be a point in M and r be the distance from o. Suppose either of the following conditions is verified:*

(1) $I_\nu(M) > 0$, *for* some $2 < \nu < \infty$;
(2) $I_\infty(M) > 0$ *and* $\inf_{x\in M} |B_1(x)| > 0$, *where* $B_a(x)$ *stands for the geodesic ball with center* x *and radius* a.

Then the Bergman distance d_B *satisfies*

$$d_B(o, x) \geq \text{const.}\, r(x).$$

Remark 3.1 (1) For the flat complex plane, one has $I_2(\mathbb{C}) > 0$, whereas $\mathbb{C}$ does not possess the Bergman metric. (2) For the punctured disc Δ^* with the Poincaré metric, one has $I_\infty(\Delta^*) > 0$, whereas the Bergman metric is not complete.

How to realize these assumptions? With respect the Poincaré metric, every uniformly perfect domain Ω has bounded geometry and $I_\infty(\Omega) > 0$ (cf. [SugawaPerfect]). Recall that a complete Riemannian manifold M has *bounded geometry* if the Ricci curvature is bounded below by a constant, and the injectivity radius is positive. Thus Ω satisfies the assumption in part (2). The point is that one may construct from Ω many open Riemannian surfaces verifying this assumption, based on the following beautiful discovery of Kanai:

Theorem 3.6 (cf. [KanaiRough]) *Let* M_1, M_2 *be complete Riemannian manifolds with bounded geometries such that they are roughly isometric to each other. Let* $\nu \geq \max\{\dim M_1, \dim M_2\}$. *Then* $I_\nu(M_1) > 0$ *if and only if* $I_\nu(M_2) > 0$.

Recall that a map $F : M_1 \to M_2$ between two Riemannian manifolds M_1 and M_2 is called a *rough isometry* if there are constants $a \geq 1$ and $b \geq 0$ such that

$$a^{-1} d_1(x, y) - b \leq d_2(F(x), F(y)) \leq a d_1(x, y) + b$$

for all $x, y \in M_1$, and F is ε*–full* for some number $\varepsilon > 0$, i.e.,

$$\bigcup_{x\in M_1} B_\varepsilon(F(x)) = M_2.$$

For instance, we learn from Kanai's theorem that every 2–dimensional jungle gym in $\mathbb{R}^n$ with $n > 2$ has a positive n–dimensional isoperimetric constant; similarly, every 2–dimensional jungle gym in a Cartan-Hadamard n–manifold ($n \geq 2$) with sectional curvature $\leq -A$ ($A > 0$) has a positive infinite-dimensional isoperimetric constant.

Problem 3.6 Let M be an open real surface. Does there always exist a complex structure on M such that the related Bergman metric is complete?

Acknowledgments The author would like to thank the organizers of KSCV10 for their invitation and the referee for pointing out many misprints in the manuscript.

References

[AhlforsSario] Ahlfors, L., Sario, L.: Riemann Sufaces. Princeton,New Jersey, Princeton University Press(1960)

[AhnKim] Ahn, T., Gaussier, H., Kim, K.-T.: Bergman and Carathéodory metrics of the Kohn-Nirenberg domains, arXiv:1406.3406

[BerndtssonLempert] Berndtsson, B., Lempert, L.: A proof of the Ohsawa-Takegoshi theorem with sharp estimates, arXiv:1407.4946v1

[Blocki93] Blocki, Z.: Estimates for the complex Monge-Ampere operator. Bull. Pol. Acad. Sci. **41**, 151–157 (1993)

[Blocki96] Blocki, Z.: The complex Monge-Ampere operator in hyperconvex domains. Annali della Scuola Norm. Sup. Pisa **23**, 721–747 (1996)

[BlockiGreen] Blocki, Z.: The Bergman metric and the pluricomplex Green function. Trans. Am. Math. Soc. **357**, 2613–2625 (2004)

[BlockiSuita] Blocki, Z.: Suita conjecture and the Ohsawa-Takegoshi extension theorem. Invent. Math. **193**, 149–158 (2013)

[BlockiBourgain] Z. Blocki, A lower bound for the Bergman kernel and the Bourgain-Milman inequality, GAFA Seminar Notes, Lect. Notes in Math., to appear

[BlockiPflug] Blocki, Z., Pflug, P.: Hyperconvexity and Bergman completeness. Nagoya Math. J. **151**, 221–225 (1998)

[Bremermann] Bremermann, H.J.: Holomorphic continuation of the kernel function and the Bergman metric in several complex variables. Lectures on Functions of a Complex Variable, Michigan (1955)

[Chen99] Chen, B.-Y.: Completeness of the Bergman metric on non-smooth pseudoconvex domains. Ann. Polon. Math. **71**, 241–251 (1999)

[Chen00] Chen, B.-Y.: A remark on the Bergman completeness. Complex Variables **42**, 11–15 (2000)

[ChenHyperconvex] Chen, B.-Y.: Bergman completeness of hyperconvex manifolds. Nagoya Math. J. **175**, 165–170 (2004)

[ChenEssay] Chen, B.-Y.: An essay on Bergman completeness. Ark. Mat. **51**, 269–291 (2013)

[ChenRiemann] Chen, B.-Y.: The Bergman metric and Isoperimetric inequalities, preprint

[ChenFu11] Chen, B.-Y., Fu, S.: Comparison of the Bergman and Szegö kernels. Adv. Math. **228**, 2366–2384 (2011)

[ChenKamimotoOhsawa] Chen, B.-Y., Kamimoto, J., Ohsawa, T.: Behavior of the Bergman kernel at infinity. Math. Z. **248**, 695–708 (2004)

[ChenZhang] Chen, B.-Y., Zhang, J.-H.: The Bergman metric on a Stein manifold with a bounded plurisubharmonic function. Trans. Am. Math. Soc. **354**, 2997–3009 (2002)

[Claudon] Claudon, B.: Non algebraicity of universal covers of Kähler surfaces. arXiv:1001.2379v1

[Conway95] Conway, J.B.: Functions of One Complex Variable II (GTM, vol. 159), Spinger, New York (1995)

[Demailly82] Demailly, J.-P.: Mesures de Monge-Ampere et mesures pluriharmoniques. Math. Z. **194**, 519–564 (1987)

[DiederichOhsawa] Diederich, K., Ohsawa, T.: An estimate for the Bergman distance on pseudoconvex domains. Ann. Math. **141**, 181–190 (1995)

[DonnellyFefferman] Donnelly, H., Fefferman, C.: L^2-cohomology and index theorem for the Bergman metric. Ann. Math. **118**, 593–618 (1983)

[GreeneWuBook] Greene, R.E., Wu, H.: Function Theory on Manifolds Which Possess a Pole, Lecture Notes in Mathematics, vol. 699. Springer, Berlin (1979)

[HarveyWells] Harvey, F.R., Wells, R.O.: Holomorphic approximation and hyperfunction theory on C^1 totally real submanifold of a complex manifold. Math. Ann. **197**, 287–318 (1972)
[HerbortHyperconvex] Herbort, G.: The Bergman metric on hyperconvex domains. Math. Z. **232**, 183–196 (1999)
[HerbortGreen] Herbort, G.: The pluricomplex Green function on pseudoconvex domains with a smooth boundary. Int. J. Math. **11**, 509–522 (2000)
[HormanderBook] Hörmander, L.: An introduction to Complex Analysis in Several Variables, North Holland, Amsterdam (1990)
[JarnickiPflug89] Jarnicki, M., Pflug, P.: Bergman completeness of complete circular domains. Ann. Pol. Math. **50**, 219–222 (1989)
[JarnickiPflug] Jarnicki, M., Pflug, P.: Invariant Distances and Metrics in Complex Analysis, De Gruyter Expositions in Math., vol. 9. Walter de Gruyter & Co., Berlin(1993)
[JarnickiPflug2] Jarnicki, M., Pflug, P.: Invariant Distances and Metrics in Complex Analysis-Revisited, Dissertationes Math., vol. 430, 192 pp (2005)
[KanaiRough] Kanai, M.: Rough isometries, and combinatorial approximations of geometries of noncompact Riemannian manifolds. J. Math. Soc. Jpn. **37**, 391–413 (1985)
[KerzmanRosay] Kerzman, N., Rosay, J.P.: Fonctions plurisousharmoniques d'exhaustion bornées et domaines taut. Math. Ann. **257**, 171–184 (1981)
[Kobayashi59] Kobayashi, S.: Geometry of bounded domains. Trans. Am. Math. Soc. **92**, 267–290 (1959)
[Kobayashi61] Kobayashi, S.: On complete Bergman metrics. Proc. Am. Math. Soc. **13**, 511–513 (1962)
[KobayashiBook98] Kobayashi, S.: Hyperbolic Complex spaces, A Series of Comprehensive Studies in Mathematics, vol. 318, Springer, Berlin (1998)
[Krushkal] Krushkal, S.L.: Strengthening pseudoconvexity of finite-dimentional Teichmüller spaces. Math. Ann. **290**, 681–687 (1991)
[NikolovPflugZwonek] Nikolov, N., Pflug, P., Zwonek, W.: Estimates for invariant metrics on $\mathbb{C}-$convex domains. Trans. Am. Math. Soc. **363**, 6245–6256 (2011)
[Ohsawa81] Ohsawa, T.: A remark on the completeness of the Bergman metric. Proc. Jpn. Acad. **57**, 238–240 (1981)
[Ohsawa84] Ohsawa, T.: Boundary behavior of the Bergman kernel function on pseudoconvex domains, vol. 20, pp. 897–902. Publ. RIMS, Kyoto Univ. (1984)
[OhsawaHyperconvex] Ohsawa, T.: On the Bergman kernel of hyperconvex domains. Nagoya Math. J. **129**, 43–52 (1993)
[Ohsawa95] Ohsawa, T.: Addendum to "On the Bergman kernel of hyperconvex domains". Nagoya Math. J. **137**, 145–148 (1995)
[OhsawaSibony] Ohsawa, T., Sibony, N.: Bounded p.s.h. functions and pseudoconvexity in Kähler manifolds. Nagoya Math. J. **149**, 1–8 (1998)
[Pflug75] Pflug, P.: Quadratintegrable holomorphe Functionen und die Serre-Vermutung. Math. Ann. **216**, 285–288 (1975)
[PflugZwonek05] Pflug, P., Zwonek, W.: Bergman completeness of unbounded Hartogs domains. Nagoya Math. J. **180**, 121–133 (2005)
[PflugZwonek03] Pflug, P., Zwonek, W.: Logarithmic capacity and Bergman functions. Arch. Math. **80**, 536–552 (2003)
[RosayRudin] Rosay, J.-P., Rudin, W.: Holomorphic maps from $\mathbb{C}^n$ to $\mathbb{C}^n$. Trans. Am. Math. Soc. **310**, 47–86 (1988)
[SiuNeighborhood] Siu, Y.-T.: Every Stein subvariety admits a Stein neighborhood. Invent. Math. **38**, 89–100 (1976)

[SugawaPerfect] Sugawa, T.: Uniformly perfect sets: analytic and geometric aspects, Sugaku Expositions (2003)

[Vajaitu] Vâjâitu, V.: On locally hyperconvex morphisms, C. R. Acad. Sci. Paris Sér. I Math. **322**, 823–828 (1996)

[XuWang] Wang, X.: Bergman completeness is not a quasiconformal invariant. Proc. Am. Math. Soc. **141**, 543–548 (2003)

[Wolpert] Wolpert, S.A.: Weil-Petersson perspectives, Proc. Sympos. Pure Math., vol. 74, Am. Math. Soc., Procidence, RI, 2006

[Yau76] Yau, S.-T.: Some function-theoretic properties of complete Riemannian manifolds and their applications to geometry. Indiana Math. J. **25**, 659–670 (1976)

[ZwonekExample] Zwonek, W.: An example concerning Bergman completeness. Nagoya Math. J. **164**, 89–101 (2001)

[ZwonekWiener] Zwonek, W.: Wiener's type criterion for Bergman exhaustiveness. Bull. Pol. Acad. Sci. Math. **50**, 297–311 (2002)

Structure Theorems for Compact Kähler Manifolds with Nef Anticanonical Bundles

Jean-Pierre Demailly

Abstract This survey presents various results concerning the geometry of compact Kähler manifolds with numerically effective first Chern class: structure of the Albanese morphism of such manifolds, relations tying semipositivity of the Ricci curvature with rational connectedness, positivity properties of the Harder-Narasimhan filtration of the tangent bundle.

Keywords Compact Kähler manifold · Anticanonical bundle · Semipositive Ricci curvature · Ricci flat manifold · Rationally connected variety · Holonomy principle

1 Introduction and Preliminaries

The goal of this survey is to present in a concise manner several recent results concerning the geometry of compact Kähler manifolds with numerically effective first Chern class. Especially, we give a rather complete sketch of currently known facts about the Albanese morphism of such manifolds, and study the relations that tie semipositivity of the Ricci curvature with rational connectedness. Many of the ideas are borrowed from [DPS96, BDPP] and the recent PhD thesis of Cao [Cao13a, Cao13b].

Recall that a compact complex manifold X is said to be rationally connected if any two points of X can be joined by a chain of rational curves. A line bundle L is said to be hermitian semipositive if it can be equipped with a smooth hermitian metric of semipositive curvature form. A sufficient condition for hermitian semipositivity is that some multiple of L is spanned by global sections; on the other hand, the hermitian semipositivity condition implies that L is numerically effective (nef) in the sense of [DPS94], which, for X projective algebraic, is equivalent to saying that $L \cdot C \geq 0$ for every curve C in X. Examples contained in [DPS94] show that all three conditions are different (even for X projective algebraic). Finally, let us recall that a line bundle

J.-P. Demailly
Institut Fourier, Université de Grenoble I, BP74, 100 rue des Maths, 38402 Saint-Martin d'Hères, France
e-mail: jean-pierre.demailly@ujf-grenoble.fr

F. Bracci et al. (eds.), *Complex Analysis and Geometry*, Springer Proceedings in Mathematics & Statistics 144, DOI 10.1007/978-4-431-55744-9_8

$L \to X$ is said to be pseudoeffective if here exists a singular hermitian metric h on L such that the Chern curvature current $T = \mathrm{i}\Theta_{L,h} = -\mathrm{i}\partial\overline{\partial}\log h$ is non-negative; equivalently, if X is projective algebraic, this means that the first Chern class $c_1(L)$ belongs to the closure of the cone of effective $\mathbb{Q}$-divisors.

The (Chern-)Ricci curvature is the curvature of the anticanonical bundle $K_X^{-1} = \det(T_X)$, and by Yau's solution of the Calabi conjecture (see [Aub76, Yau78]), a compact Kähler manifold X has a hermitian semipositive anticanonical bundle K_X^{-1} if and only if X admits a Kähler metric ω with $\mathrm{Ricci}(\omega) \geq 0$. Let us first review some classical examples of varieties with K_X^{-1} nef.

(ZFCC) *Compact Kähler manifolds with zero first Chern class*

The celebrated Bogomolov-Kobayashi-Beauville theorem yields the structure of compact Kähler Ricci-flat manifolds ([Bog74a, Bog74b, Kob81, Bea83]) which, by Yau's theorem [Yau78], are precisely compact Kähler manifolds with zero first Chern class. Recall that a *hyperkähler manifold* X is a simply connected compact Kähler manifold admitting a holomorphic symplectic 2-form σ (i.e. a holomorphic 2-form of maximal rank $n = 2p = \dim_{\mathbb{C}} X$ everywhere; in particular $K_X = \mathcal{O}_X$). A *Calabi-Yau* manifold is a simply connected projective manifold with $K_X = \mathcal{O}_X$ and $H^0(X, \Omega_X^p) = 0$ for $0 < p < n = \dim X$. Sometimes, finite étale quotient of such manifolds are also included in these classes (so that $\pi_1(X)$ is finite and possibly non trivial).

1.1 Theorem ([Bea83]) *Let (X, ω) be a compact Ricci flat Kähler manifold. Then there exists a finite étale Galois cover $\widehat{X} \to X$ such that*

$$\widehat{X} = T \times \prod Y_j \times \prod S_k$$

where $T = \mathbb{C}^q/\Lambda = \mathrm{Alb}(\widehat{X})$ is the Albanese torus of $\widehat{X}$, and Y_j, S_k are compact simply connected Kähler manifolds of respective dimensions n_j, n'_k with irreducible holonomy, Y_j being Calabi-Yau manifolds (holonomy group $= \mathrm{SU}(n_j)$) and S_k holomorphic symplectic manifolds (holonomy group $= \mathrm{Sp}(n'_k/2)$).

(RC-NAC) *Rationally connected manifolds with nef anticanonical class*

A classical example of projective surface with K_X^{-1} nef is the complex projective plane $\mathbb{P}^2_{\mathbb{C}}$ blown-up in 9 points $\{a_j\}_{1\leq j\leq 9}$. By a trivial dimension argument, there always exist a cubic curve $C = \{P(z) = 0\}$ containing the 9 points, and we assume that C is nonsingular (hence a smooth elliptic curve). Let $\mu : X \to \mathbb{P}^2$ the blow-up map, $E_j = \mu^{-1}(a_j)$ the exceptional divisors and $\widehat{C}$ the strict transform of C. One has

$$K_X = \mu^* K_{\mathbb{P}^2} \otimes \mathcal{O}_X(\textstyle\sum E_j),$$

thus

$$K_X^{-1} = \mu^* \mathcal{O}_{\mathbb{P}^2}(3) \otimes \mathcal{O}_X(-\sum E_j) = \mathcal{O}_X(\widehat{C}),$$
$$\widehat{L} := (K_X^{-1})_{|\widehat{C}} = (\mu_{|\widehat{C}})^* L$$

where $L := \mathcal{O}_C(3) \otimes \mathcal{O}_C(-\sum[a_j]) \in \mathrm{Pic}^0(C)$. As a consequence we have $K_X^{-1} \cdot \widehat{C} = (\widehat{C})^2 = 0$. For any other irreducible curve Γ in X, we find $K_X^{-1} \cdot \Gamma = \widehat{C} \cdot \Gamma \geq 0$, therefore K_X^{-1} is nef. There is a non trivial section in $H^0(\widehat{C}, \widehat{L}^{\otimes m})$ if and only if L is a m-torsion point in $\mathrm{Pic}^0(C)$ (i.e. iff L has rational coordinates with respect to the periods of $\widehat{C}$), and in that case, it is easy to see that this section extends to a section of $H^0(X, K_X^{-m})$ (cf. e.g. [DPS96]). This also means that there is an elliptic pencil $\alpha P(z)^m + \beta Q_m(z) = 0$ defined by a fibration

$$\pi_m = Q_m / P^m : X \to \mathbb{P}^1,$$

where $Q_m \in H^0(\mathbb{P}^2, \mathcal{O}(3m))$ vanishing at order m at all points a_j; the generic fiber of π_m is then a singular elliptic curve of multiplicity m at a_j, and we have $K_X^{-m} = (\pi_m)^* \mathcal{O}_{\mathbb{P}^1}(1)$, in particular K_X^{-m} is generated by its sections and possesses a real analytic metric of semipositive curvature. Now, when $L \notin \mathrm{Pic}^0(C)$ (corresponding to a generic position of the 9 points a_j on C), Ueda has analyzed the structure of neighborhoods of $\widehat{C}$ in X, and shown that it depends on a certain following diophantine condition for the point $\lambda \in H^1(C, \mathcal{O}_C)/H^1(C, \mathbb{Z})$ on the Jacobian variety of C associated with L (cf. [Ued82, p. 595], see also [Arn76]). This condition can be written

$$-\log d(m\lambda, 0) = O(\log m) \quad \text{as } m \to +\infty, \tag{1.1}$$

where d is a translation invariant geodesic distance on the Jacobian variety. Especially, (1.1) is independent of the choice of d and is satisfied on a set of full measure in $\mathrm{Pic}^0(C)$. When this is the case, Ueda has shown that $\widehat{C}$ possesses a "pseudoflat neighborhood", namely an open neighborhood U on which there exists a pluriharmonic function with logarithmic poles along $\widehat{C}$. Relying on this, Brunella [Bru10] has proven

1.2 Theorem *Let X, C, L be as above and assume that L is not a torsion point in* $\mathrm{Pic}^0(C)$. *Then*

(a) There exists on X a smooth Kähler metric with semipositive Ricci curvature if and only if $\widehat{C}$ admits a pseudoflat neighborhood in X.
(b) There does not exist on X a real analytic Kähler metric with semipositive Ricci curvature.

It seems likely (but is yet unproven) that $\widehat{C}$ does not possess pseudoflat neighborhoods when (0.2) badly fails, e.g. when the coordinates of λ with respect to periods are some sort of Liouville numbers like $\sum 1/10^{n!}$. Then, K_X^{-1} would be a nef line bundle without any smooth semipositive hermitian metric[1]. It might still be possible that there always exist singular hermitian metrics with zero Lelong numbers (and thus with trivial multiplier ideal sheaves) on such a rational surface, but this seems to be an open question as well. In general, the example of ruled surface over an elliptic curve given in [DPS94, Example 1.7] shows that such metrics with zero Lelong numbers need not always exist when K_X^{-1} is nef, but we do not know the answer when X is rationally connected. Studying in more depth the class of rationally connected projective manifolds with nef or semipositive anticanonical bundles is thus very desirable.

2 Criterion for Rational Connectedness

We give here a criterion characterizing rationally connected manifolds X in terms of positivity properties of invertible subsheaves contained in Ω_X^p or $(T_X^*)^{\otimes p}$; this is only a minor variation of Theorem 5.2 in [Pet06].

2.1 Critertion *Let X be a projective algebraic n-dimensional manifold. The following properties are equivalent.*

(a) X is rationally connected.
(b) For every invertible subsheaf $\mathscr{F} \subset \Omega_X^p := \mathscr{O}(\Lambda^p T_X^*)$, $1 \le p \le n$, $\mathscr{F}$ is not pseudoeffective.
(c) For every invertible subsheaf $\mathscr{F} \subset \mathscr{O}((T_X^*)^{\otimes p})$, $p \ge 1$, $\mathscr{F}$ is not pseudoeffective.
(d) For some (resp. for any) ample line bundle A on X, there exists a constant $C_A > 0$ such that

$$H^0(X, (T_X^*)^{\otimes m} \otimes A^{\otimes k}) = 0 \quad \text{for all } m,\ k \in \mathbb{N}^* \text{ with } m \ge C_A k.$$

Proof Observe first that if X is rationally connected, then there exists an immersion $f : \mathbb{P}^1 \subset X$ (in fact, many of them) passing through any given finite subset of X, and such that $f^* T_X$ is ample, see e.g. [Kol96, Theorem 3.9, p. 203]. It follows easily from there that 1.1 (a) implies 1.1 (d). The only non trivial implication that remains to be proved is that 1.1 (b) implies 1.1 (a). First note that K_X is not pseudoeffective, as one sees by applying the assumption 1.1 (b) with $p = n$. Hence X is uniruled by [BDPP]. We consider the quotient with maximal rationally connected fibers (rational quotient or MRC fibration, see [Cam92, KMM92])

$$f : X \dashrightarrow W$$

[1]Added in proof. In a very recent manuscript, Takayuki Koike has established the existence of such nef and non semipositive configurations, cf. arXiv:1507.00109, "Ueda theory for compact curves with nodes".

to a smooth projective variety W. By [GHS01], W is not uniruled, otherwise we could lift the ruling to X and the fibers of f would not be maximal. We may further assume that f is holomorphic. In fact, assumption 1.1 (b) is invariant under blow-ups. To see this, let $\pi : \hat{X} \to X$ be a birational morphisms from a projective manifold $\hat{X}$ and consider a line bundle $\hat{\mathscr{F}} \subset \Omega^p_{\hat{X}}$. Then $\pi_*(\hat{\mathscr{F}}) \subset \pi_*(\Omega^p_{\hat{X}}) = \Omega^p_X$, hence we introduce the line bundle

$$\mathscr{F} := (\pi_*(\hat{\mathscr{F}}))^{**} \subset \Omega^p_X.$$

Now, if $\hat{\mathscr{F}}$ were pseudoeffective, so would be $\mathscr{F}$. Thus 1.1 (b) is invariant under π and we may suppose f holomorphic. In order to show that X is rationally connected, we need to prove that $p := \dim W = 0$. Otherwise $K_W = \Omega^p_W$ is pseudoeffective by [BDPP], and we obtain a pseudo-effective invertible subsheaf $\mathscr{F} := f^*(\Omega^p_W) \subset \Omega^p_X$, in contradiction with 1.1 (b). □

2.2 Remark By [DPS94], assumptions 1.1 (b) and (c) make sense on arbitrary compact complex manifolds and imply that $H^0(X, \Omega^2_X) = 0$. If X is assumed to be compact Kähler, then X is automatically projective algebraic by Kodaira [Kod54], therefore, 1.1 (b) or (c) also characterize rationally connected manifolds among all compact Kähler ones. □

3 A Generalized Holonomy Principle

Recall that the restricted holonomy group of a hermitian vector bundle (E, h) of rank r is the subgroup $H \subset \mathrm{U}(r) \simeq U(E_{z_0})$ generated by parallel transport operators with respect to the Chern connection ∇ of (E, h), along loops based at z_0 that are contractible (up to conjugation, H does not depend on the base point z_0). The standard holonomy principle (see e.g. [BY53]) admits a generalized "pseudoeffective" version, which can be stated as follows.

3.1 Theorem *Let E be a holomorphic vector bundle of rank r over a compact complex manifold X. Assume that E is equipped with a smooth hermitian structure h and X with a hermitian metric ω, viewed as a smooth positive $(1,1)$-form $\omega = i \sum \omega_{jk}(z) dz_j \wedge d\overline{z}_k$. Finally, suppose that the ω-trace of the Chern curvature tensor $\mathrm{i}\Theta_{E,h}$ is semipositive, that is*

$$\mathrm{i}\Theta_{E,h} \wedge \frac{\omega^{n-1}}{(n-1)!} = B\,\frac{\omega^n}{n!}, \qquad B \in \mathrm{Herm}(E, E), \quad B \geq 0 \text{ on } X,$$

and denote by H the restricted holonomy group of (E, h).

(a) *If there exists an invertible sheaf $\mathscr{L} \subset \mathscr{O}((E^*)^{\otimes m})$ which is pseudoeffective as a line bundle, then $\mathscr{L}$ is flat and $\mathscr{L}$ is invariant under parallel transport by the* *connection of* $(E^*)^{\otimes m}$ *induced by the Chern connection* ∇ *of* (E,h)*; in fact, H acts trivially on* $\mathscr{L}$.

(b) *If H satisfies* $H = \mathrm{U}(r)$*, then none of the invertible subsheaves* $\mathscr{L}$ *of* $\mathscr{O}((E^*)^{\otimes m})$ *can be pseudoeffective for* $m \geq 1$.

Proof The semipositivity hypothesis on $B = \mathrm{Tr}_\omega \mathrm{i}\Theta_{E,h}$ is invariant by a conformal change of metric ω. Without loss of generality we can assume that ω is a Gauduchon metric, i.e. that $\partial\overline{\partial}\omega^{n-1} = 0$, cf. [Gau77]. We consider the Chern connection ∇ on (E,h) and the corresponding parallel transport operators. At every point $z_0 \in X$, there exists a local coordinate system $(z_1,\ldots,z_n)$ centered at z_0 (i.e. $z_0 = 0$ in coordinates), and a holomorphic frame $(e_\lambda(z))_{1\leq\lambda\leq r}$ such that

$$\langle e_\lambda(z), e_\mu(z)\rangle_h = \delta_{\lambda\mu} - \sum_{1\leq j,k\leq n} c_{jk\lambda\mu} z_j \overline{z}_k + O(|z|^3), \quad 1 \leq \lambda,\mu \leq r,$$

$$\Theta_{E,h}(z_0) = \sum_{1\leq j,k,\lambda,\mu\leq n} c_{jk\lambda\mu} dz_j \wedge d\overline{z}_k \otimes e_\lambda^* \otimes e_\mu, \quad c_{kj\mu\lambda} = \overline{c_{jk\lambda\mu}},$$

where $\delta_{\lambda\mu}$ is the Kronecker symbol and $\Theta_{E,h}(z_0)$ is the curvature tensor of the Chern connection ∇ of (E,h) at z_0.

Assume that we have an invertible sheaf $\mathscr{L} \subset \mathscr{O}((E^*)^{\otimes m})$ that is pseudoeffective. There exist a covering U_j by coordinate balls and holomorphic sections f_j of $\mathscr{L}_{|U_j}$ generating $\mathscr{L}$ over U_j. Then $\mathscr{L}$ is associated with the Čech cocycle g_{jk} in $\mathscr{O}_X^*$ such that $f_k = g_{jk} f_j$, and the singular hermitian metric $e^{-\varphi}$ of $\mathscr{L}$ is defined by a collection of plurisubharmonic functions $\varphi_j \in \mathrm{PSH}(U_j)$ such that $e^{-\varphi_k} = |g_{jk}|^2 e^{-\varphi_j}$. It follows that we have a globally defined bounded measurable function

$$\psi = e^{\varphi_j} \|f_j\|^2 = e^{\varphi_j} \|f_j\|^2_{h^{*m}}$$

over X, which can be viewed also as the ratio of hermitian metrics $(h^*)^m/e^{-\varphi}$ along $\mathscr{L}$, i.e. $\psi = (h^*)^m_{|\mathscr{L}} e^{\varphi}$. We are going to compute the Laplacian $\Delta_\omega \psi$. For simplicity of notation, we omit the index j and consider a local holomorphic section f of $\mathscr{L}$ and a local weight $\varphi \in \mathrm{PSH}(U)$ on some open subset U of X. In a neighborhood of an arbitrary point $z_0 \in U$, we write

$$f = \sum_{\alpha\in\mathbb{N}^m} f_\alpha\, e_{\alpha_1}^* \otimes \ldots \otimes e_{\alpha_m}^*, \qquad f_\alpha \in \mathscr{O}(U),$$

where (e_λ^*) is the dual holomorphic frame of (e_λ) in $\mathscr{O}(E^*)$. The hermitian matrix of (E^*,h^*) is the transpose of the inverse of the hermitian matrix of (E,h), hence we get

$$\langle e_\lambda^*(z), e_\mu^*(z)\rangle_h = \delta_{\lambda\mu} + \sum_{1\leq j,k\leq n} c_{jk\mu\lambda} z_j \overline{z}_k + O(|z|^3), \qquad 1 \leq \lambda,\mu \leq r.$$

On the open set U the function $\psi = (h^*)^m_{|\mathscr{L}} e^\varphi$ is given by

$$\psi = \Big(\sum_{\alpha\in\mathbb{N}^m} |f_\alpha|^2 + \sum_{\alpha,\beta\in\mathbb{N}^m,\ 1\le j,k\le n,\ 1\le\ell\le m} f_\alpha \overline{f_\beta}\, c_{jk\beta_\ell\alpha_\ell} z_j \overline{z}_k + O(|z|^3)|f|^2 \Big) e^{\varphi(z)}.$$

By taking $\mathrm{i}\partial\overline{\partial}(...)$ of this at $z = z_0$ in the sense of distributions (that is, for almost every $z_0 \in X$), we find

$$\begin{aligned}\mathrm{i}\partial\overline{\partial}\psi =& e^\varphi \Big(|f|^2 \mathrm{i}\partial\overline{\partial}\varphi + i\langle \partial f + f\partial\varphi, \partial f + f\partial\varphi\rangle \\ &+ \sum_{\alpha,\beta,j,k,1\le\ell\le m} f_\alpha \overline{f_\beta}\, c_{jk\beta_\ell\alpha_\ell}\, i dz_j \wedge d\overline{z}_k \Big).\end{aligned}$$

Since $\mathrm{i}\partial\overline{\partial}\psi \wedge \frac{\omega^{n-1}}{(n-1)!} = \Delta_\omega \psi \frac{\omega^n}{n!}$ (we actually take this as a definition of Δ_ω), a multiplication by ω^{n-1} yields the fundamental inequality

$$\Delta_\omega \psi \ge |f|^2 e^\varphi (\Delta_\omega \varphi + m\lambda_1) + |\nabla_h^{1,0} f + f\partial\varphi|^2_{\omega,h^{*m}}\, e^\varphi$$

where $\lambda_1(z) \ge 0$ is the lowest eigenvalue of the hermitian endomorphism $B = \mathrm{Tr}_\omega \mathrm{i}\Theta_{E,h}$ at an arbitrary point $z \in X$. As $\partial\overline{\partial}\omega^{n-1} = 0$, we have

$$\int_X \Delta_\omega \psi \,\frac{\omega^n}{n!} = \int_X \mathrm{i}\partial\overline{\partial}\psi \wedge \frac{\omega^{n-1}}{(n-1)!} = \int_X \psi \wedge \frac{\mathrm{i}\partial\overline{\partial}(\omega^{n-1})}{(n-1)!} = 0$$

by Stokes' formula. Since $\mathrm{i}\partial\overline{\partial}\varphi \ge 0$, the above inequality implies $\Delta_\omega \varphi = 0$, i.e. $\mathrm{i}\partial\overline{\partial}\varphi = 0$, and $\nabla_h^{1,0} f + f\partial\varphi = 0$ almost everywhere. This means in particular that the line bundle $(\mathscr{L}, e^{-\varphi})$ is flat. In each coordinate ball U_j the pluriharmonic function φ_j can be written $\varphi_j = w_j + \overline{w}_j$ for some holomorphic function $w_j \in \mathscr{O}(U_j)$, hence $\partial\varphi_j = dw_j$ and the condition $\nabla_h^{1,0} f_j + f_j \partial\varphi_j = 0$ can be rewritten $\nabla_h^{1,0}(e^{w_j} f_j) = 0$ where $e^{w_j} f_j$ is a local holomorphic section. This shows that $\mathscr{L}$ must be invariant by parallel transport and that the local holonomy of the Chern connection of (E, h) acts trivially on $\mathscr{L}$. Statement 2.1 (a) follows.

Finally, if we assume that the restricted holonomy group H of (E, h) is equal to $\mathrm{U}(r)$, there cannot exist any holonomy invariant invertible subsheaf $\mathscr{L} \subset \mathscr{O}((E^*)^{\otimes m})$, $m \ge 1$, on which H acts trivially, since the natural representation of $\mathrm{U}(r)$ on $(\mathbb{C}^r)^{\otimes m}$ has no invariant line on which $\mathrm{U}(r)$ induces a trivial action. Property 2.1 (b) is proved. □

4 Structure Theorem for Compact Kähler Manifolds with K_X^{-1} Semipositive

In this context, the following generalization of the Bogomolov-Kobayashi-Beauville Theorem 1.1 holds.

4.1 Structure Theorem *Let X be a compact Kähler manifold with K_X^{-1} hermitian semipositive. Then there exists a finite étale Galois cover $\widehat{X} \to X$ such that*

$$\widehat{X} \simeq \mathbb{C}^q/\Lambda \times \prod Y_j \times \prod S_k \times \prod Z_\ell$$

where $\mathbb{C}^q/\Lambda = \mathrm{Alb}(\widehat{X})$ is the Albanese torus of $\widehat{X}$, and Y_j, S_k, Z_ℓ are compact simply connected Kähler manifolds of respective dimensions n_j, n'_k, n''_ℓ with irreducible holonomy, Y_j being Calabi-Yau manifolds (holonomy $\mathrm{SU}(n_j)$), S_k holomorphic symplectic manifolds (holonomy $\mathrm{Sp}(n'_k/2)$), and Z_ℓ rationally connected manifolds with $K_{Z_\ell}^{-1}$ semipositive (holonomy $\mathrm{U}(n''_\ell)$).

Proof The proof relies on our generalized holonomy principle, combined with De Rham's splitting theorem [DR52] and Berger's classification [Ber55]. Foundational background can be found in papers by Lichnerowicz [Lic67, Lic71], and Cheeger and Gromoll [CG71, CG72].

We suppose here that X is equipped with a Kähler metric ω such that $\mathrm{Ricci}(\omega) \geq 0$, and we set $n = \dim_{\mathbb{C}} X$. We consider the holonomy representation of the tangent bundle $E = T_X$ equipped with the hermitian metric $h = \omega$. Here

$$B = \mathrm{Tr}_\omega \mathrm{i}\Theta_{E,h} = \mathrm{Tr}_\omega \mathrm{i}\Theta_{T_X,\omega} \geq 0$$

is nothing but the Ricci operator. Let $\widetilde{X} \to X$ be the universal cover of X and

$$(\widetilde{X}, \omega) \simeq \prod (X_i, \omega_i)$$

be the De Rham decomposition of $(\widetilde{X}, \omega)$, induced by a decomposition of the holonomy representation in irreducible representations. Since the holonomy is contained in $\mathrm{U}(n)$, all factors (X_i, ω_i) are Kähler manifolds with irreducible holonomy and holonomy group $H_i \subset \mathrm{U}(n_i)$, $n_i = \dim X_i$. By Cheeger and Gromoll [CG71], there is possibly a flat factor $X_0 = \mathbb{C}^q$ and the other factors X_i, $i \geq 1$, are compact and simply connected. Also, the product structure shows that each $K_{X_i}^{-1}$ is hermitian semipositive. By Berger's classification of holonomy groups [Ber55] there are only three possibilities, namely $H_i = \mathrm{U}(n_i)$, $H_i = \mathrm{SU}(n_i)$ or $H_i = \mathrm{Sp}(n_i/2)$. The case $H_i = \mathrm{SU}(n_i)$ leads to X_i being a Calabi-Yau manifold, and the case $H_i = \mathrm{Sp}(n_i/2)$ implies that X_i is holomorphic symplectic (see e.g. [Bea83]). Now, if $H_i = \mathrm{U}(n_i)$, the generalized holonomy principle 2.1 (b) shows that none of the invertible subsheaves $\mathscr{L} \subset \mathscr{O}((T^*_{X_i})^{\otimes m})$ can be pseudoeffective for $m \geq 1$. Therefore X_i is rationally connected by Criterion 2.1.

It remains to show that the product decomposition descends to a finite cover $\widehat{X}$ of X. However, the fundamental group $\pi_1(X)$ acts by isometries on the product, and does not act at all on the rationally connected factors Z_ℓ which are simply connected. Thanks to the irreducibility, the factors have to be preserved or permuted by any element $\gamma \in \pi_1(X)$, and the group of isometries of the factors S_j, Y_j are finite (since $H^0(Y, T_Y) = 0$ for such factors and the remaining discrete group $\mathrm{Aut}(Y)/\mathrm{Aut}^0(Y)$

is compact). Therefore, there is a subgroup Γ_0 of finite index in $\pi_1(X)$ which acts trivially on all factors except $\mathbb{C}^q$. By Bieberbach's theorem, there is a subgroup Γ of finite index in Γ_0 that acts merely by translations on $\mathbb{C}^q$. After taking the intersection of all conjugates of Γ in $\pi_1(X)$, we can assume that Γ is normal in $\pi_1(X)$. Then, if Λ is the lattice of translations of $\mathbb{C}^q$ defined by Γ, the quotient $\widehat{X} = \widetilde{X}/\Gamma$ is the finite étale cover of X we were looking for. □

Thanks to the exact sequence of fundamental groups associated with a fibration, we infer

4.2 Corollary *Under the assumptions of Theorem 4.1, there is an exact sequence*

$$0 \to \mathbb{Z}^{2q} \to \pi_1(X) \to G \to 0$$

where G is a finite group, namely $\pi_1(X)$ is almost abelian and is an extension of a finite group G by the normal subgroup $\pi_1(\widehat{X}) \simeq \mathbb{Z}^{2q}$.

5 Compact Kähler Manifolds with Nef Anticanonical Bundles

In this section, we investigate the properties of compact Kähler manifolds possessing a numerically effective anticanonical bundle K_X^{-1}. A simple but crucial observation made in [DPS93] is

5.1 Proposition *Let X be compact Kähler manifold and $\{\omega\}$ a Kähler class on X. Then the following properties are equivalent:*

(a) K_X^{-1} *is nef.*

(b) For every $\varepsilon > 0$, there exists a Kähler metric $\omega_\varepsilon = \omega + i\partial\overline{\partial}\varphi_\varepsilon$ in the cohomology class $\{\omega\}$ such that $\mathrm{Ricci}(\omega_\varepsilon) \geq -\varepsilon\omega$.

(c) For every $\varepsilon > 0$, there exists a Kähler metric $\omega_\varepsilon = \omega + i\partial\overline{\partial}\varphi_\varepsilon$ in the cohomology class $\{\omega\}$ such that $\mathrm{Ricci}(\omega_\varepsilon) \geq -\varepsilon\omega_\varepsilon$.

Sketch of Proof The nefness of K_X^{-1} means that $c_1(X) = c_1(K_X^{-1})$ contains a closed $(1,1)$-form ρ_ε with $\rho_\varepsilon \geq -\varepsilon\omega$, so (b) implies (a); the converse is true by Yau's theorem [Yau78] asserting the existence of Kähler metrics $\omega_\varepsilon \in \{\omega\}$ with prescribed Ricci curvature $\mathrm{Ricci}(\omega_\varepsilon) = \rho_\varepsilon$. Since $\omega_\varepsilon \equiv \omega$, (c) implies

$$c_1(X) + \varepsilon\{\omega\} \ni \rho'_\varepsilon := \mathrm{Ricci}(\omega_\varepsilon) + \varepsilon\omega_\varepsilon \geq 0,$$

hence (c) implies (a). The converse (a) ⇒ (c) can be seen to hold thanks to the solvability of Monge-Ampère equations of the form $(\omega + i\partial\overline{\partial}\varphi)^n = \exp(f + \varepsilon\varphi)$, due to Aubin [Aub76]. □

By using standard methods of Riemannian geometry such as the Bishop-Gage inequality for the volume of geodesic balls, one can then show rather easily that the fundamental group $\pi_1(X)$ has subexponential growth. This was improved by M. Păun in his PhD thesis, using more advanced tools (Gromov-Hausdorff limits and results of Cheeger and Colding [CC96, CC97], as well as the fundamental theorem of Gromov on groups of polynomial growth [Gr81a, Gr81b]).

5.2 Theorem ([Pau97, Pau98]) *Let X be a compact Kähler manifold with K_X^{-1} nef. Then $\pi_1(X)$ has polynomial growth and, as a consequence (thanks to Gromov) it possesses a nilpotent subgroup of finite index.*

We next study stability issues. Recall that the *slope* of a non zero torsion-free sheaf $\mathscr{F}$ with respect to a Kähler metric ω is

$$\mu_\omega(\mathscr{F}) = \frac{1}{\operatorname{rank}(\mathscr{F})}\int_X c_1(\mathscr{F}) \wedge \omega^{n-1}.$$

Moreover, $\mathscr{F}$ is said to be ω-stable (in the sense of Mumford-Takemoto) if $\mu_\omega(\mathscr{S}) < \mu_\omega(\mathscr{F})$ for every torsion-free subsheaf $\mathscr{S} \subset \mathscr{F}$ with $0 < \operatorname{rank}(\mathscr{S}) < \operatorname{rank}(\mathscr{F})$. In his PhD thesis [Cao13a, Cao13b], Junyan Cao observed the following important fact.

5.3 Theorem ([Cao13a, Cao13b]) *Let (X, ω) be a compact n-dimensional Kähler manifold such that K_X^{-1} is nef. Let*

$$0 = \mathscr{F}_0 \subset \mathscr{F}_1 \subset \cdots \subset \mathscr{F}_s = T_X$$

be a Harder-Narasimhan filtration of T_X with respect to ω, namely a filtration of torsion-free subsheaves such that $\mathscr{F}_i/\mathscr{F}_{i-1}$ is ω-stable with maximal slope in $T_X/\mathscr{F}_{i-1}$ [it is then well known that $i \mapsto \mu_\omega(\mathscr{F}_i/\mathscr{F}_{i-1})$ is a non increasing sequence]. Then

$$\mu_\omega(\mathscr{F}_i/\mathscr{F}_{i-1}) \geq 0 \quad \textit{for all } i.$$

Proof First consider the case where the filtration is regular, i.e., all sheaves $\mathscr{F}_i$ and their quotients $\mathscr{F}_i/\mathscr{F}_{i-1}$ are vector bundles. By the stability condition, it is sufficient to prove that

$$\int_X c_1(T_X/\mathscr{F}_i) \wedge \omega^{n-1} \geq 0 \quad \text{for all } i.$$

By 4.1 (b), for each $\varepsilon > 0$, there is a metric $\omega_\varepsilon \in \{\omega\}$ such that $\operatorname{Ricci}(\omega_\varepsilon) \geq -\varepsilon\omega_\varepsilon$. This is equivalent to the pointwise estimate

$$\mathrm{i}\Theta_{T_X,\omega_\varepsilon} \wedge \omega_\varepsilon^{n-1} \geq -\varepsilon \cdot \mathrm{Id}_{T_X}\omega_\varepsilon^n.$$

Taking the induced metric on $T_X/\mathscr{F}_i$ (which we also denote by ω_ε), the second fundamental form contributes nonnegative terms on the quotient, hence the ω_ε-trace yields

$$\mathrm{i}\Theta_{T_X/\mathscr{F}_i,\omega_\varepsilon} \wedge \omega_\varepsilon^{n-1} \geq -\varepsilon\,\mathrm{rank}(T_X/\mathscr{F}_i)\,\omega_\varepsilon^n.$$

Therefore, putting $r_i = \mathrm{rank}(T_X/\mathscr{F}_i)$, we get

$$\begin{aligned}\int_X c_1(T_X/\mathscr{F}_i) \wedge \omega^{n-1} &= \int_X c_1(T_X/\mathscr{F}_i) \wedge \omega_\varepsilon^{n-1}\\ &\geq -\varepsilon r_i \int_X \omega_\varepsilon^n = -\varepsilon r_i \int_X \omega^n,\end{aligned}$$

and we are done. In case there are singularities, they occur only on some analytic subset $S \subset X$ of codimension 2. The first Chern forms calculated on $X \smallsetminus S$ extend as locally integrable currents on X and do not contribute any mass on S. The above calculations are thus still valid. □

By the results of Bando and Siu [BS94], all quotients $\mathscr{F}_i/\mathscr{F}_{i-1}$ possess a Hermite-Einstein metric h_i that is smooth in the complement of the analytic locus S of codimension at least 2 where the $\mathscr{F}_i$ are not regular subbundles of T_X. Assuming ω normalized so that $\int_X \omega^n = 1$, we thus have

$$\Theta_{\mathscr{F}_i/\mathscr{F}_{i-1},h_i} \wedge \omega^{n-1} = \mu_i \mathrm{Id}_{\mathscr{F}_i/\mathscr{F}_{i-1}} \omega^n$$

where $\mu_i \geq 0$ is the corresponding slope. Using this, one easily obtains:

5.4 Corollary *Let (X, ω) be a compact Kähler manifold with K_X^{-1} nef, and S the analytic set of codimension ≥ 2 in X where the Harder-Narasimhan filtration of T_X with respect to ω is not regular. If a section $\sigma \in H^0(X, (T_X^*)^{\otimes m})$ vanishes at some point $x \in X \smallsetminus S$, it must vanish identically.*

Proof By dualizing the filtration of T_X and taking the m-th tensor product, we obtain a filtration

$$0 = \mathscr{G}_0 \subset \mathscr{G}_1 \subset \cdots \subset \mathscr{G}_N = (T_X^*)^{\otimes m}$$

such that all slopes $\mu_i = \mu_\omega(\mathscr{G}_i/\mathscr{G}_{i-1})$ satisfy $0 \geq \mu_1 \geq \ldots \geq \mu_N$. Now, if u is a section of a hermitian vector bundle $(\mathscr{G}, h)$ of slope $\mu \leq 0$, a standard calculation shows that

$$\Delta_\omega(\log \|u\|_h^2) = \mathrm{i}\partial\overline{\partial} \log \|u\|_h^2 \wedge \frac{\omega^{n-1}}{(n-1)!} \geq \|\nabla_h u\|_h^2 \frac{\omega^n}{n!} \geq 0.$$

By the maximum principle $\|u\|_h$ must be constant, and also u must be h-parallel, and if $\mu < 0$, the strict inequality for the trace of the curvature implies in fact $u \equiv 0$.

For $\mu = 0$ and $u \not\equiv 0$, any equality $u(x) = 0$ at a point where h does not blow up would lead to a non constant subharmonic function $\log \|u\|_h$ with a $-\infty$ pole on $X \smallsetminus S$, contradiction. From this, we conclude by descending induction starting with $i = N - 1$ that the image of σ in $H^0(X, (T_X^*)^{\otimes m}/\mathcal{G}_i)$ vanishes identically, hence σ lies in fact in $H^0(X, \mathcal{G}_i)$, and we proceed inductively by looking at its image in $H^0(X, \mathcal{G}_i/\mathcal{G}_{i-1})$. □

The next result has been first proved by Zhang [Zha96] in the projective case, and by Păun [Pau12] in the general Kähler case. We give here a different proof based on the ideas of Junyan Cao (namely, on Theorem 5.3 and Corollary 5.4).

5.5 Corollary *Let (X, ω) be a compact Kähler manifold with nef anticanonical bundle. Then the Albanese map $\alpha : X \to \mathrm{Alb}(X)$ is surjective, and smooth outside a subvariety of codimension at least 2. In particular, the fibers of the Albanese map are connected and reduced in codimension 1.*

Proof Let $\sigma_1, \ldots, \sigma_q \in H^0(X, \Omega_X^1)$ be a basis of holomorphic 1-forms. The Albanese map is obtained by integrating the σ_j's and the differential of α is thus given by $d\alpha = (\sigma_1, \ldots, \sigma_q) : T_X \to \mathbb{C}^q$. Hence α is a submersion at a point $x \in X$ if and only if no non trivial linear combination $\sigma = \sum \lambda_j \sigma_j$ vanishes at x. This is the case if $x \in X \smallsetminus S$. In particular α has generic rank equal to q, and must be surjective and smooth in codimension 1. The connectedness of fibers is a standard fact (α cannot descend to a finite étale quotient because it induces an isomorphism at the level of the first homology groups). □

A conjecture attributed to Mumford states that a projective or Kähler manifold X is rationally connected if and only if $H^0(X, (T_X^*)^{\otimes m}) = 0$ for all $m \geq 1$. As an application of the above results of J. Cao, it is possible to confirm this conjecture in the case of compact Kähler manifolds with nef anticanonical bundles.

5.6 Proposition *Let X be a compact Kähler n-dimensional manifold with nef anticanonical bundle. Then the following properties are equivalent:*

(a) X is projective and rationally connected;
(b) for every $m \geq 1$, one has $H^0(X, (T_X^*)^{\otimes m}) = 0$;
(c) for every $m = 1, \ldots, n$ and every finite étale cover $\widehat{X}$ of X, one has $H^0(\widehat{X}, \Omega^m_{\widehat{X}}) = 0$.

Proof As already seen, (a) implies (b) and (c) (apply 1.1 (d) and the fact that X is simply connected). Now, for any $p : 1$ cover $\widehat{X} \to X$, by taking a "direct image tensor product", a non zero section of $H^0(\widehat{X}, \Omega^m_{\widehat{X}})$ would yield a non zero section of

$$(\Omega_X^m)^{\otimes p} \subset (T_X^*)^{\otimes mp},$$

thus (b) implies (c). It remains to show that (c) implies (a). Assume that (c) holds. In particular $H^0(X, \Omega_X^2) = 0$ and X must be projective by Kodaira. Fix an ample line bundle A on X and look at the Harder-Narasimhan filtration $(\mathscr{F}_i)_{0\le i\le s}$ of T_X with respect to any Kähler class ω. If all slopes are strictly negative, then for any $m \gg p > 0$ the tensor product $(T_X^*)^{\otimes m} \otimes A^p$ admits a filtration with negative slopes. In this circumstance, the maximum principle then implies that Criterion 2.1 (d) holds, therefore X is rationally connected. The only remaining case to be treated is when one of the slopes is zero, i.e. for every Kähler class there is a subsheaf $\mathscr{F}_\omega \subsetneq T_X$ such that $\int_X c_1(T_X/\mathscr{F}_\omega) \wedge \omega^{n-1} = 0$. Now, by standard lemmas on stability, these subsheaves $\mathscr{F}_\omega$ live in a finite number of families. Since the intersection number $\int_X c_1(T_X/\mathscr{F}) \wedge \omega^{n-1}$ does not change in a given irreducible component of such a family of sheaves, we infer (e.g. by Baire's theorem!) that there would exist a subsheaf $\mathscr{F} \subsetneq T_X$ and a set of Kähler classes $\{\omega\}$ with non empty interior in the Kähler cone, such that $\int_X c_1(T_X/\mathscr{F}) \wedge \omega^{n-1} = 0$ for all these classes. However, by taking variations of $(\omega + t\alpha)^{n-1}$ with $t > 0$ small, we conclude that the intersection product of the first Chern class $c_1(T_X/\mathscr{F})$ with any product $\omega^{n-2} \wedge \alpha$ vanishes. The Hard Lefschetz together with Serre duality now implies that $c_1(T_X/\mathscr{F})_{\mathbb{R}} \in H^2(X, \mathbb{R})$ is equal to zero. By duality, there is a subsheaf $\mathscr{G} \subset \Omega_X^1$ of rank $m = 1, \dots, n$ such that $c_1(\mathscr{G})_{\mathbb{R}} = 0$. By taking $\mathscr{L} = \det(\mathscr{G})^{**}$, we get an invertible subsheaf $\mathscr{L} \subset \Omega_X^m$ with $c_1(\mathscr{L})_{\mathbb{R}} = 0$. Since $h^1(X, \mathscr{O}_X) = h^0(X, \Omega_X^1) = 0$, some power $\mathscr{L}^p$ is trivial and we get a finite cover $\pi : \widehat{X} \to X$ such that $\pi^*\mathscr{L}$ is trivial. This produces a non zero section of $H^0(\widehat{X}, \Omega_{\widehat{X}}^m)$, contradiction. □

The following basic question is still unsolved (cf. also [DPS96]).

5.7 Problem *Let X be a compact Kähler manifold with K_X^{-1} pseudoeffective. Is the Albanese map $\alpha : X \to \mathrm{Alb}(X)$ a (smooth) submersion? Especially, is this always the case when K_X^{-1} is nef?*

By [DPS96] or Theorem 4.1, the answer is affirmative if K_X^{-1} is semipositive. More generally, the generalized Hard Lefschetz theorem of [DPS01] shows that this is true if K_X^{-1} is pseudoeffective and possesses a singular hermitian metric of nonnegative curvature with trivial multiplier ideal sheaf. The general nef case seems to require a very delicate study of the possible degenerations of fibers of the Albanese map (so that one can exclude them in the end). In this direction, Cao and Höring [CH13] recently proved the following

5.8 Theorem ([CH13]) *Assuming X compact Kähler with K_X^{-1} nef, the answer to Problem* 4.7 *is affirmative in the following cases:*

(a) $\dim X \le 3$;
(b) $q(X) = h^0(X, \mathscr{O}_X) = \dim X - 1$;
(c) $q(X) = h^0(X, \mathscr{O}_X) \ge \dim X - 2$ *and X is projective;*
(d) the general fiber F of $\alpha : X \to \mathrm{Alb}(X)$ *is a weak Fano manifold, i.e.* K_F^{-1} *is nef and big.*

In general, a deeper understanding of the behavior of Harder-Narasimhan filtrations of the tangent bundle of a compact Kähler manifold would be badly needed.

References

[Arn76] Arnol'd, V.I.: Bifurcations of invariant manifolds of differential equations, and normal forms of neighborhoods of elliptic curves. Funct. Anal. Appl. **10**, 249–259 (1976). English translation 1977

[Aub76] Aubin, T.: Equations du type Monge-Ampère sur les variétés kählériennes compactes. C. R. Acad. Sci. Paris Ser. A **283**, 119–121 (1976); Bull. Sci. Math. **102**, 63–95 (1978)

[BS94] Bando, S., Siu, Y.-T.: Stable sheaves and Einstein-Hermitian metrics. In: Mabuchi, T., Noguchi, J., Ochiai, T. (eds.) Geometry and Analysis on Complex Manifolds, pp. 39–50. World Scientific, River Edge (1994)

[Bea83] Beauville, A.: Variétés kähleriennes dont la première classe de Chern est nulle. J. Diff. Geom. **18**, 775–782 (1983)

[Ber55] Berger, M.: Sur les groupes d'holonomie des variétés à connexion affine des variétés riemanniennes. Bull. Soc. Math. Fr. **83**, 279–330 (1955)

[Bis63] Bishop, R.: A relation between volume, mean curvature and diameter. Amer. Math. Soc. Not. **10**, 364 (1963)

[BY53] Bochner, S., Yano, K.: Curvature and Betti Numbers. Annals of Mathematics Studies, No. 32, pp. ix+190. Princeton University Press, Princeton (1953)

[Bog74a] Bogomolov, F.A.: On the decomposition of Kähler manifolds with trivial canonical class. Math. USSR Sbornik **22**, 580–583 (1974)

[Bog74b] Bogomolov, F.A.: Kähler manifolds with trivial canonical class. Izvestija Akad. Nauk **38**, 11–21 (1974)

[BDPP] Boucksom, S., Demailly, J.-P., Paun, M., Peternell, T.: The pseudo-effective cone of a compact Kähler manifold and varieties of negative Kodaira dimension. arXiv:0405285 [math.AG]; J. Alg. Geometry **22**, 201–248 (2013)

[Bru10] Brunella, M.: On Kähler surfaces with semipositive Ricci curvature. Riv. Math. Univ. Parma (N.S.), 1, 441–450 (2010)

[Cam92] Campana, F.: Connexité rationnelle des variétés de Fano. Ann. Sci. Ec. Norm. Sup. **25**, 539–545 (1992)

[Cam95] Campana, F.: Fundamental group and positivity of cotangent bundles of compact Kähler manifolds. J. Alg. Geom. **4**, 487–502 (1995)

[CPZ03] Campana, F., Peternell, Th., Zhang, Q.: On the Albanese maps of compact Kähler manifolds. Proc. Amer. Math. Soc. **131**, 549–553 (2003)

[Cao13a] Cao, J.: A remark on compact Kähler manifolds with nef anticanonical bundles and its applications. arXiv:math.AG/1305.4397

[Cao13b] Cao, J.: Vanishing theorems and structure theorems on compact Kähler manifolds. PhD thesis, Université de Grenoble, defended at institut Fourier on 18 Sep 2013. https://tel.archives-ouvertes.fr/tel-00919536/document

[CH13] Cao, J., Höring, A.: Manifolds with nef anticanonical bundle. arXiv:math.AG/1305.1018

[CC96] Cheeger, J., Colding, T.H.: Lower bounds on Ricci curvature and almost rigidity of warped products. Ann. Math. **144**, 189–237 (1996)

[CC97] Cheeger J., Colding T.H.: On the structure of spaces with Ricci curvature bounded below. J. Differ. Geom., part I: **46**, 406–480 (1997), part II: **54**, 13–35 (2000), part III: **54**, 37–74 (2000)

[CG71] Cheeger, J., Gromoll, D.: The splitting theorem for manifolds of nonnegative Ricci curvature. J. Diff. Geom. **6**, 119–128 (1971)

[CG72] Cheeger, J., Gromoll, D.: On the structure of complete manifolds of nonnegative curvature. Ann. Math. **96**, 413–443 (1972)

[DPS93] Demailly, J.-P., Peternell, T., Schneider, M.: Kähler manifolds with numerically effective Ricci class. Compositio Math. **89**, 217–240 (1993)

[DPS94] Demailly, J.-P., Peternell, T., Schneider, M.: Compact complex manifolds with numerically effective tangent bundles. J. Alg. Geom. **3**, 295–345 (1994)

[DPS96] Demailly, J.-P., Peternell, T., Schneider, M.: Compact Kähler manifolds with hermitian semipositive anticanonical bundle. Compositio Math. **101**, 217–224 (1996)

[DPS01] Demailly, J.-P., Peternell, T., Schneider, M.: Pseudo-effective line bundles on compact Kähler manifolds. Internat. J. Math. **12**, 689–741 (2001)
[DR52] de Rham, G.: Sur la reductibilité d'un espace de Riemann. Comment. Math. Helv. **26**, 328–344 (1952)
[Gau77] Gauduchon, P.: Le théorème de l'excentricité nulle. C. R. Acad. Sci. Paris **285**, 387–390 (1977)
[GHS01] Graber, T., Harris, J., Starr, J.: Families of rationally connected varieties. J. Amer. Math. Soc. **16**, 57–67 (2003)
[Gr81a] Gromov, M.: Structures métriques pour les variétés riemanniennes. Cours rédigé par J. Lafontaine et P. Pansu, Textes Mathématiques, 1, vol. VII, p. 152. Paris, Cedic/Fernand Nathan (1981)
[Gr81b] Gromov, M.: Groups of polynomial growth and expanding maps, Appendix by J. Tits. Publ. I.H.E.S. 53, 53–78 (1981)
[KMM92] Kollár, J., Miyaoka, Y., Mori, S.: Rationally connected varieties. J. Alg. **1**, 429–448 (1992)
[Kob81] Kobayashi, S.: Recent results in complex differential geometry. Jber. dt. Math.-Verein. **83**, 147–158 (1981)
[Kob83] Kobayashi, S.: Topics in complex differential geometry. In: DMV Seminar, vol. 3. Birkhäuser (1983)
[Kod54] Kodaira, K.: On Kähler varieties of restricted type. Ann. of Math. **60**, 28–48 (1954)
[Kol96] Kollár, J.: Rational Curves on Algebraic Varieties. Ergebnisse der Mathematik und ihrer Grenzgebiete, 3. Folge, Band 32, Springer (1996)
[Lic67] Lichnerowicz, A.: Variétés kähleriennes et première classe de Chern. J. Diff. Geom. **1**, 195–224 (1967)
[Lic71] Lichnerowicz, A.: Variétés Kählériennes à première classe de Chern non négative et variétés riemanniennes à courbure de Ricci généralisée non négative. J. Diff. Geom. **6**, 47–94 (1971)
[Pau97] Păun, M.: Sur le groupe fondamental des variétés kählériennes compactes à classe de Ricci numériquement effective. C. R. Acad. Sci. Paris Sér. I Math. 324, 1249–254 (1997)
[Pau98] Păun, M.: Sur les variétés kählériennes compactes classe de Ricci numériquement effective. Bull. Sci. Math. **122**, 83–92 (1998)
[Pau01] Păun, M.: On the Albanese map of compact Kähler manifolds with numerically effective Ricci curvature. Comm. Anal. Geom. **9**, 35–60 (2001)
[Pau12] Păun, M.: Relative adjoint transcendental classes and the Albanese map of compact Kähler manifolds with nef Ricci classes. arxiv:1209.2195
[Pet06] Peternell, Th: Kodaira dimension of subvarieties II. Intl. J. Math. **17**, 619–631 (2006)
[PS98] Peternell, Th, Serrano, F.: Threefolds with anti canonical bundles. Coll. Math. **49**, 465–517 (1998)
[Ued82] Ueda, T.: On the neighborhood of a compact complex curve with topologically trivial normal bundle. J. Math. Kyoto Univ. 22, 583–607 (1982/83)
[Yau78] Yau, S.T.: On the Ricci curvature of a complex Kähler manifold and the complex Monge-Ampère equation I. Comm. Pure Appl. Math. **31**, 339–411 (1978)
[Zha96] Zhang, Q.: On projective manifolds with nef anticanonical bundles. J. Reine Angew. Math. **478**, 57–60 (1996)
[Zha05] Zhang, Q.: On projective varieties with nef anticanonical divisors. Math. Ann. **332**, 697–703 (2005)

An Estimate for the Squeezing Function and Estimates of Invariant Metrics

J.E. Fornæss and E.F. Wold

Abstract We give estimates for the squeezing function on strictly pseudoconvex domains, and derive some sharp estimates for the Carathéodory, Sibony and Azukawa metrics near their boundaries.

Keywords Invariant metrics · Squeezing function · Exposing points

1 Introduction

Let Ω be a bounded domain in $\mathbb{C}^n$. The squeezing function $S_\Omega(z)$, which was introduced in [Deng] inspired by [Liu04, Liu05, Yeun09], measures how much a domain looks like the unit ball observed from a given point $z \in \Omega$. More precisely it is defined as follows: For a given injective holomorphic map $f : \Omega \to \mathbb{B}^n$ satisfying $f(z) = 0$ we set

$$S_{\Omega,f}(z) := \sup\{r > 0 : r\mathbb{B}^n \subset f(\Omega)\},$$

and then we set

$$S_\Omega(z) := \sup_f \{S_{\Omega,f}(z)\},$$

where f ranges over all injective holomorphic maps $f : \Omega \to \mathbb{B}^n$ with $f(z) = 0$. Using the the method of exposing points from [Died14] and the method from [Frid95], it was proved in [Deng] that

$$\lim_{z \to b\Omega} S_\Omega(z) = 1$$

J.E. Fornæss
NTNU, Institutt for Matematiske Fag, 7491 Trondheim, Norway
e-mail: johnefo@math.ntnu.no

E.F. Wold (✉)
UiO, Matematisk Institutt, Postboks 1053 Blindern, 0316 Oslo, Norway
e-mail: erlendfw@math.uio.no

F. Bracci et al. (eds.), *Complex Analysis and Geometry*, Springer Proceedings in Mathematics & Statistics 144, DOI 10.1007/978-4-431-55744-9_9

if Ω is a $\mathscr{C}^2$-smooth strictly pseudoconvex domain, and it was proved in [KimZ] that the squeezing function is bounded on any bounded convex domain. Our goal is to improve this estimate when the boundary has higher regularity, and to give an application to invariant metrics.

Theorem 1.1 *Let $\Omega = \{\delta < 0\} \subset \mathbb{C}^n$ be a strictly pseudoconvex domain with a defining function δ of class $\mathscr{C}^k$ for $k \geq 3$. The squeezing function $S_\Omega(z)$ for Ω satisfies the estimate*

$$S_\Omega(z) \geq 1 - C \cdot \sqrt{|\delta(z)|}$$

for a fixed constant C. If we even have $k \geq 4$, then there exists a constant $C > 0$ such that the squeezing function $S_\Omega(z)$ for Ω satisfies

$$S_\Omega(z) \geq 1 - C \cdot |\delta(z)|$$

for all z

Combining with a theorem due to D. Ma [Ma92] and a result of Deng, Guan and Zhang [Deng], an immediate consequence is a sharp estimate for invariant metrics near the boundary of a strictly pseudoconvex domain. Before we state the result, we briefly recall the definitions of some invariant metrics. Let Δ denote the unit disc, and let $\mathscr{O}(M, N)$ denote the holomorphic maps from M to N.

- Kobayashi metric $K_\Omega(p, \xi)$. We define

$$K_\Omega(p, \xi) = \inf\{|\alpha|; \exists f \in \mathscr{O}(\Delta, \Omega), f(0) = p, \alpha f'(0) = \xi\}.$$

- Carathéodory metric $C_\Omega(p, \xi)$. We define

$$C_\Omega(p, \xi) = \sup\{|f'(p)(\xi)|; \exists f \in \mathscr{O}(\Omega, \Delta), f(p) = 0\}.$$

- Sibony metric $S_\Omega(p, \xi)$. We define $S_\Omega(p, \xi) = \sup\{(\sum_{i,j} \frac{\partial^2 u(p)}{\partial z_i \partial \overline{z}_j} \xi_i \overline{\xi}_j)^{1/2}, u(p) = 0, 0 \leq u < 1, u \text{ is } \mathscr{C}^2 \text{ near } p \text{ and } \ln u \text{ is plurisubharmonic in } \Omega\}$.
- Azukawa metric $A_U(p, \xi)$. We define

$$A_\Omega(p, \xi) = \sup_{u \in P_\Omega(p)} \{\limsup_{\lambda \searrow 0} \frac{1}{|\lambda|} u(p + \lambda\xi)\}$$

where

$$\begin{aligned} P_\Omega(p) = \{u : \Omega \to [0, 1), \ln u \text{ is plurisubharmonic and} \\ \exists M_u > 0, r_u > 0 \text{ such that} \\ \mathbb{B}^n(p, r) \subset \Omega, u(z) \leq M\|z - p\|, z \in \mathbb{B}^n(p, r)\}. \end{aligned}$$

Theorem 1.2 *Let $\Omega \subset \mathbb{C}^n$ be a strictly pseudoconvex domain of class $\mathscr{C}^3$, let $p \in b\Omega$, and let δ be a defining function for Ω near p, such that $\|\nabla\delta(z)\| = 1$ for all $z \in b\Omega$. Then if $F_\Omega(z, \zeta)$ is either the Carathéodory, Sibony or Azukawa metric, there exists a constant $C > 0$ such that*

$$(1 - C\sqrt{|\delta(z)|})\left[\frac{L_{\pi(z)}(\xi_T)}{|\delta(z)|} + \frac{\|\xi_N\|}{4\delta(z)^2}\right]^{1/2} \leq F_\Omega(z, \xi)$$

$$\leq (1 + C\sqrt{|\delta(z)|})\left[\frac{L_{\pi(z)}(\xi_T)}{|\delta(z)|} + \frac{\|\xi_N\|}{4\delta(z)^2}\right]^{1/2}$$

for all z near p, and all $\xi = \xi_N + \xi_T$, where π is the orthogonal projection to $b\Omega$, ξ_N is the complex normal component of ξ at $\pi(z)$ and ξ_T is the complex tangential component, and L is the Levi form of δ.

Ma's result is the corresponding statement for the Kobayashi metric, and the result is sharp in the sense that one cannot in general do better than the square root of the boundary distance.

2 Proof of Theorem 1.2

The following was proved in [Deng], and we include the proof for the benefit of the reader.

Lemma 2.1 *Let Ω be any bounded domain in $\mathbb{C}^n$, and let $F_\Omega(z, \xi)$ be either the Carathéodory, Sibony or Azukawa metric. Then*

$$S_\Omega(z)K_\Omega(z, \xi) \leq F_\Omega(z, \xi) \leq K_\Omega(z, \xi)$$

for all $z \in \Omega$ and all $\xi \in \mathbb{C}^n$, where K denotes the Kobayashi metric.

Proof It is well known that K dominates F so we need to show the lower estimate. Let $f : \Omega \to \mathbb{B}^n$ be injective holomorphic with $f(z) = 0$, such that $B_r \subset f(\Omega)$ where $r = S_\Omega(z)$. For the existence of f see [Deng] (alternatively one can use a limiting argument). We get that

$$F_\Omega(z, \xi) = F_{f(\Omega)}(0, f_*\xi) \geq F_{\mathbb{B}^n}(0, f_*\xi) = K_{\mathbb{B}^n}(0, f_*\xi)$$
$$= S_\Omega(z)K_{B_r}(0, f_*\xi) \geq S_\Omega(z)K_{f(\Omega)}(0, f_*\xi) = S_\Omega(z)K_\Omega(z, \xi).$$

□

Proof of Theorem 1.2: By Lemma 2.1 we have that

$$S_\Omega(z)K_\Omega(z,\xi) \le F_\Omega(z,\xi) \le K_\Omega(z,\xi).$$

Then combining Theorem 1.1 with the fact that Theorem 1.2 holds with $F_\Omega(z)$ replaced by $K_\Omega(z)$ (see [Ma92]) completes the proof. □

3 Proof of Theorem 1.1

The following provides the key geometric setup for the proof. Let $k = 3$ or 4, and let Ω be a bounded strongly pseudoconvex domain of class $\mathscr{C}^k$.

Lemma 3.1 *Let $p \in b\Omega$. There exists an injective holomorphic map $\phi : \overline{\Omega} \to \mathbb{C}^n$ such that $\tilde{\Omega} = \phi(\Omega)$ satisfies the following:*

(i) $\tilde{\Omega} \subset \mathbb{B}^n$,
(ii) $\phi(p) = (1, 0, \cdots, 0) =: a$ *and* $\phi^{-1}(b\mathbb{B}^n) = \{p\}$,
(iii) *near a we have that,* $\tilde{\Omega} = \{\rho < \mu^2\}, 0 < \mu < 1$ *where*

$$\rho(z) = |z_1 - (1-\mu)|^2 + \|z'\|^2 + O(|z_1 - 1|^2) + O(\|z-a\|^k).$$

Proof By the main theorem in [Died14] there exists a map ϕ such that (i) and (ii) are satisfied. That we can achieve (iii) follows from the proof which consists of three steps. We first apply an automorphism of $\mathbb{C}^n$ to ensure that, locally near $p = 0$, our domain has a defining function

$$\rho(z) = 2Re(z_1) + \|z\|^2 + O(\|z\|^k). \tag{3.1}$$

To achieve this one approximates a local map with jet interpolation using the Andersén-Lempert theory. We next apply another automorphism of $\mathbb{C}^n$ which can be chosen to match the identity at the origin to any given order, so we still have a defining function of the form (3.1). The final exposing map is of the form $\varphi = \phi \circ \alpha$, where $\phi(z) = (f(z_1), z_2, ..., z_n)$ where f is injective holomorphic with $f'(0) > 0$, and $\alpha(z)$ can be chosen to match the identity to any given order at the origin. By a translation we assume that $\varphi(0) = 0$. We then have a defining function for $\varphi(\Omega)$ of the form

$$\begin{aligned}\rho(z) &= 2Re(c_1z_1 + c_2z_1^2 + c_3z_1^3) + |c_1|^2|z_1|^2 + \|z'\|^2 + O(|z_1|^2) + O(\|z\|^k)\\ &= 2c_1Re(z_1) + |c_1|^2|z_1|^2 + \|z'\|^2 + O(|z_1|^2) + O(\|z\|^k).\end{aligned}$$

Applying the linear change of coordinates $(z_1, z') \mapsto (z_1/c_1, z')$, we get a defining function

$$\rho(z) = 2Re(z_1) + |z_1|^2 + \|z'\|^2 + O(|z_1|^2) + O(\|z\|^k).$$

By chosing a small $0 < \mu < 1$ we have that $\mu\varphi(\Omega)$ is contained in the translated unit ball $\{2Re(z_1) + \|z\|^2 < 0\}$, with defining function

$$\rho(z) = 2\mu Re(z_1) + |z_1|^2 + \|z'\|^2 + O(|z_1|^2) + O(\|z\|^k),$$

which is the same as (iii) when translated $(z_1, z') \mapsto (z_1 + 1, z')$. □

Remark 3.1 On $b\tilde{\Omega}$ the remainder term in (iii) is actually $O(|z_1 - 1|^{k/2})$. To see this we first translate $\tilde{\Omega}$ to the origin, set $\tilde{z}_1 = z_1 - 1$, $\tilde{z} = (\tilde{z}_1, z')$ so that it is defined by

$$\tilde{\rho}(\tilde{z}) = 2Re(\tilde{z}_1) + |\tilde{z}_1|^2 + \frac{1}{\mu}\|z'\|^2 + O(|\tilde{z}_1|^2) + O(\|\tilde{z}\|^k) < 0.$$

We estimate $\|z'\|$ on $\tilde{\rho} = 0$. If $\|z'\| \leq |\tilde{z}_1|$ the remainder term is less than $C|\tilde{z}_1|^k = O(|z_1|^{k/2})$. If $|\tilde{z}_1| \leq \|z'\|$ then the remainder term is $O(\|z'\|^k)$ and we get

$$\begin{aligned}\|z'\|^2 + O(\|z'\|^k) &= \mu(-2Re(\tilde{z}_1) - |\tilde{z}_1|^2 + O(|\tilde{z}_1|^2)) \\ &= \mu|\tilde{z}_1|(-\frac{2Re(\tilde{z}_1)}{|\tilde{z}_1|} - |\tilde{z}_1| + \frac{O(|\tilde{z}_1|^2)}{|\tilde{z}_1|}).\end{aligned}$$

This implies that the remainder term is $\mathcal{O}(\|z'\|^k) = \mathcal{O}(|z_1 - 1|^{k/2})$.

From now on we assume that $\Omega = \tilde{\Omega}$ and satisfies (i)–(iii) above. Then Ω is "almost" contained in the ball $B_\mu \subset \mathbb{B}^n$ defined by

$$|z_1|^2 + \frac{1}{\mu}\|z'\|^2 < 1.$$

We will use automorphisms of the ball $\mathbb{B}^n$ of the form

$$\phi_r(z_1, z') = \left(\frac{z_1 - r}{1 - rz_1}, \frac{\sqrt{1 - r^2}}{1 - rz_1}z'\right).$$

We have that ϕ_r leaves B_μ invariant. To prove the theorem, we will estimate two things:

(a) How much $\phi_r(\Omega)$ sticks out of B_μ and
(b) the size of the largest ball in B_μ contained in $\phi_r(\Omega)$.

3.1 Estimate (a)

Lemma 3.2 *There exists a constant $C > 0$ such that for $w \in b\phi_r(\Omega)$ we have that* $|w_1|^2 + \frac{1}{\mu}\|w'\|^2 \leq 1 + C(1 - r)^{\frac{k-2}{2}}$.

Proof We would like to express the maximum of the function $\|\phi_r(z)\|$ in terms of $(1-r)$ on $b\Omega$, *i.e.*, we look at

$$\|\phi_r(z)\|^2 = \frac{|z_1-r|^2+\frac{1}{\mu}(1-r^2)\|z'\|^2}{|1-rz_1|^2} = \frac{|z_1-r|^2}{|1-rz_1|^2} + \frac{1}{\mu}\frac{(1-r^2)|z'|^2}{|1-rz_1|^2}$$

for $z \in b\Omega$. Fix any $\eta > 0$. We show first that if $z \in \mathbb{B}^n$ with $|z_1-1| > \eta$, then we have a uniform estimate

$$\|\phi_r(z)\|^2 \leq 1 + C(1-r).$$

In this case we have that the denominator of the second term stays bounded independent of r, while $|z'| \leq 1$, hence the term goes to zero like $(1-r)$. For the other term we write

$$\frac{|z_1-r|^2}{|1-rz_1|^2} = 1 + \frac{(1-r^2)(|z_1|^2-1)}{|1-rz|^2} \leq 1 + C(1-r).$$

Next we look at $|z_1-1| \leq \eta$. If η is chosen small enough, the local description (iii) is valid. Hence if $|z_1-1| < \eta$ and if $z \in b\Omega$ we have that

$$\begin{aligned}\|z'\|^2 &= -|z_1-(1-\mu)|^2 + O(|z_1-a|^{k/2}) + \mu^2 \\ &= -|z_1-1|^2 - 2\mu Re(z_1-1) + O(|z_1-a|^{k/2}),\end{aligned}$$

which gives that

$$\begin{aligned}\frac{1}{\mu}\|z'\|^2 &= -\frac{1}{\mu}|z_1-1|^2 - 2Re(z_1-1) + O(|z_1-a|^{k/2}) \\ &\leq -|z_1-1|^2 - 2Re(z_1-1) + O(|z_1-a|^{k/2}) \\ &= 1-|z_1|^2 + O(|z_1-a|^{k/2}).\end{aligned}$$

Hence

$$\begin{aligned}&\frac{|z_1-r|^2+\frac{1}{\mu}(1-r^2)\|z'\|^2}{|1-rz_1|^2} \\ &= \frac{|z_1-r|^2+(1-r^2)(1-|z_1|^2)}{|1-rz_1|^2} + \frac{(1-r^2)O(|z_1-1|^{k/2})}{|1-rz_1|^2} \\ &= 1 + \frac{(1-r^2)O(|z_1-1|^{k/2})}{|1-rz_1|^2} \leq 1 + C\frac{(1-r^2)|1-rz_1|^{k/2}}{|1-rz_1|^2} \\ &\leq 1 + C_1\frac{1-r}{|1-rz_1|^{2-(k/2)}} \leq 1 + C_1\frac{1-r}{(1-r)^{2-(k/2)}} \\ &\leq 1 + C_2(1-r)^{\frac{k-2}{2}}.\end{aligned}$$

□

3.2 Estimate (b)

We define

$$B^{\mu}_{\eta,\tilde{\eta}} = \{|z_1 - (1-\eta)|^2 + \frac{\tilde{\eta}}{\mu}|z'|^2 < \eta^2\}$$

with constants $0 < \eta \leq \tilde{\eta} < 2\eta$.

Lemma 3.3 *We set* $\tilde{\eta} = \begin{cases} \eta, & k=4 \\ \frac{\eta}{1-C\eta}, & k=3 \end{cases}$.

(i) *If* $k=4$ *then* $B^{\mu}_{\eta,\tilde{\eta}} \subset \Omega$ *for all* η *small enough*

(ii) *If* $k=3$*, and the constant* $C>0$ *is fixed large enough, then* $B^{\mu}_{\eta,\tilde{\eta}} \subset \Omega$ *for all* η *small enough.*

Proof For η small enough, the ellipsoid $B^{\mu}_{\eta,\tilde{\eta}}$ is contained in the region where the local defining function ρ is defined. Since ρ is plurisubharmonic it suffices to show that $\rho \leq 0$ on $bB^{\mu}_{\eta,\tilde{\eta}}$. We translate coordinates, by setting $\tilde{z}_1 = z_1 - 1$ and $\tilde{z} = (\tilde{z}_1, z')$. We want to show that

$$\{2\eta Re(\tilde{z}_1) + |\tilde{z}_1|^2 + \frac{\tilde{\eta}}{\mu}|z'|^2 = 0\}$$

is contained in the set

$$\{2\mu Re(\tilde{z}_1) + |\tilde{z}_1|^2 + \|z'\|^2 + O(|\tilde{z}_1|^2) + O(\|\tilde{z}\|^k) \leq 0\}.$$

Write $\tilde{z}_1 = x_1 + iy_1$. On the boundary of the ellipsoid we have that

$$2\eta x_1 + x_1^2 + y_1^2 + \frac{\tilde{\eta}}{\mu}\|z'\|^2 = 0 \Leftrightarrow \frac{\mu}{\tilde{\eta}}y_1^2 + \|z'\|^2 = -\frac{\mu}{\tilde{\eta}}(2\eta x_1 + x_1^2)$$

and consequently we get on the boundary of the ellipsoid that

$$\begin{aligned}\|\tilde{z}\|^2 &= x_1^2 + y_1^2 + \|z'\|^2 \leq x_1^2 + \frac{\mu}{\tilde{\eta}}y_1^2 + \|z'\|^2 = x_1^2 - \frac{\mu}{\tilde{\eta}}(2\eta x_1 + x_1^2) \\ &= -x_1(-x_1 + \frac{\mu}{\tilde{\eta}}(2\eta + x_1)) .\end{aligned}$$

It follows that $\|\tilde{z}\|^2 \leq C|x_1|$, and so

$$\|\tilde{z}\|^k \leq C|x_1|^{k/2} . \tag{3.2}$$

Consider again the boundary of the ellipsoid; we have

$$x_1^2 + y_1^2 + 2\eta x_1 + \frac{\tilde{\eta}}{\mu}\|z'\|^2 = 0\ .$$

Hence

$$\|z'\|^2 = -\frac{\mu}{\tilde{\eta}}(x_1^2 + y_1^2 + 2\eta x_1)\ .$$

Therefore

$$\begin{aligned}2\mu x_1 + |\tilde{z}_1|^2 + \|z'\|^2 + O(|\tilde{z}_1|^2) + O(\|\tilde{z}\|^k)\\ \leq 2\mu x_1 + |\tilde{z}_1|^2 - \frac{\mu}{\tilde{\eta}}(|\tilde{z}_1|^2 + 2\eta x_1) + C_3|x_1|^{k/2} + C_4|\tilde{z}_1|^2.\end{aligned}$$

using (3.2). It suffices therefore to show that the right side is ≤ 0. This means:

$$2\mu x_1(1 - \frac{\eta}{\tilde{\eta}}) + |\tilde{z}_1|^2(1 - \frac{\mu}{\tilde{\eta}}) + C_3|x_1|^{k/2} + C_4|\tilde{z}_1|^2 \leq 0. \tag{3.3}$$

Observe that $\frac{1}{2} \leq \frac{\eta}{\tilde{\eta}} \leq 1$, so $1 - \frac{\eta}{\tilde{\eta}} \geq 0$. Morever $x_1 \leq 0$ on the translated ellipse. Hence the first term in (3.3) is ≤ 0. It suffices therefore that

$$|\tilde{z}_1|^2(1 - \frac{\mu}{\tilde{\eta}}) + C|\tilde{z}_1|^{k/2} \leq 0, \tag{3.4}$$

where we merged the constants C_3 and C_4. When $k = 4$, this holds as soon as η is small enough. When $k = 3$, this holds when

$$C|\tilde{z}_1|^{3/2} \leq \frac{\mu}{\tilde{\eta}}|\tilde{z}_1|^{1/2}|\tilde{z}_1|^{3/2}(1 - \frac{\tilde{\eta}}{\mu})$$

or

$$C \leq \frac{|\tilde{z}_1|^{1/2}}{\tilde{\eta}}(\mu - \tilde{\eta})$$

This holds when $|\tilde{z}_1| \geq \tilde{C}\eta^2$ for large enough $\tilde{C}$. To complete the proof we need to consider the case when $k = 3$ and $|\tilde{z}_1| \leq \tilde{C}\eta^2$, and we go back to consider the full expression (3.3). Since the sum $|\tilde{z}_1|^2(1 - \frac{\mu}{\tilde{\eta}}) + C_4|\tilde{z}_1|^2$ is negative when η is small, it is enough to determine when

$$2\mu x_1(1 - \frac{\eta}{\tilde{\eta}}) + C_3|x_1|^{3/2} \leq 0.$$

or equivalently when

$$2\mu x_1(1-\frac{\eta}{\tilde{\eta}}) \leq C_3 x_1|x_1|^{1/2} \Leftrightarrow 2\mu(1-\frac{\eta}{\tilde{\eta}}) \geq C_3|x_1|^{1/2}.$$

By our assumption we now have that $C_3|x_1|^{1/2} \leq C_3(\tilde{C}\eta^2)^{1/2} = C_5\eta$, and so we need that

$$2\mu(1-\frac{\eta}{\tilde{\eta}}) \geq C_5\eta.$$

Hence the choice $\tilde{\eta} = \frac{\eta}{1-\frac{C_5}{2\mu}\eta}$ works. □

Now let $\psi(z_1, z') = (z_1, \frac{1}{\sqrt{\mu}}z')$. Then $\psi(B^{\mu}_{\eta,\tilde{\eta}})$ is the ellipsoid

$$B^1_{\eta,\tilde{\eta}} = \{|z_1-(1-\eta)|^2 + \tilde{\eta}\|z'\|^2 < \eta^2\},$$

Lemma 3.4 *Let* $0 < \eta, r < 1$ *and* $\tilde{\eta} > 0$. *If* $z \in bB^1_{\eta,\tilde{\eta}}$, *then*

$$\|\phi_r(z_1,z')\|^2 = 1 + \frac{(1-r^2)|z_1-1|^2}{|1-rz|^2} - \frac{(1-r^2)(1/\tilde{\eta})|z_1-1|^2}{|1-rz_1|^2} + \frac{(1-r^2)2(1-\frac{\eta}{\tilde{\eta}})(Re(z_1)-1)}{|1-rz_1|^2}.$$

Proof

$$\begin{aligned}
\|\phi_r(z_1,z')\|^2 &= \frac{|z_1-r|^2+(1-r^2)|z'|^2}{|1-rz_1|^2} \\
&= \frac{|z_1-r|^2+(1-r^2)(1/\tilde{\eta})(\eta^2-|z_1-(1-\eta)|^2)}{|1-rz_1|^2} \\
&= \frac{|z_1-r|^2+(1-r^2)(1/\tilde{\eta})(\eta^2-|z_1-1|^2-2\eta Re(z_1-1)-\eta^2)}{|1-rz_1|^2} \\
&= \frac{|z_1-r|^2+(1-r^2)(1/\tilde{\eta})(-2\eta Re(z_1-1)-|z_1-1|^2)}{|1-rz_1|^2} \\
&= 1+\frac{|z_1-r|^2-|1-rz_1|^2-(1-r^2)\frac{2\eta}{\tilde{\eta}}Re(z_1-1)}{|1-rz_1|^2} \\
&\quad - \frac{(1-r^2)(1/\tilde{\eta})|z_1-1|^2}{|1-rz_1|^2} \\
&= 1+\frac{|z_1|^2-2rRe(z_1)+r^2-(1-2rRe(z_1)+r^2|z_1|^2)}{|1-rz_1|} \\
&\quad - \frac{(1-r^2)\frac{2\eta}{\tilde{\eta}}Re(z-1)}{|1-rz_1|^2} - \frac{(1-r^2)(1/\tilde{\eta})|z_1-1|^2}{|1-rz_1|^2}
\end{aligned}$$

$$
\begin{aligned}
&= 1 + \frac{(1-r^2)(|z_1|^2 - \frac{2\eta}{\eta}Re(z_1) + (\frac{2\eta}{\tilde{\eta}} - 1))}{|1-rz_1|^2} \\
&\quad - \frac{(1-r^2)(1/\tilde{\eta})|z_1 - 1|^2}{|1-rz_1|^2} \\
&= 1 + \frac{(1-r^2)|z_1-1|^2}{|1-rz|^2} - \frac{(1-r^2)(1/\tilde{\eta})|z_1-1|^2}{|1-rz_1|^2} \\
&\quad + \frac{(1-r^2)2(1-\frac{\eta}{\tilde{\eta}})(Re(z_1)-1)}{|1-rz_1|^2}
\end{aligned}
$$

□

Lemma 3.5 *Let* $\psi(z) = (z_1, \frac{1}{\sqrt{\mu}}z')$. *Suppose that* $0 < \eta, r < 1, 1 - 2\eta < r$ *and* $\tilde{\eta} > 0$. *Then* $\psi(\phi_r(B^{\mu}_{\eta,\tilde{\eta}}))$ *contains the ball of radius*

$$
\sqrt{1 - 2(1-r)\frac{1}{\tilde{\eta}} - 4|1 - \frac{\eta}{\tilde{\eta}}|}.
$$

Proof Since $1 - 2\eta < r$, we have that $0 \in \psi(\phi_r(B^{\mu}_{\eta,\tilde{\eta}}))$. Hence it suffices to show that $\|\psi(\phi_r)(z)\|^2 \geq 1 - 2(1-r)\frac{1}{\tilde{\eta}} - 4|1 - \frac{\eta}{\tilde{\eta}}|$ on the boundary of $B^{\mu}_{\eta,\tilde{\eta}}$. (This is nonempty if the expression on the right is nonnegative.) Since $\psi \circ \phi_r = \phi_r \circ \psi$, it suffices to show that $\|\phi_r(z)\|^2 \geq 1 - 2(1-r)\frac{1}{\tilde{\eta}} - 4|1 - \frac{\eta}{\tilde{\eta}}|$ on the boundary of $B^{1}_{\eta,\tilde{\eta}}$. From the previous lemma we have that

$$
\begin{aligned}
\|\phi_r(z)\|^2 &\geq 1 + \frac{(1-r^2)|z_1-1|^2}{|1-rz|^2} - \frac{(1-r^2)(1/\tilde{\eta})|z_1-1|^2}{|1-rz_1|^2} \\
&\quad + \frac{(1-r^2)2(1-\frac{\eta}{\tilde{\eta}})(Re(z_1)-1)}{|1-rz_1|^2} \\
&\geq 1 - \frac{(1-r^2)(1/\tilde{\eta})|z_1-1|^2}{|1-rz_1|^2} - \frac{(1-r^2)2|1-\frac{\eta}{\tilde{\eta}}||Re(z_1)-1|}{|1-rz_1|^2} \\
&\geq 1 - \frac{(2(1-r))(1/\tilde{\eta})|rz_1-1|^2}{|1-rz_1|^2} - \frac{(2(1-r))2|1-\frac{\eta}{\tilde{\eta}}||rz_1-1|}{|1-rz_1|^2} \\
&\geq 1 - (1-r))(2/\tilde{\eta}) - \frac{4|1-rz_1|)(1-\frac{\eta}{\tilde{\eta}})}{|1-rz_1|} \\
&\geq 1 - (1-r))(2/\tilde{\eta}) - 4|1 - \frac{\eta}{\tilde{\eta}}|
\end{aligned}
$$

□

We prove Theorem 1.1

Proof We will estimate the squeezing function at points $(r, 0)$ when $r < 1$ is close to 1. That this gives the uniform constant claimed in Theorem 1.1, follows from the dependence on p as p varies over the boundary of the original domain. In particular, the constants in our estimates can be chosen independently of the point p, and the radial lines will foliate a neighborhood of the boundary so that we get an estimate for all points near the boundary. The map $\psi \circ \phi_r$ maps $(r, 0)$ to the origin. We estimate the image of Ω.

It follows from Lemma 3.2 that there exists a constant $C > 0$ such that for $w \in b\phi_r(\Omega)$ we have that $|w_1|^2 + \frac{1}{\mu}\|w'\|^2 \leq 1 + C(1-r)^{\frac{k-2}{2}}$. Since the left side is plurisubharmonic, the same estimate holds by the maximum principle on $\overline{\phi_r(\Omega)}$. Suppose that $(z_1, z') \in \psi(\overline{\phi_r(\Omega)})$. Then $(z_1, z') = \psi(w_1, w') = (w_1, \frac{1}{\sqrt{\mu}}w')$ for some $w \in \overline{\phi_r(\Omega)}$. Hence $\|z\|^2 = |w_1|^2 + \frac{1}{\mu}\|w'\|^2 \leq 1 + C(1-r)^{\frac{k-2}{2}}$. It follows that $\psi(\phi_r(\Omega))$ is contained in the ball centered at the origin of radius $1 + C(1-r)^{\frac{k-2}{2}}$.

We next estimate the radius of the largest ball contained in $\psi(\phi_r(\Omega))$. By Lemma 3.3 we have ellipsoids $B^{\mu}_{\eta,\tilde{\eta}} = \{|z_1 - (1-\eta)|^2 + \frac{\tilde{\eta}}{\mu}|z'|^2 < \eta^2\}$ contained in Ω for certain $\eta, \tilde{\eta}$: We set

$$\tilde{\eta} = \begin{cases} \eta, & k = 4 \\ \frac{\eta}{1-C\eta}, & k = 3. \end{cases}$$

(i) If $k = 4$ we have that $B^{\mu}_{\eta,\tilde{\eta}} \subset \Omega$ for all η small enough, and

(ii) if $k = 3$, and the constant $C > 0$ is fixed large enough, then $B^{\mu}_{\eta,\tilde{\eta}} \subset \Omega$ for all η small enough. We can then estimate instead the largest ball contained in $\psi(\phi_r(B^{\mu}_{\eta,\tilde{\eta}}))$.

We use Lemma 3.5: Suppose that $0 < \eta, r < 1$, $1 - 2\eta < r$ and $\tilde{\eta} > 0$. Then $\psi(\phi_r(B^{\mu}_{\eta,\tilde{\eta}}))$ contains the ball of radius

$$\sqrt{1 - 2(1-r)\frac{1}{\tilde{\eta}} - 4|1 - \frac{\eta}{\tilde{\eta}}|}.$$

We deal first with the case $k = 4$. Then we assume that $1 - 2\eta < r$ and $\tilde{\eta} = \eta$. It follows that

$\psi(\phi_r(\Omega)) \supset \psi(\phi_r\left(B^{\mu}_{\eta,\tilde{\eta}}\right)) \supset \mathbb{B}(0, \sqrt{1 - 2(1-r)\frac{1}{\eta}})$. We choose a fixed η, and let $r \to 1$. We then get that for a fixed constant C', $\psi(\phi_r(\Omega)) \supset \mathbb{B}(0, 1 - C'(1-r))$. Hence we have shown that in the case $k = 4$, $\frac{k-2}{2} = 1$,

$$\mathbb{B}(0, 1 - C'(1-r)) \subset \psi(\phi_r(\Omega)) \subset \mathbb{B}(0, 1 + C(1-r)).$$

Composing with the map $\lambda(z) = \frac{z}{1+C(1-r)}$ we obtain that $\lambda(\psi(\phi_r(r,0))) = 0$ and that

$$\mathbb{B}\left(0, \frac{1-C'(1-r)}{1+C(1-r)}\right) \subset \lambda(\psi(\phi_r(\Omega))) \subset \mathbb{B}(0,1).$$

Hence it follows that the squeezing function at $(r,0)$ is at least $1-C''(1-r)$. Since the defining function $\delta(z) = -(1-r) + \mathcal{O}((1-r)^2)$ for $z = (r,0)$ and r close to 1, we obtain Theorem 1.1 in the case when $k = 4$.

It remains to do the case $k = 3$.

It follows as above that $\psi(\phi_r(\Omega))$ is contained in the ball centered at the origin of radius $1 + C(1-r)^{\frac{k-2}{2}} = 1 + C(1-r)^{\frac{1}{2}}$.

As above we suppose that $0 < \eta, r < 1$, $1-2\eta < r$, and we have that $\psi(\phi_r(B^{\mu}_{\eta,\tilde{\eta}}))$ contains the ball of radius

$$\sqrt{1 - 2(1-r)\frac{1}{\tilde{\eta}} - 4|1 - \frac{\eta}{\tilde{\eta}}|}.$$

We have that $\frac{\tilde{\eta}}{\eta} = 1 - C\eta$, and so it follows that

$$\begin{aligned}
\psi(\phi_r(\Omega)) \supset & \psi(\phi_r(B^{\mu}_{\eta,\tilde{\eta}})) \\
\supset & \mathbb{B}\left(0, \sqrt{1 - 2(1-r)\frac{1}{\tilde{\eta}} - 4C\eta}\right) \\
\supset & \mathbb{B}\left(0, \sqrt{1 - 2(1-r)\frac{1}{\eta} - 4C\eta}\right).
\end{aligned}$$

In this case, we let η depend on r. Set $\eta = \sqrt{1-r}$. Then $r = 1 - \eta^2 > 1 - 2\eta$ if r is close enough to 1. We then get that

$$\begin{aligned}
\psi(\phi_r(\Omega)) \supset & \mathbb{B}\left(0, \sqrt{1 - 2(1-r)\frac{1}{\eta} - 4C\eta}\right) \\
= & \mathbb{B}\left(0, \sqrt{1 - 2(1-r)\frac{1}{\sqrt{1-r}} - 4C\sqrt{1-r}}\right) \\
= & \mathbb{B}\left(0, \sqrt{1 - (2+4C)\sqrt{1-r}}\right) \\
\supset & \mathbb{B}\left(0, 1 - (2+4C)\sqrt{1-r}\right)
\end{aligned}$$

Now it follows by the same scaling type argument with a map λ that we get the desired lower bound for the squeezing function in the case $k = 3$. □

4 An Example

Let Ω be the domain $\Omega := \mathbb{B}^n \setminus \frac{1}{2}\overline{\mathbb{B}}^n$. We will show that $S_\Omega(z)$ cannot approach 1 faster than $1 - C\text{dist}(z, b\Omega)$. By abuse of notation we set $r = (r, 0, ..., 0), 0 < r < 1$ and we set $a = (1/2, 0, ..., 0)$. Then the Kobayashi distance with respect to $\mathbb{B}^n$ from a to r is $\frac{1}{2}(\log(\frac{1+r}{1-r}) - \log(3))$. Now let $f : \Omega \to \mathbb{B}^n$ be an injective holomorphic map with $f(r) = 0$. Then f extends to a holomorphic map $\tilde{f} : \mathbb{B}^n \to \mathbb{B}^n$, so by the decreasing property of the Kobayashi metric we have that the Kobayashi distance between $f(r)$ and $f(a)$ is less that $\frac{1}{2}\log(\frac{1+r}{1-r})$. It follows that $S_{\Omega,f}(r) \leq r$.

References

[Deng] Deng, F., Guan, Q., Zhang, L.: Some properties of squeezing functions on bounded domains. Pac. J. Math. **257**(2), 319–341 (2012)

[Died14] Diederich, K., Fornæss, J.E., Wold, E.F.: Exposing points on the boundary of a strictly pseudoconvex or a locally convexifiable domain of finite 1-type. J. Geom. Anal. **24**(4), 2124–2134 (2014)

[Frid95] Fridman, B., Ma, D.: On exhaustion of domains. Indiana Univ. Math. J. **44**(2), 385–395 (1995)

[KimZ] Kang-Tae, K., Zhang, L.: On the uniform squeezing property and the squeezing function. http://arxiv.org/pdf/1306.2390

[Liu04] Liu, K., Sun, X., Yau, S.T.: Canonical metrics on the moduli space of Riemann Surfaces I. J. Differ. Geom. **68**, 571–637 (2004)

[Liu05] Liu, K., Sun, X., Yau, S.T.: Canonical metrics on the moduli space of Riemann Surfaces II. J. Differ. Geom. **69**, 163–216 (2005)

[Ma92] Ma, D.: Sharp estimates for the Kobayashi metric near strongly pseudoconvex points. Contemp. Math. **137**, 329–338 (1992)

[Yeun09] Yeung, S.K.: Geometry of domains with the uniform squeezing property. Adv. Math. **221**, 547–569 (2009)

Classification of Proper Holomorphic Mappings Between Generalized Pseudoellipsoids of Different Dimensions

Atsushi Hayashimoto

Abstract We give a rigidity theorem of proper holomorphic mappings between generalized pseudoellipsoids. The theorem claims that any proper holomorphic mapping which is holomorphic extendable up to the boundary between generalized pseudoellipsoids of non-equidimensions is a collections of totally geodesic embeddings up to automorphisms.

Keywords Gap theorem · Proper holomorphic mappings · Generalized pseudoellipsoids

1 The Research of Proper Holomorphic Mappings

The purpose of this article is to classify proper holomorphic mappings between certain bounded weakly pseudoconvex domains of different dimensions. Before going on this, we survey the research of proper holomorphic mappings. Let $f : D_1 \to D_2$ be a holomorphic mapping. If the inverse image of any compact subset of D_2 is compact, then f is called proper. Therefore biholomorohic mappings are the typical example of proper holomorphic mappings, and many properties on biholomophic mappings are generalized to those on proper holomorphic mappings. In this section, we review three research topics of proper holomorphic mappings. There are many topics which we shall not refer here, for example, complexity of proper holomorphic mappings, group invariant proper holomorphic mappings, the relations between proper holomorphic mappings and CR mappings, rationality of proper holomorphic mappings, and so on.

A. Hayashimoto (✉)
716 Tokuma, Nagano 381-8550, Japan
e-mail: atsushi@nagano-nct.ac.jp

F. Bracci et al. (eds.), *Complex Analysis and Geometry*, Springer Proceedings in Mathematics & Statistics 144, DOI 10.1007/978-4-431-55744-9_10

1.1 Extension Problem

In 1974, C. Fefferman proved that any biholomorphic mapping between bounded strictly pseudoconvex domains in $\mathbb{C}^n$ with smooth boundaries can be extended to a closure of the domains as a biholomorphic mapping. Motivated by this theorem, many research have been done. The central problem of this line is the following.

Problem 1.1 *Does every proper holomorphic mapping between bounded domains D_1, D_2 with smooth boundaries in $\mathbb{C}^n$ extend smoothly to the boundary of D_1?*

Bell-Catlin [BC] and Diederich-Fornaess [DF] gave important progress by introducing the notion of Condition R. A smooth bounded domain $D \subset \mathbb{C}^n$ is said to satisfy Condition R if the Bergman projection associated to D maps $C^\infty(\bar{D})$ into $C^\infty(\bar{D})$.

Theorem 1.1 *Let D_1 and D_2 be bounded pseudoconvex domains in $\mathbb{C}^n$ with smooth boundaries and let D_1 satisfy the Condition R. Then any proper holomorphic mapping $f : D_1 \to D_2$ extends smoothly to $\bar{D}_1$.*

The examples of domains satisfying Condition R are

- strongly pseudoconvex domains with smooth boundaries,
- pseudoconvex domains with real analytic boundaries,
- pseudoconvex domains of finite type with smooth boundaries.

For a long time, Condition R had been considered as the best method to getting the positive answer to the extension problem. However Barrett [B] constructed non pseudoconvex domains in $\mathbb{C}^2$ which do not satisfy Condition R. Therefore we need another tools to attack this problem. If we do not assume the pseudoconvexity, Diederich-Pinchuk proved the theorem in dimension 2 in [DP1] and then proved in the general dimension in [DP2].

Theorem 1.2 *Let D_1, D_2 be bounded domains in $\mathbb{C}^n$ with real analytic boundaries and let $f : D_1 \to D_2$ a proper holomorphic mapping. Then f extends holomorphically to an open neighborhood of D_1.*

They use Segre varieties. Segre variety is determined by a point near the boundary of D_1. Since there are only the finite number of points which determine the same Segre variety, the mapping f can be extended as a proper holomorphic correspondence. Then they show that the correspondence is actually a mapping.

1.2 Proper Holomorphic Mappings and Biholomorphic Mappings

By definition, biholomorphic mappings are proper, and in some cases, the inverse holds. The fundamental theorem of this line was shown by Alexander [A].

Theorem 1.3 *A proper holomorphic mapping $f : B^n \to B^n$ is necessarily biholomorphic if $n > 1$. Up to a unitary transformation, f has the form:*

$$f(z) = \frac{a - P_a z - s_a Q_a z}{1 - <z, a>}, \tag{1}$$

where $a \in B^n$, $P_a z = \dfrac{<z, a>}{<a, a>} a$ if $a \neq 0$ and $P_0 z = 0$ if $a = 0$, $s_a = (1 - |a|^2)^{1/2}$, and $Q_a = I - P_a$.

The same conclusion holds for wider class of domains, such as

- bounded strictly pseudoconvex domains in $\mathbb{C}^n (n > 1)$ with C^2 boundaries,
- bounded weakly pseudoconvex domains in $\mathbb{C}^n (n > 1)$ with real analytic boundaries,
- bounded smooth pseudoconvex complete Reinhardt domains whose weakly pseudoconvex boundaries points are contained in coordinate hyperplanes.

The more should be known for the special kind of domain, such as complex ellipsoids, symmetric domains and Hua domains. Let

$$E_p = \{(z_1, \dots, z_n) \in \mathbb{C}^n : \sum_{j=1}^{n} |z_j|^{2p_j} < 1, p_j \in \mathbb{R}_{>0}\} \tag{2}$$

be a complex ellipsoid. As a corollary of the theorem in [L1], we know that any proper holomorphic self mapping of E_p is an automorphism if $p_j \in \mathbb{N}$. We expect that this holds if $p_j \in \mathbb{R}_{>0}$, but it is not known yet.

The mappings between symmetric domains, refer to [T] and between Hua domains, refer to [TW].

1.3 Determination of Proper Holomorphic Mappings

The determination problem is the following.

Problem 1.2 *Let D_1, D_2 be domains in $\mathbb{C}^n$ with certain conditions. Determine the form of a proper holomorphic mapping between D_1 and D_2. For example, Taylor expansion of them, or dependency of variables, and so on. Or determine it up to automorphisms of D_1 and D_2.*

The typical answer to this problem is also Theorem 1.3. Many research of this topics has been studied for complex ellipsoids or more generally, for Reinhardt domains and we focus on these domains here. Let T_a be a transformation defined by

$$T_a : (z_1, \dots, z_n) \mapsto (a_1 z_1, \dots, a_n z_n) \tag{3}$$

for $|a_j| = 1, a = (a_1, \dots, a_n)$. Dini-Selvaggi [DS] proved the following theorem.

Theorem 1.4 *Let $R \subset \mathbb{C}^n$ be a Reinhardt domain and $f : R \to E_p$ a proper polynomial mapping. Here E_p is defined by (2) for $p_j \in \mathbb{N}$. Then there exists a linear automorphism K of E_p and a transformation T_a such that $f = K \circ (z_1^{q_1}, \dots, z_n^{q_n}) \circ T_a$ for $q_j \in \mathbb{N}$. Hence $R = T_b(E_r)$ for some b and $r = (p_1 q_1, \dots, p_n q_n)$.*

For Reinhardt domains, the forms of the mappings are not determined in general. But some information, for example, variable splitting, is obtained. Landucci-Pinchuk [LP] proved the following theorem.

Theorem 1.5 *Let $D_1, D_2 \subset \mathbb{C}^2$ be bounded pseudoconvex complete Reinhardt domains and $f = (f_1, f_2) : D_1 \to D_2$ a proper holomorphic mapping. Assume that there exist a complex variety V and an open neighborhood U of some point $p \in \partial D_1$ such that $V \cap U \subset \partial D_1$. Then f_1 and f_2 depend only on a single variable. Namely, $f(z_1, z_2) = (f_1(z_1), f_2(z_2))$ or $f(z_1, z_2) = (f_1(z_2), f_2(z_1))$.*

Recently, the forms of proper holomorphic mappings between generalized Hartogs triangles [CX, CL, Z] and between weakly spherical domains [L2] are obtained.

If we consider mappings between non equidimensional domains, more facts are known and it is the topics that we shall study in this note. We shall refer it in the next section.

This article is organized as follows. In Sect. 2, we state the main result of this note, which is a kind of gap or rigidity theorem and we shall review a history of gap theorems. In Sect. 3, we fix some notation and expansions of a mapping under consideration. In Sect. 4, we explain the main result and in Sect. 5, we consider the low dimensional case. In the last Sect. 6, some conjectures are posed.

Finally, the author would like to express his gratitude to the organizers of KSCV10 for the hospitality during the conference. He attended KSCV1 at POSTEC and it was his first time to go abroad. I am very happy to attend the memorial 10th conference.

2 Main Theorem and History of Gap Theorems

2.1 Main Theorem

Let $E(m; m_1, \dots, m_N; \alpha_1, \dots, \alpha_N) = E(m; (m_j); (\alpha_j))$ be a bounded domain in $\mathbb{C}^{m+1}$ with a real analytic boundary defined by

$$E(m; (m_j); (\alpha_j)) = \{(z, w_1, \dots, w_N) \in \mathbb{C} \times \mathbb{C}^{m_1} \times \cdots \times \mathbb{C}^{m_N}; \\ |z|^2 + ||w_1||^{2\alpha_1} + \cdots + ||w_N||^{2\alpha_N} - 1 < 0\} \quad (4)$$

where $\alpha_1, \dots, \alpha_N \in \mathbb{N}$ and $\alpha_j \geq 2$, $m_1 + \cdots + m_N = m$ and $||w_j||^{2\alpha_j} = (|w_j^1|^2 + \cdots + |w_j^{m_j}|^2)^{\alpha_j}$ for $w_j = (w_j^1, \dots, w_j^{m_j}) \in \mathbb{C}^{m_j}$. For simplicity, we use the notation $|||w|||^{2\alpha} = ||w_1||^{2\alpha_1} + \cdots + ||w_N||^{2\alpha_N}$. This domain $E(m; (m_j); (\alpha_j))$ is called a generalized pseudoellipsoid with N blocks. We say that the mapping $F : D_1 \to D_2$

is equivalent to the mapping $G : D_1 \to D_2$ if there exist automorphisms ϕ_1 of D_1 and ϕ_2 of D_2 such that $F = \phi_2 \circ G \circ \phi_1$. We classify mappings under this equivalence relation. The following is the main theorem in this article.

Theorem 2.1 *Let $E(m; (m_j); (\alpha_j))$ and $E(n; (n_j); (\beta_j))$ be generalized pseudoellipsoids with N blocks. Assume that there exists a proper holomorphic mapping $(\mathscr{F}, \mathscr{G}_1, \ldots, \mathscr{G}_N) : E(m; (m_j); (\alpha_j)) \to E(n; (n_j); (\beta_j))$ which is holomorphic up to the boundary and fixes the origin. Assume that there exists a permutation of indices σ of order N such that one of three relations holds:*

(*a*) $2 \leq m_{\sigma(j)} \leq n_j \leq 2m_{\sigma(j)} - 2$,
(*b*) $3 \leq m_{\sigma(j)}, n_j = 2m_{\sigma(j)} - 1$,
(*c*) $4 \leq m_{\sigma(j)}, 2m_{\sigma(j)} \leq n_j \leq 3m_{\sigma(j)} - 4$.

Then $\alpha_{\sigma(j)} = \beta_j$ and $(\mathscr{F}, \mathscr{G}_1, \ldots, \mathscr{G}_N)$ is equivalent to $(z, \widetilde{\mathscr{G}_1}, \ldots, \widetilde{\mathscr{G}_N})$. Here, $\widetilde{\mathscr{G}_j} = (w_{\sigma(j)}, 0, \ldots, 0)$ with the first $m_{\sigma(j)}$ components being $w_{\sigma(j)}$ and the rest being zeros.

We also show the case of $m_j = 2, n_j = 3$ in Theorem 5.1.

If all α_j are equal to one, then a generalized pseudoellipsoid $E(m; (m_j); (1))$ is a ball, and this is a well known case. Since any proper holomorphic mapping between equidimensional balls is an automorphism, we are interested in the non-equidimensional case, which is the topics that we see in the next subsection.

2.2 History of Gap Theorems

Webster [W] proved that any proper holomorphic mapping $f : B^n \to B^{n+1}, n \geq 3$, which extends C^3 up to the boundary is equivalent to a totally geodesic embedding given by $z \to (z, 0)$. In the case of $n = 2$, Faran [F1] proved that, under the assumption that it is C^2 up to the boundary, such a mapping is classified into four cases:

$$(z, w, 0),\ (z^2, \sqrt{2}zw, w^2),\ (z, zw, w^2),\ (z^3, \sqrt{3}zw, w^3). \tag{5}$$

Let $\mathscr{I}_k$ be a closed interval

$$\mathscr{I}_k = [kn + 1, (k + 1)n - \frac{k(k + 1)}{2} - 1]. \tag{6}$$

Faran [F2] proved that any proper holomorphic mapping between n dimensional ball to N dimensional ball which is holomorphic up to the boundary is equivalent to the totally geodesic embedding provided that $N \in \mathscr{I}_1$. Huang [Hu] also obtained the same conclusion under the weaker assumption that the mapping is only C^2 up to the boundary. X. Huang and S. Ji studied the case that the codimension $N - n$ is higher. They proved in [HJ] that if $N = 2n - 1$, such a mapping is equivalent to a totally geodesic embedding or Whitney mapping,

$$(z_1, \dots, z_{n-1}, z_n) \to (z_1, \dots, z_{n-1}, z_1 z_n, \dots, z_{n-1} z_n, z_n^2). \tag{7}$$

Hamada proved in [Ham] that if $N = 2n$, such a mapping is equivalent to D'Angelo mapping,

$$F_\theta(z) = (z_1, \dots, z_{n-1}, z_n \cos\theta, z_1 z_n \sin\theta, \dots, z_{n-1} z_n \sin\theta, z_n^2 \sin\theta), \tag{8}$$

for some θ with $0 \leq \theta \leq \pi/2$. Note that $F_0(z)$ is a totally geodesic embedding and $F_{\pi/2}(z)$ is a Whitney mapping. If θ differs, then $F_\theta(z)$'s are mutually non-equivalent and therefore there are infinitely many non equivalent classes. Huang et al. [HJX] proved that if a proper holomorphic mapping F between B^n and B^N that is C^3 up to the boundary is equivalent to $(F_\theta(z), 0, \dots, 0)$ provided that $N \in \mathscr{I}_2$. In the case of $3n - 3 \leq N \leq 3n$, as far as I know, classification has not obtained yet. But there are infinitely many mappings, see [HJY1]. If $N \in \mathscr{I}_3$, a similar theorem is known by X. Huang, S. Ji and W. Yin. They proved in [HJY2] that under the assumption that $n > 7$ and $N \in \mathscr{I}_3$, any proper holomorphic mapping between B^n and B^N is equivalent to a mapping $(G, 0, \dots, 0)$, where G is a proper holomorphic mapping from B^n to B^{3n}. The paper by Huang et al. [HJY1] is a good survey of these topics. These results say that if the target dimension N stays in each interval $\mathscr{I}_1, \dots, \mathscr{I}_3$, then new mapping does not appear, but it is out of intervals, new mappings appear. We call such a phenomenon a gap phenomenon. If the domains under consideration are Levi degenerate, then Ebenfelt-Son [ES] proved a kind of a rigidity theorem for a local holomorphic mapping between boundaries of complex ellipsoids of different dimensions.

3 Expansion of a Mapping

The bounded domains $E(m; (m_j); (\alpha_j))$ and $E(n; (n_j); (\beta_j))$ have unbounded representations E_1 and E_2 respectively by Cayley transformation, which are defined by

$$\begin{aligned} E_1 = \{&(z, w_1, \dots, w_N) \in \mathbb{C} \times \mathbb{C}^{m_1} \times \cdots \times \mathbb{C}^{m_N}; \\ &\operatorname{Im} z > ||w_1||^{2\alpha_1} + \cdots + ||w_N||^{2\alpha_N}\}, \end{aligned} \tag{9}$$

$$\begin{aligned} E_2 = \{&(z, w_1, \dots, w_N) \in \mathbb{C} \times \mathbb{C}^{n_1} \times \cdots \times \mathbb{C}^{n_N}; \\ &\operatorname{Im} z > ||w_1||^{2\beta_1} + \cdots + ||w_N||^{2\beta_N}\}. \end{aligned} \tag{10}$$

We denote by $(F, G) = (F, G_1, \dots, G_N)$ the mapping between E_1 and E_2. Since it is holomorphic up to the boundary, the restricted mapping on the boundary, which is denoted by the same notation, is expanded as

$$F(x, w) = \sum_{|p|+q\geq 0} a_{p,q}(w)^p(x + i|||w|||^{2\alpha})^q, \tag{11}$$

$$G_j^\lambda(x, w) = \sum_{|p|+q\geq 0} b_{j;p,q}^\lambda(w)^p(x + i|||w|||^{2\alpha})^q, \ j = 1, \ldots, N, \lambda = 1, \ldots, n_j \tag{12}$$

as a CR mapping on the boundary. Here we use a multi-index notation. For $w = (w_1, \ldots, w_m) \in \mathbb{C}^m$ and $p = (p_1, \ldots, p_m) \in \mathbb{N}^m$, we define $(w)^p = w_1^{p_1} \ldots w_m^{p_m}$. By observing the Levi degenerate set and its image, the following lemma holds.

Lemma 3.1 *Let $(F, G_1, \ldots, G_N) : E(m; (m_j); (\alpha_j)) \to E(n; (n_j); (\beta_j))$ be a proper holomorphic mapping which can be extended holomorphically up to the boundary. Then there exists a permutation of indices σ of order N such that $G_i|_{w_{\sigma(i)}=0} = 0$ holds for any $i = 1, \ldots, N$. Therefore we have $G(x, 0) = 0$.*

Let L_j^λ be a CR vector field on the boundary of E_1:

$$L_j^\lambda = \frac{\partial}{\partial w_j^\lambda} + i\alpha_j||w_j||^{2(\alpha_j-1)}\bar{w}_j^\lambda\frac{\partial}{\partial x}, \ j = 1, \ldots, N, \lambda = 1, \ldots, m_j. \tag{13}$$

In the expansion of F, it does not contain the $(w)^p$ variable.

Lemma 3.2 *We have the expansion*

$$F(x, w) = \sum_{q\geq 0} a_q(x + i|||w|||^{2\alpha})^q, \ a_q \in \mathbb{R}. \tag{14}$$

Proof By iterating CR vector fields, we have

$$L_{j_k}^{\lambda_k} \ldots L_{j_1}^{\lambda_1} = \frac{\partial^k}{\partial w_{j_k}^{\lambda_k} \ldots \partial w_{j_1}^{\lambda_1}} + \bar{w}_{j_1}^{\lambda_1}P_1 + \cdots + \bar{w}_{j_k}^{\lambda_k}P_k, \tag{15}$$

where $P_1, \ldots, P_k$ are differential operators with order smaller than or equal to k. Since we have $G(x, 0) = 0$, we obtain

$$\begin{aligned} 0 &= (L_{j_k}^{\lambda_k} \ldots L_{j_1}^{\lambda_1}|||G|||^{2\beta})(x, 0) \\ &= (L_{j_k}^{\lambda_k} \ldots L_{j_1}^{\lambda_1}\mathrm{Im}F)(x, 0) = \frac{1}{2i}(L_{j_k}^{\lambda_k} \ldots L_{j_1}^{\lambda_1}F)(x, 0) \end{aligned} \tag{16}$$

for any vector fields $L_{j_1}^{\lambda_1}, \ldots, L_{j_k}^{\lambda_k}$. Therefore $F(x, w)$ satisfies

$$\frac{\partial^k F}{\partial w_{j_k}^{\lambda_k} \ldots \partial w_{j_1}^{\lambda_1}}(x, 0) = 0. \tag{17}$$

It implies, in the expansion (11) of F, that $a_{p,q} = 0$ for $|p| > 0$. We replace $a_{0,q}$ by a_q. Since $F(x, 0)$ is real, $a_q \in \mathbb{R}$ follows. □

4 Main Result (General Case)

In this section, we shall explain how the Theorem 2.1 is proved. The proof contains four steps. The first two steps appear in [Hay].

Proof **Step 1.** In this and the next Steps, we assume that $2 \leq m_{\sigma(j)} \leq n_j \leq 2m_{\sigma(j)} - 2$. We prove it for one block case ($N = 1$). We substitute (14) and (12) with $N = 1$ into the defining function of the target domain. Pick up the terms with same degree in the variables to get the relations among the coefficients in the expansions. We continue the following arguments.

- By the method of undetermined coefficients, some coefficients of Taylor series are zero.
- We construct a unitary matrix which normalizes some coefficients of expansions.
- We construct a proper homogeneous mapping between balls, which are already classified, to get normalizations of some coefficients. The assumption $2 \leq m_1 \leq n_1 \leq 2m_1 - 2$ is posed to use the fact that any proper holomorphic mapping between B^{m_1} and B^{n_1} with such dimension condition is a totally geodesic embedding up to automorphisms.

By these methods, we have proved the one block case.

Step 2. Next the N blocks case. Let $w_j \in \mathbb{C}^{m_j}$ be a variable in the jth block in the source domain. If we put $w_1 = \cdots = w_N = 0$ except for w_j, then all components, except for F and $G_{\sigma^{-1}(j)}$, vanish by Lemma 3.1. This is the one block case, which has already proved. Since j is arbitrary, we obtain the relations of α's and β's. Assume that $\sigma = \mathrm{id}$ and $\alpha_1 = \beta_1 \leq \cdots \leq \alpha_N = \beta_N$. To get the normalization, we use the three arguments as in the one block case. The difficulty is that, after getting partially normalization of, say, G_1, next we normalize G_2 by certain change of coordinate. But, in the process of normalization of G_2, the coordinate change gives some influences the obtained form of G_1 and therefore a partial normalization of G_1 breaks. Therefore we need to check the influences seriously. By these Steps, we completes the proof.

Step 3. If $m_j \geq 3$, $n_j = 2m_{\sigma(j)} - 1$, then we use Whitney map and a totally geodesic embedding instead of a totally geodesic embedding in Steps 1 and 2. Then we obtain the result.

Step 4. If $m_j \geq 4$, $2m_{\sigma(j)} \leq n_j \leq 3m_{\sigma(j)} - 4$, then we use D'Angelo map instead of a totally geodesic embedding in Steps 1 and 2. Then we obtain the result. □

Theorem 2.1 says that, up to a permutation of indices and automorphisms, the mapping is a collection of totally geodesic embeddings between corresponding blocks. Since it holds for $n_j = m_{\sigma(j)}$, we have the following corollary.

Corollary 4.1 *The proper holomorphic mapping from a generalized pseudoellipsoid into itself is an automorphism.*

5 Main Result (Low Dimensional Case)

The theorem proved in the previous section does not cover the case of $m_j = 2$, $n_j = 3$. Let

$$E(2;2;\alpha) = \{(z, w_1, w_2) \in \mathbb{C} \times \mathbb{C}^2 : |z|^2 + ||w||^{2\alpha} - 1 < 0, \alpha \in \mathbb{N}\}, \quad (18)$$

$$E(3;3;\beta) = \{(z, w_1, w_2, w_3) \in \mathbb{C} \times \mathbb{C}^3 : |z|^2 + ||w||^{2\beta} - 1 < 0, \beta \in \mathbb{N}\} \quad (19)$$

be generalized pseudoellipsoids with one block. On the line of the previous section, this case has close relation to the proper mapping $f : B^2 \to B^3$, which is classified into four mappings.

Theorem 5.1 *Let $E(2;2;\alpha)$ and $E(3;3;\beta)$ be as defined. Assume that there exists a proper holomorphic mapping $(\mathscr{F}, \mathscr{G}_1, \mathscr{G}_2, \mathscr{G}_3) : E(2;2;\alpha) \to E(3;3;\beta)$ which is holomorphic up to the boundary and fixes the origin. Then we have the two equivalent classes:*

- $\alpha = \beta$, $(\mathscr{F}, \mathscr{G}_1, \mathscr{G}_2, \mathscr{G}_3) \sim (z, w_1, w_2, 0)$,
- $\alpha = 2\beta$, $(\mathscr{F}, \mathscr{G}_1, \mathscr{G}_2, \mathscr{G}_3) \sim (z, w_1^2, w_2^2, \sqrt{2}w_1w_2)$.

Among four mappings in the classification (5) of $f : B^2 \to B^3$, the two mappings appearing in this theorem satisfy

$$||(z, w_1, w_2, 0)||^2 = ||(z, w_1, w_2)||^2, \quad (20)$$

$$||(z, w_1^2, w_2^2, \sqrt{2}w_1w_2)||^2 = ||(z, w_1, w_2)||^4. \quad (21)$$

But other two do not satisfy $||F(z)||^2 = ||z||^{2n}$ for any n. This is the reason why only two mappings appear in the theorem.

6 Some Conjectures

There are some situations to be studied in the line of gap theorems.

6.1 The Number of Blocks

In Theorem 2.1, the number of blocks of source and target domains are the same. Assume that the conditions of the mapping are the same as Theorem 2.1.

Conjecture 6.1 (1) *Let N and N' be the numbers of blocks in the source and target domains respectively. If the inequality $N < N'$ holds, then any proper holomorphic mapping as in Theorem 2.1 is a collection of totally geodesic embeddings.*

(2) *If the opposite inequality $N > N'$ holds, then there does not exist such a mapping.*

6.2 Dimension in the Blocks

We pose assumptions on the dimensions m_j and n_j of variables in the blocks in Theorem 2.1. It is needed to use the gap theorem for balls. If the codimension $n_j - m_{\sigma(j)}$ is sufficiently large, a gap phenomenon does not occur and we can not classify the mappings as shown in [DL]. But I guess that proper holomorphic mappings between generalized pseudoellipsoids are not the cases.

Conjecture 6.2 *Let $(\mathscr{F}, \mathscr{G}_1, \ldots, \mathscr{G}_N) : E(m; (m_j); (\alpha_j)) \to E(n; (n_j); (\beta_j))$ be a proper holomorphic mapping which is extendable holomorphically up to the boundary between generalized pseudoellipsoids with N blocks. Assume that there exists a permutation of indices σ such that $2 < m_{\sigma(j)} < n_j$. Then $\alpha_{\sigma(j)} = \beta_j$ and $(\mathscr{F}, \mathscr{G}_1, \ldots, \mathscr{G}_N)$ is equivalent to $(z, \widetilde{\mathscr{G}_1}, \ldots, \widetilde{\mathscr{G}_N})$. Here, $\widetilde{\mathscr{G}_j} = (w_{\sigma(j)}, 0, \ldots, 0)$ with the first $m_{\sigma(j)}$ components being $w_{\sigma(j)}$ and the rest being zeros.*

References

[A] Alexander, H.: Proper holomorphic mappings in $\mathbb{C}^n$. Indiana Univ. Math. J. **26**, 137–146 (1977)

[B] Barrett, D.: Irregularity of the Bergman projection on a smooth bounded domain in $\mathbb{C}^2$. Ann. Math. **119**, 431–436 (1984)

[BC] Bell, S., Catlin, D.: Boundary regularity of proper holomorphic mappings. Duke Math. J. **49**(2), 385–396 (1982)

[CL] Chen, Z.H., Liu, Y.: A note on the classification of the proper mappings between some generalized Hartogs triangles. Chin. J. Contemp. Math. **24**(3), 215–222 (2003)

[CX] Chen, Z.H., Xu, D.K.: Rigidity of proper self-mapping on some kinds of generalized Hartogs triangle. Acta Math. Sin. English Ser. **18**(2), 357–362 (2002)

[DF] Diederich, K., Fornaess, J.E.: Boundary regularity of proper holomorphic mappings. Int. J. Math. **67**(3), 363–384 (1982)

[DL] D'Angelo, J.P., Lebl, J.: Complexity results for CR mappings between spheres. Int. J. Math. **20**(2), 146–166 (2009)

[DP1] Diederich, K., Pinchuk, S.: Proper holomorphic maps in dimension 2. Indiana Univ. J. Math. **44**(4), 1089–1126 (1995)

[DP2] Diederich, K., Pinchuk, S.: Reflection principle in higher dimensions. Doc. Math. **II**, 703–712 (1998)

[DS] Dini, G., Primicerio, A.S.: Proper polynomial holomorphic mappings for a class of Reinhardt domains. Boll. Unione Mat. Ital. **7**(1-A), 11–20 (1987)

[ES] Ebenfelt, P., Son, D.N.: Holomorphic mappings between pseudoellipsoids in different dimensions. arXiv:1210.4434v1[math.CV]

[F1] Faran, J.: Maps from the two-ball to the three-ball. Invent. Matt. **68**, 441–475 (1982)

[F2] Faran, J.: On the linearity of proper maps between balls in the low codimensional case. J. Diff. Geom. **24**, 15–17 (1986)

[Ham] Hamada, H.: Rational proper holomorphic maps from B^n into B^{2n}. Math. Ann. **331**, 693–711 (2005)

[Hay] Hayashimoto, A.: A gap theorem for proper holomorphic mappings between generalized pseudoellipsoids (in preparation)

[HJ] Huang, X., Ji, S.: Mapping B^n into B^{2n-1}. Invent. Math. **145**, 219–250 (2001)

[HJX] Huang, X., Ji, S., Xu, D.: A new gap phenomenon for proper holomorphic mappings from B^n into B^N. Math. Res. Lett. **13**(4), 515–529 (2006)

[HJY1] Huang, X., Ji, S., Yin, W.: A survey on the recent progress of some problems in several complex variables. ICCM **I**, 563–575 (2007)

[HJY2] Huang, X., Ji, S., Yin, W.: On the third gap for proper holomorphic maps between balls. Math. Ann. **358**(1–2), 115–142 (2014)

[Hu] Huang, X.: On the linearity problem for proper holomorphic maps between balls in complex spaces of different dimensions. J. Diff. Geom. **51**, 13–33 (1999)

[L1] Landucci, M.: On the proper holomorphic equivalence for a class of pseudoconvex domains. Trans. Am. Math. Soc. **282**, 807–811 (1984)

[L2] Landucci, M.: Proper holomorphic maps between weakly spherical domains. Adv. Geom. **12**, 515–523 (2012)

[LP] Landucci, M., Pinchuk, S.: Proper mappings between Reinhardt domains with an analytic variety on the boundary. Ann. Scuola Norm. Sup. Pisa Cl. Sci. (4) **22**(3), 363–373 (1995)

[T] Tu, Z.: Rigidity of proper holomorphic maps between equidimensional bounded symmetric domains. Proc. Amer. Math. Soc. **130**, 1035–1042 (2002)

[TW] Tu, Z., Wang, L.: Rigidity of proper holomorphic mappings between equidimensional Hua domains. arXiv:1411.3162v1[mathCV]

[W] Webster, S.: On mapping an n-ball into an $(n+1)$-ball in complex spaces. Pac. J. Math. **81**, 267–277 (1979)

[Z] Zapalowski, P.: Proper holomorphic mappings between complex ellipsoids and generalized Hartogs triangles. arXiv:1211.0786v2[math.CV]

Bergman Kernel Asymptotics and a Pure Analytic Proof of the Kodaira Embedding Theorem

Chin-Yu Hsiao

Abstract In this paper, we survey recent results in [HMA12] about the asymptotic expansion of Bergman kernel and we give a Bergman kernel proof of the Kodaira embedding theorem.

Keywords Bergman kernel asymptotics · The Kodaira embedding theorem

1 Introduction and Set up

Let L be a holomorphic line bundle over a complex manifold M and let L^k be the k-th tensor power of L. The Bergman projection P_k is the orthogonal projection onto the space of L^2-integrable holomorphic sections of L^k. The study of the large k behaviour of P_k is an active research subject in complex geometry and is closely related to topics like the structure of algebraic manifolds, the existence of canonical Kähler metrics, Toeplitz quantization, equidistribution of zeros of holomorphic sections, quantum chaos and mathematical physics. We refer the reader to the book [MM07] for a comprehensive study of the Bergman kernel and its applications and also to the survey [Ma10].

When M is compact and L is positive, Catlin [Cat97] and Zelditch [Zel98] established the asymptotic expansion of the Bergman kernel (see Theorem 4.2) by using a fundamental result by Boutet de Monvel-Sjöstrand [BouSj76] about the asymptotics of the Szegö kernel on a strictly pseudoconvex boundary. Dai et al. [DLM06] obtained the full off-diagonal asymptotic expansion and Agmon estimates of the Bergman kernel for a high power of positive line bundle on a compact complex manifold by using the heat kernel method. Ma and Marinescu [MM07, MM08a] proved the asymptotic expansion for yet another generalization of the Kodaira Laplacian, namely the renormalized Bochner-Laplacian on a symplectic manifold and also showed the existence of the estimate on a large class of non-compact manifolds. Another proof based on

C.-Y. Hsiao (✉)
Institute of Mathematics, Academia Sinica, 6F, Astronomy-Mathematics Building, No.1, Sec.4, Roosevelt Road, Taipei 10617, Taiwan
e-mail: chsiao@math.sinica.edu.tw; chinyu.hsiao@gmail.com

F. Bracci et al. (eds.), *Complex Analysis and Geometry*, Springer Proceedings in Mathematics & Statistics 144, DOI 10.1007/978-4-431-55744-9_11

microlocal analysis of the existence of the full asymptotic expansion for the Bergman kernel for a high power of a positive line bundle on a compact complex manifold was obtained by Berndtsson, Berman and Sjöstrand [BBS04].

In [HMA12], we impose a very mild semiclassical local condition on $\overline{\partial}_k$, namely the $O(k^{-N})$ small spectral gap on an open set $D \Subset M$ (see Definition 3.1), where $\overline{\partial}_k$ denotes the Cauchy-Riemann operator with values in L^k. We prove that the Bergman kernel admits an asymptotic expansion on D if $\overline{\partial}_k$ has $O(k^{-N})$ small spectral gap on D, cf. Theorem 3.1. Our approach is based on the microlocal Hodge decomposition for Kohn Laplacian established in [Hsiao08]. The distinctive feature of these asymptotics is that they work under minimal hypotheses. This allows us to apply them in situations which were up to now out of reach. We illustrate this in the study of the Bergman kernels of positive but singular Hermitian line bundles (see Theorem 3.3).

1.1 Set up

In this paper, we let M be a not necessary compact complex manifold of dimension n with a smooth positive $(1,1)$ form Θ. The form Θ induces Hermitian metrics on the complexified tangent bundle $\mathbb{C}TM$ and $T^{*0,q}M$ bundle of $(0,q)$ forms on M , $q = 0, 1, \ldots, n$. We shall denote all these Hermitian metrics by $\langle \cdot | \cdot \rangle$. Let $(L, h^L) \to M$ be a holomorpic line bundle over M, where h^L denotes the Hermitian fiber metric of L. Let R^L be the canonical curvature two form induced by h^L. Given a local trivializing section s of L on an open subset $D \subset M$ we define the associated local weight of h^L by

$$|s(x)|^2_{h^L} = e^{-2\phi(x)}, \quad \phi \in C^\infty(D, \mathbb{R}). \tag{1.1}$$

Then $R^L|_D = 2\partial\overline{\partial}\phi$. Let (L^k, h^{L^k}) be the k-th tensor power of the line bundle L. If s is a local trivializing section of L, $|s|^2_{h^L} = e^{-2\phi}$, then s^k is a local trivializing section of L^k and $\left|s^k\right|^2_{h^{L^k}} = e^{-2k\phi}$. We take $dv_M = dv_M(x)$ as the volume form on M induced by Θ. For every $q = 0, 1, 2, \ldots, n$, let $(\cdot | \cdot)$ and $(\cdot | \cdot)_{h^{L^k}}$ be the standard L^2 inner products on $\Omega^{0,q}_0(M) := C^\infty_0(M, T^{*0,q}M)$ and $\Omega^{0,q}_0(M, L^k) := C^\infty_0(M, T^{*0,q}M \otimes L^k)$ respectively induced by dv_M, $\langle \cdot | \cdot \rangle$ and h^{L^k} and we write $\|\cdot\|$ and $\|\cdot\|_{h^{L^k}}$ to denote the corresponding norms. Let $L^2_{(0,q)}(M)$ and $L^2_{(0,q)}(M, L^k)$ be the completions of $\Omega^{0,q}_0(M)$ and $\Omega^{0,q}_0(M, L^k)$ with respect to $\|\cdot\|$ and $\|\cdot\|_{h^{L^k}}$ respectively.

Let $\overline{\partial}_k : C^\infty(M, L^k) \to \Omega^{0,1}(M, L^k)$ be the Cauchy-Riemann operator with values in L^k. We extend $\overline{\partial}_k$ to $L^2(M, L^k) := L^2_{(0,0)}(M, L^k)$ by $\overline{\partial}_k : \mathrm{Dom}\,\overline{\partial}_k \subset L^2(M, L^k) \to L^2_{(0,1)}(M, L^k)$, where $\mathrm{Dom}\,\overline{\partial}_k := \{u \in L^2(M, L^k); \overline{\partial}_k u \in L^2_{(0,1)}(M, L^k)\}$. Let

$$P_k : L^2(M, L^k) \to \mathrm{Ker}\,\overline{\partial}_k$$

be the Bergman projection, i.e. P_k is the orthogonal projection onto $\operatorname{Ker}\overline{\partial}_k$ with respect to $(\cdot\,|\,\cdot)_{h^{L^k}}$ and let $P_k(x,y)\in C^\infty(M\times M,\mathscr{L}(L^k_y,L^k_x))$ be the distribution kernel of P_k.

2 Terminology in Semi-classical Analysis

In this section, we collect some definitions and notations in semi-classical analysis.

Let $B_k : L^2(M,L^k)\to L^2(M,L^k)$ be a continuous operator with smooth kernel $B_k(x,y)$. Let s, s_1 be local trivializing sections of L on $D_0\Subset M$, $D_1\Subset M$ respectively, $|s|^2_{h^L}=e^{-2\phi}$, $|s_1|^2_{h^L}=e^{-2\phi_1}$. The localized operator (with respect to the trivializing sections s and s_1) of B_k is given by

$$\begin{aligned}B_{k,s,s_1} : L^2_{\mathrm{comp}}(D_1)&\to L^2(D),\\ u&\to e^{-k\phi}s^{-k}B_k(s_1^k e^{k\phi_1}u).\end{aligned}\tag{2.1}$$

and let $B_{k,s,s_1}(x,y)\in C^\infty(D\times D_1)$ be the distribution kernel of B_{k,s,s_1}, where

$$L^2_{\mathrm{comp}}(D_1):=\left\{v\in L^2(D_1);\ \operatorname{Supp}v\Subset D_1\right\}.$$

Let D be a local coordinate patch of M and let $A_k : C^\infty_0(D)\to C^\infty(D)$ be a k-dependent continuous operator with smooth kernel $A_k(x,y)$. We write $A_k\equiv 0 \mod O(k^{-\infty})$ (on D) or $A_k(x,y)\equiv 0 \mod O(k^{-\infty})$ (on D) if $A_k(x,y)$ satisfies $\left|\partial^\alpha_x\partial^\beta_y A_k(x,y)\right|=O(k^{-N})$ uniformly on every compact set in $D\times D$, for all multi-indices $\alpha,\beta\in\mathbb{N}^{2n}$ and all $N>0$. Let $B_k : L^2(M,L^k)\to L^2(M,L^k)$ be a k-dependent continuous operator with smooth kernel. We write $B_k\equiv 0 \mod O(k^{-\infty})$ if $B_{k,s,s_1}\equiv 0 \mod O(k^{-\infty})$ for every local trivializing sections s and s_1.

Definition 2.1 Let D be a local coordinate patch of M. Let $S(1;D)=S(1)$ be the set of all $a\in C^\infty(D)$ such that for every $\alpha\in\mathbb{N}^{2n}$, there exists $C_\alpha>0$, such that $\left|\partial^\alpha_x a(x)\right|\le C_\alpha$ on D. If $a=a(x,k)$ depends on $k\in]1,\infty[$, we say that $a(x,k)\in S_{\mathrm{loc}}(1;D)=S_{\mathrm{loc}}(1)$ if $\chi(x)a(x,k)$ uniformly bounded in $S(1)$ when k varies in $]1,\infty[$, for any $\chi\in C^\infty_0(D)$. For $m\in\mathbb{R}$, we put $S^m_{\mathrm{loc}}(1;D)=S^m_{\mathrm{loc}}(1)=k^mS_{\mathrm{loc}}(1)$. If $a_j\in S^{m_j}_{\mathrm{loc}}(1)$, $m_j\searrow-\infty$, we say that $a\sim\sum_{j=0}^{\infty}a_j$ (in $S^{m_0}_{\mathrm{loc}}(1)$) if $a-\sum_{j=0}^{N_0}a_j\in S^{m_{N_0+1}}_{\mathrm{loc}}(1)$ for every N_0. For a given sequence a_j as above, we can always find such an asymptotic sum a and a is unique up to an element in $S^{-\infty}_{\mathrm{loc}}(1)=S^{-\infty}_{\mathrm{loc}}(1;D):=\bigcap_m S^m_{\mathrm{loc}}(1)$.

3 Asymptotic Expansion of Bergman Kernel

Let s, s_1 be local trivializing sections of L on $D_0 \Subset M$, $D_1 \Subset M$ respectively, $|s|^2_{h^L} = e^{-2\phi}$, $|s_1|^2_{h^L} = e^{-2\phi_1}$. Let P_{k,s,s_1} be the localized operator of P_k given by (2.1) and let $P_{k,s,s_1}(x, y) \in C^\infty(D \times D_1)$ be the distribution kernel of P_{k,s,s_1}. When $s = s_1$, $D = D_1$, we write $P_{k,s} := P_{k,s,s_1}$, $P_{k,s}(x, y) := P_{k,s,s_1}(x, y)$. When $x = y$, $P_{k,s}(x, x)$ is independent of s. We write $P_k(x) := P_{k,s}(x, x)$ and we call $P_k(x)$ Bergman kernel function. Let $f_1 \in C^\infty(M, L^k), \ldots, f_{d_k} \in C^\infty(M, L^k)$ be orthonormal frame for Ker $\overline{\partial}_k$, $d_k \in \{0\} \bigcup \mathbb{N} \bigcup \{\infty\}$. On D_0 and D_1, we write

$$f_j = s^k e^{k\phi} \widetilde{f}_j, \quad \widetilde{f}_j \in C^\infty(D), \quad j = 1, 2, \ldots, d_k,$$
$$f_j = s_1^k e^{k\phi_1} \hat{f}_j, \quad \hat{f}_j \in C^\infty(D_1), \quad j = 1, 2, \ldots, d_k.$$

We can check that

$$P_{k,s,s_1}(x, y) = \sum_{j=1}^{d_k} \widetilde{f}_j(x) \overline{\hat{f}_j}(y),$$
$$P_k(x) = \sum_{j=1}^{d_k} \left| f_j(x) \right|^2_{h^{L^k}}. \tag{3.1}$$

We recall $O(k^{-N})$ small spectral gap property introduced in [HMA12]

Definition 3.1 Let $D \subset M$. We say that $\overline{\partial}_k$ has $O(k^{-N})$ *small spectral gap on* D if there exist constants $C_D > 0$, $N \in \mathbb{N}$, $k_0 \in \mathbb{N}$, such that for all $k \geq k_0$ and $u \in C^\infty_0(D, L^k)$, we have

$$\|(I - P_k)u\|_{h^{L^k}} \leq C_D \, k^N \left\| \overline{\partial}_k u \right\|_{h^{L^k}}.$$

It should be mentioned that in [HMA12], we actually introduced $O(k^{-N})$ small spectral gap for Kodaira Laplacian. Note that $O(k^{-N})$ small spectral gap for $\overline{\partial}_k$ implies $O(k^{-N})$ small spectral gap for Kodaira Laplacian.

One of the main results in [HMA12] is the following

Theorem 3.1 *With the notations and assumptions used before, let s be a local trivializing section of L on an open set $D \subset M$, $|s|^2_{h^L} = e^{-2\phi}$, and assume that R^L is positive on D. Suppose that $\overline{\partial}_k$ has $O(k^{-N})$ small spectral gap on D. Then, $\chi_1 P_k \chi \equiv 0$* mod $O(k^{-\infty})$ *for every $\chi_1 \in C^\infty_0(M)$, $\chi \in C^\infty_0(D)$ with* Supp $\chi_1 \bigcap$ Supp $\chi = \emptyset$ *and*

$$P_{k,s}(x, y) \equiv e^{ik\Psi(x,y)} b(x, y, k) \mod O(k^{-\infty}) \text{ on } D,$$

where $b(x, y, k) \sim \sum_{j=0}^{\infty} b_j(x, y) k^{n-j}$ *in the sense of Definition 2.1,* $b_j(x, y) \in C^{\infty}(D \times D)$*,* $j = 0, 1, \ldots$*,* $b_0(x, x) = (2\pi)^{-n} \left| \det R^L(x) \right|$ *and*

$$\begin{aligned} &\Psi(x, y) \in C^{\infty}(D \times D), \quad \Psi(x, y) = -\overline{\Psi}(y, x), \\ &\exists c > 0: \ \operatorname{Im} \Psi \geq c\,|x - y|^2, \ \Psi(x, y) = 0 \Leftrightarrow x = y, \end{aligned} \tag{3.2}$$

for any $p \in D$*, take local holomorphic coordinates* $z = (z_1, \ldots, z_n)$ *vanishing at* p*, then near* (p, p)*,*

$$\begin{aligned} \Psi(z, w) =& i(\phi(z) + \phi(w)) \\ &- 2i \sum_{\substack{\alpha, \beta \in (\{0\} \bigcup \mathbb{N})^n, \\ |\alpha| + |\beta| \leq N}} \frac{\partial^{|\alpha| + |\beta|} \phi}{\partial z^{\alpha} \partial \overline{z}^{\beta}}(0) \frac{z^{\alpha} \overline{w}^{\beta}}{\alpha! \beta!} + O(|(z, w)|^{N+1}), \quad \forall N \in \mathbb{N}, \end{aligned} \tag{3.3}$$

where $\det R^L(x) = \lambda_1(x) \cdots \lambda_n(x)$*,* $\lambda_j(x)$*,* $j = 1, \ldots, n$*, are the eigenvalues of* R^L *with respect to* $\langle \cdot | \cdot \rangle$*.*

In particular, $P_k(x) \sim \sum_{j=0}^{\infty} b_j(x, x) k^{n-j}$ *in the sense of Definition 2.1.*

3.1 Big Line Bundles and Shiffman Conjecture

As an application of Theorem 3.1, we will establish Bergman kernel asymptotic expansion for a big line bundle and this yields yet another proof of the Shiffman conjecture. Until further notice, we assume that M is compact. We recall

Conjecture 3.1 (Shiffman, 1990). If h^L is a singular Hermitian metric, smooth outside a proper analytic set Σ, $R^L > 0$ in the sense of current, then L is big.

A line bundle L is said to be big if $\dim H^0(M, L^k) \approx k^n$, where

$$H^0(M, L^k) = \left\{ u \in C^{\infty}(M, L^k);\ \overline{\partial}_k u = 0 \right\}.$$

Ji and Shiffman [JS93] solved this conjecture.

Now, we assume that h^L is a singular Hermitian metric, smooth outside a proper analytic set Σ, $R^L > 0$ in the sense of current. Consider the non-compact complex manifold $M \setminus \Sigma$. We also write $\overline{\partial}_k$ to denote the Cauchy-Riemann operator on $M \setminus \Sigma$ with values in L^k. Let $P_{k, M \setminus \Sigma}$ be the associated Bergman projection on $M \setminus \Sigma$ and let $P_{k, M \setminus \Sigma}(x)$ be the associated Bergman kernel function. In [HMA12], we showed that

Theorem 3.2 $\overline{\partial}_k$ *has* $O(k^{-N})$ *small spectral gap on every* $D \Subset M \setminus \Sigma$*.*

From Theorem 3.2 and Theorem 3.1, we deduce that

Theorem 3.3 $P_{k,M\setminus\Sigma}(x) \sim (2\pi)^{-n}\big|\det R^L(x)\big|k^n + b_1(x)k^{n-1} + b_2(x)k^{n-2} + \cdots$ *locally uniformly on* $M \setminus \Sigma$*, where* $b_j(x) \in C^\infty(M \setminus \Sigma)$*,* $j = 1, 2, \ldots$.

Let $\{g_1, g_2, \ldots, g_{m_k}\}$ be an orthonormal frame for $H^0(M, L^k) \bigcap L^2(M\setminus\Sigma, L^k)$. The multiplier Bergman kernel function is defined by

$$P_{k,\mathscr{I}}(x) := \sum_{j=1}^{m_k} \big|g_j(x)\big|^2_{h^{L^k}}, \quad x \in M \setminus \Sigma.$$

The following result is essentially due to Skoda (see Lemma 7.3 of Ch. VIII in Demailly [Desps11]).

Theorem 3.4 $P_{k,M\setminus\Sigma}(x) = P_{k,\mathscr{I}}(x)$, $\forall x \in M \setminus \Sigma$.

Proof (Proof of Shiffman conjecture) From Theorem 3.4 and Theorem 3.3, we establish Bergman kernel asymptotic expansion for big line bundle:

$$P_{k,\mathscr{I}}(x) \sim (2\pi)^{-n}\big|\det R^L(x)\big|k^n + b_1(x)k^{n-1} + b_2(x)k^{n-2} + \cdots \quad \text{locally uniformly on} M \setminus \Sigma, \tag{3.4}$$

where $b_j(x) \in C^\infty(M \setminus \Sigma)$, $j = 1, 2, \ldots$. Let $K \Subset M \setminus \Sigma$. Note that $\dim H^0(M, L^k) \geq \int_K P_{k,\mathscr{I}}(x)dv_M(x)$. From this observation and (3.4), we reprove Shiffman conjecture. □

4 A Bergman Kernel Proof of the Kodaira Embedding Theorem

For a holomorphic line bundle $E \to M$, we say that E is positive if there is a Hermitian metric h^E of E such that the associated curvature R^E is positive definite on M. Let us recall the Kodaira embedding theorem first.

Theorem 4.1 *Let* M *be a compact complex manifold. If there is a positive holomorphic line bundle* E *over* M*, then* M *can be holomorphic embedded into* $\mathbb{CP}^N$*, for some* $N \in \mathbb{N}$.

We return to our situation and we will use the same notations as before. By using Hörmander's L^2 estimates [Hor90], it is easy to see that if M is compact and R^L is positive on M then $\overline{\partial}_k$ has $O(k^{-N})$ small spectral gap on M. From this observation and Theorem 3.1, we deduce

Theorem 4.2 *Assume that* M *is compact and* R^L *is positive on* M*. Then,*

$$\chi_1 P_k \chi \equiv 0 \mod O(k^{-\infty}) \tag{4.1}$$

for every $\chi_1 \in C^\infty(M)$, $\chi \in C^\infty(M)$ *with* $\operatorname{Supp}\chi_1 \bigcap \operatorname{Supp}\chi = \emptyset$. *Let* s *be a local trivializing section of* L *on an open set* $D \subset M$, $|s|^2_{h^L} = e^{-2\phi}$, *then*

$$P_{k,s}(x, y) \equiv e^{ik\Psi(x,y)} b(x, y, k) \quad \text{mod } O(k^{-\infty}) \text{ on } D, \tag{4.2}$$

where $b(x, y, k)$ *and* $\Psi(x, y)$ *are as in Theorem 3.1.*

In particular,

$$P_k(x) \sim (2\pi)^{-n} \left|\det R^L(x)\right| k^n + b_1(x)k^{n-1} + b_2(x)k^{n-2} + \cdots \text{ uniformly on } M. \tag{4.3}$$

By using Theorem 3.1, we are going to give a Bergman kernel proof of the Kodaira embedding theorem. From now on, we assume that R^L is positive on M. As before, put

$$H^0(M, L^k) := \left\{u \in C^\infty(M, L^k);\ \overline{\partial}_k u = 0\right\}$$

and let $\{f_1, \ldots, f_{d_k}\}$ be an orthonormal basis for $H^0(M, L^k)$ with respect to $(\,\cdot\,|\,\cdot\,)_{h^{L^k}}$. The Kodaira map is given by

$$\Phi_k : x \in X \to [f_1(x), f_2(x), \ldots, f_{d_k}(x)] \in \mathbb{CP}^{d_k - 1}. \tag{4.4}$$

From (4.3), we see that there is a $k_0 > 0$ such that for every $k \geq k_0$, $\sum_{j=1}^{d_k} \left|f_j(x)\right|^2_{h^{L^k}} \geq ck^n$ on M, where $c > 0$ is a constant independent of k. Hence, fix any $k \geq k_0$, for every $x \in X$, there is a f_j, $j \in \{1, 2, \ldots, d_k\}$, such that $\left|f_j(x)\right|^2_{h^{L^k}} > 0$. We conclude that Φ_k is a well-defined as a smooth map from X to $\mathbb{CP}^{d_k-1}$. We will prove

Theorem 4.3 *For* k *large,* Φ_k *is a holomorphic embedding.*

It is clearly that the Kodaira embedding theorem follows from Theorem 4.3. We recall that for a smooth map $\Phi : X \to \mathbb{CP}^N$ is an embedding if $d\Phi_x : TX \to T\mathbb{CP}^N$ is injective at each point $x \in X$ and $\Phi : X \to \mathbb{CP}^N$ is globally injective.

Let s be a local trivializing section of L on an open set $D \subset M$. Fix $p \in D$ and let $z = (z_1, \ldots, z_n) = x = (x_1, \ldots, x_{2n})$, $z_j = x_{2j-1} + ix_{2j}$, $j = 1, \ldots, n$, be local holomorphic coordinates of X defined in some small neighbourhood of p such that

$$\phi(z) = \sum_{j=1}^{n} \lambda_j \left|z_j\right|^2 + O(|z|^3), \tag{4.5}$$

where $2\lambda_1, \ldots, 2\lambda_n$ are the eigenvalues of $R^L(p)$ with respect to $\langle\,\cdot\,|\,\cdot\,\rangle$. We may assume that the local coordinates z defined on D. We also write $y = (y_1, \ldots, y_{2n})$.

Until further notice, we work on D. Take $\chi \in C_0^\infty(\mathbb{R}, [0, 1])$ with $\chi(x) = 1$ on $[-\frac{1}{2}, \frac{1}{2}]$, $\chi(x) = 0$ on $]-\infty, -1] \bigcup [1, \infty[$ and $\chi(t) = \chi(-t)$, for every $t \in \mathbb{R}$. Let

$$u_k := P_k\Big(s^k e^{k\phi}\chi(\sqrt{k}y_1)\cdots\chi(\sqrt{k}y_{2n})\Big) \in H^0(M, L^k). \tag{4.6}$$

On D, we write $u_k = s^k e^{k\phi}\widetilde{u}_k$, $\widetilde{u}_k \in C^\infty(D)$. Then, $|u_k(x)|^2_{h^{L^k}} = |\widetilde{u}_k(x)|^2$, $\forall x \in D$. We need

Lemma 4.1 *With the notations used above, there is a $k_0 > 0$ independent of k and the point p such that for all $k \geq k_0$,*

$$|u_k(p)|^2_{h^{L^k}} \geq c_0, \tag{4.7}$$

$$|u_k(x)|^2_{h^{L^k}} \leq \frac{1}{c_0 k}, \quad \forall x \notin D \tag{4.8}$$

and

$$\left|\frac{1}{\sqrt{k}}\frac{\partial \widetilde{u}_k}{\partial x_s}(p)\right| \leq \frac{1}{c_0 k}, \quad s = 1, 2, \ldots, 2n, \tag{4.9}$$

where $c_0 > 0$ is a constant independent of k and the point p.

Proof From (4.2), we can check that

$$\begin{aligned}&\widetilde{u}_k(x)\\&\equiv \int e^{ik\Psi(x,y)}b(x,y,k)\chi(\sqrt{k}y_1)\cdots\chi(\sqrt{k}y_{2n})dv_M(y) \mod O(k^{-\infty})\\&\equiv \int e^{ik\Psi(x,\frac{y}{\sqrt{k}})}k^{-n}b(x,\frac{y}{\sqrt{k}},k)\chi(y_1)\cdots\chi(y_{2n})dv_M(y) \mod O(k^{-\infty}).\end{aligned} \tag{4.10}$$

From (3.3), Theorem 4.2 and note that $\Psi(0,0) = 0$, we can check that

$$\lim_{k\to\infty}\widetilde{u}_k(p) = \frac{1}{2}\pi^{-n}\left|\det R^L_p\right|\int\chi(y_1)\cdots\chi(y_{2n})dv_M(y).$$

Similarly, it is straightforward to check that $\lim_{k\to\infty}\frac{1}{\sqrt{k}}\frac{\partial\widetilde{u}_k}{\partial x_s}(p) = 0$, $s = 1, 2, \ldots, 2n$. Hence, there is a constant $\widetilde{k}_0 > 0$ such that for every $k \geq \widetilde{k}_0$, (4.7) and (4.9) hold. Since X is compact, $\widetilde{k}_0$ can be taken to be independent of the point p.

Now, we prove (4.8). Since $x \notin D$, from (4.1), we see that $|u_k(x)|^2_{h^{L^k}} \equiv 0 \mod O(k^{-\infty})$ outside D. Thus, there is a constant $\hat{k}_0 > 0$ such that for every $k \geq \hat{k}_0$, (4.8) holds. Since X is compact, $\hat{k}_0$ can be taken to be independent of the point p. The lemma follows. □

For every $j = 1, 2, \dots, n$, let

$$u_k^j := P_k\Big(s^k e^{k\phi}\sqrt{k}(y_{2j-1} + iy_{2j})\chi(\sqrt{k}y_1)\cdots\chi(\sqrt{k}y_{2n})\Big) \in H^0(M, L^k). \quad (4.11)$$

On D, we write $u_k^j = s^k e^{k\phi}\widetilde{u}_k^j$, $\widetilde{u}_k^j \in C^\infty(D)$, $j = 1, 2, \dots, n$. The following follows from some straightforward computation and essentially the same as the proof of Lemma 4.1. We omit the details.

Lemma 4.2 *With the notations used above, there is a $k_1 > 0$ independent of k and the point p such that for all $k \geq k_1$,*

$$\begin{aligned}
&\left|\widetilde{u}_k^j(p)\right| \leq \frac{1}{c_1 k}, \quad j = 1, 2, \dots, n, \quad \left|\frac{1}{\sqrt{k}}\frac{\partial \widetilde{u}_k^j}{\partial \overline{z}_s}(p)\right| \leq \frac{1}{c_1 k}, \quad j, s = 1, 2, \dots, n,\\
&\left|\frac{1}{\sqrt{k}}\frac{\partial \widetilde{u}_k^j}{\partial z_s}(p)\right| \leq \frac{1}{c_1 k}, \quad j, s = 1, 2, \dots, n-1, \quad j \neq s,\\
&\left|\frac{1}{\sqrt{k}}\frac{\partial \widetilde{u}_k^j}{\partial z_j}(p)\right| \geq c_1, \quad j = 1, 2, \dots, n, \qquad (4.12)
\end{aligned}$$

where $c_1 > 0$ is a constant independent of k and the point p.

From now on, we take k be a large constant so that $k \gg 2(k_0 + k_1)$, where $k_0 > 0$ and $k_1 > 0$ are constants as in Lemma 4.1 and Lemma 4.2. We can prove

Theorem 4.4 *$d\Phi_k(x) : T_x X \to T_x\mathbb{CP}^{d_k-1}$ is injective at every $x \in X$.*

Proof Fix $p \in X$ and let s be a local trivializing section of L on an open set $D \subset M$, $p \in D$. Let $u_k \in H^0(M, L^k)$ and $u_k^j \in H^0(M, L^k)$, $j = 1, 2, \dots, n$, be as in Lemma 4.1 and Lemma 4.2. From Lemma 4.1 and Lemma 4.2, it is not difficult to check that $u_k, u_k^1, u_k^2 \dots, u_k^n$ are linearly independent. Take $\{u_k, u_k^1, u_k^2, \dots, u_k^n, g_1, \dots, g_{m_k}\}$ be a basis (not orthogonal) for $H^0(M, L^k)$, $m_k = d_k - n - 1$. From Lemma 4.1 and Lemma 4.2, it is easy to see that

$$\text{the differential of the map } x \to \Big(\frac{u_k^1}{u_k}, \dots, \frac{u_k^n}{u_k}, \frac{g_1}{u_k}, \dots, \frac{g_{m_k}}{u_k}\Big) \text{ is injective at } p. \quad (4.13)$$

From (4.13) and some elementary linear algebra, it is not difficult to check that $d\Phi_k(p) : T_p X \to T_p\mathbb{CP}^{d_k-1}$ is injective. We omit the detail. □

Now, we can prove

Theorem 4.5 *For k large, $\Phi_k : X \to \mathbb{CP}^{d_k-1}$ is globally injective.*

Proof We assume that the claim of the theorem is not true. We can find $x_{k_j}, y_{k_j} \in M$, $x_{k_j} \neq y_{k_j}$, $0 < k_1 < k_2 < \cdots$, $\lim_{j\to\infty} k_j = \infty$, such that $\Phi_{k_j}(x_{k_j}) = \Phi_{k_j}(y_{k_j})$, for each j. We may suppose that there are $x_k, y_k \in M$, $x_k \neq y_k$, such that $\Phi_k(x_k) = \Phi_k(y_k)$, for each k. Thus, $[f_1(x_k), \cdots, f_{d_k}(x_k)] = [f_1(y_k), \cdots, f_{d_k}(y_k)]$, for each k. We conclude that for every $g_k \in H^0(M, L^k)$, there is a $\lambda_k \in \mathbb{C}$ such that

$$g_k(x_k) = \lambda_k g_k(y_k). \tag{4.14}$$

We may assume that $|\lambda_k| \geq 1$. Hence, for every $g_k \in H^0(M, L^k)$,

$$|g_k(x_k)|^2_{h^{L^k}} \geq |g_k(y_k)|^2_{h^{L^k}}. \tag{4.15}$$

Since M is compact, we may assume that $x_k \to p \in M$, $y_k \to q \in M$, as $k \to \infty$. Suppose that $p \neq q$. In view of Lemma 4.1, we see that there is a $v_k \in H^0(M, L^k)$ with $|v_k(y_k)|^2_{h^{L^k}} \geq c_0$ and $|v_k(x_k)|^2_{h^{L^k}} \leq \frac{1}{c_0 k}$, where $c_0 > 0$ is a constant independent of k. Thus, for k large, $|v_k(x_k)|^2_{h^{L^k}} < |v_k(y_k)|^2_{h^{L^k}}$. From this and (4.15), we get a contradiction. Thus, we must have $p = q$.

Let s be a local trivializing section of L on an open subset $D \subset X$ of p, $|s|^2_{h^L} = e^{-2\phi}$. Now, we assume that $x_k \to p \in M$, $y_k \to p \in M$, as $k \to \infty$. Let $z = (z_1, \ldots, z_n) = x = (x_1, \ldots, x_{2n})$, $z_j = x_{2j-1} + ix_{2j}$, $j = 1, \ldots, n$, be local holomorphic coordinates of X defined in some small neighbourhood of p such that (4.5) hold. We may assume that $x_k, y_k \in D$ for each k and the local coordinates x defined on D. We shall use the same notations as before.

Case I: $\limsup_{k\to\infty} \sqrt{k}\,|x_k - y_k| = M > 0$ (M can be ∞).
For simplicity, we may assume that

$$\lim_{k\to\infty} \sqrt{k}\,|x_k - y_k| = M, \quad M \in]0, \infty]. \tag{4.16}$$

On D, we write $f_j = s^k e^{k\phi} \widetilde{f}_j$, $\widetilde{f}_j \in C^\infty(D)$, $j = 1, \ldots, d_k$. Put

$$v_k(x) := \sum_{j=1}^{d_k} f_j(x)\overline{\widetilde{f}_j(y_k)} \in H^0(M, L^k). \tag{4.17}$$

We can check that

$$\begin{aligned} |v_k(x_k)|^2_{h^{L^k}} &= \left|\sum_{j=1}^{d_k} \widetilde{f}_j(x_k)\overline{\widetilde{f}_j(y_k)}\right|^2 = \left|P_{k,s}(x_k, y_k)\right|^2 = \left|e^{ik\Psi(x_k,y_k)} b(x_k, y_k, k)\right|^2 \\ &\leq e^{-2k\operatorname{Im}\Psi(x_k,y_k)} \left|b(x_k, y_k, k)\right|^2 \end{aligned} \tag{4.18}$$

and

$$|v_k(y_k)|^2_{h^{L^k}} = \left|P_{k,s}(y_k, y_k)\right|^2 = \left|e^{ik\Psi(y_k,y_k)} b(y_k, y_k, k)\right|^2 = |b(y_k, y_k, k)|^2. \tag{4.19}$$

From the fact that $\operatorname{Im}\Psi(x,y) \geq c\,|x-y|^2$, where $c > 0$ is a constant, (4.16), (4.18) and (4.19), we can check that

$$\lim_{k\to\infty} k^{-2n}\,|v_k(x_k)|^2_{h^{L^k}} \leq e^{-2cM^2}\,|b_0(p,p)|^2 < |b_0(p,p)|^2 = \lim_{k\to\infty} k^{-2n}\,|v_k(y_k)|^2_{h^{L^k}}, \tag{4.20}$$

where b_0 is the leading term of $b(x,y,k)$. Note that $b_0(p,p) = (2\pi)^{-n}\left|\det R^L(p)\right| > 0$ (see Theorem 3.1). From (4.20) and (4.15), we get a contradiction.

Case II: $\limsup_{k\to\infty}\sqrt{k}\,|x_k - y_k| = 0$.

Put $f_k(t) = \frac{|v_k(tx_k+(1-t)y_k)|^2_{h^{L^k}}}{P_k(tx_k+(1-t)y_k)P_k(y_k)}$, where v_k is as in (4.17). We can check that

$$f_k(t) = \frac{\left|\sum_{j=1}^{d_k}\widetilde{f}_j(tx_k+(1-t)y_k)\overline{\widetilde{f}_j(y_k)}\right|^2}{\sum_{j=1}^{d_k}\left|\widetilde{f}_j(tx_k+(1-t)y_k)\right|^2\sum_{j=1}^{d_k}\left|\widetilde{f}_j(y_k)\right|^2} = \frac{\left|P_{k,s}(tx_k+(1-t)y_k,y_k)\right|^2}{P_k(tx_k+(1-t)y_k)P_k(y_k)}. \tag{4.21}$$

From (4.14) and (4.21), it is easy to see that $0 \leq f_k(t) \leq 1$, $\forall t \in [0,1]$ and $f_k(0) = f_k(1) = 1$. Thus, for each k, there is a $t_k \in [0,1]$ such that $f''_k(t_k) \geq 0$. Hence,

$$\liminf_{k\to\infty}\frac{f''_k(t_k)}{|x_k-y_k|^2\,k} \geq 0. \tag{4.22}$$

From (4.2), we see that

$$\begin{aligned}
&\left|P_{k,s}(tx_k+(1-t)y_k,y_k)\right|^2 = e^{-2k\operatorname{Im}\Psi(tx_k+(1-t)y_k,y_k)}\,|b(tx_k+(1-t)y_k,y_k,k)|^2,\\
&P_k(tx_k+(1-t)y_k)\\
&= b(tx_k+(1-t)y_k,tx_k+(1-t)y_k,k) \sim \sum_{j=0}^{\infty}k^{n-j}b_j(tx_k+(1-t)y_k,tx_k+(1-t)y_k).
\end{aligned} \tag{4.23}$$

From (4.23), it is straightforward to calculate that

$$\begin{aligned}
&\frac{\partial\left|P_{k,s}(tx_k+(1-t)y_k,y_k)\right|}{\partial t}\\
&= e^{-2k\operatorname{Im}\Psi(tx_k+(1-t)y_k,y_k)}\\
&\quad\Big(\langle -2k\operatorname{Im}\Psi'_x(tx_k+(1-t)y_k,y_k)\,,\,x_k-y_k\rangle\,|b(tx_k+(1-t)y_k,y_k,k)|^2\\
&\qquad + \langle\, O(k^{2n})\,,\,x_k-y_k\,\rangle\Big),\\
&\frac{\partial^2\left|P_{k,s}(tx_k+(1-t)y_k,y_k)\right|}{\partial t^2}\\
&= e^{-2k\operatorname{Im}\Psi(tx_k+(1-t)y_k,y_k)}
\end{aligned}$$

$$\Big(\big(\langle -2k\mathrm{Im}\,\Psi'_x(tx_k+(1-t)y_k,y_k)\,,\,x_k-y_k\,\rangle\big)^2\,|b(tx_k+(1-t)y_k,y_k,k)|^2$$
$$+\langle -2k\mathrm{Im}\,\Psi''_x(tx_k+(1-t)y_k,y_k)(x_k-y_k)\,,\,x_k-y_k\,\rangle\,|b(tx_k+(1-t)y_k,y_k,k)|^2$$
$$+\langle -2k\mathrm{Im}\,\Psi'_x(tx_k+(1-t)y_k,y_k)\,,\,x_k-y_k\,\rangle\langle\, O(k^{2n})\,,\,x_k-y_k\,\rangle$$
$$+\langle\, O(k^{2n})(x_k-y_k)\,,\,x_k-y_k\,\rangle\Big),$$
$$\frac{\partial P_k(tx_k+(1-t)y_k,y_k)}{\partial t}=\langle\, O(k^{2n})\,,\,x_k-y_k\,\rangle,$$
$$\frac{\partial^2 P_k(tx_k+(1-t)y_k,y_k)}{\partial t^2}=\langle\, O(k^{2n})(x_k-y_k)\,,\,x_k-y_k\,\rangle, \tag{4.24}$$

where $\mathrm{Im}\,\Psi'_x(x,y)$ and $\mathrm{Im}\,\Psi''_x(x,y)$ denote the derivative and the Hessian of $\mathrm{Im}\,\Psi(x,y)$ with respect to x respectively. Note that

$$\big|\langle -2k\mathrm{Im}\,\Psi'_x(tx_k+(1-t)y_k,y_k)\,,\,x_k-y_k\,\rangle\big|\le\frac{1}{c_0}k\,|x_k-y_k|^2\to 0 \text{ as } k\to\infty$$

and

$$\langle -2k\mathrm{Im}\,\Psi''_x(tx_k+(1-t)y_k,y_k)(x_k-y_k)\,,\,x_k-y_k\,\rangle<-c_0k\,|x_k-y_k|^2,$$

where $c_0>0$ is a constant independent of k. From this observation, (4.21) and (4.24), it is straightforward to see that $\liminf_{k\to\infty}\frac{f''_k(t_k)}{|x_k-y_k|^2k}<0$. From this and (4.22), we get a contradiction.

The theorem follows. □

From Theorem 4.4 and Theorem 4.5, we obtain Theorem 4.3 and the Kodaira embedding theorem follows then.

Acknowledgments The author was partially supported by Taiwan Ministry of Science of Technology project 103-2115-M-001-001 and the Golden-Jade fellowship of Kenda Foundation.

References

[BBS04] Berman, R., Berndtsson, B., Sjöstrand, J.: A direct approach to Bergman kernel asymptotics for positive line bundles. Ark. Math. **46**(2), 197–217 (2008)

[BouSj76] Boutet de Monvel, L., Sjöstrand, J.: Sur la singularité des noyaux de Bergman et de Szegö. Astérisque **34–35**, 123–164 (1976)

[Cat97] Catlin, D.: The Bergman kernel and a theorem of Tian. Analysis and geometry in several complex variables, Katata, Trends in Mathematics, pp. 1–23

[DLM06] Dai, X., Liu, K., Ma, X.: On the asymptotic expansion of Bergman kernel. J. Differ. Geom. **72**(1), 1–41 (2006)

[Desps11] Demailly, J.-P.: Complex analytic and algebraic geometry. www-fourier.ujf-grenoble.fr/~demailly/books.html

[Hor90] Hörmander, L.: An introduction to complex analysis in several variables, vol. 7. North-Holland mathematical library, North-Holland Publishing Co., Amsterdam (1990)

[Hsiao08] Hsiao, C.-Y.: Projections in several complex variables. Mém. Soc. Math. France Nouv. Sér. **123**, 131 (2010)

[HMA12] Hsiao, C.-Y., Marinescu, G.: Asymptotics of spectral function of lower energy forms and Bergman kernel of semi-positive and big line bundles. Commun. Anal. Geom. **22**, 1–108 (2014)

[JS93] Ji, S., Shiffman, B.: Properties of compact complex manifolds carrying closed positive currents. J. Geom. Anal. **3**(1), 36–61 (1993)

[Ma10] Ma, X.: Geometric quantization on Kähler and symplectic manifolds. In: Proceedings of the International Congress of Mathematicians (ICM 2010), vol. II, pp. 785–810. Hyderabad, India, 19–27 Aug 2010

[MM07] Ma, X., Marinescu, G.: Holomorphic Morse inequalities and Bergman kernels. In: Progress in Mathematics, vol. 254, pp. 422. Birkhäuser, Basel (2007)

[MM08a] Ma, X., Marinescu, G.: Generalized Bergman kernels on symplectic manifolds. Adv. Math. **217**(4), 1756–1815 (2008)

[Zel98] Zelditch, S.: Szegö kernels and a theorem of Tian. Int. Math. Res. Not. **6**, 317–331 (1998)

On the Density and the Volume Density Property

Shulim Kaliman and Frank Kutzschebauch

Abstract This article gives a short introduction into the notions of density property (DP) and volume density property (VDP). Moreover we develop an effective criterion of verifying whether a given X has VDP. As an application of this method we give a new proof of the basic fact that the product of two Stein manifolds with VDP admits VDP.

Keywords Stein manifolds · Density property · Flexibility · Volume density property · Holomorphic automorphisms

1 Introduction

The density property has emerged from the so called Andersén-Lempert theory of automorphisms of affine space $\mathbb{C}^n$ $n \geq 2$, which was presented by Rosay in his overview talks at the 1.KSCV Conference in 1997 [Ros]. It turns out that there are many more Stein manifolds whose automorphism group is similarly big. For a Stein manifold X the density property is a precise way of saying that X has a big automorphism group. Density property has become an important notion for the study of many geometric questions in Several Complex Variables.

Definition 1.1 (*Varolin,* [Var01]) A complex manifold X has the density property (DP) if in the compact-open topology the Lie algebra generated by complete holomorphic vector fields $\mathrm{Lie}_{\mathrm{hol}}(X)$ on X is dense in the Lie algebra of all holomorphic vector fields on X.

We remind the reader that a holomorphic vector field Θ on a complex manifold X is called complete if the ODE

S. Kaliman
Department of Mathematics, University of Miami, Coral Gables, FL 33124, USA
e-mail: kaliman@math.miami.edu

F. Kutzschebauch (✉)
Mathematisches Institut, Universität Bern, Sidlerstr. 5, 3012 Bern, Switzerland
e-mail: Frank.Kutzschebauch@math.unibe.ch

F. Bracci et al. (eds.), *Complex Analysis and Geometry*, Springer Proceedings in Mathematics & Statistics 144, DOI 10.1007/978-4-431-55744-9_12

$$\frac{d}{dt}\varphi(x,t) = \Theta(\varphi(x,t))$$

$$\varphi(x,0) = x$$

has a solution $\varphi(x, t)$ defined for all complex times $t \in \mathbb{C}$ and all starting points $x \in X$. It gives a complex one-parameter subgroup of the holomorphic automorphism group $\mathrm{Aut}_{hol}(X)$. The set of complete vector fields on X will be denoted by $\mathrm{CVF}(X)$.

In the presence of a holomorphic volume form ω on X (i.e., a nowhere vanishing holomorphic n-form, where $n = \dim X$) it is natural to consider a similar notion for vector fields θ preserving this volume form, i.e., such that $L_\theta \omega = 0$ or equivalently by the Cartan formula $\mathrm{d}i_\theta \omega = 0$. One also refers to such fields as fields of ω-divergence zero or ω-divergence free(where the ω-divergence $\mathrm{div}_\omega \theta$ of a vector field θ is defined by the formula $L_\theta \omega = \mathrm{div}_\omega \theta\ \omega$).

Definition 1.2 (*Varolin,* [Var01]) A complex manifold X has the ω-volume density property if in the compact-open topology the Lie algebra $\mathrm{Lie}^{\omega}_{\mathrm{hol}}(X)$ generated by complete holomorphic vector fields of ω-divergence zero on X is dense in the Lie algebra of all holomorphic vector fields of ω-divergence zero on X.

We will sketch in Chap. 2 some remarkable consequences these two properties have together with some very recent applications.

In Chap. 3 we give an analogous criterion for volume density property to what we developed for the algebraic volume density property in [KaKu]. We apply this criterion to give a short new proof of the fact that that the product of two Stein manifolds with VDP has again VDP. The first quite complicated proof of this fact using Grothendiecks theory of completions of tensor products has been given by the authors in [KaKu3].

We end the paper with some open problems.

2 Main Feature of Density and Volume Density Property and Some Recent Applications

The density property is a precise way of saying that the automorphism group of a manifold is big, in particular for a Stein manifold this is underlined by the main result of the theory (see [FR93] for $\mathbb{C}^n$, [Var01], a detailed proof can be found in the Appendix of [Rit13] or in [For11]).

Theorem 1 (Andersén-Lempert theorem) *Let X be a Stein manifold with the density property and let Ω be an open subset of X. Suppose that $\Phi : [0, 1] \times \Omega \to X$ is a continuous map such that*

(1) *$\Phi_t : \Omega \to X$ is holomorphic and injective for every $t \in [0, 1]$,*
(2) *$\Phi_0 = \mathrm{id} : \Omega \to X$ is the natural inclusion of Ω into X, and*

(3) *$\Phi_t(\Omega)$ is a Runge subset*[1] *of X for every $t \in [0, 1]$.*

Then for each $\varepsilon > 0$ and every compact subset $K \subset \Omega$ there is a continuous family, $\alpha : [0, 1] \to \mathrm{Aut}_{hol}(X)$ of holomorphic automorphisms of X such that

$$\alpha_0 = id \text{ and } |\alpha_t - \Phi_t|_K < \varepsilon \text{ for every } t \in [0, 1]$$

Here is a number of consequences the density property has, the proof of each of them is a certain application of the Andersén-Lempert theorem:

If X is a Stein manifold with DP, then

(I) X is covered by Fatou-Bieberbach domains, i.e., each $x \in X$ has a n-hood $\Omega_x \subset X$ biholomorphic to $\mathbb{C}^{\dim X}$, see [Var00].

(II) There is $\varphi : X \to X$, injective holomorphic not surjective (biholomorphic images of X in itself), see [Var00].

(III) If X is Stein with DP $\dim X \geq 3$ and Y is a complex manifold such that $End(X)$ and $End(Y)$ are isomorphic as abstract semigroups, then X and Y are biholomorphic or anti-biholomorphic, see [A, AW]. We believe that the same is true if the dimension of X is 2, but the known proofs do not apply.

(IV) There are complete homomorphic vector fields $\theta_1, \ldots, \theta_N \in CVF_{hol}(X)$ such that $\mathrm{span}(\theta_1(x), \ldots, \theta_N(x)) = T_x X \quad \forall x \in X$ (see [KaKu2]) and therefore

(V) X is an Oka-Forstnerič manifold which means it is an appropriate (nonlinear) target for generalizing classical Oka-Weil interpolation and Runge approximation for holomorphic functions (linear target $\mathbb{C}$) or sections of vector bundles (linear target as well). More precisely, the following is true (see [For11]). For any Stein space W, complex subspace W', compact $\mathcal{O}(X)$-convex subset $K = \widehat{K} \subset W$ and any $\varphi : W \to X$ continuous, such that the restriction to $W' \cup K$ is holomorphic, there is a homotopy of continuous maps

$$h : [0, 1] \times W \to X$$

from the continuous $h_0 = \varphi$ to a holomorphic h_1, with
interpolation: $h_t = \varphi$ on $W' \quad \forall t$
and
approximation: $|h_t - \varphi|_K$ arbitrary small $\forall t$.

(VI) $\mathrm{Aut}_{hol}(X)$ acts ∞-transitively on X, i.e., for all natural numbers N and pairs of N-tupels of distinct points $(x_1, x_2, \ldots, x_N)$ and $(y_1, y_2, \ldots, y_N)$ there is a holomorphic automorphism $\alpha \in \mathrm{Aut}_{hol}(X)$ with $\alpha(x_i) = y_i$ $\forall i = 1, 2, 3, \ldots, N$, see [Var00].

(VII) Moreover a parametrized version has been proven recently (see [KR])

---

[1]Recall that an open subset U of X is Runge if any holomorphic function on U can be approximated by global holomorphic functions on X in the compact-open topology. Actually, for X Stein (Footnote 1 continued)
by Cartan's Theorems A and B this definition implies more: for any coherent sheaf on X its section over U can be approximated in the compact-open topology by global sections.

Theorem 2 *Let W be a Stein manifold and X a Stein manifold with the density property.*
Let $x : W \to X^N \setminus \{(z^1, \dots, z^N) \in X^N; z^i = z^j \text{ for some } i \neq j\}$ be a holomorphic map. Then the parametrized points $x^1(w), \dots, x^N(w)$ are simultaneously standardizable by an automorphism lying in the path-connected component of the identity $(\mathrm{Aut}_W(X))^0$ of $\mathrm{Aut}_W(X)$ if and only if x is null-homotopic.

Here simultaneously standardizable means that given any fixed positions $\tilde{x}_1, \tilde{x}_2, \dots, \tilde{x}_N \in X$ there are holomorphic automorphisms α of X depending holomorphically on w, i.e., an element of $\mathrm{Aut}_W(X) = \{\alpha \in \mathrm{Aut}(W \times X); \alpha(w, z) = (w, \alpha^w(z))\}$, with $\alpha^w(x^j(w)) = \tilde{x}_j$ for all $w \in W$ and $j = 1, \dots, N$.

Coming back to the volume density property a theorem similar to the Andersén-Lempert theorem formulated above can be proven. In addition to the natural condition on volume preservation the open set Ω has to satisfy an additional cohomological condition, namely the map $H^{n-1}(X) \to H^{n-1}(\Omega)$ induced by the inclusion of Ω into X is surjective. For contractible X, for example $X = \mathbb{C}^n$, this is equivalent to $H^{n-1}(\Omega) = 0$. All properties mentioned above, except for (I) are also known for Stein X with VDP, if in addition we demand that $\dim X \geq 2$. Property (I) is unknown in this case. Concerning the dimension condition it is easy to see that DP implies $\dim X \geq 2$, whereas $\mathbb{C}^*$ is a (the only) 1-dimensional Stein manifold with VDP and has to be excluded for all properties except (V) to hold. In order for (VII) to hold for Stein X with VDP at the moment we need the additional assumption that X is contractible. It is unknown whether this condition can be relaxed, see [KR].

Here is the complete list of known **Examples of Stein manifolds with DP or VDP**

- A homogeneous space $X = G/H$ has DP, where G is a Linear Algebraic Group and H is a closed reductive subgroup, such that $X^0 \neq (\mathbb{C}^*)^k, \mathbb{C}$. Here $(\mathbb{C}^*)^k$ for $k \geq 2$ is unknown. For $\mathbb{C}^n$ the result is due to Andersén-Lempert see [AL92], for G semisimple with trivial center to Varolin-Toth see [VT1, VT2], for G Linear Algebraic is due to the authors see [KaKu1], the general case to Kaliman-Donzelli-Dvorsky, see [DDK].[2]
- A homogeneous space $X = G/H$ as above has VDP w.r.t. left invariant (Haar) form in case this form exists. This result is due to the authors, for G see [KaKu4], for the general case see [KaKu]. We would like to mention that before the work of the authors only very view examples of manifolds with VDP had been known. They were found by Andersén ($\mathbb{C}^n$ in [And1] and Varolin, e.g., $Sl_2(\mathbb{C})$, see [Var00, Var01],
- The manifolds (sometimes called suspensions or modifications) $\{(u, v, z) \in \mathbb{C}_u \times \mathbb{C}_v \times \mathbb{C}^n_z \ : \ uv = f(z)\}$ where $f \in \mathcal{O}(\mathbb{C}^n)$ has a smooth zero locus $Z = \{z \in \mathbb{C}^n : f(z) = 0\}$ have DP , and in case f is a polynomial and $H^{n-2}(Z) = 0$ they have VDP. Both results are due to the authors see [KaKu2] and [KaKu4].

[2]In fact in the coming paper of the authors these results are extended to **affine** homogeneous spaces of linear algebraic groups. More, precisely any such a space different from $\mathbb{C}$ or $(\mathbb{C}^*)^k$ has DP. Similarly, any such a space (including $\mathbb{C}$ or $(\mathbb{C}^*)^k$) equipped with a left invariant volume form has VDP.

- Danilov-Gizatullin surfaces have DP due to Donzelli, see [Do].
- A hypersurface S in $\mathbb{C}^3_{x,y,z}$ given by an equation

$$p(x) + q(y) + xyz = 1$$

where p and q are polynomials such that $p(0) = q(0) = 0$ and $1 - p(x)$ and $1 - q(y)$ have simple roots only, has VDP. This is a recent result of the authors, [KaKu]. The surface has only discrete algebraic automorphism group.
- New examples by Matthias Leuenberger, for instance, the Koras-Russel threefold $\{(x, y, s, t) \in \mathbb{C}^4 : x + x^2y + s^2 + t^3 = 0\}$ with both DP and VDP, see [Leu14].

Here comes a number of applications which in one or the other form among other things use the flexible behavior described by the Andersén-Lempert theorem. The precise statements can be looked up at the references.

(a) Embedding Stein spaces of dimension n into affine space $\mathbb{C}^N$ of the optimal dimension $N = [\frac{3}{2}n] + 1$ **with interpolation**, see [FIKP]. Without interpolation this is the celebrated result of Gromov-Eliashberg and Schürmann. Prezelj was able to prove almost the same interpolation result but loosing one dimension in half of the cases by adapting the methods of Gromov-Eliashberg. Instead of following the Gromov-Eliashberg proof one can simply use their result and achieve interpolation by constructing a sequence of automorphisms, each moving one point into the correct position at the time. The construction of the automorphisms involves the Andersén-Lempert theorem.
(b) Embedding many more examples of Riemann surfaces $R \hookrightarrow \mathbb{C}^2$. There is a series of results due to Fornæss-Wold and Forstnerič, see [FW1, FW2, W1, W2, W3] substantially enlarging the number of examples of open Riemann surfaces which can be properly embedded into $\mathbb{C}^2$. The general notoriously difficult problem whether all Riemann surfaces embed properly holomorphically into $\mathbb{C}^2$ remains open.
(c) Fatou-Bieberbach domains in $\mathbb{C}^2$ with boundary of Hausdorff dimension 4. It has been known from dynamics that Fatou-Bieberbach domains can have boundaries of Hausdorff dimension d for any d between 3 and 4. The case $d = 3$ is a famous result of Stensønes. Peters and Fornaess-Wold settled $d = 4$ with the help of the Andersén-Lempert theorem [HFW].
(d) Embedding $\mathbb{C}^k \hookrightarrow \mathbb{C}^n$ in many different ways. The Andersén-Lempert theorem was first used by Forstnerič, Globevnik, Rosay in [FGR], later by Buzzard and Fornæss [BF]. The most elaborated results are due to the second author and his students [LK, BK].
(e) There are (many different) non-linearizable holomorphic group actions on $\mathbb{C}^n$ as a consequence of the existence of different embeddings [DK1, DK2, LK]. This is in turn an application of an application of the DP for $\mathbb{C}^n$ but it seems worth mentioning since it solved the Holomorphic Linearization Problem to the negative.
(f) Each open Riemann surface with abelian fundamental group admits an acyclic embedding into a Stein manifold with DP (thus into an Oka-Forstnerič manifold).

This is the first result, due to Ritter, concerning the general question whether any Stein manifold embeds acyclically into an Oka-Forstnerič manifold, see [Rit13].

(g) If X is a Stein manifold with DP or VDP and W a Stein manifold such that $\dim X \geq 2 \dim W + 1$ then there is a proper holomorphic embedding $W \hookrightarrow X$. In case X is affine space (linear target) this is the classical Bishop-Remmert embedding theorem. This result by Andrist, Fornæss-Wold, Forstnerič and Ritter can be viewed as a generalization to any (non linear) Stein target with DP or VDP, [AFFWR].

(h) Constructions of Fatou-Bieberbach domains in $\mathbb{C}^2$ which are not Runge and a long $\mathbb{C}^2$ which is not $\mathbb{C}^2$. These beautiful results by Fornæss-Wold use DP for $\mathbb{C} \times \mathbb{C}^*$, see [W4, W5].

(i) Any open Riemann surface admits a proper harmonic map to $\mathbb{R}^2$ (disproving a conjecture of Schoen and Yau). This result due to Andrist and Fornæss-Wold uses VDP for $\mathbb{C}^* \times \mathbb{C}^*$, see [AW].

There are versions of the Anderseń-Lempert theorem with (very special) control on (very special) non-compact sets. One of them leads to

Theorem 3 ([KW]) *Let $\varphi : \mathbb{R}^s \to \mathbb{R}^s$ be a smooth diffeomorphism, and assume that $s < n$. Then ϕ can be approximated in the fine Whitney topology by holomorphic automorphisms of $\mathbb{C}^n$.*

One can generalize the DP to (non-smooth) complex spaces X by considering only vector fields vanishing on a subvariety which contains the singular locus. These versions of DP have the same remarkable consequences as in the Andersén-Lempert theorem, but for the automorphisms of X leaving the subset Y fixed (up to certain order). The notions have been defined in a recent work of Liendo, Leuenberger and the second author. Before coming to the definition we would also like to remark that in case of X being affine algebraic one conveniently works with the dense (in the Lie algebra of holomorphic vector fields) Lie sub algebra of algebraic vector fields on X. One defines in the straightforward manner the algebraic density property ADP and algebraic volume density property AVDP which imply DP resp. VDP. Thus these algebraic versions are considered as tools for proving DP or VDP.

Definition 2.1 Let X be an affine algebraic variety and let X^{sing} be the singular locus. Let $Y \subseteq X$ be an algebraic subvariety of X containing X^{sing} and let $I = I(Y) \subseteq \mathbb{C}[X]$ be the ideal of Y. Let $\text{VF}_{\text{alg}}(X, Y)$ be the $\mathbb{C}[X]$-module of vector fields vanishing in Y and $\text{Lie}_{\text{alg}}(X, Y)$ be the Lie algebra generated by all the complete vector fields in $\text{VF}_{\text{alg}}(X, Y)$.

Definition 2.2 X has the strong ADP relative to Y if $\text{VF}_{\text{alg}}(X, Y) = \text{Lie}_{\text{alg}}(X, Y)$.

Furthermore, we say that X has the ADP relative to Y if there exists $\ell \geq 0$ such that $I^{\ell}\, \text{VF}_{\text{alg}}(X, Y) \subseteq \text{Lie}_{\text{alg}}(X, Y)$.

If we let $Y = X^{\text{sing}}$ we simply say that X has the strong ADP or the ADP, respectively.

In connection with this relative property we have two results:

Theorem 4 ([KaKu1]) *Let $Y \subset \mathbb{C}^n$ be an algebraic subvariety with* codim$Y \geq 2$. *Then $\mathbb{C}^n$ has ADP relative to Y.*

Theorem 5 ([LLK]) *Let X be an affine toric variety of dimension n at least two for the torus $T = (\mathbb{C}^*)^n$ and let Y be a T-invariant closed subvariety of X containing X^{sing}. Then X has the ADP relative to Y if and only if $X \setminus Y \neq T$.*

Every affine non-smooth (the smooth ones are the torus $\mathbb{C}^* \times \mathbb{C}^*$, $\mathbb{C}^2$ and $\mathbb{C}^\star \times \mathbb{C}$) toric surface is obtained as a quotient of $\mathbb{C}^2$ by the action of a cyclic group. Let $d > e$ be relatively prime positive integers. We denote by $V_{d,e}$ the toric surface obtained as the quotient of $\mathbb{C}^2$ by the $\mathbb{Z}_d$-action $\zeta \cdot (u, v) = (\zeta u, \zeta^e v)$, where ζ is a primitive d-th root of unity. We can exactly characterize in these terms which toric surfaces have strong ADP.

Theorem 6 ([LLK]) *$V_{d,e}$ has the strong ADP if and only if e divides $d + 1$ and $e^2 \neq d + 1$.*

3 The Criterion and Volume Density of Products

Notation: Let X be a Stein manifold with a holomorphic volume form ω. Let $\mathscr{C}_k(X)$ be the space of holomorphic differential k-forms on X and $\mathscr{Z}_k(X)$ and $\mathscr{B}_k(X)$ be its subspaces of closed and exact k-forms respectively. If dim $X = n$ then there exists an isomorphism $\Theta : \mathrm{VF}_\omega(X) \to \mathscr{Z}_{n-1}(X)$ given by the formula $\xi \to \iota_\xi \omega$ where $\iota_\xi \omega$ is the interior product of ω and $\xi \in \mathrm{VF}_\omega(X)$.

Consider the homomorphism $D_k : \mathscr{C}_{k-1}(X) \to \mathscr{B}_k(X)$ generated by outer differentiation d and let $D = D_{n-1}$. The main theme of our new criterion is the search for a $\mathscr{O}(X)$-module in the space $D^{-1} \circ \Theta(\overline{\mathrm{Lie}^\omega_{\mathrm{hol}}(X)})$. With some additional assumptions the existence of such a module implies VDP.

Definition 3.1 Let ξ and η be nontrivial complete holomorphic vector fields on a Stein manifold X. We say that the pair (ξ, η) is semi-compatible

if the closure of the span of Ker $\xi \cdot$ Ker η *contains a nonzero ideal of $\mathscr{O}(X)$.*

The largest ideal contained in the closure of the span will be called the associate ideal of the pair (ξ, η).

The next simple observation provides a crucial connection between semi-compatibility and existence of $\mathscr{O}(X)$-modules in $D^{-1} \circ \Theta(\overline{\mathrm{Lie}^\omega_{\mathrm{hol}}(X)})$ where $D = D_{n-1}$.

Proposition 3.1 *Let ξ and η be vector fields from* $\mathrm{VF}_\omega(X)$. *Then*

$$\iota_{[\xi,\eta]}\omega = \mathrm{d}\iota_\xi \iota_\eta \omega. \tag{1}$$

Proof Recall the following relations between the outer differentiation d, Lie derivative L_ξ and the interior product ι_ξ

$$L_\xi = \mathrm{d}\iota_\xi + \iota_\xi \mathrm{d} \text{ and } [L_\xi, \iota_\eta] = \iota_{[\xi,\eta]}. \tag{2}$$

By this formula

$$\iota_\xi \mathrm{d}\iota_\eta \omega = \iota_\xi (L_\eta - \iota_\eta \mathrm{d})\omega$$

where the right-hand side is zero since $L_\eta \omega - \iota_\eta \mathrm{d}\omega = 0$ for closed ω and η of ω-divergence zero. Then another application of formula (2) in combination with the fact that $\iota_\xi \mathrm{d}\iota_\eta \omega = 0$ yields

$$[L_\xi, \iota_\eta]\omega = L_\xi \iota_\eta \omega - \iota_\eta L_\xi \omega = L_\xi \iota_\eta \omega = \mathrm{d}\iota_\xi \iota_\eta \omega + \iota_\xi \mathrm{d}\iota_\eta \omega = \mathrm{d}\iota_\xi \iota_\eta \omega.$$

Thus by formula (2) we have the desired equality

$$\iota_{[\xi,\eta]}\omega = \mathrm{d}\iota_\xi \iota_\eta \omega.$$

□

Let ξ, η be nonzero complete divergence-free holomorphic vector fields on X, $f \in \operatorname{Ker} \xi$, and $g \in \operatorname{Ker} \eta$. Replacing ξ and η in Formula (1) by $f\xi$ and $g\eta$ respectively we can see that $(fg)\iota_\xi \iota_\eta \omega \in D^{-1} \circ \Theta(\mathrm{Lie}^\omega_{\mathrm{hol}}(X))$. Hence one has the following.

Corollary 3.1 *Let X be a complex manifold equipped with a holomorphic volume form ω and let ξ and η be semi-compatible divergence-free vector fields on X. Then $D^{-1} \circ \Theta(\overline{\mathrm{Lie}^\omega_{\mathrm{hol}}(X)})$ contains a nontrivial $\mathscr{O}(X)$-submodule L of the module $\mathscr{C}_{n-2}(X)$.*

Let $\mu(x) \subset \mathscr{O}(X)$ be the maximal ideal of functions vanishing at $x \in X$ and let L be the largest $\mathscr{O}(X)$-submodule of $D^{-1} \circ \Theta(\overline{\mathrm{Lie}^\omega_{\mathrm{hol}}(X)})$. By Cartan's Theorem B equality $L = \mathscr{C}_{n-2}(X)$ holds as soon as $L/\mu(x)L = \mathscr{C}_{n-2}(X)/\mu(x)\mathscr{C}_{n-2}(X)$ for every $x \in X$. The latter is provided by Condition (A) below and we have the following.

Proposition 3.2 *Let X be a Stein manifold equipped with a holomorphic volume form ω and let $(\xi_j, \eta_j)_{j=1}^k$ be pairs of divergence-free semi-compatible vector fields. Let I_j be the ideal associated with (ξ_j, η_j), and let $I_j(x) = \{f(x) | f \in I_j\}$ for $x \in X$. Suppose that*

(A) *for every $x \in X$ the set $\{I_j(x)\xi_j(x) \wedge \eta_j(x)\}_{j=1}^k$ generates the fiber $\Lambda^2 T_x X$ of $\Lambda^2 TX$ over x.*

Then $\Theta(\overline{\mathrm{Lie}^\omega_{\mathrm{hol}}(X)})$ contains $\mathscr{B}_{n-1}(X)$.

As a consequence of Proposition 3.2 we have our criterion.

Theorem 7 *Let X be a Stein manifold equipped with a holomorphic volume form ω and pairs of divergence-free semi-compatible vector fields satisfying*

(A) *for every $x \in X$ the set $\{I_j(x)\xi_j(x) \wedge \eta_j(x)\}_{j=1}^k$ generates the fiber $\Lambda^2 T_x X$ of $\Lambda^2 TX$ over x.*

and

(B) *the image of $\Theta(\overline{\mathrm{Lie}^{\omega}_{\mathrm{hol}}(X)})$ under De Rham homomorphism $\Phi_{n-1} : \mathscr{Z}_{n-1}(X) \to H^{n-1}(X, \mathbb{C})$ coincides with the subspace $\Phi_{n-1}(\mathscr{Z}_{n-1}(X))$ of $H^{n-1}(X, \mathbb{C})$.*

Then $\Theta(\overline{\mathrm{Lie}^{\omega}_{\mathrm{hol}}(X)}) = \mathscr{Z}_{n-1}(X)$ and therefore $\overline{\mathrm{Lie}^{\omega}_{\mathrm{hol}}(X)} = \mathrm{VF}_{\omega}(X)$, i.e., X has the volume density property.

Theorem 8 *Let (X, ω_X) and (Y, ω_Y) be Stein manifolds with VDP. Then the product $(X \times Y, \omega_X \wedge \omega_Y)$ has VDP.*

Proof The proof is an application of Theorem 7. Set $n := \dim X$ and $m := \dim Y$, $\omega = \omega_X \wedge \omega_Y$.

First we will show that Condition (B) is fulfilled. Remember that the de Rham cohomology of Stein manifolds can be calculated by means of holomorphic forms [GR] p.155. Let α be a holomorphic $(n+m-1)$ form on the Stein manifold $X \times Y$. By the Künneth formula we can assume that either $\alpha = \alpha_1 \wedge \alpha_2$ where α_1 is a holomorphic $(n-1)$form on X and α_2 a holomorphic m-form on Y or α_1 is a holomorphic n-form on X and α_2 a holomorphic $(m-1)$-form on Y. By symmetry it is enough to consider the first case. Since ω_Y is a volume form we have $\alpha_2 = f(y)\omega_Y$ for some holomorphic function $f \in \mathscr{O}(Y)$. Since X has VDP there is a divergence-free holomorphic field $\theta \in \overline{\mathrm{Lie}^{\omega_X}_{hol}(X)}$ such that $i_\theta \omega_X$ is cohomologous to α_1. The field $f(y)\theta$ is divergence-free on $X \times Y$ and clearly contained in $\overline{\mathrm{Lie}^{\omega}_{\mathrm{hol}}(X \times Y)}$ (multiply one function in each iterated Lie bracket by $f(y)$). Moreover $i_{f\theta}\omega = i_\theta \omega_X \wedge \alpha_2$ represents the class of α.

In order to prove condition (A) remark that if θ is a complete holomorphic vector field on X and η a complete holomorphic vector field on Y, the pair (θ, η) is semi-compatible on $X \times Y$. Indeed, let the Stein manifolds X and Y be closed submanifolds of $\mathbb{C}^N$ resp. $\mathbb{C}^M$. The algebra P_X consisting of restrictions of polynomial functions on $\mathbb{C}^N$ to X are contained in $Ker(\eta)$. Analogously the algebra P_Y consisting of restrictions of polynomial functions on $\mathbb{C}^M$ to Y are contained in $Ker(\theta)$. Thus $Ker(\eta) \cdot Ker(\theta)$ contains all restrictions of polynomial functions on $\mathbb{C}^N \times \mathbb{C}^M$ to the closed submanifold $X \times Y$. By the Oka-Weill approximation theorem the closure of $Ker(\eta) \cdot Ker(\theta)$ is $\mathscr{O}(X \times Y)$, the associated ideal is therefore the whole algebra of holomorphic functions.

Next remark that VDP implies the existence of finitely many globally integrable divergence-free holomorphic fields which span the tangent space at every point, see e.g. [KaKu3]. Let θ_i $i = 1, 2, \ldots, k$ and η_i $i = 1, 2, \ldots, l$ be such fields for X resp. Y. The semi-compatible pairs (θ_i, η_j) span the subspace $T_x X \wedge T_y Y$ of $\Lambda^2 T_{(x,y)} X \times Y$ at a point (x, y). In order to span the remaining part of $\Lambda^2 T_{(x,y)} X \times Y$, namely the subspaces $\Lambda^2 T_x X$ and $\Lambda^2 T_y Y$ we will to our pairs of semi-compatible fields apply holomorphic volume preserving automorphisms h of $X \times Y$ fixing the point (x, y). The image of a semi-compatible pair under a volume preserving automorphism is again semi-compatible. By symmetry it's enough to show how to span $\Lambda^2 T_x X$. We will show how to get $\theta_i \wedge \theta_j(x)$.

The following is easily proved in local coordinates:

Let ν be a completely integrable vector field on a manifold Z and $f \in \mathrm{Ker}\, \nu$ be a function vanishing at a point z, i.e., $f(z) = 0$. Then the phase flow φ_t associated with the completely integrable field $f\nu$ generates an isomorphism $T_z Z \to T_z Z$ given by the formula

$$w \to w + t\,df(w)v \tag{3}$$

where $v = \nu(z)$.

The holomorphic volume preserving automorphism h of $X \times Y$ fixing the point (x, y) will be the time 1 map of the globally integrable divergence-free vector field $f\theta_2$ where $f \in \mathcal{O}(Y)$ is chosen such that $f(y) = 0$ and $df(\eta_m) = 1$. For this we first chose m such that the field η_m does not vanish at y and then use Cartans Theorem A for Stein manifolds to find the appropriate f. The evaluation of the semi-compatible pair (θ_1, η_m) transported by the automorphism h at the h-fixed point (x, y) will be by formula (3) equal to $\theta_1(x), \eta_m(y) + \theta_2(x)$. Since $\theta_1(x) \wedge \eta_m(y)$ is already spanned by the semi-compatible pair (θ_1, η_m) we are done.

Since both conditions (A) and (B) from our criterion are fulfilled the proof is complete. □

4 Open Problems

Open Problem 1: Suppose X is a Stein manifold with density property and $Y \subset X$ is a closed submanifold. Is there always another proper holomorphic embedding $\varphi : Y \hookrightarrow X$ which is not equivalent to the inclusion $i : Y \hookrightarrow X$?
Here we say that two proper holomorphic embeddings $\varphi_{1,2} : Y \hookrightarrow X$ are equivalent if there are holomorphic automorphisms $\alpha \in \mathrm{Aut}_{hol}(X)$ and $\beta \in \mathrm{Aut}_{hol}(Y)$ such that $\alpha \circ \varphi_1 = \varphi_2 \circ \beta$ or equivalently there is holomorphic automorphism $\alpha \in \mathrm{Aut}_{hol}(X)$ such that the images of $\alpha \circ \varphi_1$ and φ_2 coincide. We should remark that an affirmative answer to this question is stated in [Var00], but the author apparently had another (weaker) notion of equivalence in mind.
Open Problem 2: Is any Stein manifold X, $n = \dim X \geq 2$ with VDP covered by open subsets biholomorphic to $\mathbb{C}^n$? In particular, does there exist an open subset biholomorphic to $\mathbb{C}^2$ in $\mathbb{C}^* \times \mathbb{C}^*$?
Open Problem 3: [see, e.g., Rosay at KSCV 1 in 1997 [Ros]] Does $(\mathbb{C}^*)^k$ for $k \geq 2$ have DP?
$(\mathbb{C}^*)^k$ has VDP, see [Var01], which was crucial for disproving Schoen-Yau conjecture (see application (i) above). It is conjectured that $\mathrm{Aut}_{hol}((\mathbb{C}^*)^k)$ preserves the Haar volume form $\frac{dz_1}{z_1} \wedge \ldots \wedge \frac{dz_k}{z_k}$ up to sign. For the subgroup $\mathrm{AAut}_{hol}((\mathbb{C}^*)^k)$ of the

holomorphic automorphism group $\mathrm{Aut}_{hol}((\mathbb{C}^*)^k)$ generated by flows of complete algebraic vector fields this conjecture has been confirmed in [And].

Acknowledgments This research was started during a visit of the first author to the University of Bern and continued during a visit of the second author to the University of Miami, Coral Gables. We thank these institutions for their generous support and excellent working conditions. The research of the first author was also partially supported by NSA Grant no. H982301010185 and the second author was also partially supported by Schweizerische Nationalfonds grants No. 200020-134876/1 and 200021-140235/1

References

[And] Andersén, E.: Complete vector fields on $(\mathbb{C}^*)^n$. Proc. Am. Math. Soc. **128**(4), 1079–1085 (2000)

[And1] Andersén, E.: Volume-preserving automorphisms of $\mathbb{C}^n$. Complex Var. Theory Appl. **14**(1–4), 223–235 (1990)

[AL92] Andersén, E., Lempert, L.: On the group of holomorphic automorphisms of C^n. Invent. Math. **110**(2), 371–388 (1992)

[A] Rafael, B.: Andrist, Stein spaces characterized by their endomorphisms. Trans. Am. Math. Soc. **363**(5), 2341–2355 (2011)

[AFFWR] Andrist, R., Forstnerič, F., Ritter, T., Wold, E.F.: Proper holomorphic embeddings into Stein manifolds with the density property. arXiv: To appear in J. d'Analyse Math

[AW] Andrist, R.B., Wold, E.F.: Riemann surfaces in Stein manifolds with density property. arXiv:1106.4416

[BK] Borell, S., Kutzschebauch, F.: Non-equivalent embeddings into complex Euclidean spaces. Int. J. Math. **17**(9), 1033–1046 (2006)

[BF] Buzzard, G.T., Fornæss, J.-E.: An embedding of $\mathbb{C}$ with hyperbolic complement. Math. Ann. **306**(3), 539–546 (1996)

[DK1] Derksen, H.: Frank Kutzschebauch Nonlinearizable holomorphic group actions. Math. Ann. **311**(1), 41–53 (1998)

[DK2] Derksen, H., Kutzschebauch, F.: Global holomorphic linearization of actions of compact Lie groups on $\mathbb{C}^n$. Complex geometric analysis in Pohang (1997). Contemp. Math. **222**, 201–210 (Amer. Math. Soc., Providence, RI, 1999)

[Do] Donzelli, F.: Algebraic density property of Danilov-Gizatullin surfaces. Math. Z. **272**(3–4), 1187–1194 (2012)

[DDK] Donzelli, F., Dvorsky, A., Kaliman, S.: Algebraic density property of homogeneous spaces. Transform. Groups **15**(3), 551–576 (2010)

[For11] Forstnerič, F.: Stein Manifolds and Holomorphic Mappings, vol. 56 of Ergebnisse der Mathematik und ihrer Grenzgebiete. 3. F. Springer, Heidelberg (2011)

[FR93] Forstnerič, F., Rosay, J.-P.: Approximation of biholomorphic mappings by automorphisms of C^n. Invent. Math. **112**(2), 323–349 (1993)

[FIKP] Forstnerič, F., Ivarsson, B., Kutzschebauch, F., Prezelj, J.: An interpolation theorem for proper holomorphic embeddings. Math. Ann. **338**(3), 545–554 (2007)

[FW1] Forstnerič, F., Wold, E.F.: Embeddings of infinitely connected planar domains into $\mathbb{C}^2$. Anal. PDE **6**(2), 499–514 (2013)

[FW2] Forstnerič, F., Wold, E.F.: Bordered Riemann surfaces in $\mathbb{C}^2$. J. Math. Pures Appl. (9), 91 (2009) ((1), 100–114)

[FGR] Forstnerič, F., Globevnik, J., Rosay, J.-P.: Nonstraightenable complex lines in $\mathbb{C}^2$. Ark. Math. **34**(1), 97–101 (1996)

[GR] Grauert, H., Remmert, R.: Theory of Stein Spaces. Translated from the German by Alan Huckleberry. Reprint of the: translation. Springer, Berlin, Classics in Mathematics (1979)

[HFW] Peters, H., Wold, E.F.: Non-autonomous basins of attraction and their boundaries. J. Geom. Anal. **15**(1), 123–136 (2005)

[KaKu] Kaliman, S., Kutzschebauch, F.: On algebraic volume density property. Transform. Groups. arxiv:1201.4769 (to appear)

[KaKu1] Kaliman, S.: Frank Kutzschebauch Criteria for the density property of complex manifolds. Invent. Math. **172**(1), 71–87 (2008)

[KaKu2] Kaliman, S., Kutzschebauch, F.: Density property for hypersurfaces $uv = p(\bar{x})$. Math. Z. **258**(1), 115–131 (2008)

[KaKu3] Kaliman, S., Kutzschebauch, F.: On the present state of the Andersen-Lempert theory In: Affine Algebraic Geometry: The Russell Festschrift, pp. 85–122. Centre de Recherches Mathématiques. CRM Proceedings and Lecture Notes, vol. 54 (2011)

[KaKu4] Kaliman, S., Kutzschebauch, F.: Algebraic volume density property of affine algebraic manifolds. Invent. Math. **181**(3), 605–647 (2010)

[KR] Kutzschebauch, F., Ramos Peon, A.: An Oka Principle for a Parametric Infinite Transitivity Property. arXiv:1401.0093

[KW] Kutzschebauch, F., Wold, E.F.: Carleman approximation by holomorphic automorphisms of $\mathbb{C}^n$. J. Reine Angew, Math. (to appear) arXiv:1401.2842

[Leu14] Leuenberger, M.: Complete holomorphic vector fields on affine surfaces. Ph.D. thesis, University of Bern (2015)

[LLK] Kutzschebauch, F., Leuenberger, M., Liendo, A.: The algebraic density property for affine toric varieties. J. Pure Appl. Algebra 219(8), 3685–3700 (2015). arXiv:1402.2227

[LK] Kutzschebauch, F., Lodin, S.: Holomorphic families of nonequivalent embeddings and of holomorphic group actions on affine space. Duke Math. J. **162**(1), 49–94 (2013)

[Rit13] Ritter, T.: A strong Oka principle for embeddings of some planar domains into $\mathbb{C} \times \mathbb{C}^*$. J. Geom. Anal. **23**(2), 571–597 (2013)

[Ros] Rosay, J.-P.: Automorphisms of $\mathbb{C}^n$, a survey of Andersén-Lempert theory and applications. Complex geometric analysis in Pohang. Contemp. Math., **222**, 131–145 (1997) (Amer. Math. Soc., Providence, RI, 1999)

[Var00] Varolin, D.: The density property for complex manifolds and geometric structures. II. Int. J. Math. **11**(6), 837–847 (2000)

[Var01] Varolin, D.: The density property for complex manifolds and geometric structures. J. Geom. Anal. **11**(1), 135–160 (2001)

[Var01] Varolin, D.: A general notion of shears, and applications. Michigan Math. J. **46**(3), 533–553 (1999)

[VT1] Varolin, D.: Arpad Toth holomorphic diffeomorphisms of complex semisimple Lie groups. Invent. Math. **139**(2), 351–369 (2000)

[VT2] Varolin, D., Toth, A.: Holomorphic diffeomorphisms of semisimple homogeneous spaces. Compos. Math. **142**(5), 1308–1326 (2006)

[W1] Wold, E.F.: Embedding subsets of tori properly into $\mathbb{C}^2$. Ann. Inst. Fourier (Grenoble) **57**(5), 1537–1555 (2007)

[W2] Wold, E.F.: Embedding Riemann surfaces properly into $\mathbb{C}^2$. Int. J. Math. **17**(8), 963–974 (2006)

[W3] Wold, E.F.: Proper holomorphic embeddings of finitely and some infinitely connected subsets of $\mathbb{C}$. Math. Z. **252**(1), 1–9 (2006)

[W4] Wold, E.F.: A Fatou-Bieberbach domain in $\mathbb{C}^2$ which is not Runge. Math. Ann. **340**(4), 775–780 (2008)

[W5] Wold, E.F.: A long $\mathbb{C}^2$ which is not Stein. Ark. Mat. **48**(1), 207–210 (2010)

On Meromorphic Continuation of Local Zeta Functions

Joe Kamimoto and Toshihiro Nose

Abstract We investigate meromorphic continuation of *local zeta functions* and properties of their poles. In the real analytic case, local zeta functions can be meromorphically continued to the whole complex plane and, moreover, properties of the poles have been precisely investigated. However, in the only smooth case, the situation of meromorphic continuation is very different. Actually, there exists an example in which a local zeta function has a singularity different from poles. We give a sufficient condition for that the first finitely many poles samely appear as in the real analytic case and exactly investigate properties of the first pole.

Keywords Local zeta function · Newton polyhedron · Superadapted coordinate

1 Introduction

The purpose of this article is to announce our recent studies about *local zeta functions*, that is, integrals of the form

$$Z(s;\varphi) = \int_{\mathbb{R}^n} |f(x)|^s \varphi(x) dx \quad s \in \mathbb{C}, \tag{1.1}$$

where f and φ are real-valued (C^∞) smooth functions defined on an open neighborhood U of the origin in $\mathbb{R}^n$ and the support of φ is contained in U.

The integral in (1.1) converges locally uniformly on the region $\mathrm{Re}(s) > 0$, which implies that local zeta functions are holomorphic there. Moreover, when $f(0) \neq 0$,

J. Kamimoto
Faculty of Mathematics, Kyushu University, Motooka 744, Nishi-ku,
Fukuoka 819-0395, Japan
e-mail: joe@math.kyushu-u.ac.jp

T. Nose (✉)
Faculty of Engineering, Kyushu Sangyo University, Matsukadai 2-3-1, Higashi-ku,
Fukuoka 813-8503, Japan
e-mail: t-nose@ip.kyusan-u.ac.jp

F. Bracci et al. (eds.), *Complex Analysis and Geometry*, Springer Proceedings
in Mathematics & Statistics 144, DOI 10.1007/978-4-431-55744-9_13

if the support of φ is sufficiently small, then $Z(s;\varphi)$ is holomorphic on the whole complex plane. It is known that in many cases when $f(0)=0$, the situation of analytic continuation of $Z(s;\varphi)$ is strongly affected by the singularity theoretical property of f at the origin. In this article, we always assume that

$$f(0)=0, \quad \nabla f(0)=0$$

and will consider detailed situation of analytic continuation of local zeta functions from the viewpoint of singularity theory.

There have been many strong results about the above problem in the case when f is real analytic on U. In this case, it is known (cf. [bg69, ati70]) by using Hironaka's resolution of singularities [hir64] that if the support of φ is sufficiently small, then $Z(s;\varphi)$ can be analytically continued as a meromorphic function in the whole complex plane and its poles belong to finitely many arithmetic progressions which are constructed from negative rational numbers independent of φ. (Recently, Greenblatt [gre10jam] gives new proof of the results of [bg69, ati70] by using the elementary resolution of singularities constructed in his paper [gre08].) When $Z(s;\varphi)$ is regarded as a meromorphic function, various analytic properties of poles of $Z(s;\varphi)$ have been investigated in [gs64, bg69, ati70, var76, igu78, agv88], etc. In particular, Varchenko [var76] obtains quantitative results about the location and the order of its poles by using the theory of toric varieties based on the geometry of the Newton polyhedron of f. Since his study, it has been strongly recognized that Newton polyhedra play important roles in the analysis of local zeta functions. His results, in general dimensional case, need some nondegeneracy condition, which depends on the coordinates. Furthermore, in his same paper [var76], Varchenko more deeply investigates the two-dimensional case: the exisitence of so-called *adapted coordinates* is shown and the above quantitative results about the poles of $Z(s;\varphi)$ are obtained by using these coordinates without the nondegenracy condition. Later, the results of Varchenko about adapted coordinates have been improved and developped in [pss99, gre09, im11tams] (see also Sect. 2.4). In particular, Greenblatt [gre09] introduces special adapted coordinates, which are called *superadapeted coordinates*, and gives accurate results about the behavior of oscillatory integrals at infinity.

On the other hand, let us consider these problems about meromorphic continuation of $Z(s;\varphi)$ without the real analytic assumption on f. The authors [kn13] introduce a certain class of smooth functions containing the real analytic functions and naturally generalize the general dimensional results of Varchenko in the case when f belongs to this class under the nondegeneracy conditions. This class consists of the funtions admitting "the γ-parts" (see Sect. 2.3) for any face γ of the Newton polyhedron of f. The purpose of this article is to discuss how to generalize the above two-dimensional results due to Varchenko in the smooth case.

The difficulties of analysis in the smooth case are often caused by the nonexistence of "complete" resolution of singularities of f. But, Greenblatt [gre06] uses an idea of his resolution of singularities in [gre04], applies Van Der Corput lemma and obtains interesting results showing that the local zeta function can be analytically continued in the region $\mathrm{Re}(s) > -1/d(f)$, where $d(f)$ is the Newton distance of f with respect to the adapted coordinates. Moreover, he investigates the behavior of $Z(s;\varphi)$ at the point $s = -1/d(f)$ and shows the sharpness of his results. Our result in this article shows that $Z(s;\varphi)$ can be meromorphically continued in a wider region $\mathrm{Re}(s) > -1/d(f) - \varepsilon$, with some positive ε, under the additional assumption: in a superadapted coordinate, f admits the γ-part for the edges γ of the Newton polyhedron of f intersecting the bisectrix. Corresponding to our general dimensional results in [kn13], we emphasize that these results do not always need the assumption that f admits the γ-part for *all* edges γ of the Newton polyhedron of f. Moreover, precise properties of poles of local zeta functions are obtained. In our analysis, after using some kind of toric blowing-ups constructed in a similar fashion to the work of Varchenko [var76], we also apply Var Der Corput lemma. In order to obtain many estimates, we deeply use the properties of superadapted coordinates.

The properties of poles of local zeta functions are closely related to the asymptotic behavior of *oscillatory integrals*, that is, integrals of the form

$$I(t;\varphi) = \int_{\mathbb{R}^n} e^{itf(x)}\varphi(x)dx \quad t > 0,$$

where f and φ are as in (1.1). This relationship is explained in detail in [agv88], Chap. 6. From many kinds of motivation, the bahavior of oscillatory integrals as $t \to \infty$ has been deeply investigated (see for example [agv88, mul]). We can also obtain corresponding results of the asymptotic behavior of oscillatory integrals in two dimensions with smooth phases f. These results will appear elsewhere.

Notation and symbols.

- We denote by $\mathbb{Z}_+$, $\mathbb{R}_+$ the subsets consisting of all nonnegative numbers in $\mathbb{Z}$, $\mathbb{R}$, respectively. For $s \in \mathbb{C}$, $\mathrm{Re}(s)$ expresses the real part of s.
- For $x = (x_1, x_2)$, $y = (y_1, y_2) \in \mathbb{R}^2$, $\alpha = (\alpha_1, \alpha_2) \in \mathbb{Z}_+^2$, define

$$\langle x, y\rangle = x_1y_1 + x_2y_2, \quad x^\alpha = x_1^{\alpha_1}x_2^{\alpha_2}.$$

- For $A, B \subset \mathbb{R}^2$, we set $A + B = \{a + b \in \mathbb{R}^2 : a \in A \text{ and } b \in B\}$.

2 Preliminaries

2.1 Polyhedra

Let us explain fundamental notions in the theory of convex polyhedra in two dimensions, which are necessary for our investigation. Refer to [zie95] for general theory of convex polyhedra.

For $(a, l) \in \mathbb{R}^2 \times \mathbb{R}$, let $H(a, l)$ and $H^+(a, l)$ be a straight line and a closed halfspace in $\mathbb{R}^2$ defined by

$$H(a,l) := \{x \in \mathbb{R}^2 : \langle a, x\rangle = l\},$$
$$H^+(a,l) := \{x \in \mathbb{R}^2 : \langle a, x\rangle \geq l\},$$

respectively. A (*convex rational*) *polyhedron* is an intersection of closed halfspaces: a set $P \subset \mathbb{R}^2$ presented in the form $P = \bigcap_{j=1}^N H^+(a^j, l_j)$ for some $a^1, \dots, a^N \in \mathbb{Z}^2$ and $l_1, \dots, l_N \in \mathbb{Z}$.

Let P be a polyhedron in $\mathbb{R}^2$. A pair $(a,l) \in \mathbb{Z}^2 \times \mathbb{Z}$ is said to be *valid* for P if P is contained in $H^+(a,l)$. A *face* of P is any set of the form $F = P \cap H(a,l)$, where (a,l) is valid for P. Since $(0,0)$ is always valid, we consider P itself as a trivial face of P; the other faces are called *proper faces*. The *dimension* of a face F is the dimension of its affine hull of F (i.e., the intersection of all affine flats that contain F). The faces of dimensions 0 and 1 are called *vertices* and *edges*, respectively.

2.2 Newton Polyhedra

Let f be a real-valued smooth function defined on a neighborhood of the origin in $\mathbb{R}^2$, which has the Taylor series at the origin:

$$f(x) \sim \sum_{\alpha \in \mathbb{Z}_+^2} c_\alpha x^\alpha. \tag{2.1}$$

The *Newton polyhedron* $\Gamma_+(f)$ of f is defined by the convex hull of the set $\bigcup\{\alpha + \mathbb{R}_+^2; c_\alpha \neq 0\}$. Of course, the Newton polyhedron is a polyhedron. We say that f is *flat* if $\Gamma_+(f) = \emptyset$ (i.e., all derivatives of f vanish at the origin). The *Newton distance* of f is given by the coordinate d of the point (d,d) at which the bisectrix $\alpha_1 = \alpha_2$ intersects the boundary of the Newton polyhedron of f, which is denoted by $d(f)$. Of course, this distance depends on the coordinates. In order to make clear the chosen coordinate x, we sometimes write this distance as $d_x(f)$. The *principal face* γ_* of the Newton polyhedron of f is the smallest face of $\Gamma_+(f)$ containing the point $(d(f), d(f))$. The *multiplicity* of the Newton distance is given by the codimension of γ_*, which is denoted by $m(f)$.

2.3 The γ-Part

Let f be a nonflat real-valued smooth function defined on an open neighborhood V of the origin in $\mathbb{R}^2$ with the Taylor series (2.2).

Definition 2.1 Let γ be a face of $\Gamma_+(f)$. We say that f *admits the* γ *-part* on an open neighborhood $U \subset V$ of the origin if for any x in U the limit:

$$\lim_{t \to 0} \frac{f(t^{a_1}x_1, t^{a_2}x_2)}{t^l} \tag{2.2}$$

exists for *all* valid pairs $(a, l) = ((a_1, a_2), l) \in \mathbb{Z}_+^2 \times \mathbb{Z}_+$ defining γ. When f admits the γ-part, it is known in [kn13], Proposition 5.2 (iii), that the above limits take the same value for any valid pair $(a, l) \in \mathbb{Z}_+^2 \times \mathbb{Z}_+$ defining γ, which is denoted by $f_\gamma(x)$. Let us consider f_γ as a function on U, which is called the γ-part of f on U.

Remark 2.1 We summarize important properties of the γ-part. See [kn13] for the details.

(i) The γ-part f_γ is a smooth function defined on U.
(ii) If f admits the γ-part f_γ on U, then f_γ has the quasihomogeneous property:

$$f_\gamma(t^{a_1}x_1, t^{a_2}x_2) = t^l f_\gamma(x) \text{ for } 0 < t < 1 \text{ and } x \in U,$$

where $(a, l) = ((a_1, a_2), l) \in \mathbb{Z}_+^2 \times \mathbb{Z}_+$ is a valid pair defining γ.
(iii) For a compact face γ of $\Gamma_+(f)$, f always admits the γ-part near the origin. Then $f_\gamma(x)$ is the same as the γ-part of f defined in [var76, agv88], i.e., $f_\gamma(x) = \sum_{\alpha \in \gamma \cap \mathbb{Z}_+^2} c_\alpha x^\alpha$.
(iv) If f is real analytic, then f always admits the γ-part on U for any face γ of $\Gamma_+(f)$. Moreover, $f_\gamma(x)$ is real analytic and is equal to a convergent power series $\sum_{\alpha \in \gamma \cap \mathbb{Z}_+^2} c_\alpha x^\alpha$ on some neighborhood of the origin.
(v) Let f be a smooth function and γ a noncompact edge of $\Gamma_+(f)$. Then, f does not admit the γ-part in general. If f admits the γ-part, then the Taylor series of $f_\gamma(x)$ at the origin is $\sum_{\alpha \in \gamma \cap \mathbb{Z}_+^2} c_\alpha x^\alpha$.
(vi) When a noncompact edge γ of $\Gamma_+(f)$ is contained in some coordinate axis, f always admits the γ-part on U. Indeed, for every valid pair (a, l) defining γ, we have $l = 0$ and so the limit (2.2) exists.
(vii) When f is smooth and γ is a noncompact edge, there are many examples in which f does not admit the γ-part. For example, consider the case when $f(x_1, x_2) = x_1^2 + e^{-1/x_2^2}$ and the face γ defined by $\{(\alpha_1, \alpha_2) : \alpha_1 = 2, \alpha_2 \geq 0\}$.

2.4 Adapted Coordinates and Superadapted Coordinates

Let f be a nonflat real-valued smooth function defined near the origin in $\mathbb{R}^2$ with $f(0) = 0$ and $\nabla f(0) = 0$. The *height* of real analytic (resp. smooth) function f is defined by

$$h(f) := \sup_x d_x(f),$$

where the supremum is taken over all local analytic (resp. smooth) coordinate systems x at the origin and $d_x(f)$ is the Newton distance of f in the coordinates x.

Definition 2.2 A coordinate x is *adapted* to f (or f is in an *adapted* coordinate x) if $h(f) = d_x(f)$.

When f is real analytic, the existence of adapted coordinates is shown by Varchenko [var76] by means of two-dimensional resolution of singularities and by Phong-Stein-Sturm [pss99] by means of the Puiseux series expansion of roots of f. Moreover, Ikromov and Müller [im11tams] apply Varchenko's algrithm for the construction of the coordinates to the method of Phong-Stein [ps97] and give stronger results for the existence and the criterion for the adaptedness. Indeed, they shows the existence in the case when f is smooth. We remark that in dimension higher than two, adapted coordinates may not exist, as Varchenko shows in [var76].

Remark 2.2 We gives some remarks on adapted coordinates. See [im11tams] for the details.

(i) When γ_* is a vertex or a noncompact edge of $\Gamma_+(f)$ in a coordinate, this coordinate is adapted to f.
(ii) A coordinate is adapted to f if and only if for any compact edge γ of $\Gamma_+(f)$ intersecting the bisectrix, any zero of the functions $f_\gamma(\pm 1, \cdot)$ or $f_\gamma(\cdot, \pm 1)$ has order less than or equal to $d(f)$.
(iii) When f is in adapted coordinates, if a compact face γ of $\Gamma_+(f)$ does not intersect with the bisectrix, then any zero of $f_\gamma(\cdot, \pm 1)$ and $f_\gamma(\pm 1, \cdot)$ has order less than $d(f)$, where f_γ is the γ-part of f.
(iv) The multiplicity $m(f)$ of $d(f)$ depends on taking adapted coordinates.

Greenblatt [gre09] introduces the following special adapted coordinates, called *superadapted coordinates*. Though his coordinates are slightly different from the adapted coordinates (compare to Remark 2.4 (ii)), they are much more useful for the analysis of local zeta functions.

Definition 2.3 A coordinate x is *superadapted* to f (or f is in a *superadapted* coordinate x) if for any compact edge γ of $\Gamma_+(f)$ intersecting the bisectrix, any zero of the functions $f_\gamma(\pm 1, \cdot)$ or $f_\gamma(\cdot, \pm 1)$ has order less than $d_x(f)$.

For any smooth function f, the existence of superadapted coordinates is shown by Greenblatt [gre09].

Remark 2.3 We give some remarks on superadapted coordinates. See [gre09] for the details.

(i) Any superadapted coordinate system is adapted.
(ii) If the principal face of the Newton polyhedron $\Gamma_+(f)$ is a noncompact edge, then the function f is in superadapted coordinates.
(iii) For any superadapted coordinates, the multiplicity $m(f)$ of $d(f)$ is uniquely determined. (i.e., The multiplicity $m(f)$ does not depend on taking superadapted coordinates.)

3 Main Results

Now, let us explain our results. In this section, the following two conditions are assumed: Let U be an open neighborhood of the origin in $\mathbb{R}^2$.

(A) f is a nonflat real-valued smooth function defined on U satisfying that $f(0,0)=0$ and $\nabla f(0,0)=(0,0)$;
(B) φ is a real-valued smooth function whose support is contained in U.

Let us consider the local zeta function:

$$Z(s,\varphi)=\int_{\mathbb{R}^2}|f(x_1,x_2)|^s\varphi(x_1,x_2)dx_1dx_2 \quad s\in\mathbb{C}.$$

The main theorem in [gre06] due to Greenblatt implies that $Z(s;\varphi)$ can be analytically continued in the region $\mathrm{Re}(s)>-1/h(f)$, where $h(f)$ is the height of f defined in Sect. 2.4. Adding some assumption, we can see better properties of $Z(s;\varphi)$ about the analytic continuation as follows.

Theorem 3.1 *Suppose that a coordinate x is superadapted to f. We assume that if the bisectrix intersects with a noncompact edge γ of $\Gamma_+(f)$, then $f(x_1,x_2)$ admits the γ-part on U. For simplicity, we denote $h=h(f)$ and $m=m(f)$. If the support of φ is contained in a sufficiently small neighborhood of the origin, then the following hold:*

(i) *There exists a positive number ε independent of φ such that the function $Z(s;\varphi)$ can be analytically continued as a meromorphic function to the region* $\mathrm{Re}(s)>-1/h-\varepsilon$.
(ii) *The poles of the function $Z(s;\varphi)$ in the region* $\mathrm{Re}(s)>-1/h-\varepsilon$ *belong to finitely many arithmetic progressions which are precisely obtained by using the theory of toric varieties based on the geometry of the Newton polyhedron of f.*
(iii) *When $Z(s;\varphi)$ has a pole at $s=-1/h$, its order is at most m. More precisely, the coefficient of the pole of $Z(s;\varphi)$ at $s=-1/h$:*

$$C(\varphi):=\lim_{s\to-1/h}(1+1/h)^m Z(s;\varphi)$$

is explicitly given as follows:

(a) *Suppose that the principal face γ_* of $\Gamma_+(f)$ is a compact edge defined by a valid pair $(a,l)=((a_1,a_2),l)\in\mathbb{Z}_+^2\times\mathbb{Z}_+$. Then*

$$C(\varphi)=\frac{\varphi(0,0)}{h(a_2/a_1+1)}\int_{\mathbb{R}}\left(|f_{\gamma_*}(1,u)|^{-1/h}+|f_{\gamma_*}(-1,u)|^{-1/h}\right)du.$$

(b) *Suppose that the principal face* γ_* *of* $\Gamma_+(f)$ *is a vertex. Let* $((a_1, a_2), l_1)$ *and* $((b_1, b_2), l_2)$ *be valid pairs in* $\mathbb{Z}_+^2 \times \mathbb{Z}_+$ *defining the two edges of* $\Gamma_+(f)$ *containing* γ_**, where* $0 \leq a_2/a_1 \leq b_2/b_1 \leq \infty$*. Then*

$$C(\varphi) = \frac{4\varphi(0,0)|f_{\gamma_*}(1,1)|^{-1/h}}{h^2}\left(\frac{1}{a_2/a_1+1} - \frac{1}{b_2/b_1+1}\right).$$

(c) *Suppose that the principal face* γ_* *is a noncompact vertical edge. Then*

$$C(\varphi) = \frac{1}{h}\int_{\mathbb{R}} \left(|f_{\gamma_*}(1,u)|^{-1/h} + |f_{\gamma_*}(-1,u)|^{-1/h}\right)\varphi(0,u)du.$$

In the case where the principal face γ_* *is a horizontal edge, we have the analogous formula of* $C(\varphi)$*.*

In particular, when $\varphi(0,0) > 0$ *and* φ *is nonnegative on* U*,* $C(\varphi)$ *is always positive.*

Remark 3.1 The above meromorphic continuation needs the additional assumption of the admission of the γ-part. Indeed, consider the following example discovered by Greenblatt in [gre06]:

$$f(x_1, x_2) = x_1^a x_2^b + x_1^a x_2^{b-2} e^{|-x_1|^{-1/(2b)}},$$

where $a, b \in \mathbb{N}$ satisfy $a < b$ and $b \geq 2$. In this case, it is easy to see that the height of f is b, the principal face γ_* is $\{(\alpha_1, \alpha_2) : \alpha_1 \geq a, \alpha_2 = b\}$ and f does not admit the γ_*-part. Moreover, the limit of $Z(s;\varphi)$ as $s \in \mathbb{R}$ tends to $-1/b$ from the right hand side exists, as is shown in [gre06]. This means that $s = -1/b$ cannot become a pole of $Z(s;\varphi)$.

References

[agv88] Arnold, V.I., Gusein-Zade, S.M., Varchenko, A.N.: Singularities of Differentiable Maps II. Birkhäuser (1988)

[ati70] Atiyah, M.F.: Resolution of singularities and division of distributions. Comm. Pure Appl. Math. **23**, 145–150 (1970)

[bg69] Bernstein, I.N., Gel'fand, S.I.: Meromorphy of the function P^λ. Funktsional. Anal. Prilozhen. **3**, 84–85 (1969)

[gs64] Gel'fand, I.M., Shilov, G.E.: Generalized Functions I. Academic Press, New York (1964)

[gre04] Greenblatt, M.: A direct resolution of singularities for functions of two variables with applications to analysis. J. Anal. Math. **92**, 233–257 (2004)

[gre06] Greenblatt, M.: Newton polygons and local integrability of negative powers of smooth functions in the plane. Trans. Am. Math. Soc. **358**, 657–670 (2006)

[gre08] Greenblatt, M.: An elementary coordinate-dependent local resolution of singularities and applications. J. Funct. Anal. **255**, 1957–1994 (2008)

[gre09] Greenblatt, M.: The asymptotic behavior of degenerate oscillatory integrals in two dimensions. J. Funct. Anal. **257**, 1759–1798 (2009)

[gre10jam] Greenblatt, M.: Resolution of singularities, asymptotic expansions of integrals and related phenomena. J. Anal. Math. **111**, 221–245 (2010)

[hir64] Hironaka, H.: Resolution of singularities of an algebraic variety over a field of characteristic zero I, II. Ann. Math. **79**, 109–326 (1964)

[im11tams] Ikromov, I.A., Müller, D.: On adapted coordinate systems. Trans. Am. Math. Soc. **363**, 2821–2848 (2011)

[im11jfaa] Ikromov, I.A., Müller, D.: Uniform estimates for the Fourier transform of surface carried measures in $\mathbb{R}^3$ and an application to Fourier restriction. J. Fourier Anal. Appl. **17**, 1292–1332 (2011)

[igu78] Igusa, J.: Forms of Higher Degree. Tata Institute of Fundamental Research Lectures on Mathematics and Physics, vol. 59. New Delhi (1978)

[kn13] Kamimoto, J., Nose, T.: Toric resolution of singularities in a certain class of C^∞ functions and asymptotic analysis of oscillatory integrals (Preprint). arXiv:1208.3924

[kn] Kamimoto, J., Nose, T.: On local zeta functions in two dimensions (in preparation)

[mul] Müller, D.: Problems of harmonic analysis related to finite-type hypersurfaces in $\mathbb{R}^3$, and Newton polyhedra. In: C. Fefferman et al. (eds.) Advances in Analysis: The Legacy of Elias M. Stein, 301–345. Princeton University Press, Princeton (2014)

[ps97] Phong, D.H., Stein, E.M.: The Newton polyhedron and oscillatory integral operators. Acta Math. **179**, 105–152 (1997)

[pss99] Phong, D.H., Stein, E.M., Sturm, J.A.: On the growth and stability of real-analytic functions. Am. J. Math. **121–3**, 519–554 (1999)

[var76] Varchenko, A.N.: Newton polyhedra and estimation of oscillating integrals. Funct. Anal. Appl. **10–3**, 175–196 (1976)

[zie95] Ziegler, G.M.: Lectures on Polytopes, Graduate Texts in Mathematics, vol. 152. Springer, New York (1995)

Themes on Non-analytic Singularities of Plurisubharmonic Functions

Dano Kim

Abstract We survey some results and questions on the singularity of psh functions with non-analytic singularities. Also we show that the Demailly approximation sequence of a psh function does not contain a monotone singularity-increasing linear subsequence, in general.

Keywords Plurisubharmonic function · Lelong number · Multiplier ideal sheaf

1 Introduction

A plurisubharmonic (psh for short) function is a fundamental object in several complex variables and complex geometry. It also plays an important role in algebraic geometry of complex projective varieties since it appears as a local weight function of a singular hermitian metric of a holomorphic line bundle.

A psh function can be considered as limit objects of those nice psh functions of the form $c \log |f|$ where f is a holomorphic function and $c > 0$. More precisely, the following definition gives the class of 'nice' psh functions.

Definition 1.1 [D] We say that a psh function φ on a complex manifold *has analytic singularities* if it can be locally written as $\varphi = c \log(\sum_{i=1}^{N} |f_i|^{\alpha_i}) + v$ where f_i are local holomorphic functions, $c > 0$ a real number, v a locally bounded function and $\alpha_i > 0$ rational numbers.

It is easy to see that we get an equivalent definition if we require $\alpha_i = 2$. We need α_i to be at least rational since otherwise φ may not have a log-resolution (see Example 4.1). Despite the name 'analytic' singularities, a psh funciton with analytic singularities can be considered as an algebro-geometric object since at least its singularity is completely described in terms of its log-resolution.

D. Kim (✉)
Department of Mathematical Sciences, Seoul National University, 1 Gwanak-ro, Seoul, Gwanak-gu 151-747, Korea
e-mail: kimdano@snu.ac.kr

F. Bracci et al. (eds.), *Complex Analysis and Geometry*, Springer Proceedings in Mathematics & Statistics 144, DOI 10.1007/978-4-431-55744-9_14

On the other hand, the singularity of a general psh function φ with non-analytic singularities can be highly complicated and mysterious. In particular, its pole set $\varphi^{-1}(-\infty)$ can be far from an analytic subset, unlike the case of φ with analytic singularities. Nevertheless, a psh function with non-analytic singularities is expected to be equally well-behaved in some important questions. One of them is the openness conjecture of Demailly and Kollár, which was recently proved in general by Guan and Zhou [GZ].

In this paper, we introduce two other unsolved questions of this nature and survey related ideas. They are on the existence of attenuation of psh singularity (Question 4.1) and on the coherence of analytic adjoint ideal sheaves (Question 5.1), respectively. On the other hand, one original result in this paper is Theorem 3.1 (2) on the monotone linear subsequence of approximation of a psh function, which complements [K13].

This paper is organized as follows. In Sect. 2, we recall some basic notions and mention recent progresses on the openness conjecture for multiplier ideal sheaves. In Sect. 3, we discuss our recent work on the monotone subsequences of Demailly approximation of psh functions. In Sect. 4, we discuss the question on the attenuation of psh singularities with a simple example. In Sect. 5, we discuss the question on coherence of analytic adjoint ideal sheaves. Sections 3, 4 and 5 are largely independent from each other.

Although a psh function with non-analytic singularities is a transcendental object, its various aspects we describe in this paper are closely related to algebraic geometry: in Siu-type hermitian metrics, in log resolutions and in multiplier and adjoint ideal sheaves.

2 Multiplier Ideal Sheaves and the Openness Conjecture

2.1 Basic Notions and Some Examples

We refer to [D] for the definition and basic properties of a psh function and Lelong numbers.

Since a psh function often appears as a local weight function of a singular hermitian metric in a given geometric context, our discussion often refers to the singular hermitian metric instead of the psh function. Of course, a psh function on a complex manifold X itself can be always considered as defining a singular hermitian metric of a trivial line bundle on X.

We define the notion of equivalence of psh singularities. Let L be a line bundle on a complex manifold X and $h_1 = e^{-\varphi_1}$ and $h_2 = e^{-\varphi_2}$ two singular hermitian metrics of L. Following the usual convention, we often use φ_1 to refer to the metric h_1.

Definition 2.1 [D14, Definition 0.5] We say $h_1 = e^{-\varphi_1}$ is *less singular* than $h_2 = e^{-\varphi_2}$ and write $\varphi_1 \preceq \varphi_2$ and $h_1 \preceq h_2$ if the local weight functions satisfy $\varphi_2 \leq$

$\varphi_1 + O(1)$. We say φ_1 and φ_2 have *equivalent singularities* and write $\varphi_1 \sim \varphi_2$ if $\varphi_1 \preceq \varphi_2$ and $\varphi_2 \preceq \varphi_1$.

Now we introduce some important classes of psh functions. Let $\mathcal{T}_0$ be the class of psh functions with analytic singularities. Another nice class of psh functions is $\mathcal{T}_1$ of *toric psh* functions: those locally given as $\varphi(z_1, \cdots, z_n)$ which depends only on $|z_1|, \cdots, |z_n|$ in a polydisk $(|z_i| < r)$ $(i = 1, \cdots n)$ in $\mathbf{C}^n$. (See [G, 1.3] and [R13g] for more about toric psh functions.) Note that $\mathcal{T}_0 \cap \mathcal{T}_1$ contains psh functions with analytic singularities coming from monomials (in the place of f_i in Definition 1.1).

Let $\mathcal{T}_2$ be the class of psh functions φ such that e^{φ} is locally Hölder continuous (Definition [DK]).

As an example, we consider a specific type of psh function of the form

$$\varphi_S := \log(\sum_{m=1}^{\infty} \varepsilon_m \sum_{j=1}^{k_m} |s_{j,m}|^{2\alpha_m})$$

where $s_{j,m}$ are holomorphic functions (or holomorphic sections of a line bundle), $\alpha_m > 0$ and the coefficients $\varepsilon_m > 0$ are such that the series converges.

If $s_{j,m}$ are holomorphic sections of the m-th tensor power mL of a holomorphic line bundle L on a complex manifold X and $\alpha_m = \frac{1}{m}$, then this expression defines a singular hermitian metric of L (whose local weight functions are psh functions of the same form). Such a singular metric (which we will refer to as of Siu type) was first defined and used successfully in Siu's proof of invariance of plurigenera [S98] in the case of general type.

Based on the openness theorem [GZ], a recent work [K14] showed that if X is a projective manifold and L a big line bundle, then φ_S has analytic singularities if and only if the section ring of L is finitely generated. Therefore for each instance of L whose section ring is not finitely generated, we have an example of a psh function φ_S which does not have analytic singularities. Note that such φ_S typically belongs to the class $\mathcal{T}_2 \setminus (\mathcal{T}_0 \cup \mathcal{T}_1)$.

2.2 *Multiplier Ideal Sheaves and the Openness Conjecture*

Given a psh function φ on a complex manifold X, its multiplier ideal sheaf $\mathcal{J}(\varphi)$ is defined to be the ideal sheaf of holomorphic function germs u for which $|u|^2 e^{-\varphi}$ is locally integrable.

The term *multiplier ideal sheaf* in this context was first introduced in [N89, N90] though the version in [D93, N90a] is the standard definition we are using here.

Let us define $\mathcal{J}_+(\varphi) := \lim_{\varepsilon \to 0} \mathcal{J}((1+\varepsilon)\varphi)$. The openness conjecture $\mathcal{J}_+(\varphi) = \mathcal{J}(\varphi)$ was recently proved by [GZ] in all dimensions, after which [H] presented a simplified proof. Before [GZ], it was proved in dimension 2 case by [FJ1, FJ2], in the toric psh case by [G] and more recently in the case when $\mathscr{J}(\varphi)$ is a trivial ideal by [B]. Also [JM, GZ1, Le] are related to the recent activity on the conjecture.

A consequence of the openness theorem [GZ, H] in relation to the local algebraic characterization of multiplier ideal sheaves ([LL]) is as follows. A multiplier ideal is integrally closed ([D, L]). Let $\mathscr{S}_0$ be the class of integrally closed ideal sheaves and $\mathscr{S}_1$ the class of multiplier ideal sheaves. Also let $\mathscr{S}_2$ be the class of multiplier ideal sheaves $\mathscr{J}(\varphi)$ where φ is a psh with analytic singularities.

Then clearly we have the inclusion $\mathscr{S}_2 \subset \mathscr{S}_1 \subset \mathscr{S}_0$. Lazarsfeld and Lee [LL] showed that $\mathscr{S}_2 \neq \mathscr{S}_0$. Subsequently $\mathscr{S}_1 \neq \mathscr{S}_0$ was shown by [K10]. Now that the openness conjecture is proved, we know $\mathscr{S}_2 = \mathscr{S}_1$ by equisingular approximation of psh functions (see [D92, D, D14]). In the next section, we discuss the usual approximation of psh functions.

3 Demailly Approximation of Psh Singularity

A fundamental theorem of Demailly [D92, Proposition 3.1] states that given a plurisubharmonic function φ on a domain, there always exists a sequence $\{\varphi_m\}$ of plurisubharmonic functions with analytic singularities converging to φ. Moreover, the approximating function φ_m is given in a very natural form: $\varphi_m = \frac{1}{2m} \log \sum |\sigma_l|^2$ where (σ_l) is an orthonormal basis of the Hilbert space of holomorphic functions that are square integrable with respect to the weight $e^{-2m\varphi}$.

It was further proved that the subsequence $\{\varphi_{2^k}\}$ is increasing in its singularity:

$$\varphi_{2^k} \preceq \varphi_{2^{k+1}} \tag{1}$$

in [DPS, Step 3, Proof of Theorem 2.3] using a subadditivity property of the sequence φ_m's. (Note that in [K13], such sequence was referred to as *decreasing*, rather than increasing, in terms of its values. This discrepancy arises from Definition 2.1.)

It remained a natural question, raised explicitly in [B, p.134], to ask whether the entire sequence $\{\varphi_m\}$ is increasing in singularities. In [K13], we showed by an example of φ that the Demailly approximation sequence of a plurisubharmonic function is not necessarily increasing, thus answering the above question negatively. The example φ was given as a plurisubharmonic function with analytic singularities, for which we can compute the multiplier ideal sheaf of each $m\varphi$ and determine the singularities of φ_m using a finite number of local generators of $\mathscr{J}(m\varphi)$.

Then there still remained a question at the end of [K13] (asked by J.-P. Demailly) asking whether there always exists an increasing singularity subsequence of φ_m with linear indices $m = ak + b$ $(a, b \in \mathbf{Z})$.

Note that the existence of an increasing singularity subsequence with exponential indices (1) is a consequence of subadditivity of multiplier ideal sheaves (which in turn uses Ohsawa-Takegoshi extension theorem). Existence of such subsequence with linear indices should require an even stronger general property for multiplier ideal sheaves than subadditivity. However, we answer the question negatively in the second statement of the following

Theorem 3.1 *(1) If φ has analytic singularities with the coefficient c rational (in Definition 1.1), then its Demailly approximation has an increasing singularity subsequence with linear indices.*

(2) If $\varphi = c \log |z_1|^2$ with the coefficient c irrational, then its Demailly approximation does not have an increasing singularity subsequence with linear indices.

The proof of (1) is exactly as in the proof for the special case of the example in [K13]. The statement (2) follows immediately from the next arithmetic proposition, which may be of independent interest.

Proposition 3.1 *For $c > 0$ an irrational number, the sequence*

$$\{q_m := \frac{\lfloor mc \rfloor}{m}\}_{m \geq 1}$$

does not have a monotone increasing subsequence of linear indices $\{q_{ak+b}\}_{k \geq k_0}$ where $a, b, k_0 > 0$ are integers. (Here $\lfloor x \rfloor$ refers to the greatest integer ℓ such that $\ell \leq x$.)

First note that $c - \frac{1}{m} \leq \frac{\lfloor mc \rfloor}{m} \leq c$ for $m \geq 1$.

Proof Suppose that such subsequence $\{q_{ak+b}\}_{k \geq k_0}$ exists for some $a, b, k_0 > 0$. Then we have

$$\frac{\lfloor (ak+b)c \rfloor}{ak+b} \leq \frac{\lfloor (ak+b)c + ac \rfloor}{ak+b+a} \tag{2}$$

for every $k \geq k_0$.

Define $\{x\} := x - \lfloor x \rfloor$. We will use the following well-known classical fact.

Lemma 3.1 *Let $c > 0$ be an irrational number. For any open interval $J := (e_1, e_2) \subset (0, 1)$, there exists an integer $n \geq 1$ such that $\{nc\} \in (e_1, e_2)$.*

Consider $(ak + b)c + ac = \lfloor (ak + b)c \rfloor + \lfloor ac \rfloor + \{(ak + b)c\} + \{ac\}$. Since $0 < \{ac\} < 1$, there exists $k = k_1 \geq k_0$ such that

$$\{(ak+b)c\} < \min(\{ac\}, 1 - \{ac\}) \tag{3}$$

from Lemma 3.1. Hence for $k = k_1$, we get $\lfloor (ak + b)c + ac \rfloor = \lfloor (ak + b)c \rfloor + \lfloor ac \rfloor$.

For $k = k_1$, from the previous line and (2), we get

$$\begin{aligned} \lfloor ac \rfloor (ak+b) &\geq \lfloor (ak+b)c \rfloor a = (ak+b)ca - \{(ak+b)c\}a \\ &\geq \lfloor ac \rfloor (ak+b) + \{ac\}(ak+b) - \{ac\}a \end{aligned}$$

where we used (3) in the last inequality. This is contradiction since $\{ac\} > 0$ and $a, b, k \geq 1$. □

4 Resolution of Psh Singularity

In contrast to approximating a psh function by ones with analytic singularities, one may try to resolve the psh singularity by blow ups. More precisely

Definition 4.1 Let φ be a psh function on a complex manifold X. A *log resolution* of φ is a proper modification $\pi : X' \to X$ from a complex manifold X' such that $\pi^*\varphi = \psi + v$ where v is a locally bounded function and ψ is locally equal to a function of the form $\sum_{i=1}^n a_i \log|w_i|^2$ (for $a_i \geq 0$ and local analytic coordinates $(w_1, \cdots, w_n)$ of X with dimension n).

In other words, ψ is the psh weight associated to a simple normal crossing (SNC for short) effective divisor on X'. If φ has analytic singularities as in Definition 1.1, then φ has a log-resolution π which one can take as the log resolution of the ideal sheaf (or its integral closure) of the functions $f_1, \cdots, f_N$ (see [L, Definition 9.1.12]).

For example, near the origin of $\mathbf{C}^2$, one can take a log resolution π of $\varphi = \log(|x^3| + |y^2|)$ to be the composition of three blow-ups which gives a log resolution of the ideal $\mathfrak{a} = (x^3, y^2)$ [L, Example 9.1.13].

Obviously not every psh function can have a log resolution. For a simple example, consider a psh function with zero Lelong number at every point, but not locally bounded. Instead of log resolution, it is natural to raise the following

Question 4.1 Let φ be a psh function on X. For arbitrary $\varepsilon > 0$, does there exist a proper modification $\pi = \pi_\varepsilon : X' \to X$ from a complex manifold X' such that $\pi^*\varphi = \psi + v$ where ψ is as in Definition 4.1 and v is a psh function whose Lelong number is less than ε at every point of X'?

Such π_ε is called an *attenuation* of the psh singularity of φ. This question can be considered as generalization of Hironaka's celebrated theorem [Hi] on the existence of a log resolution of an ideal.

Question 4.1 is answered affirmatively in dimension 2 by [FJ1] (also by [Gu] in the compact case). We would like to illustrate the question with a probably simplest nontrivial example in dimension 2.

Example 4.1 Let $\varphi = \log(|x| + |y|^\alpha)$ near the origin of $\mathbf{C}^2$. Suppose that $\alpha > 1$ is an irrational number. Consider the blow up π of the origin as given by $(u, v) \to (x, y) := (u, uv)$ and $(u, v) \to (x, y) := (uv, v)$. The 'strict transform' of φ is given by $\log(|u| + |v|^{\alpha-1})$ which is of the same form as the original φ and again the origin $(u, v) = (0, 0)$ is the only point to be blown up again. (For general psh φ, one may have a countable number of points to be blown up at each step, as was explained in [DH1].)

From this observation, we see that after $\lfloor\alpha\rfloor$ times of point blow ups, the strict transform psh function is of the form $\psi := \log(|x| + |y|^{\alpha - \lfloor\alpha\rfloor})$ whose Lelong number at the origin is $\alpha - \lfloor\alpha\rfloor < 1$. Let $\beta = \frac{1}{\alpha - \lfloor\alpha\rfloor} > 1$. Up to equivalence of singularities, we can retake $\psi = \log(|x| + |y|^{\frac{1}{\beta}}) \sim \frac{1}{\beta}\log(|y| + |x|^\beta)$. Then similarly as before, we have $\lfloor \beta \rfloor$ times of point blow ups and at each time the Lelong number of the strict transform psh function at the only singular point is $\frac{1}{\beta}$.

Now we will see that attenuation of φ indeed exists: let $a_1 := \alpha$ and $a_2 := \beta$. Define $a_m := \frac{1}{a_{m-1} - \lfloor a_{m-1} \rfloor}$ $(m \geq 2)$ which is well-defined and $a_m > 1$ since $\alpha \notin \mathbf{Q}$. Given $\varepsilon > 0$, the attenuation π_ε is the composition $\mu_k \circ \cdots \circ \mu_1$ for some $k \geq 1$ sufficiently large:

$$\mathbf{C}^2 \longleftarrow_{\mu_1} X_1 \longleftarrow_{\mu_2} X_2 \longleftarrow_{\mu_3} X_3 \longleftarrow \cdots$$

where μ_i is the composition of $\lfloor a_i \rfloor$ times of point blow ups, each of which has Lelong number of the strict transform at the origin as $b_i := \frac{1}{a_2 \cdots a_i}$ (for $i \geq 2$) or as 1 (for $i = 1$). It is easy to see that the following equality of an infinite series holds from the definition of a_m:

$$\alpha = \underbrace{1^2 + \cdots + 1^2}_{\lfloor a_1 \rfloor} + \frac{1}{a_2^2} \left(\underbrace{1^2 + \cdots + 1^2}_{\lfloor a_2 \rfloor} + \frac{1}{a_3^2} \left(\underbrace{1^2 + \cdots + 1^2}_{\lfloor a_3 \rfloor} + \frac{1}{a_4^2} (\cdots\cdots) \right) \right) \tag{4}$$

The sequence b_i is decreasing and it should converge to zero as $i \to \infty$, since the RHS of (4) converges. This answers Question 4.1 positively in this special case.

Note that (4) also confirms the conjectured equality (**) of [DH1, p.26] in this special case since the second Lelong number $e_2(\varphi) = \alpha$ by [D, Corollary III (7.4)].

5 Analytic Adjoint Ideal Sheaves

The adjoint ideal sheaf is a variant of the multiplier ideal sheaf in algebraic geometry (see [L, T10, E] and the references therein for its algebraic definition and its applications). Guenancia [G] gave an analytic definition of an adjoint ideal sheaf of a psh function with respect to an SNC divisor $H = \sum H_i$. For simplicity of exposition, we will assume in this section that H has only one irreducible component. But the discussion in this section makes sense in the general case as well.

Let X be a complex manifold and $H \subset X$ a smooth irreducible hypersurface. Let φ be a psh function on X (or a singular hermitian metric of a line bundle). For $\alpha > 1$, let $Adj^{\alpha}_{H,*}(\varphi) \subset \mathscr{O}_X$ be the ideal sheaf of holomorphic function germs u for which

$$|u|^2 \frac{1}{|h|^2 (-\log |h|)^\alpha} e^{-\varphi}$$

is locally integrable where h is a local equation of H. It is easy to see that the definition is independent of the choice of h (as far as $|h| < 1$ around a point on H). Note that [G] was using $\alpha = 2$ in the definition, but we allow $\alpha > 1$ for a reason to be explained later on. The weight function $\frac{1}{|h|^2(-\log|h|)^\alpha}$ appears in the norm of the *extended* section in many versions of Ohsawa-Takegoshi extension theorem with $\alpha = 2$, but also with $\alpha > 1$ in [MV].

Guenancia [G] gave the following adjusted definition as the correct one to generalize the algebraic adjoint ideal sheaf.

Definition 5.1 The ideal sheaf $Adj^{\alpha}_{H}(\varphi) = \cup_{\varepsilon>0} Adj^{\alpha}_{H,*}((1+\varepsilon)\varphi)$ is called the *analytic adjoint ideal sheaf* of φ with respect to H (for $\alpha > 1$).

This generalizes the algebraic adjoint ideal sheaf in the sense of the following

Proposition 5.1 *Let X be a smooth complex variety and $\mathfrak{a} \subset \mathcal{O}_X$ an ideal sheaf. Then the algebraic adjoint ideal sheaf associated to $\mathfrak{a}^c$ $(c > 0)$ is equal to the analytic adjoint ideal sheaf associated to $c\varphi_{\mathfrak{a}}$ where $\varphi_{\mathfrak{a}}$ is the psh function defined by $\mathfrak{a}$.*

Proof This is [G, Proposition 2.11] for $\alpha = 2$. The proof works for $\alpha > 1$ when one replaces the first identity in [G, Lemma 2.12] by

$$\int_{(0,\delta)^2} \frac{x^a y^{-1}}{(-\log(xy))^{\alpha}} dydx = \frac{\delta^{-1-a}}{\alpha - 1} \int_0^{\delta^2} x^a(-\log x)^{-\alpha+1} dx. \qquad \square$$

From the coherence of analytic multiplier ideal sheaves and algebraic adjoint ideal sheaves, it is natural to raise the following

Question 5.1 ([G]) For $\alpha > 1$ and φ psh, is the analytic adjoint ideal sheaf $Adj^{\alpha}_{H}(\varphi)$ coherent ?

This was answered affirmatively for those φ with e^{φ} locally Hölder continuous by [G, Corollary 2.19] using adjunction exact sequence.

It is reasonable to try using L^2 estimates for a $\overline{\partial}$ equation ([Ho, D]) for Question 5.1 as in the proof of coherence of multiplier ideal sheaves [D]. The problem is that the weight $e^{\psi} := \frac{1}{(-\log|h|)}$ appears with the 'opposite' sign: ψ is psh and we need L^2 existence for $\overline{\partial}$ for the weight of the form $e^{\alpha\psi-\eta}$ where $\alpha > 1$ and η psh.

Such L^2 estimates for $\overline{\partial}$ equations result for $\alpha < 1$ was given by Berndtsson [B01] under the crucial condition $\partial\psi \wedge \overline{\partial}\psi = \partial\overline{\partial}\psi$ which is satisfied by the above ψ. Recent results of this type of Błocki in the case $\alpha = 1$ were given and applied toward Suita conjecture in [Bl12, Bl13] (also see the references therein related to this progress).

For Question 5.1, one would need a result of L^2 estimates for $\overline{\partial}$ for $\alpha > 1$ and that is our motivation in replacing $\alpha = 2$ with $\alpha > 1$ in Definition 5.1. Note that in this context, the psh function η in the weight can be taken as $\varphi + \log|h|^2$ so the weight $e^{\alpha\psi-\eta}$ at hand is more special than general, as in [B01].

Therefore, besides having the intrinsic interest on the coherence of the naturally defined ideal sheaf, Question 5.1 motivates further investigation into the fundamental method of L^2 estimates for $\overline{\partial}$ equations.

Acknowledgments This work was supported by the National Research Foundation of Korea grants NRF-2012R1A1A1042764 and No.2011-0030795, funded by the Republic of Korea government. The author would like to thank J.-P. Demailly for sharing many of his insights on the topic of this paper. He also thanks B. Berndtsson, Z. Błocki, S. Boucksom, V. Guedj, M. Păun and N. Sibony for valuable comments and discussions.

References

[B] Berndtsson, B.: The openness conjecture for plurisubharmonic functions. arXiv:1305.5781

[B01] Berndtsson, B.: Weighted estimates for the $\bar{\partial}$-equation. Complex Analysis and Geometry, vol. 9, pp. 43 - 57. Ohio State University Mathematical Research Institute Publications, Columbus. de Gruyter, Berlin (2001)

[Bl] Błocki, Z.: Some applications of the Ohsawa-Takegoshi extension theorem. Expo. Math. **27**(2), 125–135 (2009)

[Bl12] Błocki, Z.: On the Ohsawa-Takegoshi extension theorem. Univ. Iagel. Acta Math. **50**, 53–61 (2012)

[Bl13] Błocki, Z.: Suita conjecture and the Ohsawa-Takegoshi extension theorem. Invent. Math. **193**, 149–158 (2013)

[D92] Demailly, J.-P.: Regularization of closed positive currents and intersection theory. J. Alg. Geom. **1**, 361–409 (1992)

[D93] Demailly, J.-P.: A numerical criterion for very ample line bundles. J. Differ. Geom. **37**(2), 323–374 (1993)

[D97] Demailly, J.-P.: Complex Analytic and Differential Geometry. http://www-fourier.ujf-grenoble.fr/~demailly/manuscripts/agbook.pdf

[D] Demailly, J.-P.: Analytic Methods in Algebraic Geometry, Surveys of Modern Mathematics, International Press, Somerville. Higher Education Press, Beijing (2012)

[D14] Demailly, J.-P.: On the cohomology of pseudoeffective line bundles. arXiv:1401.5432

[DH] Demailly, J.-P., Hiep, P.H.: A sharp lower bound for the log canonical threshold. Acta Math. **212**(1), 1–9 (2014)

[DH1] Demailly, J.-P., Hiep, P.H.: A sharp lower bound for the log canonical threshold. Slides of the talk given at the NORDAN 2012 Conference. http://www-fourier.ujf-grenoble.fr/~demailly/documents.html

[DK] Demailly, J.-P., Kollár, J.: Semi-continuity of complex singularity exponents and Kähler-Einstein metrics on Fano orbifolds. Ann. Sci. École Norm. Sup. **34**(4), 525-556 (2001)

[DPS] Demailly, J.-P., Peternell, T., Schneider, M.: Pseudo-effective line bundles on compact Kähler manifolds. Int. J. Math. **12**(6), 689–741 (2001)

[E] Eisenstein, E.: Inversion of adjunction in high codimension. Ph. D. thesis, University of Michigan, 118 pp (2011)

[FJ1] Favre, C., Jonsson, M.: Valuative analysis of planar plurisubharmonic functions. Invent. Math. **162**(2), 271–311 (2005)

[FJ2] Favre, C., Jonsson, M.: Valuations and multiplier ideals. J. Am. Math. Soc. **18**(3), 655–684 (2005)

[Gu] Guedj, V.: Desingularization of quasiplurisubharmonic functions. Int. J. Math. **16**(5), 555–560 (2005)

[G] Guenancia, H.: Toric plurisubharmonic functions and analytic adjoint ideal sheaves. Mathematische Zeitschrift **271**, 1011–1035 (2012)

[GZ] Guan, Q., Zhou, X.: Strong openness conjecture and related problems for plurisubharmonic functions. arXiv:1401.7158. Also Strong openness conjecture for plurisubharmonic functions. arXiv:1311.3781

[GZ1] Guan, Q., Zhou, X.: Effectiveness of Demailly's strong openness conjecture and related problems

[H] Hiep, P.H.: The weighted log canonical threshold. C. R. Math. Acad. Sci. Paris **352**(4), 283–288 (2014)
[Hi] Hironaka, H.: Resolution of singularities of an algebraic variety over a field of characteristic zero. I, II. Ann. Math. **79**(2), 109–326 (1964)
[Ho] Hörmander, L.: An Introduction to Complex Analysis in Several Variables, 3rd edn. North-Holland Publishing Co., Amsterdam (1990)
[JM] Jonsson, M., Mustaţă, M.: An algebraic approach to the openness conjecture of Demailly and Kollár. J. Inst. Math. Jussieu **13**(1), 119–144 (2014)
[K10] Kim, D.: The exactness of a general Skoda complex. Michigan Math. J. **63**(1), 3–18 (2014)
[K13] Kim, D.: A remark on the approximation of plurisubharmonic functions. C. R. Math. Acad. Sci. Paris **352**(5), 387–389 (2014)
[K14] Kim, D.: Equivalence of plurisubharmonic singularities and Siu-type metrics. arXiv:1407.6474 (to appear in Monatshefte für Mathematik)
[L] Lazarsfeld, R.: Positivity in Algebraic Geometry, Ergebnisse der Mathematik und ihrer Grenzgebiete, vol. 3. Folge, pp. 48–49. Springer, Berlin (2004)
[LL] Lazarsfeld, R., Lee, K.: Local syzygies of multiplier ideals. Invent. Math. **167**(2), 409–418 (2007)
[Le] Lempert, L.: Modules of square integrable holomorphic germs. arXiv:1404.0407
[MV] McNeal, J., Varolin, D.: Analytic inversion of adjunction: L^2 extension theorems with gain. Ann. Inst. Fourier (Grenoble) **57**(3), 703–718 (2007)
[N89] Nadel, A.: Multiplier ideal sheaves and existence of Kähler-Einstein metrics of positive scalar curvature. Proc. Nat. Acad. Sci. USA **86**(19), 7299–7300 (1989)
[N90] Nadel, A.: Multiplier ideal sheaves and Kähler-Einstein metrics of positive scalar curvature. Ann. Math. **132**(2, 3), 549-596 (1990)
[N90a] Nadel, A.: Relative Bounds for Fano Varieties of the Second Kind. Einstein Metrics and Yang-Mills Connections (Sanda, 1990), pp. 181–191. Lecture Notes in Pure and Applications and Mathematics, vol. 145. Dekker, New York (1993)
[R13g] Rashkovskii, A.: Multi-circled singularities, Lelong numbers, and integrability index. J. Geom. Anal. **23**(4), 1976–1992 (2013)
[S98] Siu, Y.-T.: Invariance of plurigenera. Invent. math. **134**, 661–673 (1998)
[S11] Siu, Y.-T.: Section extension from hyperbolic geometry of punctured disk and holomorphic family of flat bundles. Sci. China Math. **54**(8), 1767–1802 (2011)
[T10] Takagi, S.: Adjoint ideals along closed subvarieties of higher codimension. J. Reine Angew. Math. **641**, 145–162 (2010)

Proper Holomorphic Maps Between Bounded Symmetric Domains

Sung-Yeon Kim

Abstract We consider rigidity problem of proper holomorphic maps between bounded symmetric domains. We give an introduction to differential geometric techniques on rigidity problems, based on the similar phenomenon for local CR maps between arbitrary boundary components of two bounded symmetric domains of Cartan type I.

Keywords Bounded symmetric domains · Proper holomorphic map · CR map · Totally geodesic embedding

1 Introduction

Rigidity of holomorphic maps was first studied by Poincaré [P07] and later by Alexander [A74] for maps sending one open piece of the sphere into another. Then Webster [W79] obtained rigidity for holomorphic maps between pieces of spheres of different dimensions, proving that any such map between spheres in $\mathbb{C}^n$ and $\mathbb{C}^{n+1}$ is totally geodesic. Further rigidity results are obtained by Faran [Fa86], Cima-Suffridge [CS83, CS90], Forstneric [Fo86, Fo89] and Huang [H99] for CR maps between pieces of spheres in $\mathbb{C}^{n+1}$ and $\mathbb{C}^{n'+1}$ under the assumption $n' < 2n$. Beyond this bound, rigidity fails to hold as one can see Whitney map as a counterexample [HJ01]. Rigidity for CR maps between real hypersurfaces and hyperquadrics are studied by Ebenfelt-Huang-Zaitsev [EHZ04, EHZ05], Baouendi-Huang [BH05], Baouendi-Ebenfelt-Huang [BEH08, BEH09].

On the other hand, since the work of Bochner [Bo47] and Calabi [Ca53], rigidity phenomena for the quotients of bounded symmetric domains have been widely studied. The reader is referred to the survey by Mok [M11]. Among rigidity phenomena, problems such as metric rigidity and the characterization of totally geodesic complex submanifolds are formulated and studied by differential geometric methods. Though

S.-Y. Kim (✉)
Department of Mathematics Education, Kangwon National University,
123 Hyoja-dong, Kangwon-do Chuncheon 200-701, Korea
e-mail: sykim87@kangwon.ac.kr

F. Bracci et al. (eds.), *Complex Analysis and Geometry*, Springer Proceedings in Mathematics & Statistics 144, DOI 10.1007/978-4-431-55744-9_15

they are formulated using metric, they may be deduced from which concern primarily the complex structure, such as rigidity results on holomorphic mappings. Remarkably, many rigidity properties survive when the isometry condition is replaced by purely topological conditions such as properness [MN12, MNT10, Ng12, Ng13].

In this direction, Mok introduced a method to incorporate the study of proper holomorphic maps into the study of germs of holomorphic mappings preserving some form of geometric structures [M89]. Then he proposed a question of using properness only to verify a condition on the preservation of certain geometric structures such as variety of minimal rational tangents, etc.

The aim of this article is to introduce a differential geometric method used in [KZ13, KZ14] for the study of proper holomorphic maps between bounded symmetric domains. We use CR structure on the boundary components as geometric structures preserved by proper holomorphic maps extending smoothly to an open piece of a boundary component. Then we follow Cartan's moving frame method which was first adopted by Webster in the study of rigidity of locally defined CR maps between spheres [W79]. In sphere case, moving frame method on local CR rigidity rely heavily on Tanaka-Chern-Moser approach [Ta62, CM74] and many of them also on Tanaka-Webster connection, which is unavailable for boundaries of higher rank bounded symmetric domain. To compensate for the lack of the power of Tanaka-Chern-Moser normalization, we introduce a sequence of several subsequent adjustments of moving frames reaching further and further normalization conditions.

By slightly modifying the statements in [KZ14], we prove the following theorems.

Theorem 1.1 *Let f be a smooth CR map between connected open pieces of boundary components $S_{p,q,r}$ and $S_{p',q',r'}$ of rank $r < q$ and r' respectively of bounded symmetric domains $D_{p,q}$ and $D_{p',q'}$ with $q, q' > 1$, such that $df(\xi) \in T' \setminus T'^c$ for any tangent vector $\xi \in T \setminus T^c$. Assume that*

$$q' - r' < \min(p - r, 2(q - r)).$$

Then $r \leq r'$ and after composing with suitable automorphisms of $D_{p,q}$ and $D_{p',q'}$, f takes the block matrix form

$$f(z) = \begin{pmatrix} z & 0 & 0 \\ 0 & I_{r'-r} & 0 \\ 0 & 0 & h(z) \end{pmatrix},$$

where $h\colon S_{p,q,r} \to \mathbb{C}^{[(q'-r')-(q-r)]\times[(p'-r')-(p-r)]}$ is a CR map satisfying

$$Id - h(z)^* h(z) > 0.$$

Theorem 1.2 *Let $f\colon D_{p,q} \to D_{p',q'}$ $(p \geq q > 1)$ be a proper holomorphic map which extends smoothly to a neighborhood of a smooth boundary point. Assume that*

$$1 < q' < \min(p, 2q - 1).$$

Then $p' \geq p$, $q' \geq q$ *and after composing with suitable automorphisms of* $D_{p,q}$ *and* $D_{p',q'}$, f *takes the block matrix form*

$$f : D_{p,q} \to D_{p',q'}, \quad z \mapsto \begin{pmatrix} z & 0 \\ 0 & h(z) \end{pmatrix},$$

where $h(z)$ *is arbitrary holomorphic matrix-valued function satisfying*

$$I_{q'-q} - h(z)^* h(z) \text{ is positive definite}, \quad z \in D_{p,q}.$$

In §1, we introduce a CR structure on the boundary components of bounded symmetric domain of Cartan type I. In §2, we formulate a Pfaffian differential system for CR mappings. Then we construct second fundamental forms and Gauss formulae. With these, we sketch the proof of Theorem 1.1. Theorem 1.2 follows from Theorem 1.1 by letting $r = 1$. Notice that the condition in Theorem 1.2 is a condition imposed only on the rank of the target manifold(=q'), while the condition in [KZ14] is on both p' and q'.

Throughout this paper, we adopt Einstein summation convention unless stated otherwise.

2 Geometry of Boundary Components

Recall that a bounded symmetric domain $D_{p,q}$ of Cartan type I has the standard realization in the space $\mathbb{C}^{p\times q}$ of $p \times q$ matrices, given by

$$D_{p,q} := \{z \in \mathbb{C}^{p\times q} : I_q - z^* z \text{ is positive definite}\},$$

where I_q is the identity $q \times q$ matrix and $z^* = \bar{z}^t$. We shall always assume $p \geq q$ so that the rank of $D_{p,q}$ is q. Then each boundary of $D_{p,q}$ is given by

$$S_{p,q,r} = \{z \in \mathbb{C}^{p\times q} : I_q - z^* z \text{ has } (q-r) \text{-positive and } r \text{-zero eigenvalues}\},$$

for $r = 1, \ldots, q$. In particular, $S_{p,q,1}$ is the hypersurface boundary and $S_{p,q,q}$ is the Shilov boundary. For $q = r = 1$, $S_{p,1,1}$ is the unit sphere in $\mathbb{C}^p$.

We shall consider the standard inclusion $D_{p,q} \subset \mathbb{C}^{p\times q} \subset G_{p,q}$, where $G_{p,q}$ is the Grassmanian of all q-dimensional subspaces (q-planes) of $\mathbb{C}^{p+q}$. We equip the space $\mathbb{C}^{p+q}$ with the nondegenerate hermitian form

$$\langle z, w\rangle = \sum_j \varepsilon_j z_j \bar{w}_j, \quad \varepsilon_j = \begin{cases} -1, & j = 1, \dots, q, \\ 1, & j = q+1, \dots, q+p, \end{cases}$$

called the *basic form*.

In this identification, $D_{p,q}$ is represented by all q-planes $V \subset \mathbb{C}^{p+q}$ such that the restriction $\langle\cdot,\cdot\rangle|_V$ is negative definite, and the boundary component $S_{p,q,r} \subset \partial D_{p,q}$ by all q-planes $V \subset \mathbb{C}^{p+q}$ such that restriction $\langle\cdot,\cdot\rangle|_V$ has $(q-r)$-negative and r-zero eigenvalues. For $V \in S_{p,q,r}$, denote by $V_0 \subset V$ the r-dimensional kernel of $\langle\cdot,\cdot\rangle|_V$. The connected identity component G of the biholomorphic automorphism group $\mathsf{Aut}\,(D_{p,q})$ is now identified with the group of all linear transformations of $\mathbb{C}^{p+q}$ preserving $\langle\cdot,\cdot\rangle$, and each $S_{p,q,r}$ is a G-orbit.

2.1 Adapted Frames

An adapted $S_{p,q,r}$-frame is a set of vectors

$$Z_1, \dots, Z_r, Z'_1, \dots, Z'_{q-r}, X_1, \dots, X_{p-r}, Y_1, \dots, Y_r,$$

for which the basic form is given by the matrix

$$\begin{pmatrix} 0 & 0 & 0 & I_r \\ 0 & -I_{q-r} & 0 & 0 \\ 0 & 0 & I_{p-r} & 0 \\ I_r & 0 & 0 & 0 \end{pmatrix}.$$

Thus we have

$$V_0 = \mathrm{span}\,\{Z_1, \dots, Z_r\}, \quad V = V_0 \oplus \mathrm{span}\,\{Z'_1, \dots, Z'_{q-r}\}$$

and denote

$$V' := \mathrm{span}\,\{Z'_1, \dots, Z'_{q-r}\}, \quad X := \mathrm{span}\,\{X_1, \dots, X_{p-r}\}, \quad Y := \mathrm{span}\,\{Y_1, \dots, Y_r\}.$$

2.2 The Connection Matrix Form

Write $S := S_{p,q,r}$ and denote by $\mathscr{B} \to S$ the adapted $S_{p,q,r}$-frame bundle and by π the Maurer-Cartan (connection) form on $\mathscr{B}$ satisfying the structure equation $d\pi = \pi \wedge \pi$. Then we can write

$$\begin{pmatrix} dZ_\alpha \\ dZ'_u \\ dX_k \\ dY_\alpha \end{pmatrix} = \pi \begin{pmatrix} Z_\beta \\ Z'_v \\ X_j \\ Y_\beta \end{pmatrix} = \begin{pmatrix} \psi_\alpha{}^\beta & \theta_\alpha{}^v & \theta_\alpha{}^j & \phi_\alpha{}^\beta \\ \sigma_u{}^\beta & \omega_u{}^v & \delta_u{}^j & \theta_u{}^\beta \\ \sigma_k{}^\beta & \delta_k{}^v & \omega_k{}^j & \theta_k{}^\beta \\ \xi_\alpha{}^\beta & \sigma_\alpha{}^v & \sigma_\alpha{}^j & \widehat{\psi}_\alpha{}^\beta \end{pmatrix} \begin{pmatrix} Z_\beta \\ Z'_v \\ X_j \\ Y_\beta \end{pmatrix},$$

where the matrix π satisfies the symmetry relation

$$\begin{pmatrix} \psi_\alpha{}^\beta & \theta_\alpha{}^v & \theta_\alpha{}^j & \phi_\alpha{}^\beta \\ \sigma_u{}^\beta & \omega_u{}^v & \delta_u{}^j & \theta_u{}^\beta \\ \sigma_k{}^\beta & \delta_k{}^v & \omega_k{}^j & \theta_k{}^\beta \\ \xi_\alpha{}^\beta & \sigma_\alpha{}^v & \sigma_\alpha{}^j & \widehat{\psi}_\alpha{}^\beta \end{pmatrix} = - \begin{pmatrix} \widehat{\psi}_{\bar\beta}{}^{\bar\alpha} & \varepsilon_v\theta_{\bar v}{}^{\bar\alpha} & \varepsilon_j\theta_{\bar j}{}^{\bar\alpha} & \phi_{\bar\beta}{}^{\bar\alpha} \\ \varepsilon_u\sigma_{\bar\beta}{}^{\bar u} & \varepsilon_u\varepsilon_v\omega_{\bar v}{}^{\bar u} & \varepsilon_u\varepsilon_j\delta_{\bar j}{}^{\bar u} & \varepsilon_u\theta_{\bar\beta}{}^{\bar u} \\ \varepsilon_k\sigma_{\bar\beta}{}^{\bar k} & \varepsilon_k\varepsilon_v\delta_{\bar v}{}^{\bar k} & \varepsilon_k\varepsilon_j\omega_{\bar j}{}^{\bar k} & \varepsilon_k\theta_{\bar\beta}{}^{\bar k} \\ \xi_{\bar\beta}{}^{\bar\alpha} & \varepsilon_v\sigma_{\bar v}{}^{\bar\alpha} & \varepsilon_j\sigma_{\bar j}{}^{\bar\alpha} & \psi_{\bar\beta}{}^{\bar\alpha} \end{pmatrix},$$

where

$$\varepsilon_u := \langle Z'_u, Z'_u \rangle = -1, \quad u = 1, \dots, q-r, \quad \varepsilon_j := \langle X_j, X_j \rangle = 1, \quad j = 1, \dots, p-r.$$

The defining equations of $S_{p,q,r}$ can be written as

$$S_{p,q,r} = \{[V] \in G_{p,q} : \langle \cdot, \cdot \rangle|_{V_0} = 0\}$$

and hence their differentiation yields

$$\langle dZ_\alpha, Z_\beta \rangle + \langle Z_\alpha, dZ_\beta \rangle = \varphi_\alpha{}^\beta + \overline{\varphi_\beta{}^\alpha} = 0. \tag{2.1}$$

By substituting $dZ_\Lambda = \pi_\Lambda{}^\Gamma Z_\Gamma$ into $(1,0)$ component of (2.1) we obtain, in particular,

$$\varphi_\alpha{}^\gamma \langle Y_\gamma, Z_\beta \rangle \; = \; \varphi_\alpha{}^\beta \; = \; 0,$$

when restricted to the $(1,0)$ tangent space. Comparing the dimensions, we conclude that the kernel of $\{\varphi_\alpha{}^\beta, \alpha, \beta = 1, \dots, q\}$ forms the CR bundle of $S_{p,q,r}$, i.e.,

$$\ker(\varphi|_Z) = T_Z^{1,0} S_{p,q,r} \oplus T_Z^{0,1} S_{p,q,r}.$$

In other words, $\varphi = (\varphi_\alpha{}^\beta)$ span the space of contact forms on $S_{p,q,r}$. Since

$$dZ_\alpha = \psi_\alpha{}^\beta Z_\beta + \theta_\alpha{}^v Z'_v + \theta_\alpha{}^j X_j + \varphi_\alpha{}^\beta Y_\beta,$$
$$dZ'_u = \sigma_u{}^\beta Z_\beta + \omega_u{}^v Z'_v + \delta_u{}^j X_j + \theta_u{}^\beta Y_\beta$$

and $\varphi = (\varphi_\alpha{}^\beta)$ is a contact form, we conclude that the upper right block forms

$$\begin{pmatrix} \theta_{\alpha}{}^{j} & \varphi_{\alpha}{}^{\beta} \\ \delta_{u}{}^{k} & \theta_{u}{}^{\beta} \end{pmatrix}$$

give together a basis in the space of all $(1, 0)$ forms on $S_{p,q,r}$.

2.3 The Tangent Space of $S_{p,q,r}$

The tangent space to the Grassmanian $G_{p,q}$ at the element V is isomorphic to $\mathsf{Hom}\,(V, \mathbb{C}^{p+q}/V)$. Hence, given an adapted frame (Z, Z', X, Y), it is isomorphic to

$$T_V G_{p,q} = \mathsf{Hom}\,(V, X \oplus Y).$$

By taking into account the splitting $V = V_0 \oplus V'$, the elements of

$$\mathsf{Hom}\,(V, X \oplus Y) = \mathsf{Hom}\,(V_0 \oplus V', X \oplus Y)$$

are given by block 2×2 matrices decomposed as

$$R \in \begin{pmatrix} \mathsf{Hom}\,(V_0, X) & \mathsf{Hom}\,(V_0, Y) \\ \mathsf{Hom}\,(V', X) & \mathsf{Hom}\,(V', Y) \end{pmatrix}. \tag{2.2}$$

Then the real tangent space $T_V S_{p,q,r}$ to $S_{p,q,r}$ is

$$T = T_V S_{p,q,r} = \begin{pmatrix} * & \widetilde{R} \\ * & * \end{pmatrix}, \quad \widetilde{R} = -\widetilde{R}^*,$$

the complex tangent subspace is

$$T^c = \begin{pmatrix} * & 0 \\ * & * \end{pmatrix}.$$

The complex tangent space T^c contains further two invariantly defined subspaces

$$T^- := \{R \in T^c : R(V_0) \subset V\} = \begin{pmatrix} 0 & 0 \\ * & * \end{pmatrix},$$

$$T^+ := \{R \in T^c : \langle R(V), V_0 \rangle = 0\} = \begin{pmatrix} * & 0 \\ * & 0 \end{pmatrix},$$

such that

$$T^+ + T^- = T^c.$$

For a change of frame given by

$$\begin{pmatrix} \widetilde{Z} \\ \widetilde{Z}' \\ \widetilde{X} \\ \widetilde{Y} \end{pmatrix} := U \begin{pmatrix} Z \\ Z' \\ X \\ Y \end{pmatrix},$$

π changes via

$$\widetilde{\pi} = dU \cdot U^{-1} + U \cdot \pi \cdot U^{-1}.$$

We shall employ several types of frame changes.

Definition 2.1 We call a change of frame

(i) change of position if

$$\widetilde{Z}_\alpha = W_\alpha{}^\beta Z_\beta, \quad \widetilde{Z}'_u = W_u{}^\beta Z_\beta + W_u{}^\nu Z'_\nu, \quad \widetilde{Y}_\alpha = V_\alpha{}^\beta Y_\beta + V_\alpha{}^\nu Z'_\nu, \quad \widetilde{X}_j = X_j,$$

where $W_0 = (W_\alpha{}^\beta)$ and $V_0 = (V_\alpha{}^\beta)$ are $r \times r$ matrices satisfying $V_0^* W_0 = I_r$, $W' = (W_u{}^\nu)$ is a $(q-r) \times (q-r)$ matrix satisfying $W'^* W' = I_{q-r}$ and $V_\alpha{}^\beta W^*{}_\beta{}^\gamma + V_\alpha{}^\nu W^*{}_\nu{}^\gamma = 0$;

(ii) change of real vectors if

$$\widetilde{Z}_\alpha = Z_\alpha, \quad \widetilde{Z}'_u = Z'_u, \quad \widetilde{X}_j = X_j, \quad \widetilde{Y}_\alpha = Y_\alpha + H_\alpha{}^\beta Z_\beta,$$

where $H = (H_\alpha{}^\beta)$ is a skew hermitian matrix;

(iii) dilation if

$$\widetilde{Z}_\alpha = \lambda_\alpha^{-1} Z_\alpha, \quad \widetilde{Z}'_u = Z'_u, \quad \widetilde{Y}_\alpha = \lambda_\alpha Y_\alpha, \quad \widetilde{X}_j = X_j,$$

where $\lambda_\alpha > 0$;

(iv) rotation if

$$\widetilde{Z}_\alpha = Z_\alpha, \quad \widetilde{Z}'_u = Z'_u, \quad \widetilde{Y}_\alpha = Y_\alpha, \quad \widetilde{X}_j = U_j{}^k X_k,$$

where $(U_j{}^k)$ is a unitary matrix.

The remaining change of frame is given by

$$\widetilde{Z}_\alpha = Z_\alpha, \quad \widetilde{Z}'_u = Z'_u, \quad \widetilde{X}_j = X_j + C_j{}^\beta Z_\beta, \quad \widetilde{Y}_\alpha = Y_\alpha + A_\alpha{}^\beta Z_\beta + B_\alpha{}^j X_j,$$

or

$$\begin{pmatrix} \widetilde{Z}_\alpha \\ \widetilde{Z}'_u \\ \widetilde{X}_j \\ \widetilde{Y}_\alpha \end{pmatrix} = \begin{pmatrix} I_r & 0 & 0 & 0 \\ 0 & I_{q-r} & 0 & 0 \\ C_j^\beta & 0 & I_{p-r} & 0 \\ A_\alpha^\beta & 0 & B_\alpha^j & I_r \end{pmatrix} \begin{pmatrix} Z_\beta \\ Z'_u \\ X_k \\ Y_\beta \end{pmatrix},$$

such that

$$C_j^{\ \alpha} + B_j^{\ \alpha} = 0$$

and

$$(A_\alpha^{\ \beta} + \overline{A_\beta^{\ \alpha}}) + B_\alpha^{\ j} B_j^{\ \beta} = 0,$$

where

$$B_j^{\ \alpha} := \overline{B_\alpha^{\ j}}.$$

2.4 Structure Identities

The structure equations yield

$$d\phi_\alpha^{\ \beta} = \theta_\alpha^{\ j} \wedge \theta_j^{\ \beta} + \theta_\alpha^{\ u} \wedge \theta_u^{\ \beta} \mod \phi,$$

$$d\theta_\alpha^{\ j} = \theta_\alpha^{\ v} \wedge \delta_v^{\ j} \mod \{\theta_\beta^{\ k}, \phi\},$$

$$d\theta_u^{\ \alpha} = \delta_u^{\ k} \wedge \theta_k^{\ \alpha} \mod \{\theta_v^{\ \beta}, \phi\},$$

where ϕ stands for the span of all $\phi_\alpha^{\ \beta}$. The first equation via Cartan's formula

$$d\tau(R_1, R_2) = R_1\tau(R_2) - R_2\tau(R_1) - \tau([R_1, R_2]),$$

determines the invariant tensor

$$\mathscr{L} = \mathscr{L}_1 \colon T^{1,0} \times T^{1,0} \to \frac{\mathbb{C}T}{T^{1,0} + T^{0,1}}, \quad (R_1, R_2) \mapsto [R_1, \overline{R}_2] \mod T^{1,0} + T^{0,1},$$

which represents the Levi form of $S_{p,q,r}$ up to imaginary constant. In particular,

$$T^0 := T^+ \cap T^- = \begin{pmatrix} 0 & 0 \\ * & 0 \end{pmatrix} \subset T^c$$

is the kernel of the Levi form of $S_{p,q,r}$. Levi kernel is always integrable. In fact, since T^0 is given by

$$\phi_\alpha^{\ \beta} = \theta_u^{\ \beta} = \theta_\alpha^{\ j} = 0,$$

on the integral manifold of Levi kernel, we obtain

$$dZ_\alpha = \psi_\alpha^{\ \beta} Z_\beta,$$

i.e., the space $(Z_1, \ldots, Z_r)$ is fixed. Hence up to automorphisms of $D_{p,q}$, the integral manifold of the Levi kernel is given by

$$\begin{pmatrix} I_r & 0 \\ 0 & z \end{pmatrix},$$

where I_r is the $r \times r$ identity matrix and z is a $(p-r) \times (q-r)$ complex matrix satisfying

$$I_{q-r} - z^* z > 0.$$

From this we can see that $S_{p,q,r}$ is foliated by $D_{p-r,q-r}$.

Similarly, the second and the third equations of the structure equations determine together the invariant tensor

$$\mathscr{L}_2\colon K^{1,0} \times T^{1,0} \to \frac{T^{1,0}}{K^{1,0}} \cong \frac{T^{1,0}+T^{0,1}}{K^{1,0}+T^{0,1}}, \quad (R_1, R_2) \mapsto [R_1, \overline{R}_2] \mod K^{1,0} + T^{0,1},$$

where $K^{1,0}$ is in the complexified Levi kernel. Note that for $R_1 \in K^{1,0}$, one always has $[R_1, \overline{R}_2] \subset T^{1,0} + T^{0,1}$. The tensor $\mathscr{L}_2$ can be regarded as the "second order Levi form" that comes naturally into consideration along with the (first order) Levi form $\mathscr{L}_1$ to gain the "missing nondegeneracy". In the decomposition (2.2), $\mathscr{L}_2$ takes the form

$$\left(\begin{pmatrix} 0 & 0 \\ c_1 & 0 \end{pmatrix}, \begin{pmatrix} a_2 & 0 \\ c_2 & d_2 \end{pmatrix}\right) \mapsto (-c_1 d_2^*) \oplus a_2^* c_1 \in \mathsf{Hom}\,(V_0, X) \oplus \mathsf{Hom}\,(V', Y).$$

3 Differential Equations for CR Maps

Let f be a local CR map from an open set $M \subset S_{p,q,r}$ to an open set $M' \subset S_{p',q',r'}$. We shall consider the connection forms $\phi_\alpha{}^\beta, \theta_\alpha{}^j, \psi_\alpha{}^\beta, \omega_j{}^k, \sigma_j{}^\beta, \xi_\alpha{}^\beta$ on M and denote by capital letters $\Phi_a{}^b, \Theta_a{}^J, \Psi_a{}^b, \Omega_J{}^K, \Sigma_K^{\ b}, \Xi_a{}^b$ their corresponding counterparts on M'.

Since $\phi = (\phi_\alpha{}^\beta)$ and $\Phi = (\Phi_a{}^b)$ are contact forms on M and M', respectively, the pull back of Φ via f is a linear combination of $\phi = (\phi_\alpha{}^\beta)$. Choose a diagonal contact form of M' and say $\Phi_1{}^1$. Since contact forms are spanned by $\phi_\alpha{}^\beta$, we can write

$$\Phi_1{}^1 = c_\alpha^\beta \phi_\beta{}^\alpha$$

for some smooth functions $c_\alpha{}^\beta$. At generic points, after a change of position vector Z on M, we may assume

$$\Phi_1{}^1 = \sum_{\alpha=1}^{r} c_\alpha \phi_\alpha^\alpha$$

for smooth functions c_α. Then using the structure equation for ϕ and its analogue for M', we obtain

$$\Theta_1{}^J \wedge \Theta_J{}^1 + \Theta_1{}^U \wedge \Theta_U{}^1 = \sum_\alpha c_\alpha(\theta_\alpha{}^j \wedge \theta_j{}^\alpha + \theta_\alpha{}^u \wedge \theta_u{}^\alpha) \mod \phi,$$

Since the Levi form is nonnegative, we conclude $c_\alpha \geq 0$. If $\Phi_1{}^1 \not\equiv 0$, then after dilation of Φ, we may assume that $c_1 = 1$. Moreover, since f sends the Levi kernel of M given by $\phi = \theta = 0$ into the Levi kernel of M' given by $\Phi = \Theta = 0$, we can write

$$\begin{aligned} \Theta_1{}^J &= h_j^{J,\alpha}\theta_\alpha{}^j + g_\alpha^{J,u}\theta_u{}^\alpha \mod \phi, \\ \Theta_U{}^1 &= \eta_{U,j}^\alpha\theta_\alpha{}^j + \xi_{U,\alpha}^u\theta_u{}^\alpha \mod \phi. \end{aligned}$$

The same argument in §3 of [KZ14] using $\Theta_U{}^1$ in place of $\Theta_1{}^J$ will yield the following lemma.

Lemma 3.1 *Suppose $q' - r' < p - r$. Then*

$$g_\alpha^{J,u} = \eta_{U,j}^\alpha = 0,$$

i.e.,

$$f_*(T^+) \subset T'^+, \ f_*(T^-) \subset T^-.$$

Furthermore, if $q' - r' < 2(q - r)$, then

$$\Phi_1{}^1 = \phi_1{}^1.$$

By this lemma and the same argument in §3 and §4 of [KZ14], we obtain the second fundamental forms

$$\begin{aligned} \Omega_k{}^J &= A_k{}^J{}_v\theta_\alpha{}^v + B_k{}^J{}_j\theta_\alpha{}^j \mod \phi, \quad J > p - r, \\ \Omega_u{}^U &= A_u{}^U{}_k\theta_\alpha{}^k + B_u{}^U{}_v\theta_\alpha{}^v \mod \phi, \quad U > q - r, \end{aligned} \tag{3.1}$$

on T^+ and T^-, respectively for each fixed $\alpha = 1, \ldots, r$. Since each θ_α^j, $\theta_\alpha{}^u$ is independent, if $r \geq 2$, then we conclude that the second fundamental forms are trivial modulo ϕ. If $r = 1$, then we analyze the Gauss equations for T^+;

$$\begin{aligned} B_k{}^K{}_l B^j{}_K{}^m &= g_l{}^m\hat{\delta}_k{}^j + g_l{}^j\hat{\delta}_k{}^m + g_k{}^m\hat{\delta}_l{}^j + g_k{}^j\hat{\delta}_l{}^m, \\ A_k{}^K{}_u B^j{}_K{}^m &= \eta_u{}^m\hat{\delta}_k{}^j + \eta_u{}^j\hat{\delta}_k{}^m, \end{aligned}$$

for T^-;

$$B_{u\ w}^{\ V} B_{\ V}^{v\ y} = g_w^{\ y} \hat{\delta}_u^{\ v} + g_w^{\ v} \hat{\delta}_u^{\ y} + g_u^{\ y} \hat{\delta}_w^{\ v} + g_u^{\ v} \hat{\delta}_w^{\ y}, \tag{3.2}$$

$$A_{u\ k}^{\ V} B_{\ V}^{v\ w} = \eta_k^{\ v} \hat{\delta}_u^{\ w} + \eta_k^{\ w} \hat{\delta}_u^{\ v}, \tag{3.3}$$

and mixed terms;

$$A_{u\ k}^{\ V} A_{\ V}^{v\ j} + A_{k\ u}^{\ K} A_{\ K}^{j\ v} = g_u^{\ v} \hat{\delta}_k^{\ j} + g_k^{\ j} \hat{\delta}_u^{\ v},$$

where $\hat{\delta}$ is the Kronecker delta, $\eta_u^{\ k} = -\overline{\eta_k^{\ u}}$, and $g_j^{\ k}$, $g_u^{\ v}$ satisfy

$$\begin{aligned} \Delta_u^{\ j} - \delta_u^{\ j} - \eta_u^{\ j} \phi_\alpha^{\ \alpha} &= 0, \\ \hat{\delta}_j^{\ k} \left(\Psi_\alpha^{\ \alpha} - \psi_\alpha^{\ \alpha}\right) - \left(\Omega_j^{\ k} - \omega_j^{\ k}\right) + g_j^{\ k} \phi_\alpha^{\ \alpha} &= 0, \\ \hat{\delta}_v^{\ u} \left(\Psi_\alpha^{\ \alpha} - \psi_\alpha^{\ \alpha}\right) - \left(\Omega_v^{\ u} - \omega_v^{\ u}\right) + g_v^{\ u} \phi_\alpha^{\ \alpha} &= 0 \end{aligned}$$

with relation

$$\hat{\delta}_k^{\ j} g_u^{\ v} + \hat{\delta}_u^{\ v} g_k^{\ j} = \eta_{u;k}^{\ j;v} \tag{3.4}$$

such that

$$d\eta_u^{\ j} - \eta_v^{\ j} \omega_u^{\ v} + \eta_u^{\ k} \omega_k^{\ j} = \eta_{u;k}^{\ j;v} \delta_v^{\ k} \qquad \text{mod } \theta, \bar{\theta}, \phi.$$

Lemma 5.3 from [EHZ05] applied to (3.2) implies that

$$B_{u\ w}^{\ V} = g_u^{\ v} = 0$$

provided $q' - r' < 2(q - r)$. Then differentiation of (3.1) as in [KZ14] implies

$$A_{u\ k}^{\ V} = 0.$$

By (3.3), we obtain

$$\eta_u^{\ k} = 0$$

and by (3.4), we obtain

$$g_j^{\ k} = 0.$$

Then Gauss equations for T^+ and mixed terms will imply

$$B_{k\ \ell}^{\ J} = A_{k\ u}^{\ J} = 0,$$

i.e., the second fundamental forms are trivial modulo ϕ. This will lead to the following key proposition.

Proposition 3.1 *There exist vector subspaces* $V_0, V_1, V_2 \subset \mathbb{C}^{p'+q'}$ *of dimensions*

$$\dim V_0 = p+q, \quad \dim V_1 = r'-r, \quad \dim V_2 = p'-r'+q'-r'-(p-r)-(q-r)$$

that form a direct sum, such that the basic form $\langle \cdot, \cdot \rangle$ *is null when restricted to* V_1*, nondegenerate of signature* (p, q) *when restricted to* V_0*, and nondegenerate of signature* $(p'-r'-(p-r), q'-r'-(q-r))$ *when restricted to* V_2*, and such that whenever* $x \in S_{p,q,r}$ *and* $f(x)$ *is defined, we have*

$$f(x) = W_0 \oplus V_1 \oplus W_2 \in Gr(V_0, q) \oplus V_1 \oplus Gr(V_2, (q'-r')-(q-r)), \quad (3.5)$$

such that the basic form restricted to W_0 *has rank* r.

With this proposition and the same argument in §6 of [KZ14], we can prove the theorems.

Acknowledgments This research was supported by Basic Science Research Program through the National Research Foundation of Korea (NRF) funded by the Ministry of Education, Science and Technology (grant number 2012R1A1B5003198).

References

[A74] Alexander, H.: Holomorphic mappings from the ball and polydisc. Math. Ann. **209**, 249–256 (1974)

[BEH08] Baouendi, M.S., Ebenfelt, P., Huang, X.: Super-rigidity for CR embeddings of real hypersurfaces into hyperquadrics. Adv. Math. **219**(5), 1427–1445 (2008)

[BEH09] Baouendi, M.S., Ebenfelt, P., Huang, X.: Holomorphic mappings between hyperquadrics with small signature difference. Am. J. Math. **133**(6), 1633–1661 (2011)

[BH05] Baouendi, M.S., Huang, X.: Super-rigidity for holomorphic mappings between hyperquadrics with positive signature. J. Differ. Geom. **69**(2), 379–398 (2005)

[Bo47] Bochner, S.: Curvature in Hermitian metric. Bull. Am. Math. Soc. **53**, 179–195 (1947)

[Ca53] Calabi, E.: Isometric imbedding of complex manifolds. Ann. Math. **58**(2), 1–23 (1953)

[CM74] Chern, S.S., Moser, J.K.: – Real hypersurfaces in complex manifolds. Acta Math. **133**, 219–271 (1974)

[CS83] Cima, J., Suffridge, T.J.: A reflection principle with applications to proper holomorphic mappings. Math. Ann. **265**, 489–500 (1983)

[CS90] Cima, J., Suffridge, T.J.: Boundar behavior of rational proper maps. Duke Math. **60**, 135–138 (1990)

[EHZ04] Ebenfelt, P., Huang, X., Zaitsev, D.: Rigidity of CR-immersions into spheres. Commun. Anal. Geom. **12**(3), 631–670 (2004)

[EHZ05] Ebenfelt, P., Huang, X., Zaitsev, D.: The equivalence problem and rigidity for hypersurfaces embedded into hyperquadrics. Am. J. Math. **127**(1), 169–191 (2005)

[Fa86] Faran, V.: On the linearity of proper maps between balls in the lower codimensional case. J. Differ. Geom. **24**, 15–17 (1986)

[Fo86] Forstnerič, F.: Proper holomorphic maps between balls. Duke Math. J. **53**, 427–440 (1986)

[Fo89] Forstnerič, F.: Extending proper holomorphic mappings of positive codimension. Invent. Math. **95**, 31–61 (1989)

[H99] Huang, X.: On a linearity problem for proper holomorphic maps between balls in complex spaces of different dimensions. J. Differ. Geom. **51**, 13–33 (1999)
[HJ01] Huang, X., Ji, S.: Mapping B^n into B^{2n-1}. Invent. Math. **145**(2), 219–250 (2001)
[KZ13] Kim, S.Y., Zaitsev, D.: Rigidity of CR maps between Shilov boundaries of bounded symmetric domains. Invent. Math. **193**(2), 409–437 (2013)
[KZ14] Kim, S.Y., Zaitsev, D.: Rigidity of proper holomorphic maps between bounded symmetric domains. Math. Ann. 362(2), 639–677 (2015)
[M89] Mok, N.: Metric Rigidity Theorems on Hermitian Locally Symmetric Spaces. Series in Pure Mathematics, vol. 6. World Scientific, Singapore (1989)
[M11] Mok, N.: Geometry of holomorphic isometries and related maps between bounded domains. In: Geometry and Analysis, vol. II, ALM 18. Higher Education Press and International Press, Beijing, pp. 225–270 (2011)
[MN12] Mok, N., Ng, S.-C.: Germs of measure-preserving holomorphic maps from bounded symmetric domains to their Cartesian products. J. Reine Angew. Math. **669**, 47–73 (2012)
[MNT10] Mok, N., Ng, S.-C., Tu, Z.: Factorization of proper holomorphic maps on irreducible bounded symmetric domains of rank ≥ 2. Sci. Chi. Math. **53**(3), 813–826 (2010)
[Ng12] Ng, S.-C.: Cycle spaces of flag domains on Grassmannians and rigidity of holomorphic mappings. Math. Res. Lett. **19**(6), 1219–1236 (2012)
[Ng13] Ng, S.-C.: Holomorphic double fibration and the mapping problems of classical domains. Int. Math. Res. Not. (to appear)
[P07] Poincaré, H.: Les fonctions analytiques de deux variables et la représentation conforme. Rend. Circ. Mat. Palermo **23**(2), 185–220 (1907)
[Ta62] Tanaka, N.: On the pseudo-conformal geometry of hupersurfaces of the space of n complex variables. J. Math. Soc. Jpn **14**, 397–429 (1962)
[W79] Webster, S.M.: The rigidity of C-R hypersurfaces in a sphere. Indiana Univ. Math. J. **28**, 405–416 (1979)

Characterizations of Strongly Pseudoconvex Models in Almost Complex and CR Geometries

Kang-Hyurk Lee

Abstract In this paper, we introduce the Wong-Rosay theorem, R. Schoen's theorem and its generalization in almost complex geometry.

Keywords Almost CR manifolds · Pseudo-Hermitian manifolds · Infinitesimal automorphism

1 Introduction

The aim of this paper is to introduce (1) the Wong-Rosay theorem, a characterization of the unit ball by its holomorphic automorphism group, (2) Schoen's theorem, a characterization of the unit sphere and the Heisenberg group by their CR automorphism groups, and (3) their generalizations to the almost complex and CR manifolds, respectively.

1.1 The Characterization of the Unit Ball

The Riemann mapping theorem says that a simply connected proper domain in the complex plane $\mathbb{C}$ is biholomorphic to the unit disc Δ. Hence in Complex Analysis of one variable, it is important to understand the nature of the unit disc. But in multi-dimensional complex Euclidean spaces, the Riemann mapping theorem fails as H. Poincaré showed that the unit ball $\mathbb{B}^2 = \left\{z \in \mathbb{C}^2 : \|z\| < 1\right\}$ and the bidisc $\Delta^2 = \Delta \times \Delta$ are biholomorphically distinct. Moreover as showed in [BU78, GE82], the biholomorphic equivalence classes of simply connected domains in $\mathbb{C}^n$ $(n \geq 2)$ forms indeed an infinite dimensional space. Therefore it has been a fundamental problem in Several Complex Variables to classify bounded domains which can play

K.-H. Lee (✉)
Department of Mathematics and Research Institute of Natural Science,
Gyeongsang National University, Jinju, Gyeongnam 660-701, Korea
e-mail: nyawoo@gnu.ac.kr

F. Bracci et al. (eds.), *Complex Analysis and Geometry*, Springer Proceedings in Mathematics & Statistics 144, DOI 10.1007/978-4-431-55744-9_16

the same rôle of model objects as the unit disc. A precondition for the rôle in Complex Analysis and Complex Geometry is to admit a noncompact automorphism group. While it is not possible to classify simply connected domains in $\mathbb{C}^n$, the classification of domains with noncompact automorphism groups seems to be possible since a generic bounded domain has no automorphism except the identity (see [GE82]). A typical classification is B. Wong's characterization of the unit ball $\mathbb{B}^n = \{z \in \mathbb{C}^n : \|z\| < 1\}$.

Theorem 1.1 ([WO77]) *A bounded strongly pseudoconvex domain in $\mathbb{C}^n$ with noncompact automorphism group is biholomorphic to the unit ball $\mathbb{B}^n$.*

For a bounded domain Ω, the noncompactness of the automorphism group of Ω, denoted by $\mathrm{Aut}(\Omega)$, is equivalent to the existence of an automorphism orbit $\{\varphi_k(p)\}$ for some $\varphi_k \in \mathrm{Aut}(\Omega)$ and $p \in \Omega$ which is accumulating at a bounded point. In his paper [RO79], J. P. Rosay strengthened Wong's theorem as following:

Theorem 1.2 ([RO79, EF95, GA02]) *A domain in a complex manifold which admits an automorphism orbit accumulating at a strongly pseudoconvex boundary point is biholomorphic to the unit ball.*

Theorems 1.1 and 1.2 are usually called the Wong-Rosay theorem.

1.2 The Characterization of the Unit Sphere

In the confomal geometry, the Euclidean space $\mathbb{R}^n$ and the Euclidean sphere $S^n = \{x \in \mathbb{R}^{n+1} : \|x\| = 1\}$ are characterized as global homogeneous models as showed in [AL72, SC95, FE96]:

Theorem 1.3 *The conformal group of the Riemannian manifold (M^n, g) is essential if and only if M is conformally equivalent to either $\mathbb{R}^n$ or S^n.*

Here essential means that the conformal group can not be reduced to an isometry group of a metric in the conformal class. As in [AL72], if the conformal group of M is essential, then it acts improperly on M (A topological group G *acts improperly* on M if there is a compact subset K of M such that $G_K = \{\varphi \in G : \varphi(K) \cap K \neq \emptyset\}$ is noncompact). The main proof of Theorem 1.3 is to confirm D. V. Alekseevskiĭ's assertion: *if the conformal group acts improperly on M, then M is conformally equivalent to $\mathbb{R}^n$ or S^n.*

A strongly pseudoconvex real hypersurface in a complex manifold, especially a boundary of a strongly pseudoconvex domain, has a similar geometric structure to the conformal geometry, usually called the pseudo-conformal structure. A real hypersurface M in a complex manifold X admits a CR structure inherited by the complex structure of X. If M is strongly pseudoconvex, then its CR structure is determined by the conformal structure of its pseudo-hermitian metric. R. Schoen also gave the CR version of Theorem 1.3 in case of strongly pseudoconvex CR manifolds:

Theorem 1.4 ([SC95]) *Suppose that M^{2n+1} is a strongly pseudoconvex CR manifold whose CR automorphism group acts on M improperly. Then M is CR equivalent to either the unit sphere $S^{2n+1} = \{z \in \mathbb{C}^{n+1} : \|z\| = 1\}$ if M is compact or the Heisenberg group if M is noncompact.*

This is a CR counterpart of the Wong-Rosay theorem. In case of a bounded strongly pseudoconvex domain Ω, Fefferman's extension theorem ([FE74]) implies that each automorphism of Ω extends to a CR automorphism of the boundary $\partial\Omega$ which is a compact strongly pseudoconvex CR manifold. Thus the noncompactness of $\mathrm{Aut}(\Omega)$ implies that the CR automorphism group of $\partial\Omega$ is also noncompact, equivalently, it acts improperly (for a compact manifold, the improper action by a topological group G is the same as the noncompactness of G). In case of an unbounded domain, consider the *Siegel half plane*, $\mathbb{H}^{n+1} = \{(z^0, z^1, \dots, z^n) \in \mathbb{C}^{n+1} : \mathrm{Re}\, z^0 + \sum_{\alpha=1}^{n} |z^\alpha|^2 < 0\}$ which is biholomorphic to the unit ball $\mathbb{B}^{n+1}$ by the Cayley transform. The group of affine automorphisms of $\mathbb{H}^{n+1}$ coincides with the CR automorphism group of the *Heisenberg group* $\partial\mathbb{H}^{n+1}$. Since $\mathscr{D}_s$ in (2.1) belongs to the isotropy subgroup at the origin, the CR automorphism group of $\partial\mathbb{H}^{n+1}$ is noncompact and moreover acts improperly.

1.3 Generalizations

Gaussier and Sukhov ([GA03]) showed that the Wong-Rosay theorem is also valid in almost complex manifolds of complex dimension 2. But in higher dimensional case, there is an exotic model (called a pseudo-Siegel domain) which admits an automorphism orbit accumulating at a strongly pseudoconvex boundary point and whose almost complex structure is non-integrable, so which is not biholomorphic to the unit ball with the standard complex structure. Thus the local version (Theorem 1.2) fails in almost complex manifolds. Gaussier and Sukhov [GA06] and the author [LK06] characterized the pseudo-Siegel domains: *a domain in almost complex manifold which admits an automorphism orbit accumulating at a strongly pseudoconvex boundary point is biholomorphic to a pseudo-Siegel domain* (Theorem 2.1). However as in [BY09], the global version (Theorem 1.1) is also valid in any dimension: *a relatively compact, strongly pseudoconvex domain in an almost complex manifold with a noncompact automorphism group is biholomorphic to the unit ball with the standard complex structure* (Theorem 2.2).

As in Sect. 2, a pseudo-Siegel domain is the Siegel half plane with a certain almost complex structure, so its boundary is noncompact. And its automorphism group is the same as the CR automorphism group of the boundary which acts improperly. Therefore the relationship between the Wong-Rosay theorem and Schoen's theorem makes us to expect:

Conjecture 1.1 A strongly pseudoconvex almost CR manifold M whose CR automorphism group action is improper is CR equivalent to either the standard sphere if M is compact or a boundary of a pseudo-Siegel domain if M is noncompact.

In this paper, we introduce the basic technique to get the Wong-Rosay theorem in almost complex structure as in [GA06, LK06] and a partial confirmation of the conjecture by the collaboration work with Joo [JO15].

Convention: Throughout this paper, Greek indices indicating coefficients of complex tensor run from 1 to n and Latin indices for real tensors run from 1 to $2n$. For Greek indices, the summation convention is always assumed. We will take the bar on Greek indices to denote the complex conjugation of the corresponding tensor coefficients: $\overline{Z}_\alpha = Z_{\bar\alpha}$, $\bar\omega^\alpha = \omega^{\bar\alpha}$, $\overline{R}_\beta{}^\alpha{}_{\lambda\bar\mu} = R_{\bar\beta}{}^{\bar\alpha}{}_{\bar\lambda\mu}$.

2 The Wong-Rosay Theorem in the Almost Complex Manifold

Let X be an almost complex manifold with an almost complex structure J. By an (holomorphic) *automorphism* of (X, J), we mean a biholomorphism of X onto itself with respect to J. The *automorphism group* $\mathrm{Aut}(X, J)$ of (X, J) is the topological group of automorphisms of (X, J) with the composition law and the compact-open topology.

Let us define the pseudo-Siegel domain as in [LK08]:

Definition 2.1 Consider the complex Eulidean space $\mathbb{C}^{n+1}$ with the standard coordinates $(z^0, z^1, \ldots, z^n)$. Let $P = (P_{\alpha\beta})_{\alpha,\beta=1,\ldots,n}$ be a $n \times n$ skew-symmetric complex matrix. The *model structure* J_P is the almost complex structure of $\mathbb{C}^{n+1}$ defined by the following $(1, 0)$-vector fields:

$$Z_0 = \frac{\partial}{\partial z^0}, \quad Z_\alpha = \frac{\partial}{\partial z^\alpha} - i P_{\alpha\beta} z^\beta \frac{\partial}{\partial z^{\bar 0}} \quad (\alpha = 1, \ldots, n) .$$

The pair $(\mathbb{H}^{n+1}, J_P)$ is called a *pseudo-Siegel domain* for the Siegel half plane $\mathbb{H}^{n+1}$ and the model structure J_P.

2.1 Automorphisms of the Pseudo-Siegel Domain

As mentioned, the Siegel-half plane $\mathbb{H} = \mathbb{H}^{n+1}$ with the standard complex structure J_{st} (the case of $P = 0$) is biholomorphic to the unit ball $(\mathbb{B}^{n+1}, J_{\mathrm{st}})$; thus the pseudo-Siegel domains can be considered as a deformation of the unit ball. The matrix P represents the torsion for the integrability of the structure, in the sense of $[Z_\alpha, Z_\beta] = -2i P_{\alpha\beta} \partial/\partial z^{\bar 0}$ so that J_P is always non-integrable except $P = 0$. For any choice of P, the boundary of $\mathbb{H}$ is always strongly pseudoconvex and $\mathbb{H}$ has the non-isotropic dilation

$$\mathscr{D}_s : (z^0, z^1, \ldots, z^n) \mapsto (e^s z^0, e^{s/2} z^1, \ldots, e^{s/2} z^n) \quad (s \in \mathbb{R}) \tag{2.1}$$

as its automorphism. This means that Theorem 1.2 fails in almost complex setting.

Moreover any pseudo-Siegel domain is homogeneous since it has the Heisenberg group as its holomorphic transformation group. The *Heisenberg group* is the group $\mathscr{H}_P = (\partial\mathbb{H}, *_P)$ whose binary operation $*_P$ is defined by

$$\zeta *_P \xi = \left(\zeta^0 + \xi^0 - 2\delta_{\alpha\bar{\beta}}\xi^\alpha\zeta^{\bar{\beta}} + iP_{\alpha\beta}\xi^\alpha\zeta^\beta + iP_{\bar{\alpha}\bar{\beta}}\xi^{\bar{\alpha}}\zeta^{\bar{\beta}}, \zeta' + \xi'\right), \tag{2.2}$$

for $\zeta = (\zeta^0, \zeta'), \xi = (\xi^0, \xi') \in \partial\mathbb{H}$. Each element $\zeta \in \mathscr{H}_P$ generates an automorphism by $z \mapsto \zeta *_P z$; hence $\mathscr{H}_P$ can be considered as a subgroup of $\mathrm{Aut}(\mathbb{H}, J_P)$. Then one can easily see that the transformation group generated by $\mathscr{H}_P$ and $\{\mathscr{D}_s : s \in \mathbb{R}\}$ acts on $\mathbb{H}$ transitively.

In [LK08], the automorphism groups and the bihomorphic equivalence of pseudo-Siegel domains are completely described.

2.2 The Scaling Method in Almost Complex Manifold

Here, we introduce the scaling method to the almost complex manifold due to Gaussier and Sukhov [GA03, GA06].

Let Ω be a domain in an almost complex manifold (X, J) of complex dimension $n + 1$. Suppose that there are $\varphi_k \in \mathrm{Aut}(\Omega, J)$ and $p \in \Omega$ such that

$$\varphi_k(p) \to q \in \partial\Omega \quad \text{as} \quad k \to \infty\,,$$

where $\partial\Omega$ is smooth near q and strongly J-pseudoconvex at q.

Step 1 (a local coordinate system): Choosing a local coordinate system $\Phi : U \subset \mathbb{C}^{n+1} \to M$ about q with $\Phi(0) = q$, we can identity $q = 0$, $\Phi(U) = U$ and $d\Phi^{-1} \circ J \circ d\Phi = J$. For a suitable Φ, we may assume that

1. $J(0) = J_{\mathrm{st}}$ where J_{st} is the standard complex structure of $\mathbb{C}^{n+1}$,
2. $U \cap \Omega = \{z : \rho(z) < 0\}$ where $\rho(z) = \mathrm{Re}\, z^0 + \sum_{\alpha=1}^n |z^\alpha|^2 + o(\|z\|^2)$.

Step 2 (centering): We shall only consider sufficiently large k with $\varphi_k(p) \in U$. For each k, take $p_k^* \in U \cap \partial\Omega$ that realizes the Euclidean distant τ_k from $p_k = \varphi_k(p)$ to $U \cap \partial\Omega$. Then we consider a rigid motion L_k of $\mathbb{C}^{n+1}$ with $L_k(p_k^*) = 0$ and $L_k(p_k) = (-\tau_k, 0, \ldots, 0)$.

Step 3 (dilating): Now we let

$$\Lambda_k(z) = \left(\frac{z^0}{\tau_k}, \frac{z^1}{\sqrt{\tau_k}}, \ldots, \frac{z^n}{\sqrt{\tau_k}}\right).$$

For $A_k = \Lambda_k \circ L_k$, the sequence $A_k(U \cap \Omega)$ of domains converges to the Siegel half plane $\mathbb{H}^{n+1} = \{\mathrm{Re}\, z^0 + \|z'\|^2 < 0\}$ in the sense of the Hausdorff set convergence.

Simultaneously, the sequence $dA_k \circ J \circ dA_k^{-1}$ of induced almost complex structures on $A_k(U \cap \Omega)$ converges to an almost complex structure J' of $\mathbb{H}$ for which $(\mathbb{H}, J')$ is biholomorphic to a pseudo-Siegel domain $(\mathbb{H}, J_P)$.

Finally one can get that $A_k \circ \varphi_k : \varphi_k^{-1}(U \cap \Omega) \subset \Omega \to \mathbb{C}^{n+1}$ has a subsequential limit F defined on the whole of Ω which is biholomorphism (Ω, J) to $(\mathbb{H}, J')$.

Theorem 2.1 ([GA06, LK06]) *Let Ω be a domain in an almost complex manifold (X, J). If Ω admits an automorphism orbit accumulating at a strongly J-pseudoconvex boundary point, then (Ω, J) is biholomorphic to a pseudo-Siegel domains.*

2.3 Bounded Realization of the Pseudo-Siegel Domain

For any non-integrable model structure, the induced structure by the Cayley transform on the unit ball has a singularity at the boundary point corresponding to the point at infinity. Thus it is natural to ask whether there is biholomorphism from the non-integrable pseudo-Siegel domain to a relatively compact domain in an almost complex manifold.

Let Ω be a relatively compact, strongly pseudconvex domain in an almost complex manifold (X, J). If $\mathrm{Aut}(\Omega, J)$ is noncompact, then by Theorem 2.1, there is a biholomorphism $F : (\Omega, J) \to (\mathbb{H}, J_P)$. Consider the point $-\mathbf{1} = (-1, 0, \ldots, 0) \in \mathbb{H}$ and the automorphism $\mathscr{D}_k$ as in (2.1) for $k = 1, 2, \ldots$. Since the automorphism orbit $\{\mathscr{D}_k(-\mathbf{1}) : k = 1, 2, \ldots\}$ is noncompact in $\mathbb{H}$, there is a subsequential limit $q \in \partial\Omega$ of the sequence $F^{-1}(\mathscr{D}_k(-\mathbf{1}))$. Applying the scaling method again to the automorphism orbit $\{F^{-1}(\mathscr{D}_k(-\mathbf{1}))\}$ with certain local coordinates about q, we can obtain a biholomorphism $G : (\Omega, J) \to (\mathbb{H}, J_P)$ with $G(q) = 0$ in the limit sense. Then $F^{-1} \circ G$ is the automorphism of $(\mathbb{H}, J_P)$ with $(F^{-1} \circ G)(0) = \infty$. But every automorphism of $(\mathbb{H}, J_P)$ is affine if $P \neq 0$ ([LK08]); thus $P = 0$ so J_P is integrable.

Theorem 2.2 (Byun et al. [BY09]) *A relative compact and strongly pseudoconvex domain in an almost complex manifold with a noncompact automorphism group is biholomorphic to the unit ball with the standard complex structure.*

3 Schoen's Theorem in Almost CR Manifolds

The scaling method in the Wong-Rosay theorem allows to rescale a given domain and its complex structure to a biholomorphically equivalent model. But in order to get the CR equivalence to a model, the local equivalence problem of CR structures must be considered since the CR structure is a local structure. For the CR equivalence in Theorem 1.4, R. Schoen used the pseudo-hermitian equivalence of Webster [WE78] via the CR Yamabe problem. In this section, we introduce the pseudo-hermitian equivalence problem, the Yamabe type problem and generalization of Theorem 1.4 in strongly pseudoconvex almost CR manifolds as studied in Joo and Lee [JO15].

3.1 Pseudo-hermitian Structure Equations

Let us consider an almost CR manifold M of real dimension $2n+1$ with a CR structure (H, J), that is, $H = \bigcup_{p \in M} H_p \subset TM$ is a hyperplane bundle with a smooth field of bundle isomorphisms $J : H \to H$ such that $J \circ J = -I$. By a *CR automorphism* of M, we mean a diffeomorphism φ of M onto itself with $d\varphi(H) = H$ and $J \circ d\varphi = d\varphi \circ J$. The *CR automorphism group* of M, simply denoted by $\mathrm{Aut}_{\mathrm{CR}}(M)$, is the topological group of CR automorphisms of M with the composition law and the compact-open topology.

The tensor field J decomposes the complexified bundle $\mathbb{C}H = \mathbb{C} \otimes_{\mathbb{R}} H$ by $\mathbb{C}H = H^{1,0} \oplus H^{0,1}$ where $H^{1,0} = \{v - iJv : v \in H\}$ and $H^{0,1} = \overline{H^{1,0}}$. The CR manifold is *strongly peudoconvex* if for an 1-form θ annihilating H, the *Levi form* L_θ defined by $L_\theta(Z, W) = 2id\theta(Z, \overline{W})$ for $Z, W \in H^{1,0}$ is positively or negatively definite. This is independent of the choice of θ. Let $(Z_\alpha) = (Z_1, \ldots, Z_n)$ be a $(1,0)$-frame, a local frame filed to $H^{1,0}$. Then there is an *admissible* $(1,0)$-*coframe* $(\omega^\alpha) = (\omega^1, \ldots, \omega^n)$, a $\mathbb{C}^n$-valued 1-form which is dual to (Z_α) and satisfies

$$d\theta = 2ig_{\alpha\bar{\beta}}\omega^\alpha \wedge \omega^{\bar{\beta}} + p_{\alpha\beta}\omega^\alpha \wedge \omega^\beta + p_{\bar{\alpha}\bar{\beta}}\omega^{\bar{\alpha}} \wedge \omega^{\bar{\beta}} . \tag{3.1}$$

Here $(g_{\alpha\bar{\beta}})$ stands for the Levi form and $(p_{\alpha\beta})$ is uniquely determined by $p_{\alpha\beta} = -p_{\beta\alpha}$. We will use the Levi form $(g_{\alpha\bar{\beta}})$ and its inverse $(g^{\bar{\beta}\alpha})$ to lower and raise indices (e.g. ${\omega_\beta}^\gamma g_{\gamma\bar{\alpha}} = \omega_{\beta\bar{\alpha}}$). Then we can define the *pseudo-hermitian connection form* $({\omega_\beta}^\alpha)$, uniquely determined by $dg_{\alpha\bar{\beta}} - \omega_{\alpha\bar{\beta}} - \omega_{\bar{\beta}\alpha} = 0$ and

$$d\omega^\alpha = \omega^\beta \wedge {\omega_\beta}^\alpha + {T_\beta}^\alpha{}_\gamma \omega^\beta \wedge \omega^\gamma + {N_{\bar{\beta}}}^\alpha{}_{\bar{\gamma}} \omega^{\bar{\beta}} \wedge \omega^{\bar{\gamma}} + A^\alpha{}_{\bar{\beta}}\theta \wedge \omega^{\bar{\beta}} + B^\alpha{}_\beta \theta \wedge \omega^\beta . \tag{3.2}$$

The functions ${T_\beta}^\alpha{}_\gamma$, ${N_{\bar{\beta}}}^\alpha{}_{\bar{\gamma}}$, $A^\alpha{}_{\bar{\beta}}$, $B^\alpha{}_\beta$ are also fixed by ${T_\beta}^\alpha{}_\gamma = -{T_\gamma}^\alpha{}_\beta$, ${N_{\bar{\beta}}}^\alpha{}_{\bar{\gamma}} = -{N_{\bar{\gamma}}}^\alpha{}_{\bar{\beta}}$, $B_{\beta\bar{\alpha}} = B_{\bar{\alpha}\beta}$. The J-linear connection defined by $\nabla Z_\alpha = {\omega_\alpha}^\beta \otimes Z_\beta$ is the *pseudo-hermitian connection*. Then we have the *pseudo-hermitian curvature tensor* $({R_\beta}^\alpha{}_{\lambda\bar{\mu}})$ defined by

$$ {\Omega_\beta}^\alpha \equiv {R_\beta}^\alpha{}_{\lambda\bar{\mu}}\omega^\lambda \wedge \omega^{\bar{\mu}} \mod \{\theta, \omega^\lambda \wedge \omega^\mu, \omega^{\bar{\lambda}} \wedge \omega^{\bar{\mu}}\}$$

for the curvature form ${\Omega_\beta}^\alpha = d{\omega_\beta}^\alpha - {\omega_\beta}^\gamma \wedge {\omega_\gamma}^\alpha$.

3.2 Pseudo-hermitian Equivalence Problem

Now we characterize a pseudo-hermitian structure of the boundary of the Siegel domain. First, we introduce an intrinsic form of the boundary.

Let $(t, z) = (t, z^1, \dots, z^n)$ be the standard coordinates of $\mathbb{R} \times \mathbb{C}^n$. A $n \times n$ skew-symmetric complex matrix $P = (P_{\alpha\beta})$ gives the Lie group structure $*_P$ to $\mathbb{R} \times \mathbb{C}^n$ by

$$(t, z) *_P (s, w) = (t + s + 2\mathrm{Im}\ \langle z, w\rangle - 2\mathrm{Re}\, P(z, w), z + w)$$

where $\langle z, w\rangle = \delta_{\alpha\bar{\beta}} z^\alpha w^{\bar{\beta}}$ and $P(z, w) = P_{\alpha\beta} z^\alpha w^\beta$. This is the induced operation from (2.2) under the natural projection $\pi : \partial\mathbb{H}^{n+1} \to \mathbb{R} \times \mathbb{C}^n$. We call $\mathscr{H}_P = (\mathbb{R} \times \mathbb{C}^n, *_P)$ a *Heisenberg group* associated to P. In fact all Heisenberg groups are Lie group isomorphic to each others (see [BY09]).

Each Heisenberg group has the contact distribution H_P annihilated by

$$\theta_P = dt + i\delta_{\alpha\bar{\beta}} z^\alpha dz^{\bar{\beta}} - i\delta_{\alpha\bar{\beta}} z^{\bar{\beta}} dz^\alpha + P_{\alpha\beta} z^\alpha dz^\beta + P_{\bar{\alpha}\bar{\beta}} z^{\bar{\alpha}} dz^{\bar{\beta}} \tag{3.3}$$

and the strongly pseudoconvex CR structure J_P on H_P whose the global $(1, 0)$-frame $(Z_1, \dots, Z_n)$ is defined by

$$Z_\alpha = \frac{\partial}{\partial z^\alpha} + \left(i\delta_{\alpha\bar{\beta}} z^{\bar{\beta}} + P_{\alpha\beta} z^\beta\right) \frac{\partial}{\partial t}, \quad \alpha = 1, \dots, n.$$

Then the Heisenberg group $\mathscr{H}_P$ acts transitively on itself as a CR transformation group of (H_P, J_P). We call the CR manifold $\mathbb{R} \times \mathbb{C}^n$ with the CR structure (H_P, J_P) a *Heisenberg group*, simply denoted by $\mathscr{H}_P$.

Since $[Z_\alpha, Z_\beta] = -2P_{\alpha\beta}\partial/\partial t$, the CR structure of $\mathscr{H}_P$ is non-integrable except $P = 0$. Each Heisenberg group $\mathscr{H}_P$ is CR equivalent to $(\partial\mathbb{H}^{n+1}, J_P)$ and admits the *CR dilation*,

$$\mathscr{D}_s : (t, z^1, \dots, z^n) \mapsto (e^s t, e^{s/2} z^1, \dots, e^{s/2} z^n) \quad (s \in \mathbb{R}) \tag{3.4}$$

as its CR automorphism. Therefore the CR automorphism group of $\mathscr{H}_P$ acts improperly, so Theorem 1.4 is not valid in the almost CR setting.

Let us consider the pseudo-hermitian structure equations of $\mathscr{H}_P$. For the contact form θ_P, we have $d\theta_P = 2i\delta_{\alpha\bar{\beta}} dz^\alpha \wedge dz^{\bar{\beta}} + P_{\alpha\beta} dz^\alpha \wedge dz^\beta + P_{\bar{\alpha}\bar{\beta}} dz^{\bar{\alpha}} \wedge dz^{\bar{\beta}}$, so that $g_{\alpha\bar{\beta}} \equiv \delta_{\alpha\bar{\beta}}$, $p_{\alpha\beta} \equiv P_{\alpha\beta}$ for (3.1), and $(dz^1, \dots, dz^n)$ is the admissible coframe for θ_P. Since dz^α is closed, one can see that the connection form $({\omega_\beta}^\alpha)$ of (dz^α) vanishes identically. So all torsion tensors except $p_{\alpha\beta} \equiv P_{\alpha\beta}$ and curvature tensors are vanishing identically. This characterizes the Heisenberg model with θ_P:

Proposition 3.1 *Let (M, θ) be a pseudo-hermitian manifold. Suppose that there is an admissible coframe $(\omega^1, \dots, \omega^n)$ with the following vanishing tensors:*

$$p_{\alpha\beta;\gamma} \equiv T_{\beta}{}^{\alpha}{}_{\gamma} \equiv N_{\bar{\beta}}{}^{\alpha}{}_{\bar{\gamma}} \equiv A^{\alpha}{}_{\bar{\beta}} \equiv R_{\beta}{}^{\alpha}{}_{\lambda\bar{\mu}} \equiv 0\,. \tag{3.5}$$

Then (M, θ) is locally pseudo-hermitian equivalent to a Heisenberg group model $(\mathscr{H}_P, \theta_P)$.

Here $p_{\alpha\beta;\gamma}$ stands for the coefficient of the covariant derivative of the tensor $(p_{\alpha\beta})$ by Z_γ's: $p_{\alpha\beta;\gamma} = Z_\gamma p_{\alpha\beta} - p_{\alpha\lambda}\omega_{\beta}{}^{\lambda}(Z_\gamma) - \omega_{\alpha}{}^{\lambda}(Z_\gamma)p_{\lambda\beta}$.

3.3 Sub-Riemannian Yamabe Problem

In order to use the pseudo-hermitian equivalence (Theorem 3.1), we need to find a contact form for which (3.5) holds. In his paper [SC95], R. Schoen uses the CR Yamabe problem for the Webstar scalar curvature, $R = R_{\alpha\bar{\beta}\lambda\bar{\mu}}g^{\alpha\bar{\beta}}g^{\lambda\bar{\mu}}$. Unlike the integrable case, the transformation formula of the Webster scalar curvature is much more complicated. It is not possible to be simplified as the CR Yamabe equation in the integrable pseudo-hermitian geometry. Thus in [JO15] we studied an auxiliary contact sub-Riemannian structure and its Yamabe problem to find a desired contact form.

Let (M^{2n+1}, H) be a contact manifold and θ be a contact form ($H = \ker\theta$). A positive quadratic form g on the contact distribution H is called a *sub-Riemannian metric* and the pair (θ, g) is called a *contact sub-Riemannian structure* of M.

For an orthonormal frame $(X_1, \ldots, X_{2n})$ to H with respect to g, we have a $2n \times 2n$ skew-symmetric matrix (h_{ij}) defined by

$$h_{ij} = d\theta(X_i, X_j)\,.$$

Let X_0 be the *characteristic vector field* of the contact form θ, that is, the vector field uniquely determined by $\theta(X_0) = 1$ and $X_0 \lrcorner\, d\theta = 0$. For the dual coframe $(\theta, \theta^1, \ldots, \theta^{2n})$ of $(X_0, X_1, \ldots, X_{2n})$, we have $d\theta = h_{ij}\theta^i \wedge \theta^j$. Then we can define the *contact sub-Rimannian connection form* (θ_j^i) for (θ, g) (see [FV93, FV07]) which is uniquely determined by

$$d\theta^i = \theta^j \wedge \theta_j^i + \theta \wedge \tau^i\,, \quad \theta_j^i = -\theta_i^j\,, \quad \sum_i \tau^i \wedge \theta^i = 0\,.$$

Moreover the *curvature form* (Θ_j^i) and the *curvature tensor* (R_{jkl}^i) for (θ, g) are defined by

$$\Theta_j^i = d\theta_j^i - \theta_j^k \wedge \theta_k^i \equiv R_{jkl}^i \theta^k \wedge \theta^l \mod \theta.$$

We call $R = \sum_{i,j} R_{jij}^i$ a *sub-Riemannian scalar curvature* of (θ, g). When we let $R_h = \sum_{i,j,k,l} R_{jkl}^i h^{ik}h^{jl}$ for the inverse (h^{ji}) of (h_{ij}), we call the amount

$$S = (2n+1)R - R_h$$

a *twisted scalar curvature* for (θ, g).

Assume that the contact sub-Riemannian structure (θ, g) is *orthogonal*, that is, $h = (h_{ij})$ is the orthogonal matrix. Then we get the Yamabe type transformation formula for the twisted scalar curvature:

Theorem 3.1 ([JO15]) *Let (M^{2n+1}, θ, g) be an orthogonal contact sub-Riemannian manifold. For a subconformal change $(\theta', g') = (u^{2/n}\theta, u^{2/n}g)$, let S and S' be the twisted scalar curvatures for (θ, g) and (θ', g'), respectively. Then u satisfies*

$$S' u^{\frac{2}{n}+1} = Lu\ , \tag{3.6}$$

where $L = 4(n+1)\Delta_b + S$ and Δ_b is the sub-laplacian operator *defined by $\Delta_b u = -\sum_i X_i(X_i u) + \sum_{i,j}(X_j u)\theta_i^j(X_i)$.*

For $p = 2 + 2/n$ and the volume form $dV = (1/n!)\theta \wedge d\theta^n$, the *sub-conformal Yamabe invariant* $Q(M)$ is defined by

$$Q(M) = \inf\left\{ \int_M u\, Lu\, dV : \int_M u^p dV = 1,\ u \in C_c^\infty(M) \text{ and } u \geq 0 \right\}$$

which is independent from the subconformal change of the contact sub-Riemannian structure. Then we can solve the subconformal Yamabe problem.

Theorem 3.2 ([JO15]) *Let (M, θ, g) be an orthogonal contact sub-Riemannian manifold.*

(1) *If M is compact and $Q(M) < Q(S^{2n+1})$, then there is a sub-conformal change $(\theta', g') = (u^{2/n}\theta, u^{2/n}g)$ whose twisted scalar curvature S' of (M, θ', g') is the constant $Q(M)$.*
(2) *If M is noncompact and $Q(M) \geq 0$ or $Q(M) < 0$, then there exists a sub-conformal change whose twisted scalar curvature is the constant 0 or -1, respectively.*

This is a generalization of the CR Yamabe problem for the Webster scalar curvature as in Jerison and Lee [JE87], Schoen [SC95].

3.4 A Generalization of Schoen's Theorem

Let M be a strongly pseudoconvex almost CR manifold with a CR structure (H, J). Suppose that there is a contact sub-Riemannian structure (θ, g) on (M, H) which is *associated to the almost CR structure of M*, that is, every CR automorphism of (M, J) is a subconformal transformation of (M, θ, g). Note that if (θ, g) is associated to (M, H, J), then a sub-conformal change $(u^{2/n}\theta, u^{2/n}g)$ for positive u is also associated to the almost CR structure.

Suppose that M is noncompact. Then by (2) of Theorem 3.2, we may assume that $S \equiv -1$ or $S \equiv 0$ for (θ, g).

Case 1 ($S \equiv -1$): For each CR automorphism φ of M, let u_φ be the positive function with $(\varphi^*\theta, \varphi^*g) = (u_\varphi^{2/n}\theta, u_\varphi^{2/n}g)$ satisfies $(4(n+1)\Delta_b - 1)u_\varphi = -u_\varphi^{2/n+1}$ from (3.6) since φ is the isometry from $(\varphi^*\theta, \varphi^*g)$ to (θ, g). Using the self-adjoint property of Δ_b to non-negative test functions, we get that $\int u_\varphi^{(n+2)/n}$ is locally bounded uniformly for $\varphi \in \mathrm{Aut}_{\mathrm{CR}}(M)$. Using the mean value inequality for sub-elliptic operator $4(n+1)\Delta_b - 1$, one can conclude that u_φ is locally bounded uniformly for $\varphi \in \mathrm{Aut}_{\mathrm{CR}}(M)$; so the CR automorphism group of M acts properly by the Arzela-Ascoli theorem.

Case 2 ($S \equiv 0$): Let θ be a contact form with $S \equiv 0$. If $\mathrm{Aut}_{\mathrm{CR}}(M)$ acts improperly, then there are a compact subset K of M and a sequence $\varphi_k \in \mathrm{Aut}_{\mathrm{CR}}(M)$ such that $\varphi_k(K) \cap K \neq \emptyset$ and $\sup_K u_k \to \infty$ where $\varphi_k^*\theta = u_k^{2/n}\theta$. Equiation (3.6) to each u_k is

$$\Delta_b u_k = 0\,. \tag{3.7}$$

Using the normal coordinates for the orthogonal contact sub-Rimannian structure (θ, g) about a point $p_k \in K$ with $\varphi_k(p_k) \in K$, we take a small open neighborhood V_k of p_k such that $\inf_{V_k} u_k \to \infty$ by the sub-elliptic Harnack Principle to (3.7), so $\varphi_k(V_k)$ increasingly exhausts M by passing a subsequence.

Now consider $T = (T_\beta{}^\alpha{}_\gamma)$, the torsion tensors in (3.2) for θ and let $T_k = \varphi_k^*T$ be the corresponding one for $\varphi_k^*\theta$. Since φ_k is the pseudo-hermitian isometry from $\varphi_k^*\theta$ to θ, we have $\left\| T_k \circ \varphi_k^{-1} \right\|_{\varphi_k^*\theta} = \|T\|_\theta$, where $\|\cdot\|_\theta$ and $\|\cdot\|_{\varphi_k^*\theta}$ stand for tensor norms with respect to the pseudo-hermitian metrices of θ and $\varphi_k^*\theta$, respectively.

Take any point $q \in M$. We shall consider sufficiently large k such that $q \in \varphi_k(V_k)$. For $q_k = \varphi_k^{-1}(q) \in V_k$, we have $\|T_k(q_k)\|_{\varphi_k^*\theta} = \|T(q)\|_\theta$. The transformation formula for T under the pseudo-conformal change $\varphi_k^*\theta = u_k^{2/n}\theta$ (Proposition 4.12 in [JO15]) gives

$$\|T_k(q_k)\|^2_{\varphi_k^*\theta} \le C u_k(q_k)^{-2/n}\left(\|T(q_k)\|^2_\theta + \frac{1}{(nu_k)^2}\|d_b u(q_k)\|^2_\theta\right)$$

where $\|d_b u_k\|_\theta$ is the holomorphic gradient norm of u_k with respect to θ. By the sub-elliptic Schauder estimates for (3.7), we have a uniform bound of $\|d_b u_k\|_\theta / nu_k$ on the relatively compact subset $\cup_k V_k$ of M. Since $u_k(q_k)^{-2/n} \le (\inf_{V_k} u_k)^{-2/n} \to 0$, we have that $\|T(q_k)\|_{\varphi_k^*\theta} \to 0$, so $\|T(q)\|_\theta = 0$. This means that $T_\beta{}^\alpha{}_\gamma = 0$ at q. Following the same manner, we have Condition (3.5) for θ, so get a local pseudo-hermitian equivalence to $\mathscr{H}_P$ by Theorem 3.1. Taking a local CR diffeomorphism F_k from V_k to $\mathscr{H}_P$ and a CR dilation $\Lambda_k(t, z) = (\tau_k t, \sqrt{\tau_k} z)$ of $\mathscr{H}_P$ for some $\tau_k \to \infty$, we have a global CR diffeomorphism $F : M \to \mathscr{H}_P$ as a subsequential limit of $\Lambda_k \circ F_k \circ \varphi_k^{-1}$.

Theorem 3.3 ([JO15]) *Let M be a noncompact, strongly pseudoconvex, almost CR manifold with an associated orthogonal contact sub-Riemannian structure. If the CR automorphism group of M acts on M improperly, then M is CR equivalent to a Heisenberg group $\mathscr{H}_P$.*

If M is compact and $\mathrm{Aut}_{CR}(M)$ acts improperly, then by the same way of Schoen [SC95], we have a point $p \in M$ such that there is a CR diffeomorphism $F : M \setminus \{p\} \to \mathscr{H}_P$. Then we show that the CR automorphism $\mathscr{D}_s$ of $M \setminus \{p\} \simeq \mathscr{H}_P$ as in (3.4) extends to the CR automorphism of the whole M. Since $F^{-1}(0)$ is a fixed point of each $\mathscr{D}_s$, $\{\mathscr{D}_s : s \in \mathbb{R}\}$ acts improperly on $M \setminus \{F^{-1}(0)\}$ which contains p. Form Therorem 3.3 and the homogeneity of $\mathscr{H}_P$, there is a CR diffeomorphism $G : M \setminus \{F^{-1}(0)\} \to \mathscr{H}_P$ with $G(p) = 0$. Thus the CR automorphism $G \circ F^{-1}$ of $\mathscr{H}_P \setminus \{0\}$ which can not extend on $\mathscr{H}_P$. This contacts to Proposition 3.3 in [JO15] if $P \neq 0$.

Theorem 3.4 ([JO15]) *Let M^{2n+1} be a compact, strongly pseudoconvex, almost CR manifold with an associated orthogonal contact sub-Riemannian structure. If the CR automorphism group of M is noncompact, then M is CR equivalent to the standard sphere S^{2n+1}.*

If $\dim M = 5$ or 7, M always admits an associated orthogonal contact sub-Riemannian structure. Thus we can partially confirm Conjecture 1.1.

Acknowledgments The research of the author was supported by Basic Science Research Program through the National Research Foundation of Korea (NRF) funded by the Ministry of Science, ICT & Future Planning (NRF 2012R1A1A1004849).

References

[AL72] Alekseevskiĭ, D.V.: Groups of conformal transformations of Riemannian spaces. Mat. Sb. (N.S.) **89**(131), 280–296, 356 (1972)

[BU78] Burns Jr, D., Shnider, S., Wells Jr, R.O.: Deformations of strictly pseudoconvex domains. Invent. Math. **46**, 237–253 (1978)

[BY09] Byun, J., Gaussier, H., Lee, K.-H.: On the automorphism group of strongly pseudoconvex domains in almost complex manifolds. Ann. Inst. Fourier (Grenoble) **59**, 291–310 (2009)

[EF95] Efimov, A.M.: A generalization of the Wong-Rosay theorem for the unbounded case. Mat. Sb. **186**, 41–50 (1995)

[FV07] Falbel, E., Veloso, J.M.: A bilinear form associated to contact sub-conformal manifolds. Differ. Geom. Appl. **25**, 35–43 (2007)

[FV93] Falbel, E., Veloso, J.M., Verderesi, J.A.: Constant curvature models in sub-Riemannian geometry. Mat. Contemp. **4**, 119–125 (1993). VIII School on Differential Geometry (Portuguese) (Campinas, 1992)

[FE74] Fefferman, C.: The Bergman kernel and biholomorphic mappings of pseudoconvex domains. Invent. Math. **26**, 1–65 (1974)

[FE96] Ferrand, J.: The action of conformal transformations on a Riemannian manifold. Math. Ann. **304**, 277–291 (1996)
[GA02] Gaussier, H., Kim, K.-T., Krantz, S.G.: A note on the Wong-Rosay theorem in complex manifolds. Complex Var. Theory Appl. **47**, 761–768 (2002)
[GA03] Gaussier, H., Sukhov, A.: Wong-Rosay theorem in almost complex manifolds, preprint, arXiv:math/0307335
[GA06] Gaussier, H., Sukhov, A.: On the geometry of model almost complex manifolds with boundary. Math. Z. **254**, 567–589 (2006)
[GE82] Greene, R.E., Krantz, S.G.: Deformation of complex structures, estimates for the $\bar{\partial}$ equation, and stability of the Bergman kernel. Adv. Math. **43**, 1–86 (1982)
[JE87] Jerison, D., Lee, J.M.: The Yamabe problem on CR manifolds. J. Differ. Geom. **25**, 167–197 (1987)
[JO15] Joo, J.-C., Lee, K.-H.: Subconformal Yamabe equation and automorphism groups of almost CR manifolds, J. Geom. Anal., to appear
[LK06] Lee, K.-H.: Domains in almost complex manifolds with an automorphism orbit accumulating at a strongly pseudoconvex boundary point. Mich. Math. J. **54**, 179–205 (2006)
[LK08] Lee, K.-H.: Strongly pseudoconvex homogeneous domains in almost complex manifolds, J. Reine Angew. Math. **623**, 123–160 (2008)
[RO79] Rosay, J.-P.: Sur une caractérisation de la boule parmi les domaines de $\mathbf{C}^n$ par son groupe d'automorphismes. Ann. Inst. Fourier (Grenoble), **29**, ix, 91–97 (1979)
[SC95] Schoen, R.: On the conformal and CR automorphism groups. Geom. Funct. Anal. **5**, 464–481 (1995)
[WE78] Webster, S.M.: Pseudo-Hermitian structures on a real hypersurface. J. Differ. Geom. **13**, 25–41 (1978)
[WO77] Wong, B.: Characterization of the unit ball in C^n by its automorphism group. Invent. Math. **41**, 253–257 (1977)

Compact Smooth but Non-complex Complements of Complete Kähler Manifolds

Xu Liu

Abstract We modify the techniques developed by Diederich and Fornaess, and construct compact smooth submanifolds of arbitrary real codimension ≥ 3, which are non-complex as the complements of complete Kähler manifolds.

Keywords Complements of complete Kähler domains · Complete pluripolar sets

1 Introduction

It was observed by Grauert that not every complex manifold M, $\dim_{\mathbb{C}} M > 1$, carrying a complete Kähler metric is Stein. Instead, for any closed analytic subvariety A of M, there exists a complete Kähler metric on $M \setminus A$ (Satz A in [Grau1956]).

One question arises from the above observation: what kind of condition can force the complement of a complete Kähler manifold to be complex-analytic?

The real codimension 2 case was considered by Ohsawa [Ohsa1980]: Assume M is a complex manifold, and $A \subset M$ is a closed C^1 submanifold of real codimension 2. If $M \setminus A$ admits a complete Kähler metric, then A is complex-analytic.

Later Diederich and Fornaess [Die1982] considered the higher codimensional case and showed: Assume M is a complex manifold, and $A \subset M$ is a closed real-analytic submanifold of real codimension ≥ 3. If $M \setminus A$ admits a complete Kähler metric, then A is complex-analytic.

Notice that in Ohsawa's result, the C^1 regularity condition is sufficient. In the contrary, in higher codimensional case, even smoothness is not able to guarantee the analyticity. In other words, real-analyticity is necessary, due to the following: For any $k \in \mathbb{N}$, $k \geq 3$, there exists a closed C^∞ submanifold A of real codimension k in a ball B, such that A is not complex-analytic and $B \setminus A$ admits a complete Kähler metric.

The above examples were constructed on open manifolds. After some modifications, we generalize their result to the compact case and obtain the main result:

X. Liu (✉)
Graduate School of Mathematics, Nagoya University, 464-8602 Nagoya, Japan
e-mail: x10001c@math.nagoya-u.ac.jp

F. Bracci et al. (eds.), *Complex Analysis and Geometry*, Springer Proceedings in Mathematics & Statistics 144, DOI 10.1007/978-4-431-55744-9_17

Theorem 1.1 *For any $k \in \mathbb{N}, k \geq 3$, there exists a compact C^∞ submanifold A of real codimension k in $\mathbb{P}^n$, such that A is not complex-analytic and $\mathbb{P}^n \setminus A$ admits a complete Kähler metric.*

2 Compact Counterexamples

Here we are going to directly construct compact examples of non-complex submanifolds of arbitrary real codimension ≥ 3 in the complements of complete Kähler manifolds.

Firstly, we construct a real dimension 2 submanifold $A \subset \mathbb{C}^3$ as a graph over $S^1 \times S^1$ (instead of a graph over $\mathbb{R}^2$ cf. [Die1982]) together with a complete Kähler metric near A. It is based on the following key lemma.

Lemma 2.1 *Assume $f = F|_{S^1 \times S^1}$ where $F(z_1, z_2)$ is a polynomial on $\mathbb{C}^2$. Γ_f is the graph of f. Fix a point $p \notin \Gamma_f$ with $\mathrm{dist}(p, \Gamma_f) \geq 1$. For given $n \in \mathbb{N}, \varepsilon > 0$, there exists a C^∞ strictly plurisubharmonic function ϕ on $\mathbb{C}^3$ and $h = H(z_1, z_2)|_{S^1 \times S^1}$ where H is a polynomial on $\mathbb{C}^2$ such that*

(*A*) $|\mathrm{D}^\alpha(h - f)| < \varepsilon$ *on* $S^1 \times S^1, |\alpha| \leq n$;
(*B*) $|\phi| < \varepsilon$ *on* $B_{2n} := \{z||z| \leq 2n\}$;
(*C*) $|\mathrm{D}^\alpha \phi| < \varepsilon$ *on* $B_{2n} \cap \{z|\mathrm{dist}(z, \Gamma_f) \geq \varepsilon\}, |\alpha| \leq n$;
(*D*) *the distance from p to $\Gamma_h \cap B_{2n}$ measured in B_{2n} with respect to the metric induced by $\partial\overline{\partial}\phi$ is at least n.*

Proof It is known that there exists a continuous subharmonic function on $\mathbb{C}$ such that ϕ is radially symmetric, i.e., $\phi(z) = \phi(|z|)$; ϕ is smooth on $\mathbb{C}^*$; $\partial\overline{\partial}\phi$ gives a complete Kähler metric on $\mathbb{C}^*$. A detailed constructive proof is contained in [Die1982]. There are also other approaches. Consider, e.g., $\phi_0(z) := \frac{1}{\log(-\log|z|)}$. Then $\partial\overline{\partial}\phi_0$ gives a complete Kähler metric at the origin on the punctured disk $\{0 < |z| < \frac{1}{2e}\}$. Using a cut-off function to combine ϕ_0 with a convex smooth function increasing rapidly enough will extend it to the whole plane.

Let Z_F be the graph of F and $\phi_F := \phi(z_3 - F(z_1, z_2))$. Then ϕ_F is continuous plurisubharmonic on $\mathbb{C}^3$. Then we can use Richberg's regularization [Rich1968] to $\phi_F + |z|^2$ to get required smoothness and scale the result to satisfy the conditions (B-C). Then we get a C^∞ strictly plurisubharmonic function ϕ_1 such that for any curve $\gamma : [0, 1] \mapsto B_{2n}$ going from p to $q \in \Gamma_f$ with $\gamma((0, 1)) \subset B_{2n} \backslash \Gamma_f$, γ has length at least $n+1$ with respect to $\partial\overline{\partial}\phi_1$ unless it satisfies the following condition ($\star$):

$$\gamma(\tau) \notin A := \{z \mid |z_3 - F(x_1, x_2)| \geq \frac{\varepsilon}{16}, 1 - \eta \leq |z_j| \leq 1 + \eta, j = 1, 2\}$$
$$\text{for all } \tau \geq t := \sup\{\tau \in [0, 1] \mid \mathrm{dist}(\gamma(\tau), \Gamma_f) \geq \frac{\varepsilon}{8}\},$$

where η is independent of γ and small enough such that $Z_F \cap A \cap B_{2n} = \emptyset$.

The condition ($\star$) means that γ approaches Γ_f along Z_F. It makes sense because due to the construction of ϕ, any curve going into Z_F transversely in B_{2n} has $+\infty$ length with respect to the metric induced by $\partial\overline{\partial}\phi_F$ and therefore $\partial\overline{\partial}\phi_1$. It can also be stated as: There exists $\delta_1 > 0$, $\delta_1 \ll \varepsilon$, such that any curve $\gamma : [0,1] \mapsto B_{2n} \setminus \Gamma_f$ going from p to q with $\mathrm{dist}(q, \Gamma_f) \le \delta_1$ has length at least n with respect to the metric induced by $\partial\overline{\partial}\phi_1$ unless γ satisfies ($\star$).

Next we can choose a polynomial $P(z_1)$ and let $G(z_1, z_2) := F(z_1, z_2) + P(z_1)$ such that $|P^{(k)}(z_1)| < \delta_1$ for all $|z_1| = 1$, $k \le n$ and $\mathrm{dist}((z_1, z_2, G(z_1, z_2))), \Gamma_f) > \varepsilon$ for $|z_1| = 1 \pm \frac{\eta}{2}$. Let $g := G|_{S^1\times S^1}$ and repeat the above process to $\phi_G + |z|^2$ to obtain a smooth strictly plurisubharmonic fucntion ϕ_2 such that ϕ_2 satisfies condition (B-C) and there exists $\delta_2 > 0$, $0 < \delta_2 \ll \delta_1$, $\delta_2 < \frac{\eta}{2}$ such that any curve $\gamma : [0,1] \mapsto B_{2n} \setminus \Gamma_g$ going from p to any q with $\mathrm{dist}(q, \Gamma_g) \le \delta_2$ has length at least n with respect to the metric induced by $\partial\overline{\partial}(\phi_1 + \phi_2)$ unless γ satisfies ($\star$) and $1 - \frac{\eta}{2} \le |z_1| \le 1 + \frac{\eta}{2}$ for all $\tau \ge t$.

Similarly, we can choose a polynomial $Q(z_2)$ satisfying the same conditons as $P(z_1)$ and repeat the same process to $\phi_H + |z|^2$ where $H := G + Q$ to get a smooth strictly plurisubharmonic fucntion ϕ_3 such that ϕ_3 satisfies condition (B-C) and any curve $\gamma : [0,1] \mapsto B_{2n} \setminus \Gamma_h$ going from p to any q with $\mathrm{dist}(q, \Gamma_h) \le \delta_2$ has length at least n with respect to the metric induced by $\partial\overline{\partial}(\phi_1 + \phi_2 + \phi_3)$ unless it satisfies ($\star$) and $1 - \frac{\eta}{2} \le |z_j| \le 1 + \frac{\eta}{2}$, $j = 1, 2$ for all $\tau \ge t$.

However, this is impossible since A defined in ($\star$) and $\{|z_j| < 1 - \frac{\eta}{2}$ or $|z_j| > 1 + \frac{\eta}{2}\}$, $j = 1, 2$ wrap up Γ_h, which means any curve going to Γ_h has to intersects either of these three sets. □

Remark 2.1 In fact, since any periodic function on $\mathbb{R}$ can be considered as a function defined on S^1, if we allow the variable to take complex values, we get $\mathbb{C}^*$ as the complexification of S^1 and a function defined on $\mathbb{C}^*$. Therefore, in the statement of the lemma, we can take F as such extension of f from $S^1 \times S^1$ to $\mathbb{C}^{*2}$ conversely, where f can be chosen to be rational functions on $\mathbb{C}^2$ with a permissible pole at the origin.

Proposition 2.1 *There exists $A \subset \mathbb{P}^3$ given by the graph of a C^∞ function over $S^1 \times S^1$ such that $\mathbb{P}^3 \setminus A$ admits a complete Kähler metric.*

Proof Let $f_1 \equiv 0$, $p = (\frac{3}{2}i, 0, 0) \in \mathbb{C}^3$. Apply the lemma inductively to find a series of smooth strictly plurisubharmonic functions $\phi_1, \ldots, \phi_{k-1}$ and polynomials $f_1, \ldots, f_k$ satisfying the following conditions:

(A') $|\mathrm{D}^\alpha(f_j - f_{j-1})| < \frac{1}{2^{j-1}}$ on $S^1 \times S^1$, $|\alpha| \le j - 1$, $j = 2, \ldots, k$;
(B') $|\phi_j| < \frac{1}{2^j}$ on B_{2^j}, $j = 1, \ldots, k-1$;
(C') $|\mathrm{D}^\alpha\phi_j| < \frac{1}{2^j}$ on $B_{2^j} \cap \{z|\mathrm{dist}(z, \Gamma_{f_j}) \ge \frac{1}{2^{j+1}}\}$, $|\alpha| \le j$, $j = 1, \ldots, k-1$;
(D') the distance from p to $\Gamma_{f_k} \cap B_{2^{\alpha(j,k)}}$ in $B_{2^{\alpha(j,k)}}$ with respect to $\partial\overline{\partial}\phi_j \ge \alpha(j,k) := j - 1 + \frac{1}{2^{k-1}}$, $j = 1, \ldots, k-1$.

It follows from (A') that $f_k \to f_\infty$ in the C^∞-topology on $S^1 \times S^1$ and from (B') that $|z|^2 + \sum_{j=1}^k \phi_j \to \phi$ uniformly on every compact subset such that ϕ is continuous on $\mathbb{C}^3$ and C^∞ on $\mathbb{C}^3 \setminus \Gamma_{f_\infty}$ where $\partial\overline{\partial}\phi$ induces a complete Kähler metric. It remains to use a cut-off function to combine it with a Fubini–Study metric to get the desired complete Kähler metric on $\mathbb{P}^3 \setminus \Gamma_{f_\infty}$. □

Proof (*Proof of Theorem 1.1*) In the above construction, by restricting the function f_∞ to $\mathbb{C} \times \{0\}$, we get its graph as a smooth curve in $\mathbb{C} \times \{0\} \times \mathbb{C} \cong \mathbb{C}^2$ (the real codimension 3 case). It is also easily seen that Lemma 2.1 and Proposition 2.1 can be generalized to higher dimensions, i.e., we can construct the graph $\Gamma_f \subset \mathbb{C}^n$ of a smooth function f over $\underbrace{S^1 \times ... \times S^1}_{n-1}$ and a continuous plurisubharmonic function ϕ, smooth outside Γ_f. In every case, $\partial\overline{\partial}\phi$ combined with a Fubini–Study metric will give complete Kähler metrics on $\mathbb{P}^n \setminus \Gamma_{f_\infty}$. □

3 Remarks

In the last section, the existence of complete Kähler metrics is shown by constructing their potentials which are plurisubharmonic functions. It is also closely related to the propeties of pluripolar sets.

Using similar techniques in [Die1982], Diederich and Fornaess [DieFor1982] proved that complete pluripolar sets are not necessarily complex, even when they are closed C^∞ real submanifolds.

It is known that if φ is the defining function of a closed complete pluripolar set $A \subset \mathbb{C}^n$ and smooth outside A, then

$$\mathrm{d}s^2 := \partial\overline{\partial}(|z|^2 - \log(-\varphi))$$

gives a complete Kähler metric on $\mathbb{C}^n \setminus A$. Here $f(t) = -\log(-t)$ should be understood as a function extended to be defined on $\mathbb{R}$ such that it keeps increasing and convex . This provides another approach to Theorem 1.1, if we can prove:

For any $k \in \mathbb{N}$, $k \geq 3$, there exists a compact C^∞ submanifold A of real codimension k in $\mathbb{C}^n$, such that A is complete pluripolar but not complex-analytic.

This problem was studied by Edlund [Edl2004] and answered affirmatively.

Sometimes, curvature conditions are considered when one studies the complements of complete Kähler manifolds. For example, Anchouche [Anc2009] used additional curvature conditions to reduce the compact complements of complete Kähler manifolds into finite point sets. It is shown that in general there exist nontrivial examples.

Acknowledgments I would like to thank Professor Nikolay V. Shcherbina for pointing out Edlund's result and for helpful discussion during KSCV 10.

References

[Anc2009] Anchouche, B.: Analyticity of compact complements of complete Kähler manifolds. Proc. AMS **137**, 3037–3044 (2009)

[Die1982] Diederich, K., Fornaess, J.E.: Thin complements of complete Kähler domains. Math. Ann. **259**, 331–341 (1982)

[DieFor1982] Diederich, K., Fornaess, J.E.: Smooth, but not complex-analytic pluripolar sets. Manuscr. Math. **37**, 121–125 (1982)

[Edl2004] Edlund, T.: Complete pluripolar curves and graphs. Ann. Polon. Math. **84**(1), 75–86 (2004)

[Grau1956] Grauert, H.: Charakterisierung der Holomorphiegebiete durch die vollständige Kählersche Metrik (German). Math. Ann. **131**, 38–75 (1956)

[Ohsa1980] Ohsawa, T.: Analyticity of complements of complete Kähler domains. Proc. Japan Acad. **56**(Ser. A), 484-487 (1980)

[Rich1968] Richberg, R.: Stetige streng pesudokonvexe Funktionen. Math. Ann. **175**, 257–286 (1968)

Injectivity Theorems with Multiplier Ideal Sheaves and Their Applications

Shin-ichi Matsumura

Abstract The purpose of this survey is to present analytic versions of the injectivity theorem and their applications. The proof of our injectivity theorems is based on a combination of the L^2-method for the $\overline{\partial}$-equation and the theory of harmonic integrals. As applications, we obtain Nadel type vanishing theorems and extension theorems for pluri-canonical sections of log pairs. Moreover, we give some results on semi-ampleness related to the abundance conjecture in birational geometry (the minimal model program).

Keywords Injectivity theorems · Vanishing theorems · Singular metrics · Multiplier ideal sheaves · The theory of harmonic integrals · L^2-methods · Extension theorems · Abundance conjecture

1 Introduction

The Kodaira vanishing theorem is one of the most celebrated results in complex geometry, and such results play an important role when we consider certain fundamental problems in algebraic geometry and the theory of several complex variables, including asymptotics of linear systems, extension problems of holomorphic sections, the minimal model program, and so on. According to these objectives, the study of vanishing theorems has been a constant subject of attention in the last decades. This paper is a survey of recent results in [Mat13b] and [GM13], whose purpose is to present generalizations of the Kodaira vanishing to pseudo-effective line bundles equipped with singular metrics and their applications, from the viewpoint of the theory of several complex variables and differential geometry.

S. Matsumura (✉)
Mathematical Institute, Tohoku University, 6-3, Aramaki Aza-Aoba, Aoba-ku, Sendai 980-8578, Japan
e-mail: mshinichi@m.tohoku.ac.jp; mshinichi0@gmail.com

F. Bracci et al. (eds.), *Complex Analysis and Geometry*, Springer Proceedings in Mathematics & Statistics 144, DOI 10.1007/978-4-431-55744-9_18

1.1 Analytic Versions of the Injectivity Theorem

In this subsection, we introduce analytic versions of the injectivity theorem. The injectivity theorem is a generalization of the vanishing theorem to "semi-positive" line bundles, and it has been studied by several authors, for example, Tankeev [Tan71], Kollár [Kol86], Enoki [Eno90], Esnault-Viehweg [EV92], Ohsawa [Ohs04], Fujino [Fuj12, Fuj13a], Ambro [Amb03, Amb12], and so on. In his paper [Kol86], Kollár gave the following injectivity theorem for semi-ample line bundles, whose proof depends on the Hodge theory. In [Eno90], Enoki relaxed his assumption by a different method depending on the theory of harmonic integrals, which enables us to approach the injectivity theorem from the viewpoint of complex differential geometry.

Theorem 1.1 ([Kol86] resp. [Eno90]) *Let F be a semi-ample (resp. semi-positive) line bundle on a smooth projective variety (resp. a compact Kähler manifold) X. Then for a (non-zero) section s of a positive multiple F^m of the line bundle F, the multiplication map induced by the tensor product with s*

$$\Phi_s : H^q(X, K_X \otimes F) \xrightarrow{\otimes s} H^q(X, K_X \otimes F^{m+1})$$

is injective for any q. Here K_X denotes the canonical bundle of X.

The above theorem can be regarded as a generalization of the Kodaira vanishing theorem to semi-ample (semi-positive) line bundles. On the other hand, the Kodaira vanishing theorem has been generalized by Nadel [Nad89, Nad90]. This generalization uses singular metrics with positive curvature and corresponds to the Kawamata-Viehweg vanishing theorem in algebraic geometry [Kaw82, Vie82]. Therefore, in the same direction as this generalization, it is natural and of interest to study injectivity theorems for line bundles equipped with "singular metrics".

The Kodaira vanishing	$\xrightarrow{\text{semi-positivity}}$	Kollár's injectivity theorem.
{ cpx. geom: smooth positive metrics alg. geom: ample line bundles		{ cpx. : smooth semi-positive metrics alg. : semi-ample line bundles
$\downarrow$ singular metrics		$\downarrow$ singular metrics
The Nadel, Kawamata-Viehweg vanishing	$\xrightarrow{\text{semi-positivity}}$	**Theorem 1.2**
{ cpx. : singular positive metrics alg. : big line bundles		{ cpx. : singular semi-positive metrics alg. : pseudo-effective line bundles

The following theorem is one of the main results, which can be seen as a generalization of the injectivity theorem and the Nadel vanishing theorem.

Theorem 1.2 ([Mat13b, Theorem 1.3]) *Let F be a line bundle on a compact Kähler manifold X and h be a singular metric with semi-positive curvature on F. Then for* *a (non-zero) section s of a positive multiple* F^m *satisfying* $\sup_X |s|_{h^m} < \infty$, *the multiplication map*

$$\Phi_s : H^q(X, K_X \otimes F \otimes \mathscr{I}(h)) \xrightarrow{\otimes s} H^q(X, K_X \otimes F^{m+1} \otimes \mathscr{I}(h^{m+1}))$$

is (well-defined and) injective for any q. *Here* $\mathscr{I}(h)$ *denotes the multiplier ideal sheaf associated to the singular metric* h.

Remark 1.1 The multiplication map is well-defined thanks to the assumption of $\sup_X |s|_{h^m} < \infty$. When h is a metric with minimal singularities on F, this assumption is automatically satisfied for any section s of F^m (see [Dem] for the concept of metrics with minimal singularities).

When we consider the problem of extending (holomorphic) sections from subvarieties to the ambient space, we need to refine the above formulation (see Theorem 2.1). Our injectivity theorem can be seen as an improvement of [Eno90, Fuj12, Kol86, Mat14]. For the proof, we take an analytic approach for the cohomology groups with coefficients in $K_X \otimes F \otimes \mathscr{I}(h)$, which includes techniques of [Eno90, Fuj12, Mat13a, Mat14, Ohs04, Tak97]. The proof is based on a technical combination of the L^2-method for the $\overline{\partial}$-equation and the theory of harmonic integrals. To handle transcendental (non-algebraic) singularities, after regularizing a given singular metric, we investigate the asymptotic behavior of the harmonic forms with respect to a family of the regularized metrics. Moreover we establish a method to obtain L^2-estimates of solutions of the $\overline{\partial}$-equation by using the Čech complex. See Sect. 2.1 for more details.

1.2 Applications to the Vanishing Theorem

Theorem 1.2 is formulated for singular metrics with transcendental (non-algebraic) singularities, which is one of the advantages of our injectivity theorem. For example, metrics with minimal singularities are important objects, but they do not always have algebraic singularities. By applying Theorem 1.2 to them, we can obtain an injectivity theorem for nef and abundant line bundles (Corollary 1.1) and Nadel type vanishing theorems (Theorem 1.3 and Corollary 1.2).

It is natural to expect the same conclusion as in Theorem 1.1 under the weaker assumption that F is nef. However there is a counterexample to the injectivity theorem for nef line bundles (see for example [Fuj13a, Example 5.1]). If F is nef and abundant (that is, the numerical dimension agrees with the Kodaira dimension), the line bundle F admits a metric $h_{\min}$ with minimal singularities satisfying $\mathscr{I}(h^m_{\min}) = \mathscr{O}_X$ for any $m > 0$. This follows from [Kaw85, Proposition 2.1]. Hence Theorem 1.2 leads to the following corollary.

Corollary 1.1 ([Mat13b, Corollary 1.5]) *Let F be a nef and abundant line bundle on a compact Kähler manifold X. Then for a (non-zero) section s of a positive multiple F^m of the line bundle F, the multiplication map induced by the tensor product with s*

$$\Phi_s : H^q(X, K_X \otimes F) \xrightarrow{\otimes s} H^q(X, K_X \otimes F^{m+1})$$

is injective for any q.

The same statement was proved in [Fuj12], and a similar conclusion was proved in [EP08, EV92] by different methods when X is a projective variety. It is worth pointing out that Theorem 1.1 is not sufficient to obtain Corollary 1.1. This is because the above metric $h_{\min}$ may not be smooth and not have algebraic singularities even if F is nef and abundant (see for example [Fuj13a, Example 5.2]).

As another application of Theorem 1.2, we obtain a Nadel type vanishing theorem (Theorem 1.3) and its corollary (Corollary 1.2).

Theorem 1.3 ([Mat13b, Theorem 3.21] cf. [Mat14, Theorem 5.2]) *Let F be a line bundle on a smooth projective variety X and h be a singular metric with semi-positive curvature on F. Then*

$$H^q(X, K_X \otimes F \otimes \mathscr{I}(h)) = 0 \quad \textit{for any}\, q > \dim X - \kappa_{\mathrm{bdd}}(F, h).$$

See Sect. 2.2 or [Mat14, Definition 5.1] for the definition of the bounded Kodaira dimension $\kappa_{\mathrm{bdd}}(F, h)$.

Corollary 1.2 ([Mat13b, Corollary 1.6] cf. [Mat13a, Theorem 1.2]) *Let F be a line bundle on a smooth projective variety X and $h_{\min}$ be a singular metric with minimal singularities on F. Then*

$$H^q(X, K_X \otimes F \otimes \mathscr{I}(h_{\min})) = 0 \quad \textit{for any}\, q > \dim X - \kappa(F).$$

Here $\kappa(F)$ denotes the Kodaira dimension of F.

Since multiplier ideal sheaves are coherent ideal sheaves, the family of multiplier ideal sheaves $\{\mathscr{I}(h^{1+\delta})\}_{\delta>0}$ has the maximal element, which we denote by $\mathscr{I}_+(h)$ (see [DEL00] for more details). In [Cao15], Cao gave striking results on the Nadel vanishing theorem for the cohomology groups with coefficients in $K_X \otimes F \otimes \mathscr{I}_+(h)$. However, the Nadel vanishing theorem for $K_X \otimes F \otimes \mathscr{I}(h_{\min})$ is non-trivial even if F is big. In fact, it was first proved in [Mat13a] when F is big.

It is of interest to ask whether or not $\mathscr{I}_+(\varphi)$ agrees with $\mathscr{I}(\varphi)$ for a plurisubharmonic (psh for short) function φ, which was first posed in [DEL00]. We can easily see that $\mathscr{I}_+(\varphi) = \mathscr{I}(\varphi)$ holds if φ has algebraic singularities, but $h_{\min}$ unfortunately does not always have algebraic singularities. It is a natural problem related to the (strong) openness conjecture of Demailly-Kollár (see [DK01]), but it had been an open problem. Recently, Guan-Zhou affirmatively solved the openness conjecture in [GZ15], and Hiep gave another proof in [Hie14]. Although their celebrated results imply Theorem 1.3, we believe that our techniques are still of interest, since they bring a quite different viewpoint and have further applications.

1.3 Applications to the Extension Theorem

In this subsection, we give an extension theorem for pluri-canonical sections of log pairs. Our motivation is the abundance conjecture, which is one of the most important conjectures in the classification theory of algebraic varieties. From now on, we freely use the standard notation in [BCHM10, KaMM87, KM] and further we interchangeably use the words "Cartier divisors", "line bundles", "invertible sheaves".

Conjecture 1.1 (Generalized abundance conjecture) Let X be a normal projective variety and Δ be an effective $\mathbb{Q}$-divisor such that (X, Δ) is a klt pair. Then $\kappa(K_X + \Delta) = \kappa_\sigma(K_X + \Delta)$. In particular, if $K_X + \Delta$ is nef, then it is semi-ample. See [Nak] for the definition of $\kappa(\cdot)$ and $\kappa_\sigma(\cdot)$.

Toward the abundance conjecture, we need to study the non-vanishing conjecture and the extension conjecture (see [DHP13], [Fuj00, Introduction], [FG14, Sect. 5]). We study the following extension conjecture for dlt pairs formulated in [DHP13, Conjecture 1.3]:

Conjecture 1.2 (Extension conjecture for dlt pairs) Let X be a normal projective variety and $S + B$ be an effective $\mathbb{Q}$-divisor satisfying the following assumptions:

- $(X, S + B)$ is a dlt pair.
- $\lfloor S + B \rfloor = S$.
- $K_X + S + B$ is nef.
- $K_X + S + B$ is $\mathbb{Q}$-linearly equivalent to an effective divisor D with $S \subseteq \mathrm{Supp} D \subseteq \mathrm{Supp}\,(S + B)$.

Then the restriction map

$$H^0(X, \mathscr{O}_X(m(K_X + S + B))) \to H^0(S, \mathscr{O}_S(m(K_X + S + B)))$$

is surjective for sufficiently divisible integers $m \geq 2$.

When S is a normal irreducible variety (that is, $(X, S+B)$ is a plt pair), Demailly-Hacon-Păun proved the above conjecture in [DHP13] by using technical methods based on a version of the Ohsawa-Takegoshi L^2-extension theorem. This result can be seen as an extension theorem for plt pairs.

By applying Theorem 2.1 instead of the Ohsawa-Takegoshi theorem to the extension problem, we prove the following extension theorem for *dlt pairs*. Thanks to the injectivity theorem, we can obtain some extension theorems for not only plt paris but also dlt pairs. This is an advantage of our approach. Even if $K_X + \Delta$ is semi-positive (that is, it admits a smooth metric with semi-positive curvature), it seems to be very impossible to prove the extension theorem for dlt pairs by the Ohsawa-Takegoshi theorem at least in the current situation, and thus we need our injectivity theorem (Theorem 2.1).

Theorem 1.4 ([GM13, Corollary 4.5]) *Let X be a compact Kähler manifold and $S + B$ be an effective $\mathbb{Q}$-divisor with the following assumptions:*

- *$S + B$ is a simple normal crossing divisor with $0 \leq S + B \leq 1$ and $\lfloor S + B \rfloor = S$.*
- *$K_X + S + B$ is $\mathbb{Q}$-linearly equivalent to an effective divisor D with $S \subseteq \operatorname{Supp} D$.*
- *$K_X + S + B$ admits a singular metric h with semi-positive curvature.*
- *The Lelong number $\nu(h, x)$ is equal to 0 at every point $x \in S$.*

Then, for an integer $m \geq 2$ with Cartier divisor $m(K_X + S + B)$, every section $u \in H^0(S, \mathscr{O}_S(m(K_X+S+B)))$ can be extended to a section in $H^0(X, \mathscr{O}_X(m(K_X+S+B)))$.

In particular, Conjecture 1.2 is affirmatively solved under the assumption that $K_X + \Delta$ admits a singular metric whose curvature is semi-positive and Lelong number is identically zero. This assumption is stronger than the assumption that $K_X + \Delta$ is nef, but weaker than the assumption that $K_X + \Delta$ is semi-positive. Let us observe that Verbitsky proved the non-vanishing conjecture on hyperKähler manifolds (holomorphic symplectic manifolds) under the same assumption (see [Ver10]).

As compared to Conjecture 1.2, one of our advances has been to remove the condition $\operatorname{Supp} D \subseteq \operatorname{Supp}(S+B)$. As a benefit of removing the condition $\operatorname{Supp} D \subseteq \operatorname{Supp}(S + B)$ in Conjecture 1.2, we can run the minimal model program while preserving a good condition for metrics (cf. [DHP13, Sect. 8], [FG14, Theorem 5.9]). By applying the above theorem and techniques of the minimal model program, we obtain results related to the abundance conjecture (see [GM13] for more details).

2 Proof of the Main Results

2.1 *Proof of Theorem 2.1*

In this subsection, we give a proof of the following theorem, which is an improvement of Theorem 1.2 to obtain Theorem 1.4.

Theorem 2.1 *Let (F, h_F) and (L, h_L) be (singular) hermitian line bundles with semi-positive curvature on a compact Kähler manifold X. Assume that there exists an effective $\mathbb{R}$-divisor Δ with*

$$h_F = h_L^a \cdot h_\Delta,$$

where a is a positive real number and h_Δ is the singular metric defined by Δ.

Then for a (non-zero) section s of L satisfying $\sup_X |s|_{h_L} < \infty$, *the multiplication map*

$$\Phi_s : H^q(X, K_X \otimes F \otimes \mathscr{I}(h_F)) \xrightarrow{\otimes s} H^q(X, K_X \otimes F \otimes L \otimes \mathscr{I}(h_F h_L))$$

is (well-defined and) injective for any q.

Remark 2.1 (1) The case of $\Delta = 0$ corresponds to Theorem 1.2.
(2) If h_L and h_F are smooth on a Zariski open set, the same conclusion holds under the weaker assumption of $\sqrt{-1}\Theta_{h_F}(F) \geq a\sqrt{-1}\Theta_{h_L}(L)$ (see [Mat14, Theorem 1.5]).

Proof We give here only the strategy of the proof. See [Mat13b, GM13] for the precise proof. First of all, we recall Enoki's method to generalize Kollar's injectivity theorem, which gives a proof of the special case where h_L is smooth and $\Delta = 0$. In this case, the cohomology group $H^q(X, K_X \otimes F)$ is isomorphic to the space of the harmonic forms with respect to h_F

$$\mathscr{H}^{n,q}(F)_{h_F} := \{u \mid \text{u is a smooth } F\text{-valued } (n,q)\text{-form on } X \text{ such that } \overline{\partial} u = D''^{*}_{h_F} u = 0\},$$

where $D''^{*}_{h_F}$ is the adjoint operator of the $\overline{\partial}$-operator. For an arbitrary harmonic form $u \in \mathscr{H}^{n,q}(F)_{h_F}$, we can conclude that $D''^{*}_{h_F h_L} su = 0$ from the semi-positivity of the curvature and $h_F = h_L^a$. This step heavily depends on the semi-positivity of the curvature. This implies that the multiplication map Φ_s induces the map from $\mathscr{H}^{n,q}(F)_{h_F}$ to $\mathscr{H}^{n,q}(F \otimes L)_{h_F h_L}$, and thus the injectivity is obvious.

When h_L is smooth on a Zariski open set, the cohomology group $H^q(X, K_X \otimes F \otimes \mathscr{I}(h))$ is isomorphic to the space of harmonic forms on the Zariski open set. Therefore we can give a proof similar to Enoki's proof thanks to the semi-positivity of the curvature (see [Mat14, Theorem 1.5]).

In our situation, we must consider singular metrics with transcendental (non-algebraic) singularities. It is quite difficult to directly handle transcendental singularities, and thus, in Step 1, we approximate a given singular metric h_F by metrics $\{h_\varepsilon\}_{\varepsilon>0}$ that are smooth on a Zariski open set. Then we represent a given cohomology class in $H^q(X, K_X \otimes F \otimes \mathscr{I}(h_F))$ by the associated harmonic form u_ε with respect to h_ε on the Zariski open set. We want to show that su_ε is also harmonic by using the same method as Enoki. However, the same argument as in [Eno90] fails since the curvature of h_ε is not semi-positive. For this reason, in Step2, we investigate the asymptotic behavior of the harmonic forms u_ε with respect to a family of the regularized metrics $\{h_\varepsilon\}_{\varepsilon>0}$. Then we show that the L^2-norm $\| D''^{*}_{h_\varepsilon h_{L,\varepsilon}} su_\varepsilon \|$ converges to zero as ε tends to zero, where $h_{L,\varepsilon}$ is a suitable approximation of h_L. Further, in Step 3, we construct solutions γ_ε of the $\overline{\partial}$-equation $\overline{\partial}\gamma_\varepsilon = su_\varepsilon$ such that the L^2-norm $\|\gamma_\varepsilon\|$ is uniformly bounded, by applying the Čech complex with the topology induced by the local L^2-norms. In Step 4, we see that

$$\|su_\varepsilon\|^2 = \langle\langle su_\varepsilon, \bar{\partial}\gamma_\varepsilon \rangle\rangle \le \|D^{''*}_{h_\varepsilon h_{L,\varepsilon}} su_\varepsilon\| \|\gamma_\varepsilon\| \to 0 \quad \text{as } \varepsilon \to 0.$$

From these observations, we conclude that u_ε converges to zero in a suitable sense. This completes the proof.

Step 1 (The equisingular approximation of h_F**)**

Throughout the proof, we fix a Kähler form ω on X. For the proof, we want to apply the theory of harmonic integrals, but the metric h_F may not be smooth. For this reason, we approximate h_F by metrics $\{h_\varepsilon\}_{\varepsilon>0}$ that are smooth on a Zariski open set. By [DPS01, Theorem 2.3], we can obtain metrics $\{h_\varepsilon\}_{\varepsilon>0}$ on F satisfying the following properties:

(a) h_ε is smooth on $Y := X \setminus Z$, where Z is a subvariety independent of ε.
(b) $h_{\varepsilon_2} \le h_{\varepsilon_1} \le h_F$ holds for any $0 < \varepsilon_1 < \varepsilon_2$.
(c) $\mathscr{I}(h_F) = \mathscr{I}(h_\varepsilon)$.
(d) $\sqrt{-1}\Theta_{h_\varepsilon}(F) \ge -\varepsilon\omega$.

See [Mat13b, Theorem 2.3] for property (a). By [Fuj12, Lemma 3.1], we obtain a Kähler form $\widetilde{\omega}$ on Y satisfying the following properties:

(A) $\widetilde{\omega}$ is a complete Kähler form on Y.
(B) There exists a bounded function Ψ such that $\widetilde{\omega} = dd^c\Psi$ on a neighborhood of $z \in Z$.
(C) $\widetilde{\omega} \ge \omega$.

In the proof, we mainly consider harmonic forms on Y with respect to h_ε and $\widetilde{\omega}$. Let $L^{n,q}_{(2)}(Y, F)_{h_\varepsilon,\widetilde{\omega}}$ be the space of L^2-integrable F-valued (n, q)-forms α with respect to the inner product $\|\cdot\|_{h_\varepsilon,\widetilde{\omega}}$ defined by

$$\|\alpha\|^2_{h_\varepsilon,\widetilde{\omega}} := \int_Y |\alpha|^2_{h_\varepsilon,\widetilde{\omega}}\, \widetilde{\omega}^n.$$

Then we have the following orthogonal decomposition:

$$L^{n,q}_{(2)}(Y, F)_{h_\varepsilon,\widetilde{\omega}} = \operatorname{Im}\bar{\partial} \oplus \mathscr{H}^{n,q}(F)_{h_\varepsilon,\widetilde{\omega}} \oplus \operatorname{Im} D^{''*}_{h_\varepsilon}.$$

Here the operator $D^{'*}_{h_\varepsilon}$ (resp. $D^{''*}_{h_\varepsilon}$) denotes the closed extension of the formal adjoint of the $(1,0)$-part $D^{'}_{h_\varepsilon}$ (resp. $(0,1)$-part $D^{''}_{h_\varepsilon} = \bar{\partial}$) of the Chern connection $D_{h_\varepsilon} = D^{'}_{h_\varepsilon} + D^{''}_{h_\varepsilon}$. Further $\mathscr{H}^{n,q}(F)_{h_\varepsilon,\widetilde{\omega}}$ denotes the space of harmonic forms with respect to h_ε and $\widetilde{\omega}$, namely

$$\mathscr{H}^{n,q}(F)_{h_\varepsilon,\widetilde{\omega}} := \{\alpha \mid \alpha \text{ is an } F\text{-valued}(n, q)\text{-form with } \bar{\partial}\alpha = D^{''*}_{h_\varepsilon}\alpha = 0\}.$$

A harmonic form in $\mathcal{H}^{n,q}(F)_{h_\varepsilon,\widetilde{\omega}}$ is smooth by the regularity theorem for elliptic operators. These results are known to specialists. The precise proof of them can be found in [Fuj12, Claim 1].

Take an arbitrary cohomology class $\{u\} \in H^q(X, K_X \otimes F \otimes \mathcal{I}(h_F))$ represented by an F-valued (n, q)-form u with $\|u\|_{h_F,\omega} < \infty$. In order to prove that the multiplication map Φ_s is injective, we assume that the cohomology class of su is zero in $H^q(X, K_X \otimes F \otimes L \otimes \mathcal{I}(h_F h_L))$. Our goal is to show that the cohomology class of u is actually zero under this assumption.

By the inequality $\|u\|_{h_\varepsilon,\widetilde{\omega}} \le \|u\|_{h_F,\omega} < \infty$, we can obtain $u_\varepsilon \in \mathcal{H}^{n,q}(F)_{h_\varepsilon,\widetilde{\omega}}$ and $v_\varepsilon \in L^{n,q-1}_{(2)}(Y, F)_{h_\varepsilon,\widetilde{\omega}}$ such that

$$u = u_\varepsilon + \overline{\partial} v_\varepsilon.$$

Note that the component of $\mathrm{Im} D''^{*}_{h_\varepsilon}$ is zero since u is $\overline{\partial}$-closed.

At the end of this step, we explain the strategy of the proof. In Step 2, we show that $\| D''^{*}_{h_\varepsilon h_{L,\varepsilon}} s u_\varepsilon \|_{h_\varepsilon h_{L,\varepsilon},\widetilde{\omega}}$ converges to zero as ε tends to zero. Here $h_{L,\varepsilon}$ is the singular metric on L defined by

$$h_{L,\varepsilon} := h_\varepsilon^{1/a} h_\Delta^{-1/a}.$$

Since the cohomology class of su is zero, there are solutions γ_ε of the $\overline{\partial}$-equation $\overline{\partial}\gamma_\varepsilon = su_\varepsilon$. For the proof, we need to obtain L^2-estimates of them. In Step 3, we construct solutions γ_ε of the $\overline{\partial}$-equation $\overline{\partial}\gamma_\varepsilon = su_\varepsilon$ such that the norm $\|\gamma_\varepsilon\|_{h_\varepsilon h_{L,\varepsilon},\widetilde{\omega}}$ is uniformly bounded. Then we have

$$\|su_\varepsilon\|^2_{h_\varepsilon h_{L,\varepsilon},\widetilde{\omega}} \le \| D''^{*}_{h_\varepsilon h_{L,\varepsilon}} s u_\varepsilon \|_{h_\varepsilon h_{L,\varepsilon},\widetilde{\omega}} \|\gamma_\varepsilon\|_{h_\varepsilon h_{L,\varepsilon},\widetilde{\omega}}.$$

By Step 2 and Step 3, we can conclude that the right hand side goes to zero as ε tends to zero. In Step 4, from this convergence, we prove that u_ε converges to zero in a suitable sense, which implies that the cohomology class of u is zero.

Step 2 (A generalization of Enoki's proof)

By generalizing Enoki's method, in Step 2, we prove the following proposition:

Proposition 2.1 *As ε tends to zero, the norm $\| D''^{*}_{h_\varepsilon h_{L,\varepsilon}} s u_\varepsilon \|_{h_\varepsilon h_{L,\varepsilon},\widetilde{\omega}}$ converges to zero.*

The same argument as in [Eno90] fails since the curvature of h_ε is not semi-positive, and further property (d) is not sufficient for the proof of the proposition since there is counterexample to the injectivity theorem for nef line bundles. To overcome these difficulties, we first see the following inequality:

$$\|u_\varepsilon\|_{h_\varepsilon,\widetilde{\omega}} \le \|u\|_{h_\varepsilon,\widetilde{\omega}} \le \|u\|_{h,\omega}. \tag{1}$$

This inequality and properties (b), (c) imply the proposition. This step can be considered as a generalization of Enoki's method.

Step 3 (A construction of solutions of the $\overline{\partial}$-equation via the Čech complex)

In Step 3, we construct solutions of the $\overline{\partial}$-equation with suitable L^2-norm by using the Čech complex.

Proposition 2.2 *There exist F-valued $(n, q-1)$-forms α_ε on Y satisfying the following properties:*

$$(1)\ \overline{\partial}\alpha_\varepsilon = u - u_\varepsilon. \quad (2)\ \textit{The norm}\ \|\alpha_\varepsilon\|_{h_\varepsilon,\widetilde{\omega}}\ \textit{is uniformly bounded.}$$

Remark 2.2 We have already known that there exist solutions α_ε of the $\overline{\partial}$-equation $\overline{\partial}\alpha_\varepsilon = u - u_\varepsilon$ since $u - u_\varepsilon \in \mathrm{Im}\overline{\partial}$. However, for the proof of the main theorem, we need to construct solutions with uniformly bounded L^2-norm.

The strategy of the proof is as follows: The main idea of the proof is to convert the $\overline{\partial}$-equation $\overline{\partial}\alpha_\varepsilon = u - u_\varepsilon$ to the equation $\delta V_\varepsilon = S_\varepsilon$ of the coboundary operator δ in the space of cochains $C^\bullet(K_X \otimes F \otimes \mathscr{I}(h_\varepsilon))$, by using the Čech complex and pursuing the De Rham-Weil isomorphism. Here the q-cochain S_ε is constructed from $u - u_\varepsilon$. In this construction, we locally solve the $\overline{\partial}$-equation. The important point is that the space $C^\bullet(K_X \otimes F \otimes \mathscr{I}(h_\varepsilon))$ is independent of ε thanks to property (c) of h_ε although the L^2-space $L^{n,q}_{(2)}(Y, F)_{h_\varepsilon,\widetilde{\omega}}$ depends on ε. Since $\|u - u_\varepsilon\|_{h_\varepsilon,\widetilde{\omega}}$ is uniformly bounded, we can observe that S_ε converges to some q-coboundary in $C^q(K_X \otimes F \otimes \mathscr{I}(h))$ with the topology induced by the local L^2-norms with respect to h. Further we can observe that the coboundary operator δ is an open map. Then by these observations we construct solutions V_ε of the equation $\delta V_\varepsilon = S_\varepsilon$ with uniformly bounded norm. Finally, by using a partition of unity, we conversely construct $\alpha_\varepsilon \in L^{n,q-1}_{(2)}(Y, F)_{h_\varepsilon,\widetilde{\omega}}$ from S_ε satisfying the properties in Proposition 2.2. This proof gives a new method to obtain L^2-estimates of solutions of the $\overline{\partial}$-equation.

Step 4 (The limit of the harmonic forms)

In Step 4, we investigate the limit of u_ε and complete the proof. By Step 2 and Step 3, we have

$$\|su_\varepsilon\|^2_{h_\varepsilon h_{L,\varepsilon},\widetilde{\omega}} \leq \|D''^*_{h_\varepsilon h_{L,\varepsilon}} su_\varepsilon\|_{h_\varepsilon h_{L,\varepsilon},\widetilde{\omega}} \|\gamma_\varepsilon\|_{h_\varepsilon h_{L,\varepsilon},\widetilde{\omega}} \to 0 \quad \text{as}\ \varepsilon \to 0.$$

From this convergence, we can show that u_ε converges to zero in a suitable sense, which implies that the cohomology class $\{u\}$ of u is zero in $H^q(X, K_X \otimes F \otimes \mathscr{I}(h_\varepsilon))$. By property (c), we obtain the conclusion of Theorem 2.1. □

2.2 Proof of Theorem 1.3

In this subsection, we give a proof of Theorem 1.3 by using Theorem 1.2 and [Mat14, Theorem 4.1].

Proof of Theorem 1.3 We consider the space of sections with bounded norm defined by

$$H^0_{\mathrm{bdd},h^m}(X, F^m) := \{s \in H(X, F^m) \mid \sup_X |s|_{h^m} < \infty\}.$$

The *bounded Kodaira dimension* $\kappa_{\mathrm{bdd}}(F, h)$ of (F, h) is defined to be $-\infty$ if $H^0_{\mathrm{bdd},h^m}(X, F^m) = 0$ for any $m > 0$. Otherwise, $\kappa_{\mathrm{bdd}}(F, h)$ is defined by

$$\kappa_{\mathrm{bdd}}(F, h) := \sup\{k \in \mathbb{Z} \mid \limsup_{m\to\infty} \dim H^0_{\mathrm{bdd},h^m}(X, F^m)/m^k > 0\}.$$

For a contradiction, we assume that there exists a non-zero cohomology class $\alpha \in H^q(X, K_X \otimes F \otimes \mathscr{I}(h))$. If sections $\{s_i\}_{i=1}^N$ in $H^0_{\mathrm{bdd},h^m}(X, F^m)$ are linearly independent, then $\{s_i\alpha\}_{i=1}^N$ are also linearly independent in $H^q(X, K_X \otimes F^{m+1} \otimes \mathscr{I}(h^{m+1}))$. Indeed, if $\sum_{i=1}^N c_i s_i \alpha = 0$ for some $c_i \in \mathbb{C}$, then we know $\sum_{i=1}^N c_i s_i = 0$ by Theorem 1.2. Since $\{s_i\}_{i=1}^N$ are linearly independent, we have $c_i = 0$ for any $i = 1, 2, \ldots N$. This yields

$$\dim H^0_{\mathrm{bdd},h^m}(X, F^m) \le \dim H^q(X, K_X \otimes F^{m+1} \otimes \mathscr{I}(h^{m+1})).$$

On the other hand, by [Mat14, Theorem 4.1], we have

$$\dim H^q(X, K_X \otimes F^m \otimes \mathscr{I}(h^m)) = O(m^{\dim X - q}) \quad \text{as } m \to \infty,$$

for any $q \ge 0$ (cf. [Dem, (6.18) Lemma]). If $q > \dim X - \kappa_{\mathrm{bdd}}(F, h)$, this is a contradiction. □

2.3 *Proof of Theorem 1.4*

In this subsection, we give a proof of Theorem 1.4.

Proof of Theorem 1.4 For simplicity, we put $\Delta := S + B$ and $G := m(K_X + \Delta)$. We may assume the additional assumption of $h \le h_D$, where h_D is the singular metric on $\mathscr{O}_X(K_X + \Delta)$ defined by the effective divisor D. Indeed, for a smooth metric g on $\mathscr{O}_X(K_X + \Delta)$ and an L^1-function φ (resp. φ_D) with $h = g\, e^{-\varphi}$ (resp. $h_D = g\, e^{-\varphi_D}$), the metric defined by $g\, e^{-\max(\varphi, \varphi_D)}$ satisfies the assumptions again.

Consider the following exact sequence:

$$0 \to \mathscr{O}_X(G - S) \otimes I(h^{m-1}h_B) \to \mathscr{O}_X(G) \otimes \mathscr{I}(h^{m-1}h_B) \to \mathscr{O}_S(G) \otimes \mathscr{I}(h^{m-1}h_B) \to 0.$$

We first prove the induced homomorphism

$$H^q(X, \mathscr{O}_X(G - S) \otimes I(h^{m-1}h_B)) \to H^q(X, \mathscr{O}_X(G) \otimes I(h^{m-1}h_B))$$

is injective by our injectivity theorem. By the assumption on the support of D, we can take an integer $a > 0$ such that aD is a Cartier divisor and $S \leq aD$. Then we have the following commutative diagram:

$$\begin{array}{ccc} & & H^q(X, \mathscr{O}_X(G) \otimes I(h^{m-1}h_B)) \supseteq \mathrm{Im}\,(+S) \\ & \overset{+S}{\nearrow} & \big\downarrow {\scriptstyle +(aD-S)} \\ H^q(X, \mathscr{O}_X(G-S) \otimes I(h^{m-1}h_B)) & \xrightarrow[+aD]{} & H^q(X, \mathscr{O}_X(G-S+aD) \otimes I(h^{a+m-1}h_B)), \end{array}$$

with a map $+S : H^q(X, \mathscr{O}_X(G-S) \otimes I(h^{m-1}h_B)) \to H^q(X, \mathscr{O}_X(G) \otimes I(h^{m-1}h_B))$. In order to show that the upper map on right is injective, we prove that the horizontal map is injective as an application of Theorem 2.1.

By the definition of G, we have

$$G - S = m(K_X + \Delta) - S = K_X + (m-1)(K_X + \Delta) + B.$$

Then the line bundle $F := \mathscr{O}_X((m-1)(K_X + \Delta) + B)$ equipped with the metric $h_F := h^{m-1}h_B$ and the line bundle $L := \mathscr{O}_X(aD)$ equipped with the metric $h_L := h^a$ satisfy the assumptions in Theorem 2.1. Indeed, we have $h_F = h_L^{(m-1)/a} h_B$ by the construction, and further the point-wise norm $|s_{aD}|_{h_L}$ is bounded on X by the inequality $h \leq h_D$, where s_{aD} is the natural section of aD. Therefore the horizontal map is injective by Theorem 2.1. By the assumption on the Lelong number of h, we can conclude that $\mathscr{O}_S \otimes \mathscr{I}(h^{m-1}h_B) = \mathscr{O}_S$. This follows from Skoda's lemma and Hölder's inequality. This completes the proof. □

3 Open Problems

In this section, we summarize and give open problems related to the topics mentioned in this survey.

It is of interest to consider the injectivity theorem in the relative situation. The following problem is a relative version of Theorem 1.2. For relative versions of the injectivity theorem and their applications, we refer the reader to [Fuj13a]. In his paper [Fuj13a], Fujino affirmatively solved this problem under the assumption on the regularity of singular metrics, whose proof is based on the Ohsawa-Takegoshi twisted version of the Bochner-Kodaira-Nakano identity. To remove this assumption, it seems to be needed to use a combination of his method and the techniques of Theorem 1.2.

Probelm 3.1 (cf. [Fuj13a, Problem 1.8]) Let $\pi : X \to Y$ be a surjective holomorphic map from Kähler manifold X to a complex manifold Y, and F be a line bundle on X with a singular metric h whose curvature is semi-positive. Then for a (non-zero) section s of a positive multiple F^m satisfying $\sup_X |s|_{h^m} < \infty$, the multiplication map

$$\Phi_s : R^q\pi_*(K_X \otimes F \otimes \mathscr{I}(h)) \xrightarrow{\otimes s} R^q\pi_*(K_X \otimes F^{m+1} \otimes \mathscr{I}(h^{m+1}))$$

is injective for any q. Here $R^q\pi_*(\mathscr{F})$ denotes the higher direct image of a sheaf $\mathscr{F}$.

Theorem 2.1 can be expected to hold under the weaker assumption made in the following problem. Indeed, this problem was affirmatively solved in [Mat14] under the regularity assumption on singular metrics. It is also an interesting problem to consider the relative version of this problem in the same direction as Problem 3.1.

Probelm 3.2 (cf. [Mat14, Theorem 1.5], [Fuj12, Theorem 1.2]) Let (F, h_F) and (L, h_L) be (singular) hermitian line bundles with semi-positive curvature on a compact Kähler manifold X. Assume there exists a positive real number a such that $\sqrt{-1}\Theta_{h_F}(F) \geq a\sqrt{-1}\Theta_{h_L}(L)$. Then the same conclusion as in Theorem 2.1 holds.

Fujino proposed the following problem, which asks whether one can generalize the injectivity theorem for lc pairs proved by him. The main difficulty in studying this problem is that one must handle lc singularities by analytic methods.

Probelm 3.3 (cf. [Fuj11, Theorem 6.1]) Let D be a simple normal crossing divisor and F be a semi-positive line bundle on a compact Kähler manifold X. Then, for a (non-zero) section s of a positive multiple F^m whose zero locus $s^{-1}(0)$ contains no lc centers of (X, D), the multiplication map

$$\Phi_s : H^q(X, K_X \otimes F \otimes \mathscr{O}_X(D)) \xrightarrow{\otimes s} H^q(X, K_X \otimes F^{m+1} \otimes \mathscr{O}_X(D))$$

is injective for any q.

For a nef line bundle F on a smooth projective variety X, it can be proven that

$$\dim H^q(X, F^m) = O(m^{\dim X - q}) \quad \text{as } m \to \infty.$$

When X is merely supposed to be a compact Kähler manifold, the same conclusion can be expected. This was first posed by Demailly, and proved by Berndtsson under the stronger assumption that F is semi-positive in [Ber12]. The following problem was also proved in [Mat14] when X is a smooth projective variety.

Probelm 3.4 (cf. [Mat14, Theorem 4.1]) Let F be a line bundle on a compact Kähler manifold X and h be a singular metric with semi-positive curvature on F. Then, for any vector bundle $(orlinebundle)$ M, we have

$$\dim H^q(X, M \otimes F^m \otimes \mathscr{I}(h^m)) = O(m^{\dim X - q}) \quad \text{as } m \to \infty.$$

Acknowledgments The author obtained an opportunity of discussion on the injectivity theorem and extension problem when he attended the conference "The 10th Korean Conference in Several Complex Variables". He is grateful to the organizers. He would also like to thank the referee for carefully reading the paper and for suggestions. He is partially supported by the Grant-in-Aid for Young Scientists (B) ♯25800051 from JSPS.

References

[Amb03] Ambro, F.: Quasi-log varieties. Tr. Mat. Inst. Steklova **240** (2003), Biratsion. Geom. Linein. Sist. Konechno Porozhdennye Algebry, 220–239; translation in Proc. Steklov Inst. Math. **1**(240), 214–233 (2003)

[Amb12] Ambro, F.: An injectivity theorem. Compos. Math. **150**(6), 999–1023 (2014)

[BCHM10] Birkar, C, Cascini, P., Hacon, C. D., McKernan, J.: Existence of minimal models for varieties of log general type. J. Amer. Math. Soc. **23**, 405–468 (2010)

[Ber12] Berndtsson, B.: An eigenvalue estimate for the $\overline{\partial}$-Laplacian. J. Differ. Geom. **60**(2), 295–313 (2002)

[Ber13] Berndtsson, B.: The openness conjecture for plurisubharmonic functions. Preprint, arXiv:1305.5781v1

[Cao15] Cao, J.: Numerical dimension and a Kawamata-Viehweg-Nadel type vanishing theorem on compact Kähler manifolds. Compos. Math. **150**(11), 1869–1902 (2014)

[DEL00] Demailly, J.-P., Ein, L., Lazarsfeld, R.: A subadditivity property of multiplier ideals. Mich. Math. J. **48**, 137–156 (2000)

[Dem] Demailly, J.-P.: Analytic methods in algebraic geometry. Surveys of Modern Mathematics vol. 1. International Press, Somerville; Higher Education Press, Beijing (2012)

[Dem-book] Demailly, J.-P.: Complex analytic and differential geometry. Lecture Notes on the web page of the author

[Dem82] Demailly, J.-P.: Estimations L^2 d'un fibré vectoriel holomorphe semi-positif au-dessus d'une variété kählérienne complète. Ann. Sci. École Norm. **15**(4), 457–511 (1982)

[DHP13] Demailly, J.-P., Hacon, C.D., Păun, M.: Extension theorems, non-vanishing and the existence of good minimal models. Acta Math. **210**, 203–259 (2013)

[DK01] Demailly, J.-P., Kollár, J.: Semicontinuity of complex singularity exponents and Kähler-Einstein metrics on Fano orbifolds. Ann. Sci. École Norm. **34**(4), 525–556 (2001)

[DPS01] Demailly, J.-P., Peternell, T., Schneider, M.: Pseudo-effective line bundles on compact Kähler manifolds. Int. J. Math. **6**, 689–741 (2001)

[Eno90] Enoki, I.: Kawamata-Viehweg vanishing theorem for compact Kähler manifolds. Einstein metrics and Yang-Mills connections, pp. 59–68 (Sanda, 1990)

[EP08] Ein, L., Popa, M.: Global division of cohomology classes via injectivity. Special volume in honor of Melvin Hochster. Mich. Math. J. **57**, 249–259 (2008)

[EV92] Esnault, H., Viehweg, E.: Lectures on vanishing theorems. In: DMV Seminar, vol. 20. Birkhäuser Verlag, Basel (1992)

[Fuj00] Fujino, O.: Abundance theorem for semi log canonical threefolds. Duke Math. J. **102**(3), 513–532 (2000)

[Fuj11] Fujino, O.: Fundamental theorems for the log minimal model program. Publ. Res. Inst. Math. Sci. **47**(3), 727–789 (2011)

[Fuj12] Fujino, O.: A transcendental approach to Kollár's injectivity theorem. Osaka J. Math. **49**(3), 833–852 (2012)

[Fuj13a] Fujino, O.: A transcendental approach to Kollár's injectivity theorem II. J. Reine Angew. Math. **681**, 149–174 (2013)

[Fuj13b] Fujino, O.: Injectivity theorems. Preprint, arXiv:1303.2404v1

[FG14] Fujino, O., Gongyo, Y.: Log pluricanonical representations and the abundance conjecture. Compositio Math. **150**, 593–620 (2014)

[GM13] Gongyo, Y. Matsumura, S.: Versions of injectivity and extension theorems. Preprint, arXiv:1406.6132v2

[GZ15] Guan, Q., Zhou, X.: Effectiveness of Demailly's strong openness conjecture and related problems. Invent. Math

[Hie14] Hiep, P.H.: The weighted log canonical threshold. C. R. Math. Acad. Sci. Paris **352**(4), 283–288 (2014)

[Kaw82] Kawamata, Y.: A generalization of Kodaira-Ramanujam's vanishing theorem. Math. Ann. **261**(1), 43–46 (1982)
[Kaw85] Kawamata, Y.: Pluricanonical systems on minimal algebraic varieties. Invent. Math. **79**(3), 567–588 (1985)
[Kaw92] Kawamata, Y.: Abundance theorem for minimal threefolds. Invent. Math. **108**(2), 229–246 (1992)
[KaMM87] Kawamata, Y., Matsuda, K., Matsuki, K.: Introduction to the minimal model problem. Algebraic Geometry, Sendai, pp. 283–360 (1985), Adv. Stud. Pure Math., **10**, North-Holland, Amsterdam (1987)
[KeMMc04] Keel, S., Matsuki, K., McKernan J.: Log abundance theorem for threefolds. Duke Math. J. **75**(1), 99–119 (1994), Corrections to: "Log abundance theorem for three-folds." Duke Math. J. **122**(3), 625–630 (2004)
[KM] Kollár, J., Mori, S.: Birational geometry of algebraic varieties. Cambridge Tracts in Math. **134** (1998)
[Kol86] Kollár, J.: Higher direct images of dualizing sheaves. I. Ann. Math. (2) **123**(1), 11–42 (1986)
[Laz] Lazarsfeld, R.: Positivity in Algebraic Geometry I-II. A Series of Modern Surveys in Mathematics, vol. 48, 49. Springer Verlag, Berlin (2004)
[Mat13a] Matsumura, S.: A Nadel vanishing theorem for metrics with minimal singularities on big line bundles. Preprint, arXiv:1306.2497v2
[Mat13b] Matsumura, S.: An injectivity theorem with multiplier ideal sheaves of singular metrics with transcendental singularities. Adv. in Math. **280**, 188–207 (2015)
[Mat14] Matsumura, S.: A Nadel vanishing theorem via injectivity theorems. Math. Ann. **359**(4), 785–802 (2014)
[Nad89] Nadel, A.M.: Multiplier ideal sheaves and existence of Kähler-Einstein metrics of positive scalar curvature. Proc. Nat. Acad. Sci. U.S.A. **86**(19), 7299–7300 (1989)
[Nad90] Nadel, A.M.: Multiplier ideal sheaves and Kähler-Einstein metrics of positive scalar curvature. Ann. Math. (2) **132**(3), 549–596 (1990)
[Nak] Nakayama, N.: Zariski decomposition and abundance. MSJ Memoirs, vol. 14. Mathematical Society of Japan, Tokyo (2004)
[Ohs84] Ohsawa, T.: Vanishing theorems on complete Kähler manifolds. Publ. Res. Inst. Math. Sci. **20**(1), 21–38 (1984)
[Ohs04] Ohsawa, T.: On a curvature condition that implies a cohomology injectivity theorem of Kollár-Skoda type. Publ. Res. Inst. Math. Sci. **41**(3), 565–577 (2005)
[Tak97] Takegoshi, K.: On cohomology groups of nef line bundles tensorized with multiplier ideal sheaves on compact Kähler manifolds. Osaka J. Math. **34**(4), 783–802 (1997)
[Tan71] Tankeev, S.G.: On n-dimensional canonically polarized varieties and varieties of fundamental type. Math. USSR-Izv. **5**(1), 29–43 (1971)
[Ver10] Verbitsky, M.: HyperKähler SYZ conjecture and semipositive line bundles. Geom. Funct. Anal. **19**(5), 1481–1493 (2010)
[Vie82] Viehweg, E.: Vanishing theorems. J. Reine Angew. Math. **335**, 1–8 (1982)

Amoebas of Cuspidal Strata for Classical Discriminant

E.N. Mikhalkin, A.V. Shchuplev and A.K. Tsikh

Abstract An amoeba of an analytic set is the real part of its image in a logarithmic scale. Among all hypersurfaces A-discriminantal sets have the most simple amoebas. We prove that any singular cuspidal stratum of the classical discriminant can be transformed by a monomial change of variables into an A-discriminantal set and compute the contours of the amoebas of these strata.

Keywords Amoeba · A-discriminant · Cuspidal stratum

The notion of the amoeba of an algebraic hypersurface was introduced in 1994 in the book [GKZ94]. The study of the structure of amoebas began with the papers [FPT00, Mik00], and by now there are many interesting results related both to the description of amoebas [MR01, BT12], and to their applications in the study of dimer configurations [KOS06], extensions of non-Archimedean fields [EKL06], to mention but a few. The interest in amoebas is partly stimulated by the connections to real algebraic geometry [Mik00] and tropical arithmetic [EKL06, Stu02]. The extension of the notion of amoeba to non-algebraic complex analytic sets allows to use this language in thermodynamics and statistical physics in general, for example in problems with several Hamiltonians for a given physical system [PPT13, PT09]. In statistical physics amoebas appear when using asymptotical methods for studying integrals with integration over cycles on analytic sets [LPT08, BKT14].

Denote by $\mathbb{T}^n$ the complex algebraic torus $(\mathbb{C} \setminus 0)^n$ and consider the mapping $\mathrm{Log} : \mathbb{T}^n \to \mathbb{R}^n$ given by

E.N. Mikhalkin · A.V. Shchuplev · A.K. Tsikh (✉)
Institute of Mathematics and Computer Science, Siberian Federal University,
pr. Svobodny 79, 660041 Krasnoyarsk, Russia
e-mail: atsikh@sfu-kras.ru

E.N. Mikhalkin
e-mail: mikhalkin@bk.ru

A.V. Shchuplev
e-mail: ashchuplev@sfu-kras.ru

F. Bracci et al. (eds.), *Complex Analysis and Geometry*, Springer Proceedings in Mathematics & Statistics 144, DOI 10.1007/978-4-431-55744-9_19

$$\mathrm{Log} z = (\log |z_1|, \dots, \log |z_n|).$$

The *amoeba* of an algebraic set $V \subset \mathbb{T}^n$ is its image $\mathrm{Log} V \subset \mathbb{R}^n$. The amoeba of the set V will be denoted by $\mathscr{A}_V$, while its complement $\mathbb{R}^n \setminus \mathscr{A}_V$ will be denoted by ${}^c\mathscr{A}_V$. Since the map Log is proper, the complement of the amoeba is open. For a hypersurface V, i.e. for a set of codimension 1, the complement ${}^c\mathscr{A}_V$ consists of a finite number of connected components, each is open and convex. Indeed, if a hypersurface V is the zero set of a polynomial P, then for every connected component E of ${}^c\mathscr{A}_V$ the set $\mathrm{Log}^{-1} E$ is a domain of convergence for some Laurent series for $1/P$ centered at the origin:

$$\frac{1}{P(z)} = \sum_{\alpha \in \mathbb{Z}^n} a_\alpha z^\alpha,$$

and such domains are logarithmically convex.

In the case of arbitrary codimension $k = \mathrm{codim}_{\mathbb{C}} V$ the complement ${}^c\mathscr{A}_V$ has the property of being $(k-1)$-convex (the 0-convexity is the usual convexity) [BT12, Hen04].

In comparison to the case of hypersurfaces, amoebas of surfaces of codimension $k > 1$ are studied to a less extent. One of the reasons behind that is the absence of a simple analog of the Jensen-Ronkin function [PR04]. This paper deals with amoebas of singular strata of cuspidal type for the classical discriminant. An important role in this study is played by the Horn-Kapranov parametrization for the discriminant set (see Sect. 2). The implicit function theorem yields that singularities of an algebraic function, given by a polynomial equation, appear only in those points where the discriminant of the polynomial vanishes. It turns out that a general algebraic function, i.e. given by a polynomial with independent variable coefficients, is a hypergeometric function in the sense of Horn [Hor89]. The hallmark of this property is that one can explicitly parametrize the boundary of the domain of convergence of a hypergeometric series, i.e. parametrize the set of conjugate radii of convergence. This parametrization was obtained by J.Horn in 1889, and a hundred years later M. Kapranov noticed a miraculous fact: if in Horn's parametrization we omit the absolute value signs and let the parameters be complex it becomes the parametrization of the singular set of a hypergeometric function [Kap91]. The ideas of Horn and Kapranov were further developed in [AT12] to parametrize discriminantal sets for polynomial transformations of $\mathbb{C}^n$.

We proceed as follows. In Sect. 1 we define the contour of an amoeba and the logarithmic Gauss map and formulate a theorem that establishes a relationship between them (Theorem 1). In Sect. 2 we consider cuspidal strata for the classical discriminant and find their place in the hierarchy of all A-discriminantal sets (Theorem 2). Theorems 3 and 4 are necessary steps to justify the fact that amoebas of cuspidal strata have non-empty contours, which admit explicit parameterizations (Theorem 5).

1 The Contour of Amoeba and the Logarithmic Gauss Map

Definition 1 The *contour* of the amoeba $\mathscr{A}_V$ is the set $\mathscr{C}_V$ of critical values of the logarithmic mapping Log restricted to V, i.e. of the mapping $\mathrm{Log} : V \to \mathbb{R}^n$.

The structure of the contour of an amoeba can be described in terms of the logarithmic Gauss map. This mapping, introduced by Kapranov in [Kap91] for hypersurfaces, extends naturally to the case of surfaces V of any codimension k.

Definition 2 Let $\mathrm{Gr}(n, k)$ be the Grassmanian of k-dimensional complex subspaces in $\mathbb{C}^n$. *The logarithmic Gauss map* $\gamma : V \to \mathrm{Gr}(n, k)$ sends a smooth point $z \in \mathrm{reg}V$ to the normal subspace $\gamma(z)$ to $\mathrm{Log}_{\mathbb{C}} V$ at $\mathrm{Log}_{\mathbb{C}}(z)$, where $\mathrm{Log}_{\mathbb{C}}$ is the complex logarithm $\mathrm{Log}_{\mathbb{C}} : (z_1, \ldots, z_n) \to (\log z_1, \ldots, \log z_n)$.

If V is a hypersurface

$$V = \{z \in \mathbb{T}^n : P(z) = 0\}$$

(i.e. if $k = 1$ and $\mathrm{Gr}(n, 1) = \mathbb{CP}_{n-1}$) the logarithmic Gauss map $\gamma : V \to \mathbb{CP}_{n-1}$ has the following analytic expression

$$(z_1, \ldots, z_n) \to \left(z_1 \frac{\partial P}{\partial z_1} : \ldots : z_n \frac{\partial P}{\partial z_n}\right).$$

In this case it is known [Mik00, The02] that a point $z \in \mathrm{reg}V$ is critical for the map $\mathrm{Log}|_V$ if and only if its image $\gamma(z)$ under the logarithmic Gauss map lies in the real projective subspace $\mathbb{RP}_{n-1} \subset \mathbb{CP}_{n-1}$. So the contour $\mathscr{C}_V$ of the amoeba $\mathscr{A}_V$ of a hypersurface is the set $\mathrm{Log}(\gamma^{-1}(\mathbb{RP}_{n-1}))$.

Consider now an algebraic surface $V \subset \mathbb{T}^n$, $n > 1$. Assume that V is of pure complex dimension d, i.e. all irreducible components of V have the same dimension d. Denote by $k = n - d$ the codimension of V.

In a neighborhood of any its smooth point z_0 the set V is given by the system $P_1(z) = \ldots = P_k(z) = 0$ with the Jacobian matrix of rank k. Then the logarithmic Gauss map at this point is defined by the matrix

$$\gamma(z) = \begin{pmatrix} z_1 \frac{\partial P_1}{\partial z_1} & \cdots & z_n \frac{\partial P_1}{\partial z_n} \\ \vdots & & \vdots \\ z_1 \frac{\partial P_k}{\partial z_1} & \cdots & z_n \frac{\partial P_k}{\partial z_n} \end{pmatrix}.$$

The rows of this matrix form a basis for the normal space to the image $\mathrm{Log}_{\mathbb{C}} V$ at $\mathrm{Log}_{\mathbb{C}}(z_0)$.

Theorem 1 ([BT12]) *A point $z \in \operatorname{reg} V$ is critical for the mapping* Log *if and only if the image $\gamma(z)$ of the logarithmic Gauss map contains*

- *at least $n - 2d + 1$ linearly independent real vectors if $2d \leq n$,*
- *at least one real vector if $2d \geq n$.*

In particular, if V is a hypersurface or a curve, i.e. $d = n - 1$ or $d = 1$, a point z is critical if and only if $\gamma(z)$ is real.

Let us say some words on the essence of this statement. The mapping $\operatorname{Log}|_V : V \to \mathbb{R}^n$ is the composition of the complex logarithm

$$\operatorname{Log}_{\mathbb{C}}(z) = \operatorname{Log}(z) + i\operatorname{Arg}(z) : \ V \to \mathbb{R}^n \oplus i\mathbb{R}^n$$

and the projection onto the real part $\mathbb{R}^n$:

$$\operatorname{Log}|_V = \pi_{\mathbb{R}^n} \circ \operatorname{Log}_{\mathbb{C}}\big|_V .$$

The complex logarithm does not have critical point on $\operatorname{reg} V$ (it is locally biholomorphic in $\mathbb{T}^n$), therefore the critical points of $\operatorname{Log}|_V$ appear only as critical point of the projection

$$\pi_{\mathbb{R}^n} : \ \operatorname{Log}_{\mathbb{C}} V \to \mathbb{R}^n .$$

But the critical point of this projection are defined by the properties of its tangent map

$$d(\pi_{\mathbb{R}^n})|_{\operatorname{Log}_{\mathbb{C}} V} : \ T_w(\operatorname{Log}_{\mathbb{C}} V) \to T_{\operatorname{Re}(w)}(\mathbb{R}^n), \ \ w = \operatorname{Log}_{\mathbb{C}}(z).$$

As a matter of fact, the criterion for $\operatorname{Log}|_V$ to be critical at z can be formulated as follows

- if $2d \leq n$, the tangent map of the projection $\pi_{\mathbb{R}^n}$ is *not injective*,
- if $2d \geq n$, the tangent map of the projection $\pi_{\mathbb{R}^n}$ is *not surjective*.

The conditions of being non-injective or non-surjective are related to whether the normal space to $\operatorname{Log}_{\mathbb{C}} V$ is real or not (in some sense Fig. 1 clarifies that: in critical points of the projection $\pi_{\mathbb{R}^n}$ the normal subspace $\gamma(z)$ becomes 'horizontal' and does not have a real part, i.e. $\gamma(z)$ is real).

As an example, let us examine whether an amoeba of a complex line has a contour. Let the complex line V in $\mathbb{C}^n$ be given by

$$\begin{cases} z_2 = a_2 z_1 + b_2, \\ \dots \\ z_n = a_n z_1 + b_n, \end{cases} \tag{1}$$

where all $a_j, b_j \neq 0$. The logarithmic projection of V has the form

$$\operatorname{Log}(z)|_V = (\log|z_1|, \log|a_2 z_1 + b_2|, \dots, \log|a_n z_1 + b_n|).$$

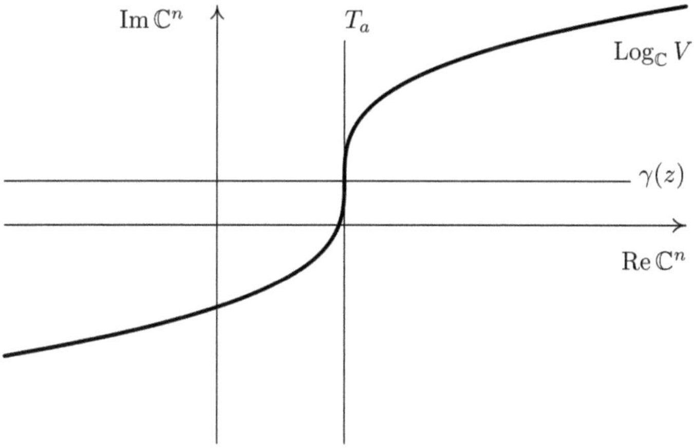

Fig. 1 An illustration to Theorem 1

Its Jacobian matrix equals

$$\frac{\partial(\mathrm{Log})}{\partial(z,\overline{z})} = \frac{1}{2}\begin{pmatrix} \frac{1}{z_1} & \frac{1}{\overline{z}_1} \\ \frac{a_2}{z_1} & \frac{\overline{a}_2}{\overline{z}_1} \\ \cdots & \\ \frac{a_n}{z_1} & \frac{\overline{a}_n}{\overline{z}_1} \end{pmatrix}.$$

Denote

$$z_1 = x + iy, \qquad \frac{b_j}{a_j} = c_j + id_j,$$

then the condition for $z_1 = x + iy$ to be critical for the mapping $\mathrm{Log}|_V$ (i.e. when the rank of the Jacobian matrix is not maximal) can be written as

$$\begin{vmatrix} x & y \\ c_j & d_j \end{vmatrix} = 0, \quad j = 2, \ldots, n,$$

$$\begin{vmatrix} c_k & d_k \\ x & y \end{vmatrix} + \begin{vmatrix} x & y \\ c_l & d_l \end{vmatrix} + \begin{vmatrix} c_k & d_k \\ c_l & d_l \end{vmatrix} = 0, \quad k,\ l = 2, \ldots, n.$$

This system is consistent if and only if $c_k d_l = c_l d_k$ for all $k, l = 2, \ldots, n$, but this condition is equivalent to

$$\frac{a_k b_l}{a_l b_k} \in \mathbb{R}, \quad k,\ l = 2, \ldots, n. \tag{2}$$

Thus, we arrive at

Proposition *For $n \geq 3$ the contour of the amoeba of a complex line (1) is not empty if and only if the conditions (2) hold. In such case the contour of the amoeba is the image of the real line $d_2 x = c_2 y$ under the mapping* Log.

Consider two examples of lines in $\mathbb{T}^3$.

Example 1 For the complex line given by

$$\begin{cases} z_2 = z_1 + 1, \\ z_3 = z_1 + 1 + i, \end{cases}$$

the conditions (2) do not hold, therefore the contour of its amoeba is empty. The logarithmic projection of this line does not have critical points, and the line is diffeomorphic to its amoeba (see Fig. 2, left). In this case we say that the amoeba is not degenerate. At each point of the line the value $\gamma(z)$ of the logarithmic Gauss map has only one real vector (see Fig. 2, right).

Example 2 For the complex line

$$\begin{cases} z_2 = z_1 + 1, \\ z_3 = z_1 + 2 \end{cases}$$

the condition (2) holds: $\dfrac{a_2 b_3}{a_3 b_2} = 2 \in \mathbb{R}$. The amoeba is a surface with a corner in $\mathbb{R}^3$, each its interior point has two preimages on the line. Namely, for every non-real $z_1 = x + iy$ the images of

$$\text{Log}(z_1, z_1 + 1, z_1 + 2) \text{ and } \text{Log}(\overline{z}_1, \overline{z}_1 + 1, \overline{z}_1 + 2)$$

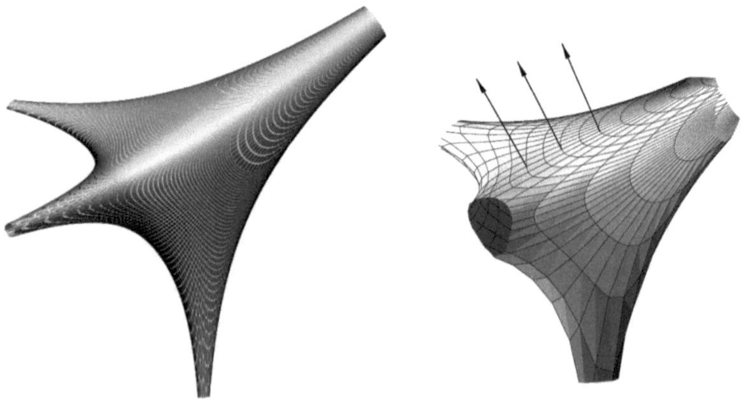

Fig. 2 The amoeba of the complex line of Example 1

Fig. 3 The amoeba of the complex line of Example 2

coincide. The real line $z_1 = x_1$ is mapped to the contour of the amoeba (its topological boundary), and the amoeba itself is the result of collapsing of a non-degenerate amoeba (see Fig. 3, left). At the points of the contour the logarithmic Gauss map $\gamma(z)$ contains a plane of real normal vectors (see Fig. 3, right).

2 Cuspidal Strata for Classical Discriminant

By a general algebraic equation we understand the equation

$$f(y) := a_0 + a_1 y + \ldots + a_{n-1} y^{n-1} + a_n y^n = 0 \tag{3}$$

with variable complex coefficients $a = (a_0, a_1, \ldots, a_n)$.

The classical discriminant is the polynomial $D(a)$ that vanishes if and only if the Eq. (3) has multiple roots. The zero set of the discriminant $D(a)$ we denote by ∇ and call *the discriminantal set* of the Eq. (3) or of the polynomial f.

Define subsets $\mathscr{M}^j \subset \nabla$ that comprise all $a \in \mathbb{C}^{n+1}$ for which the Eq. (3) has roots of multiplicity $\geq j$. They form a sequence of nested subsets

$$\nabla = \mathscr{M}^2 \supset \mathscr{M}^3 \supset \ldots \supset \mathscr{M}^n.$$

Each $\mathscr{M}^{j+1}$ is a subset of singular points sng$\mathscr{M}^j$, and the stratum $S^j = \mathscr{M}^j \setminus \mathscr{M}^{j+1}$ consists of points where either $\mathscr{M}^j$ is smooth or self intersects with its smooth components.Therefore we call $\mathscr{M}^j$ the *cuspidal strata*. Note that certain properties of these strata were studied in [Kat03].

Our recent result from the forthcoming paper [MT] states the following.

Theorem 2 *There exist monomial transformations that turn the strata* $\mathscr{M}^2$, $\mathscr{M}^3, \ldots, \mathscr{M}^n$ *into some A-discriminantal sets* $\nabla_{A_2}, \nabla_{A_3}, \ldots, \nabla_{A_n}$.

Recall the definition of an A-discriminantal set (see [GKZ94], Chap. 9). Instead of Eq. (3) in one unknown y we consider an equation in k unknowns $y = (y_1, \ldots, y_k)$:

$$f(y_1, \ldots, y_k) := \sum_{\alpha=(\alpha_1,\ldots,\alpha_k)\in A} a_\alpha {y_1}^{\alpha_1} \ldots {y_k}^{\alpha_k} = 0, \tag{4}$$

where $A \subset \mathbb{Z}^k$ is a fixed set of exponents that generate the lattice $\mathbb{Z}^k$ as an additive group, and the coefficients a_α are variables. The set of coefficients (same as the set of Eq. (4) and the set of Laurent polynomials f with exponents $\alpha \in A$) is $\mathbb{C}^A$, whose dimension is equal to the cardinality of A.

Definition 3 Let ∇° be the set of all $(a_\alpha) \in \mathbb{C}^A$ for which the Eq. (4) has critical roots $y \in (\mathbb{C} \setminus 0)^k$, i.e. the roots where the gradient of f vanishes. The closure $\overline{\nabla}^\circ$ of this set is called *an A-discriminantal set* and is denoted by ∇_A.

In the case $k = 1,\ A = \{0, 1, 2 \ldots, n\} \subset \mathbb{Z}$ the set ∇_A is the classical discriminantal set ∇ of the Eq. (3). In Theorem 2 each ∇_{A_j} is an A_j-discriminantal set of an equation in $j-1$ unknowns. Moreover, the cardinality of A_j is $n+1$ and $\nabla_{A_2} = \nabla$.

For the proof of Theorem 2, the crucial thing is the Horn-Kapranov parametrization (see [PT04]) for the discriminantal set of a reduced equation

$$f(y) = 1 + x_1 y + \ldots + x_{n-1} y^{n-1} + y^n = 0. \tag{5}$$

This parametrization $x = \Psi(s) : \mathbb{CP}^{n-2}_s \to \mathbb{C}^{n-1}_x$ is given by the formula

$$x_k = -\frac{n s_k}{\langle \alpha, s \rangle} \left(\frac{\langle \alpha, s \rangle}{\langle \beta, s \rangle} \right)^{\frac{k}{n}}, \quad k = 1, \ldots, n-1, \tag{6}$$

where $\alpha,\ \beta$ are vectors of integers

$$\alpha = (n-1, \ldots, 2, 1),\ \beta = (1, 2, \ldots, n-1).$$

Notice that the Eq. (3) can be reduced differently, fixing coefficients of any pair of monomials y^p and y^q. The parametrization of the corresponding reduced discriminantal set ∇_{pq} will differ from formula (6) (i.e. the parametrization of ∇_{0n}), it will depend on different vectors α and β, and the root in the formula will be of degree $p-q$ instead of n [PT04].

Define the sequence of critical strata $\mathscr{C}^j$ of the parametrization (6). The first stratum $\mathscr{C}^1$ is defined as the set of critical values of the parametrization Ψ. It turns out that the critical points of Ψ constitute a hyperplane $L_1 \subset \mathbb{CP}^{n-2}$, consequently, the first critical stratum $\mathscr{C}^1$ is parametrized by the restriction of Ψ to L_1. Analogously, we define the stratum $\mathscr{C}^2$ of critical values of that restriction and proceed by induction. To formulate the result, introduce the following hyperplanes in $\mathbb{CP}^{n-2}$:

$$L_j = \left\{ s : \sum_{i=j}^{n-1} i(i-1)\cdots(i-(j-1))(n-i)s_i = 0 \right\},$$

where $s = (s_1 : \ldots : s_{n-1})$ is the homogeneous coordinates. The following theorem is proved by the direct computations.

Theorem 3 *The strata $\mathscr{C}^j$ are parametrized by the restrictions $\Psi\Big|_{L^j}$ on the planes $L^j = L_1 \cap \ldots \cap L_j$.*

The next theorem shows the relationship between the critical strata of Ψ with the reduced singular strata $\mathscr{M}^j_{0n}$ obtained from $\mathscr{M}^j$ by intersecting with the plane $a_0 = a_n = 1$.

Theorem 4 *The reduced singular strata $\mathscr{M}^{j+2}_{0n} \subset \nabla_{0n}$ coincide with the critical strata $\mathscr{C}^j$ of the parametrization Ψ.*

The proof of Theorem 4 goes as follows. First, we notice that the expression

$$t(s) = \left(\frac{\langle \beta, s\rangle}{\langle \alpha, s\rangle}\right)^{\frac{1}{n}}$$

involved in (6) is a root of the Eq. (5) of multiplicity ≥ 2 for $x = \Psi(s)$.

Let t be a root of the Eq. (5) of multiplicity $\geq \mu$, i.e.

$$f(y) = (y - t)^{\mu} f_{n-\mu}(y), \tag{7}$$

where

$$f_{n-\mu}(y) := \sum_{k=0}^{n-\mu} x_k^{(n-\mu)} y^k$$

is the result of division of f by $(y - t)^{\mu}$. Computing the coefficients $x_k^{(n-\mu)}$ in terms of the root t and the coefficients of x_k of the initial polynomial f, we prove that

$$f_{n-\mu}(t(s)) = 0 \Longleftrightarrow s \in L^{\mu-1}.$$

So, if $x = \Psi(s)$ then $y = t(s)$ is a root of multiplicity $\geq \mu - 2$ if and only if $s \in L^{\mu}$. From there, it is easy to finish the proof of Theorem 4.

To explain the proof of Theorem 2, recall the Horn-Kapranov parametrization for a reduced A-discriminantal set. In order to do that, with the set of exponents $\alpha^j \in A$ of (5) we associate the matrix

$$A = \begin{pmatrix} 1 & 1 & \cdots & 1 \\ \alpha_{11} & \alpha_{21} & \cdots & \alpha_{N1} \\ \vdots & \vdots & \ddots & \vdots \\ \alpha_{1k} & \alpha_{2k} & \cdots & \alpha_{Nk} \end{pmatrix}$$

(we denote this matrix by A too). For the Eq. (3) we have

$$A = \begin{pmatrix} 1 & 1 & \cdots & 1 & 1 \\ 0 & 1 & \cdots & n-1 & n \end{pmatrix}$$

Let B be an integer right annulator of A of rank $m = N - k$. There are many such annulators and the choice of B gives a *reduction* of the Eq. (4) (see [GKZ94] or [Kap91]). Write this annulator in the form

$$B = \begin{pmatrix} b_{11} & \cdots & b_{1m} \\ \vdots & \ddots & \vdots \\ b_{N1} & \cdots & b_{Nm} \end{pmatrix}.$$

The matrix B defines the mapping

$$\Psi_B : \mathbb{CP}^{m-1} \to (\mathbb{C}^*)^m, \quad s \to z = (Bs)^B, \tag{8}$$

where $s = (s_1 : \ldots : s_m)$ is the homogeneous coordinates in $\mathbb{CP}^{m-1}$. Coordinate-wise the mapping Ψ_B has the form

$$z_k = \prod_{j=1}^{N} \langle b_j, s \rangle^{b_{jk}}, \quad k = 1, \ldots, m,$$

where $b_j = (b_{j1}, \ldots, b_{jm})$ are the rows of the matrix B. Since the first row of A is orthogonal to each column of B, the degree of homogeneity of these expressions in s is zero, therefore $(Bs)^B$ are correctly defined on $\mathbb{CP}^{m-1}$. The mapping $\Psi_B(s)$ defined by (8) is called *the Horn-Kapranov parametrization.* The importance of this mapping follows from Kapranov's theorem [Kap91] stating that

- The mapping $\Psi_B(s)$ is a parametrization of the reduced A-discriminantal set $\widetilde{\nabla}_A$.
- If $\widetilde{\nabla}_A$ is a hypersurface then $\Psi_B(s)$ is a birational isomorphism that coincide with the inversion of the logarithmic Gauss map for $\widetilde{\nabla}_A$.

Example 3 Let us sketch the idea of the proof of Theorem 2 by the example of the stratum $\mathscr{M}^3_{01}$ for the equation of degree 4. Consider a reduced equation of fourth degree

$$1 + y + z_2 y^2 + z_3 y^3 + z_4 y^4 = 0.$$

According to Kapranov's theorem the reduced discriminantal set ∇_{01} is parametrized by the mapping $\Psi : \mathbb{CP}^2 \to \mathbb{C}^3$ by

$$\begin{aligned} z_2 &= s_2(s_2 + 2s_3 + 3s_4)^1(-2s_2 - 3s_3 - 4s_4)^{-2} \\ z_3 &= s_3(s_2 + 2s_3 + 3s_4)^2(-2s_2 - 3s_3 - 4s_4)^{-3} \\ z_4 &= s_4(s_2 + 2s_3 + 3s_4)^3(-2s_2 - 3s_3 - 4s_4)^{-4}. \end{aligned}$$

The line $L^1 \subset \mathbb{CP}^2$ of its critical points has the equation $s_2 + 3s_3 + 6s_4 = 0$. Therefore, the reduced stratum $\mathscr{M}^3_{01}$ defined by the restriction $\Psi\big|_{L^1}$, in the homogeneous coordinates $s' = (s_3 : s_4)$ of this line is given by the formulas

$$z_2 = (-3s_3 - 6s_4)(-s_3 - 3s_4)^1(3s_3 + 8s_4)^{-2}$$
$$z_3 = s_3(-s_3 - 3s_4)^2(3s_3 + 8s_4)^{-3}$$
$$z_4 = s_4(-s_3 - 3s_4)^3(3s_3 + 8s_4)^{-4}.$$

The coefficients of five linear functions involved here define the matrix

$$B = \begin{pmatrix} -1 & -3 \\ 3 & 8 \\ -3 & -6 \\ 1 & 0 \\ 0 & 1 \end{pmatrix}.$$

The monomial change of variables $M\colon (z_1, z_2, z_3) \to (w_3, w_4)$ given by

$$w_3 = z_3 z_2^{-3}, \; w_4 = z_4 z_2^{-6},$$

transforms the parametrization of $\mathcal{M}_{01}^3$ into

$$w_3 = s_3^1 s_4^0(-3s_3 - 6s_4)^{-3}(-s_3 - 3s_4)^{-1}(3s_3 + 8s_4)^3$$
$$w_4 = s_3^0 s_4^1(-3s_3 - 6s_4)^{-6}(-s_3 - 3s_4)^{-3}(3s_3 + 8s_4)^8,$$

which has the form $w = (Bs')^B$. By Kapranov's theorem such a mapping parametrizes some reduced A-discriminantal set. In order to determine the set A it is enough to find a left integer annulator of B of the size 3×5 such that all elements of its first row are 1 and its columns generate $\mathbb{Z}^3$. In this case we can take

$$A = \begin{pmatrix} 1 & 1 & 1 & 1 & 1 \\ 1 & 0 & 0 & 1 & 3 \\ 0 & 0 & 1 & 3 & 6 \end{pmatrix}.$$

Therefore, $w = (Bs')^B$ parametrizes a reduced A-discriminantal set of the equation

$$a_{10}y_1 + a_{01}y_2 + a_{00} + a_{31}y_1^3 y_2 + a_{63}y_1^6 y_2^3 = 0,$$

whose exponents are columns of the matrix A without the first row. The corresponding reduction of the equation is obtained if we fix $a_{10} = a_{01} = a_{00} = 1$ and denote $a_{31} =: w_3$, $a_{63} =: w_4$.

Notice that the chosen annulator A of B has all its elements non-negative. Following the general scheme (see Lemma below), we would have chosen the matrix

$$\begin{pmatrix} 1 & 1 & 1 & 1 & 1 \\ 1 & 0 & 0 & 1 & 3 \\ 0 & 1 & 0 & -3 & -8 \end{pmatrix},$$

which is obtained from A by multiplication by a unimodular (3×3)-matrix.

The following lemma is a generalization of the observations of this example, it is an essential ingredient of the proof of Theorem 2. Here $p, q, i_0, \ldots, i_{j-1}$ is an arbitrary sequence of pair-wise distinct integers from $\{0, 1, \ldots, n\}$, and $1 \le j \le n-2$.

Lemma *Consider two sets of variables*

$$z = (z_i), i \neq p, q,$$
$$w = (w_k), k \neq p, q, i_0, \ldots, i_{j-1}.$$

The map $M : (\mathbb{C}^*)_z^{n-1} \to (\mathbb{C}^*)_w^{n-1-j}$ *defined by*

$$w_k = z_k \prod_{\nu=0}^{j-1} z_{i_\nu}^{-\frac{(k-p)(k-q)}{(i_\nu-p)(i_\nu-q)} \prod\limits_{m\neq\nu} \frac{k-i_m}{i_\nu-i_m}}, \quad k \neq p, q, i_0, \ldots, i_{j-1}, \tag{9}$$

transforms the parametrization of the stratum $\mathscr{M}_{pq}^{j+2}$ *to the form* $w = (Bs')^B$, *where* B *is a rational* $(n+1) \times (n-1-j)$*-matrix of rank* $(n-1-j)$ *such that the sum of elements in a row is zero, and* $s' = (s_k)$, $k \neq p, q, i_0, \ldots, i_{j-1}$.

For the proof of Theorem 2 it is convenient to take as p, q, $i_0, \ldots, i_{j-1}$ the sequence $0,\ 1,\ 2, \ldots,\ j+1$.

3 Amoebas of Reduced Cuspidal Strata for Classical Discriminant

Let us turn back to the reduced Eq. (5) and let $n = 4$. Consider the reduced discriminantal set ∇_{04} for this equation. According to Theorems 2 and 3, its stratum $\mathscr{M}_{04}^3$ is parametrized by the restriction of

$$\Psi : \mathbb{CP}^2 \to \nabla_{04} \subset \mathbb{C}^3$$

to the complex line of its critical points

$$L^1 = L_1 = \{(s_1 : s_2 : s_3) :\ 1 \cdot 3 \cdot s_1 + 2 \cdot 2 \cdot s_2 + 3 \cdot 1 \cdot s_3 = 0\},$$

where Ψ is defined by formula (6) for $n = 4$. Choosing s_1 as an affine coordinate in L_1, we see that the restriction $\Psi\big|_{L^1}$ is

$$x_1 = -\frac{8s_1}{3s_1-1}\left(\frac{3s_1-1}{-s_1+3}\right)^{\frac{1}{4}}$$
$$x_2 = 2\frac{3s_1+3}{3s_1-1}\left(\frac{3s_1-1}{-s_1+3}\right)^{\frac{1}{2}}$$
$$x_3 = -\frac{8}{3s_1-1}\left(\frac{3s_1-1}{-s_1+3}\right)^{\frac{3}{4}}.$$

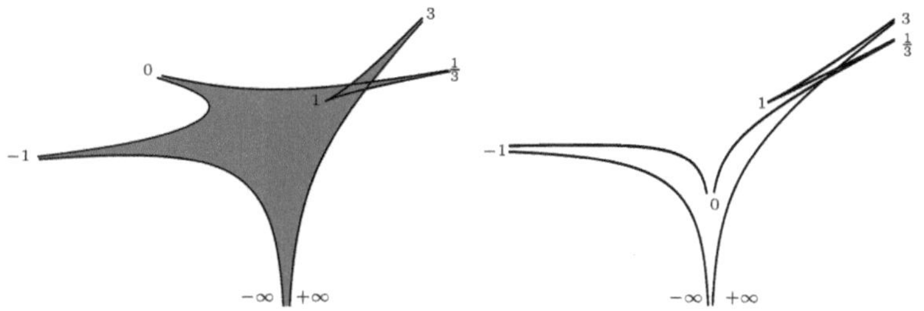

Fig. 4 The amoeba for the reduced stratum $\mathscr{M}^3_{04}$ (*left*) and its contour from a different angle (*right*)

The amoeba and its contour for the stratum $\mathscr{M}^3$, which admits this parametrization, is depicted on Fig. 4. One can see that the tentacles of the amoeba correspond to the values

$$s_1 = -\infty, \ -1, \ 0, \ \frac{1}{3}, \ 3.$$

The value $s_1 = 1$ corresponds to the zero-dimensional stratum $\mathscr{M}^4_{04}$; this is a critical point of the parametrization. Thus, the contour of the amoeba for the zero-dimensional stratum $\mathscr{M}^4_{04}$ is a cuspidal point for the contour of the amoeba for the one-dimensional stratum $\mathscr{M}^3_{04}$ attached to it.

One has to be subtle when studying the attachement of the contours of the amoebas for the strata $\mathscr{M}^2_{04} = \nabla_{04}$ and $\mathscr{M}^3_{04}$. In the affine coordinates s_1, s_2 of $\mathbb{CP}^2$ the parametrization Ψ for $\mathscr{M}^2 = \tilde{\nabla}$ looks like

$$x_1 = \frac{-4s_1}{3s_1+2s_2+1}\left(\frac{3s_1+2s_2+1}{s_1+2s_2+3}\right)^{\frac{1}{4}}$$
$$x_2 = \frac{-4s_2}{3s_1+2s_2+1}\left(\frac{3s_1+2s_2+1}{s_1+2s_2+3}\right)^{\frac{1}{2}}$$
$$x_3 = \frac{-4}{3s_1+2s_2+1}\left(\frac{3s_1+2s_2+1}{s_1+2s_2+3}\right)^{\frac{3}{4}}.$$

To draw the contour of the amoeba we need to compute the image of $\mathbb{R}^2 \subset \mathbb{RP}^2$ under the map $\mathrm{Log} \circ \Psi$. This map has four polar singularities on four lines (the fifth line $s_3 = 0$ lies at infinity of the chosen affine space):

$$s_1 = 0, \ s_2 = 0, \ 3s_1 + 2s_2 + 1 = 0, \ s_1 + 2s_2 + 3 = 0.$$

The contour of the amoeba for the stratum $\mathscr{M}^3_{04}$ is a curve of cuspidal points for the contour of the amoeba for the whole ∇_{04}, as shown on Fig. 5 (left). In a neighborhood of the edge of the contour of the amoeba for $\mathscr{M}^3_{04}$, which corresponds to $s_1 = 1$, the attachment of the contours forms 'the swallowtail'. It should be noticed that the contour of the amoeba of the discriminantal set contains the logarithmic image of the real part of the discriminantal set, which is the object of study in singularity theory.

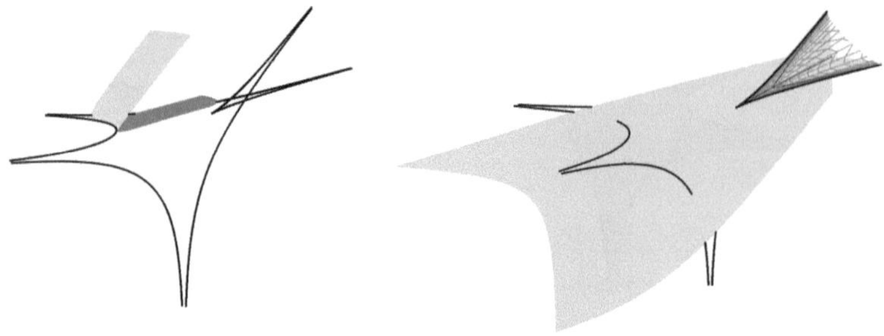

Fig. 5 Attachment of the contour of the amoeba for ∇_{04} to the contour of the amoeba for $\mathscr{M}_{04}^{3}$

The contour of the amoeba, however, is significantly larger, and its stratification is more complex.

Let us make now some observations based on studying the equation of degree 4. The contours of the amoebas for strata $\mathscr{M}_{04}^{3}$ and $\mathscr{M}_{04}^{2} = \nabla_{04}$ are parametrized by the restrictions of parameterizations $\Psi\Big|_{L^1}$ and $\Psi\Big|_{L^0} = \Psi$ (here $L^0 = \mathbb{CP}^2$) on the real parts of the planes L^1 and L^0. The mapping Ψ behaves continuously as the parameter $s \in L^0$ approach $L^1 \setminus L^2$ (where L^2 is the zero-dimensional subspace corresponding to the stratum $\mathscr{M}_{04}^{4}$). Note a sharp contrast of such a 'nice' behavior with the fact that the inverse $\Psi^{-1} : \nabla_{04} \to \mathbb{CP}^2$, which coincides with the logarithmic Gauss map, is not defined at singular points $\mathscr{M}_{04}^{3} \subset \nabla_{04}$. In general, similar arguments prove the following theorem.

Theorem 5 *The contours of the amoebas of all strata* $\mathscr{M}_{0n}^{2} \supset \mathscr{M}_{0n}^{3} \supset \cdots \supset \mathscr{M}_{0n}^{n}$ *are not empty and their preimages under the Log-projection are parametrized by the restrictions of the parametrization (6) to the real parts of the complex planes* $L^0 \supset L^1 \supset \cdots \supset L^{n-2}$.

In conclusion, we would like to raise a question about the distribution of values of the classical Gauss map for amoebas of complex curves $V \in \mathbb{T}^3$. In its smooth points the curve V admits a holomorphic parametrization $z = z(t)$, therefore the mapping $\mathrm{Log}z(t)$, which parametrizes the amoeba $\mathscr{A}_V$, is given by a triple of harmonic functions. If $t = u + iv$ were an isothermal coordinate for $\mathscr{A}_V$, the amoeba would be a minimal surface. According to the result of Fujimoto [Fuj97], the Gauss map for a minimal surface can not omit more than 4 points. In the case of amoebas the situation is quite different.

The line from Example 2 is parametrized by $t = z_1$ and the Gauss map is given by the formula

$$t \mapsto \frac{(-|t|^2,\ 2|t+1|^2,\ -|t+2|^2)}{\sqrt{|t|^4 + 4|t+1|^4 + |t+2|^4}}.$$

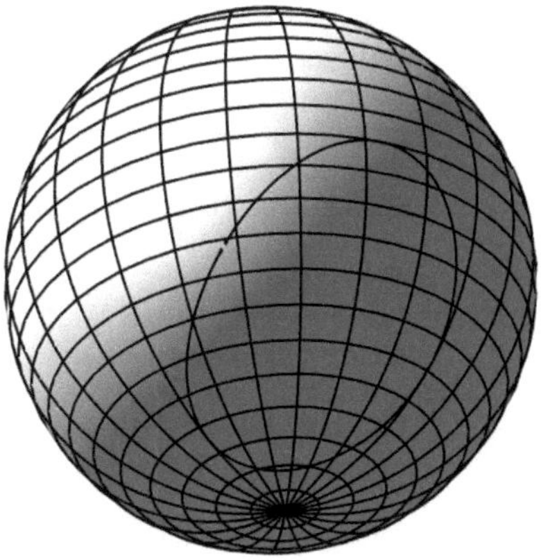

Fig. 6 An oval on the *sphere* is the image of the boundary of the amoeba

The image of the boundary of the amoeba $\mathscr{A}_V$ under this map is shown on Fig. 6. It is a smooth curve on the sphere, and the rest of the amoeba is mapped into the smaller spherical cap bounded by this curve. The Gauss map omits here a dense set of points of S^2.

Acknowledgments The first author is supported by the RFBR, research project no. 14-01-31265 mol_a. The second author is supported by the RFBR, research project no. 14-01-31239 mol_a. The third author is supported by the RFBR, research project no. 14-01-00544_a.
The research for this paper was carried out in Siberian Federal University within the framework of the research project 'Multidimensional Complex Analysis and Differential Equations' funded by the grant of the Russian Federation Government to support scientific research under the supervision of leading scientist, no. 14.Y26.31.0006.

References

[AT12] Antipova, I.A., Tsikh, A.K.: The discriminant locus of a system of n Laurent polynomials in n variables. Izv. Math. **76**(5), 881–906 (2012)
[BKT14] Bushueva, N.A., Kuzvesov, K., Tsikh, A.K.: On the asymptotics of homological solutions to linear multidimensional difference equations. J. Siberian Fed. Univ. Math. Phys. **7**(4), 417–430 (2014)
[BT12] Bushueva, N.A., Tsikh, A.K.: On amoebas of algebraic sets of higher codimension. Proc. Steklov Inst. Math. **279**(1), 52–63 (2012)
[EKL06] Einsiedler, M., Kapranov, M., Lind, D.: Non-archimedean amoebas and tropical varieties. J. Reine Angew. Math. **601**, 139–157 (2006)
[FPT00] Forsberg, M., Passare, M., Tsikh, A.: Laurent determinants and arrangements of hyperplane amoebas. Adv. Math. **151**, 54–70 (2000)
[Fuj97] Fujimoto, H.: Nevanlinna theory and minimal surfaces. Geom. V. Encyclopaedia Math. Sci. **90**, 95–151 (1997)
[GKZ94] Gelfand, I., Kapranov, M., Zelevinsky, A.: Discriminants, Resultants and Multidimensional Determinants. Birkhäuser, Boston (1994)
[Hen04] Henriques, A.: An analogue of convexity for complements of amoebas of varieties of higher codimension, an answer to a question asked by B. Sturmfels. Adv. Geom. **4**(1), 61–73 (2004)

[Hor89] Horn, J.: Über die Konvergenz der hypergeometrischen Riehen zweier und dreier Veranderlichen. Math. Ann. **34**, 544–600 (1889)

[Kap91] Kapranov, M.M.: A characterization of A-discriminantal hypersurfaces in terms of the logarithmic Gauss map. Math. Ann. **290**, 277–285 (1991)

[Kat03] Katz, G.: How tangents solve algebraic equations, or a remarkable geometry of discriminant varieties. Expo. Math. **21**, 219–261 (2003)

[KOS06] Kenyon, R., Okounkov, A., Sheffield, S.: Dimers and amoebae. Ann. Math. **163**, 1019–1056 (2006)

[LPT08] Leinartas, E.K., Passare, M., Tsikh, A.K.: Multidimensional versions of Poincare's theorem for difference equations. Sb. Math. **199**(10), 1505–1521 (2008)

[MT] Mikhalkin, E.N., Tsikh, A.K.: Singular strata of cuspidal type for classical discriminant. Sb. Math. **206**, 282–310 (2015)

[Mik00] Mikhalkin, G.: Real algebraic curves, the moment map and amoebas. Ann. Math. **151**, 309–326 (2000)

[MR01] Mikhalkin, G., Rullgård, H.: Amoebas of maximal area. Internat. Math. Res. Not. **9**, 441–451 (2001)

[PPT13] Passare, M., Pochekutov, D., Tsikh, A.: Amoebas of complex hypersurfaces in statistical thermodynamics. Math. Phys. Anal. Geomy. **16**, 89–108 (2013)

[PR04] Passare, M., Rullgård, H.: Amoebas, Monge-Ampére measures, and triangulations of the Newton polytope. Duke Math. J. **121**, 481–507 (2004)

[PT04] Passare, M., Tsikh, A.: Algebraic equations and hypergeometric series. In the book 'The legacy of Niels Henrik Abel', pp. 653–672 (2004)

[PT05] Passare, M., Tsikh, A.: Amoebas: their spines and their contours. Contemp. Math. **377**, 275–288 (2005)

[PT09] Pochekutov, D., Tsikh, A.K.: On the asymptotic of Laurent coefficients and its application in statistical mechanics. J. Siberian Fed. Univ. Math. Phys. **2**(4), 483–493 (2009)

[Stu02] Sturmfels, B.: Solving systems of polynomial equations. In: CBMS Regional Conferences Series, No. 97. American Mathematical Society, Providence, Rhode Island (2002)

[The02] Theobald, T.: Computing amoebas. Exp. Math. **11**, 513–526 (2002)

A Remark on Hörmander's Isomorphism

Takeo Ohsawa

To the memory of Lars Hörmander

Abstract A finiteness theorem on the bundle-valued L^2 $\bar{\partial}$-cohomology groups is recalled and reproved with some refinement by employing the method of Hörmander [H]. A new connection between the $\bar{\partial}$-cohomology of noncompact manifolds and the problem of extending analytic objects is remarked.

Keywords Weakly 1-complete · L^2 $\bar{\partial}$-cohomology · Extension

1 Introduction

This is a continuation of the author's master thesis [Oh] where the following was proved:

Theorem 1.1 *Let X be a weakly 1-complete manifold of dimension n and let $B \to X$ be a holomorphic line bundle. Assume that B admits a fiber metric whose curvature form is positive outside some compact subset of X. Then*

$$\dim H^{n,q}(X, B) < \infty \quad \textit{for}\, q \geq 1,$$

where $H^{n,q}(X, B)$ denotes the B-valued $\bar{\partial}$-cohomology group of X of type (n, q).

This result is an extension of Grauert's theorem in [G] asserting that the q-th cohomology groups of strongly pseudoconvex domains with coefficients in coherent

---

T. Ohsawa
Graduate School of Mathematics, Nagoya University, Nagoya, Japan
e-mail: ohsawa@math.nagoya-u.ac.jp

F. Bracci et al. (eds.), *Complex Analysis and Geometry*, Springer Proceedings in Mathematics & Statistics 144, DOI 10.1007/978-4-431-55744-9_20

analytic sheaves are finite dimensional for all $q \geq 1$. A significance of the finite-dimensionality lies in that it implies the existence of holomorphic sections of the bundles with prescribed singularities at infinity. Nakano and Rhai [N-R] generalized Theorem 1.1 for holomorphic vector bundles of higher rank, where the positivity assumption on the curvature form is in the sense of Nakano [N]. Recall that X is called weakly 1-complete if there exists a C^∞ plurisubharmonic function $\varphi : X \to \mathbb{R}$ such that the sublevel sets $X_c := \{x \in X; \varphi(x) < c\}$ $(c \in \mathbb{R})$ of φ are relatively compact (i.e. φ is also an exhaustion function on X). In what follows the pair (X, φ) will be referred to as a weakly 1-complete manifold.

For the proof of Theorem 1.1 and its generalization in [N-R], the method of Hörmander [H] is applied. For the case of Theorem 1.1, given a fiber metric of B say h as above, $H^{n,q}(X, B)$ is approximated by the L^2 $\bar{\partial}$-cohomology groups $H^{n,q}_{(2)}(X, B)$ with respect to a fixed Hermitian metric ω which satisfies $\omega = i\Theta_h$ on $X \setminus X_c$ for the curvature form Θ_h of h. More precisely $H^{n,q}_{(2)}(X, B)$ are considered with respect to a family of fiber metrics. Namely, letting $H^{n,q}_{(2),\lambda}(X, B)$ be the L^2 $\bar{\partial}$-cohomology groups with respect to $(\omega, he^{-\lambda(\varphi)})$, Theorem 1.1 is a corollary of the following more intricate assertion.

Theorem 1.2 *Let (X, φ) be a weakly 1-complete manifold of dimension n, let ω be a Hermitian metric on X, and let B be a holomorphic line bundle over X with fiber metric h whose curvature form Θ_h satisfies $\omega = i\Theta_h$ on $X \setminus X_c$ for some $c \in \mathbb{R}$. Then, for any nonconstant C^∞ convex increasing function λ, one can find a positive number $\mu(\lambda)$ such that for all $\mu \geq \mu(\lambda)$ the following are true.*

(1) $\dim H^{n,q}_{(2),\mu\lambda}(X, B) < \infty, \quad q \geq 1.$
(2) The natural restriction homomorphisms

$$H^{n,q}_{(2),\mu\lambda}(X, B) \longrightarrow H^{n,q}_{(2)}(X_c, B), \quad q \geq 1$$

with respect to the restrictions of h and ω are bijective.
(3) The natural inclusion homomorphisms

$$\alpha^q_\mu : H^{n,q}_{(2),\mu\lambda}(X, B) \to H^{n,q}(X, B), \quad q \geq 1$$

are bijective.

Theorem 1.2, which is essentially what we have proved in [Oh], is a higher dimensional analogue of Mittag-Leffler's theorem and Runge's approximation theorem in function theory of one variable. We would like to note that this way of unified generalization is what Oka has meant in [O-1, O-2]. For the reader's convenience, let us mention Hörmander's original result which is closer to Oka's idea, in a slightly modified but equivalent form:

Theorem 1.3 *Let X be a complex manifold of dimension n and let $\phi : X \to \mathbb{R}$ be an exhaustion function of class C^2. Suppose that the Levi form of ϕ has at least*

$n - q + 1$ positive eigenvalues everywhere on $X \setminus X_c$. Then, for any holomorphic vector bundle $E \to X$, the following hold.

(1) For any fiber metric h on E, there exist a Hermitian metric ω on X, a convex increasing function λ and $\mu_0 > 0$ such that

$$\dim H^{0,p}_{(2)}(X, E) < \infty$$

for all $p \geq q$ with respect to ω and $he^{-\mu\lambda(\phi)}$ for any $\mu > 0$, and

$$H^{0,p}_{(2)}(X, E) \cong H^{0,p}(X, E)$$

holds with respect to ω and $he^{-\mu\lambda(\phi)}$ if $\mu \geq \mu_0$ and $p \geq q$.

(2) $H^{0,p}(X, E) \cong H^{0,p}(X_c, E) \cong H^{0,p}_{(2)}(X_c, E)$ for all $p \geq q$.

(3) The image of

$$H^{0,q-1}(X, E) \to H^{0,q-1}(X_c, E)$$

is dense.

Theorem 1.3 is due to Andreotti and Grauert [A-G] except for the assertions on the L^2 cohomology. An advantage of Theorems 1.2 and 1.3 is that one can see a relation between them and the works of Kodaira [K] and Serre [S] on compact complex manifolds. Namely, roughly speaking, the stability of weighted L^2 cohomology groups for sufficiently large μ as above can be regarded as the vanishing of cohomology groups along the divisor at infinity, under some curvature conditions. The purpose of the present article is to make this similarily more explicit by establishing an extension theorem on compact complex manifolds. Let M be a compact complex manifold of dimension n, let D be an effective divisor on M, and let (E, h) be a Hermitian holomorphic vector bundle over M. Let $\mathscr{I}_D$ be the ideal sheaf of D, and put $\mathscr{O}_D = \mathscr{O}_M/\mathscr{I}_D$, where $\mathscr{O}_M$ denotes the structure sheaf of M. By $(|D|, \mathscr{O}_D)$ we denote the complex space whose underlying space is the support $|D|$ of D and structure sheaf is $\mathscr{O}_D$. By K_M and $[D]$ we denote the canonical line bundle of M and the line bundle associated to D, respectively. Then our remark here on Theorem 1.3 is the following.

Theorem 1.4 *Let M, E and D be as above. Suppose that $[D]$ is semipositive and $E|_{|D|}$ is Nakano positive. Then there exists $\mu_0 \in \mathbb{N}$ such that*

(1) $H^0(M, \mathscr{O}_M(K_M \otimes E \otimes [D]^\mu)) \twoheadrightarrow H^0(|D|, \mathscr{O}_D(K_M \otimes E \otimes [D]^\mu))$ if $\mu \geq \mu_0$

and

(2) $H^{n,q}(M, E \otimes [D]^\mu) \cong H^{n,q}(M \setminus |D|, E)$ if $q \geq 1$ and $\mu \geq \mu_0$.

For the reader's convenience, the proof of Theorem 1.2 will be given at first.

2 An L^2 Isomorphism Theorem

Let (X, φ), (B, h), ω and Θ_h be as in the assumption of Theorem 1.2, and let λ_0 be any C^∞ convex increasing function such that

$$\omega_\varepsilon := \omega + \varepsilon i \partial\bar{\partial}(\lambda_0 \circ \varphi) \tag{2.1}$$

is a complete metric on X for any $\varepsilon > 0$. (One may take $\lambda_0(t) = e^t$ for instance.) Here ∂ (resp. $\bar{\partial}$) denotes the complex exterior derivative of type $(1,0)$ (resp. $(0,1)$). Let λ be any nonconstant C^∞ convex increasing function and let $L^{n,q}_{\lambda,\varepsilon}(X, B)(= L^{n,q}_{(2),\lambda,\varepsilon}(X, B))$ denote the Hilbert space of square integrable ($=L^2$) B-valued (n, q)-forms on X with respect to $(\omega_\varepsilon, he^{-(\lambda+\varepsilon\lambda_0)\circ\varphi})$.

By $\bar{\partial} : L^{n,q}_{\lambda,\varepsilon}(X, B) \to L^{n,q+1}_{\lambda,\varepsilon}(X, B)$ we denote the closed extension of $\bar{\partial}$ whose domain of definition is

$$Dom\bar{\partial} \cap L^{n,q}_{\lambda,\varepsilon}(X, B) := \{u \in L^{n,q}_{\lambda,\varepsilon}(X, B);\, \bar{\partial}u \in L^{n,q+1}_{\lambda,\varepsilon}(X, B)\}.$$

We put

$$H^{n,q}_{(2),\lambda,\varepsilon}(X, B) = \frac{\mathrm{Ker}(\bar{\partial} : L^{n,q}_{\lambda,\varepsilon}(X, B) \to L^{n,q+1}_{\lambda,\varepsilon}(X, B))}{\mathrm{Im}(\bar{\partial} : L^{n,q-1}_{\lambda,\varepsilon}(X, B) \to L^{n,q}_{\lambda,\varepsilon}(X, B))}.$$

Let $L^{n,q}_{(2)}(X_c, B)$ and $H^{n,q}_{(2)}(X_c, B)$ be defined similarly with respect to the restrictions of ω and h. Let $\bar{\partial}^*(= \bar{\partial}^*_{\lambda,\varepsilon}) : L^{n,q+1}_{\lambda,\varepsilon}(X, B) \to L^{n,q}_{\lambda,\varepsilon}(X, B)$ denote the adjoint of $\bar{\partial}$ and put

$$\mathscr{H}^{n,q}_{\lambda,\varepsilon} = \mathrm{Ker}\bar{\partial} \cap \mathrm{Ker}\bar{\partial}^* \cap L^{n,q}_{\lambda,\varepsilon}(X, B).$$

Similarly we define

$$\bar{\partial} : L^{n,q}_{(2)}(X_c, B) \to L^{n,q+1}_{(2)}(X_c, B),$$

$$\bar{\partial}^*(= \bar{\partial}^*_c) : L^{n,q+1}_{(2)}(X_c, B) \to L^{n,q}_{(2)}(X_c, B)$$

and

$$\mathscr{H}^{n,q}_c = \mathrm{Ker}\bar{\partial} \cap \mathrm{Ker}\bar{\partial}^* \cap L^{n,q}_{(2)}(X_c, B).$$

Finite-dimensionality theorems by Hörmander's method are based on the following.

Proposition 2.1 *Under the above situation, there exists a constant* $C > 0$ *which does not depend on the choices of* λ_0, λ *and* ε *such that*

$$\|u\|^2 \leq C\{\|\bar{\partial}u\|^2 + \|\bar{\partial}^*u\|^2 + \int_{X_{c+1}} |u|^2 e^{-(\lambda+\varepsilon\lambda_0)\circ\varphi}\omega_\varepsilon^n\} \tag{2.2}$$

holds for any $u \in L^{n,q}_{\lambda,\varepsilon}(X, B) \cap \mathrm{Dom}\bar{\partial} \cap \mathrm{Dom}\bar{\partial}^*$ *with* $q \geq 1$. *Here* $\|\cdot\|$ *denotes the* L^2 *norm with respect to* $(\omega_\varepsilon, he^{-(\lambda+\varepsilon\lambda_0)(\varphi)})$ *and* $|u|$ *the length of* u *with respect to* ω_ε *and* h.

Corollary 1 $\dim \mathscr{H}^{n,q}_{\lambda,\varepsilon} < \infty$ *and* $H^{n,q}_{(2),\lambda,\varepsilon}(X, B) \cong \mathscr{H}^{n,q}_{\lambda,\varepsilon}$ *hold if* $q \geq 1$.

Corollary 2 *There exists* μ_0 *such that for any* $\mu' \geq \mu \geq \mu_0$ *and* $\varepsilon > 0$ *the natural inclusion homomorphisms*

$$H^{n,q}_{(2),\mu\lambda,\varepsilon}(X, B) \to H^{n,q}_{(2),\mu'\lambda,\varepsilon}(X, B), \quad q \geq 1$$

are injective.

Proof of Proposition 2.1. By the assumption on the curvature form of h, for any compactly supported C^∞ B-valued (n, q)-form u such that $\mathrm{supp}u \Subset X \setminus X_c$ and $q \geq 1$,

$$\|u\|^2 \leq \|\bar{\partial}u\|^2 + \|\bar{\partial}^*u\|^2$$

holds with respect to ω_ε and $he^{-(\lambda+\varepsilon\lambda_0)\circ\varphi}$, by Kodaira-Nakano's identity on Kähler manifolds. Hence, for any u in $L^{n,q}_{\lambda,\varepsilon}(X, B) \cap \mathrm{Dom}\bar{\partial} \cap \mathrm{Dom}\bar{\partial}^*$ with $q \geq 1$, by multiplying a C^∞ cut-off function χ with $\mathrm{supp}\chi \subset X \setminus X_c$ and $\mathrm{supp}(1-\chi) \subset X_{c+1}$ to u and approximating χu by C^∞ forms, one has (2.2) by the completeness of ω_ε. C depends only on the gradient of χ. □

Proof of Corollary 1. From (2.2) and Rellich's lemma, it follows that, for any sequence u_μ ($\mu = 1, 2, \ldots$) in $L^{n,q}_{\lambda,\varepsilon}(X, B)$ ($q \geq 1$) such that $u_\mu \perp \mathscr{H}^{n,q}_{\lambda,\varepsilon}$, $\|u_\mu\| = 1$, $\|\bar{\partial}u_\mu\| \to 0$ and $\|\bar{\partial}^*u_\mu\| \to 0$, one can choose a strongly convergent subsequence, once we fix λ and ε. Therefore, there has to exist a constant C_0 such that

$$\|u\|^2 \leq C_0\{\|\bar{\partial}u\|^2 + \|\bar{\partial}^*u\|^2\}$$

holds for any $u \in (L^{n,q}_{\lambda,\varepsilon}(X, B) \ominus \mathscr{H}^{n,q}_{\lambda,\varepsilon}) \cap \mathrm{Dom}\bar{\partial} \cap \mathrm{Dom}\bar{\partial}^*$ with $q \geq 1$, where $\ominus$ denotes the orthocomplement. Indeed, if there were no such C_0, one would have a sequence $u_\mu \in (L^{n,q}_{\lambda,\varepsilon}(X, B) \ominus \mathscr{H}^{n,q}_{\lambda,\varepsilon}) \cap \mathrm{Dom}\bar{\partial} \cap \mathrm{Dom}\bar{\partial}^*$, $q \geq 1$, such that $\|u_\mu\| = 1$, $\|\bar{\partial}u_\mu\| \to 0$ and $\|\bar{\partial}^*u_\mu\| \to 0$. Then the limit, say u, of a subsequence of u_μ would satisfy $\|u\| = 1$, $\bar{\partial}u = \bar{\partial}^*u = 0$ and $u \perp \mathscr{H}^{n,q}_{\lambda,\varepsilon}$, which is an absurdity. Hence the assertion follows by a standard argument using Hahn-Banach's theorem. □

Proof of Corollary 2. Let λ_0 be as in (2.1). We may assume in advance that $\mathrm{supp}\lambda_0 = \mathrm{supp}\lambda = (-\infty, c]$ because for any convex increasing function λ_1 one can find a convex increasing function λ_2 such that $\mathrm{supp}\lambda_2 = (-\infty, c]$ and $|\lambda_1 - \lambda_2|$ is bounded. Then, similarly as in the proof of Corollary 1, one can find $\mu_0 = \mu_0(\lambda) > 0$ and a constant C_1 such that, for any $\varepsilon > 0$ and $\mu \geq \mu_0$,

$$\|u\|^2 \leq C_1\{\|\bar{\partial}u\|^2 + \|\bar{\partial}^* u\|^2\}$$

holds for any $u \in L^{n,q}_{\mu\lambda,\varepsilon}(X, B) \cap \mathrm{Dom}\bar{\partial} \cap \mathrm{Dom}\bar{\partial}^*$ satisfying

$$\int_{X_c} \langle u, v\rangle \omega^n = 0$$

for all $v \in \mathscr{H}_c^{n,q}$, where $\langle\cdot, \cdot\rangle$ denotes the pointwise inner product with respect to ω and h. Hence, by Hahn-Banach again, we obtain the desired conclusion. □

Proof of Theorem 1.2. Since this part is not used in the proof of Theorem 1.4, and moreover the argument is essentially the repetition of the above, we shall only give a sketch. Let μ_0 be as above. Then, similarly as above, the natural homomorphisms

$$\beta^q_{\mu,\varepsilon} : H^{n,q}_{(2),\mu\lambda,\varepsilon}(X, B) \to H^{n,q}_{(2)}(X_c, B), \ q \geq 1$$

are injective for all $\mu \geq \mu_0$, $\dim H^{n,q}_{(2)}(X_c, B) < \infty$ for $q \geq 1$, and

$$\overline{(\cup_{\mu=1}^{\infty}\mathrm{Im}\beta^q_{\mu,\varepsilon})} = H^{n,q}_{(2)}(X_c, B)$$

for all $q \geq 0$.

Therefore, there exists $\mu_1 \geq \mu_0$ such that $\beta^q_{\mu,\varepsilon}$ are bijective if $q \geq 1, \mu \geq \mu_1$ and $\varepsilon > 0$. Since C_1 is independent of μ and ε, we are allowed to let $\varepsilon \to 0$ and obtain the bijectivity of $\beta^q_{\mu,0}$ for $\mu \geq \mu_1$ as well. Similarly, it is easy to see that α^q_μ are bijective if $\mu \geq \mu_1$. Hence it suffices to put $\mu(\lambda) = \mu_1$. □

As is easily seen from the above proof, given a Hermitian holomorphic vector bundle (E, h_E) over X, the conclusions of Theorem 1.2 are also valid for the E-valued L^2 $\bar{\partial}$-cohomology groups, provided that there exists a Hermitian metric ω on X with $\mathrm{supp}(d\omega) \subset X_c$ such that the curvature form Θ_{h_E} of h_E satisfies

$$i\Theta_{h_E} - Id_E \otimes \omega \geq 0$$

on $X \setminus X_c$ in the sense of Nakano, since Proposition 2.1 holds for (E, h_E) then. This is the reason why Theorem 1.4 is stated for vector bundles of any rank. In the proof of Theorem 1.4 below, we shall use the above notations replacing (B, h) by (E, h_E).

3 Extension and Isomorphism on Compact Manifolds

Let M, E and D be as in the assumption of Theorem 1.4, and let s be a canonical section of $[D]$. Since $[D]$ is semipositive, there exists a fiber metric of $[D]$ for which the function $-\log|s|$ is plurisubharmonic on $M \setminus |D|$. Here $|s|$ denotes the length of s. Since E is Nakano positive along $|D|$, there exists a fiber metric h_E of E and a neighborhood $U \supset |D|$ such that (E, h_E) is Nakano positive on U. In this situation, one can find a Hermitian metric ω_M on M such that

$$\|u\|^2 \leq \|\bar{\partial}u\|^2 + \|\bar{\partial}^* u\|^2$$

holds for any compactly supported C^∞ E-valued (n, q)-form u on $U \setminus |D|$, in virtue of Nakano's identity (cf. [N]). Hence, similarly as in the proof of Corollary 2, one can find a complete Hermitian metric ω_ε on $M \setminus |D|$ and $\mu_0 \in \mathbb{N}$ such that the homomorphisms

$$H^{n,1}_{(2),-\mu\log|s|,\varepsilon}(M \setminus |D|, E) \to H^{n,1}_{(2),-\mu'\log|s|,\varepsilon}(M \setminus |D|, E),\ \ \mu' \geq \mu \geq \mu_0 - 1$$

are injective. Given any $f \in H^0(|D|, K_M \otimes [D]^\mu \otimes E)$, let $\tilde{f}$ be any extension of f to M as a C^∞ section of $K_M \otimes [D]^\mu \otimes E$ such that $\bar{\partial}\tilde{f}|_{|D|} = 0$. Then

$$\frac{\bar{\partial}\tilde{f}}{s^\mu} \in L^{n,1}_{-(\mu-1)\log|s|,\varepsilon}(M \setminus |D|, E) \cap \mathrm{Ker}\bar{\partial}.$$

Since $\bar{\partial}(\tilde{f}/s^\mu) = \bar{\partial}\tilde{f}/s^\mu$ and $\tilde{f}/s^\mu \in L^{n,0}_{-\mu\log|s|,\varepsilon}(M \setminus |D|, E)$, the above injectivity implies that there exists $g \in L^{n,0}_{-(\mu-1)\log|s|,\varepsilon}(M \setminus |D|, E)$ such that $\bar{\partial}g = \bar{\partial}\tilde{f}/s^\mu$ holds on $M \setminus |D|$. Then $\tilde{f} - s^\mu g$ is the desired extension of f in $H^0(M, K_M \otimes [D]^\mu \otimes E)$, which is the end of the proof of 1). The proof of 2) is similar as in Theorem 1.2. □

4 Additional Remarks

Corollary of Theorem 1.4. Let M be a compact complex manifold. If there exist an effective divisor D on M and a holomorphic line bundle $B \to M$ such that $[D] \geq 0$ and $B|_{|D|} > 0$. Then M is a Moishezon manifold.

In particular, by a theorem of Chow and Kodaira [C-K], M is projective algebraic if moreover $\dim M = 2$. If there exists a compact Riemann surface S of genus $g \geq 2$ and a holomorphic embedding $p : S \to M$ such that $[p(S)] \geq 0$, one may take K_M or $[p(S)]$ as B. In other words, *a compact complex surface containing a semipositively embedded irreducible smooth curve of genus ≥ 2 is algebraic*. Therefore, in the case

$\deg[p(S)]|_{p(S)} = 0$, it might be an interesting question whether or not there eixsts a constant μ_2 independent of the choice of M, S and p such that the restriction maps

$$H^0(M, \mathcal{O}_M(K_M^{\mu})) \to H^0(p(S), \mathcal{O}_{p(S)}(K_M^{\mu}))$$

are surjective for all $\mu \geq \mu_2$.

References

[A-G] Andreotti, A., Grauert, H.: Théorème de finitude pour la cohomologie des espaces complexes. Bull. Soc. Math. France **90**, 193–259 (1962)

[C-K] Chow, W.-L., Kodaira, K.: On analytic surfaces with two independent meromorphic functions. Proc. Natl. Acad. Sci. U. S. A. **38**, 319–325 (1952)

[G] Grauert, H.: On Levi's problem and the embedding of real-analytic manifolds. Ann. Math. **68**, 460–472 (1958)

[H] Hörmander, L.: L^2 estimates and existence theorems for the $\bar{\partial}$ operator. Acta Math. **113**, 89–152 (1965)

[K] Kodaira, K.: On a differential-geometric method in the theory of analytic stacks. Proc. Natl. Acad. Sci. U. S. A. **39**, 1268–1273 (1953)

[N] Nakano, S.: On complex analytic vector bundles. J. Math. Soc. Jpn. **7**, 1–12 (1955)

[N-R] Nakano, S., Rhai, T.S.: Vector bundle version of Ohsawa's finiteness theorems. Math. Jpn. **24**, 657–664 (1980)

[Oh] Ohsawa, T.: Finiteness theorems on weakly 1-complete manifolds. Publ. RIMS, Kyoto Univ. **15**, 853–870 (1979)

[O-1] Oka, K.: Domaines convexes par rapport aux fonctions rationelles. J. Sci. Hiroshima Univ. **6**, 245–255 (1936)

[O-2] Oka, K.: Domaines pseudoconvexes. Tôhoku Math. J. **49**, 15–52 (1942)

[S] Serre, J.-P.: Géométrie algébrique et géométrie analytique, Université de Grenoble. Annales de l'Institut Fourier **6**, 1–42 (1956)

The Julia-Wolff-Carathéodory Theorem and Its Generalizations

Jasmin Raissy

Abstract This note is a short introduction to the Julia-Wolff-Carathéodory theorem, and its generalizations in several complex variables, up to very recent results for infinitesimal generators of semigroups.

Keywords Infinitesimal generators · Semigroups of holomorphic mappings · Julia-Wolff-Carathéodory theorem · Boundary behaviour

1 The Classical Julia-Wolff-Carathéodory Theorem

One of the classical result in one-dimensional complex analysis is Fatou's theorem:

Theorem 1.1 (Fatou [Fa]) *Let $f\colon \Delta \to \Delta$ be a holomorphic self-map of the unit disk $\Delta \subset \mathbb{C}$. Then f admits non-tangential limit at almost every point of $\partial\Delta$.*

This result however does not give precise information about the behavior at a specific point σ of the boundary. Of course, to obtain a more precise statement in this case some hypotheses on f are needed. In fact, as it was found by Julia [Ju1] in 1920, the right hypothesis is to assume that $f(\zeta)$ approaches the boundary of Δ at least as fast as ζ, in a weak sense. More precisely, we have the classical *Julia's lemma:*

Theorem 1.2 (Julia [Ju1]) *Let $f\colon \Delta \to \Delta$ be a bounded holomorphic function such that*

$$\liminf_{\zeta\to\sigma} \frac{1-|f(\zeta)|}{1-|\zeta|} = \alpha < +\infty \tag{1.1}$$

for some $\sigma \in \partial\Delta$. Then f has non-tangential limit $\tau \in \partial\Delta$ at σ. Moreover, for all $\zeta \in \Delta$ one has

---

J. Raissy(✉)
Institut de Mathématiques de Toulouse; UMR5219,
Université de Toulouse, CNRS,UPS IMT, 31062 Toulouse Cedex 9, France
e-mail: jraissy@math.univ-toulouse.fr

F. Bracci et al. (eds.), *Complex Analysis and Geometry*, Springer Proceedings in Mathematics & Statistics 144, DOI 10.1007/978-4-431-55744-9_21

$$\frac{|\tau - f(\zeta)|^2}{1 - |f(\zeta)|^2} \le \alpha \frac{|\sigma - \zeta|^2}{1 - |\zeta|^2} . \tag{1.2}$$

The latter statement admits an interesting geometrical interpretation. The *horocycle* $E(\sigma, R)$ contained in Δ of *center* $\sigma \in \partial\Delta$ and *radius* $R > 0$ is the set

$$E(\sigma, R) = \left\{ \zeta \in \Delta \;\middle|\; \frac{|\sigma - \zeta|^2}{1 - |\zeta|^2} < R \right\} .$$

Geometrically, $E(\sigma, R)$ is an euclidean disk of radius $R/(R+1)$ internally tangent to $\partial\Delta$ at σ. Therefore (1.2) becomes $f\big(E(\sigma, R)\big) \subseteq E(\tau, \alpha R)$ for all $R > 0$, and the existence of the non-tangential limit more or less follows from (1.2) and from the fact that horocycles touch the boundary in exactly one point.

A horocycle can be thought of as the limit of Poincaré disks of fixed euclidean radius and centers going to the boundary; so it makes sense to think of horocycles as Poincaré disks centered at the boundary, and of Julia's lemma as a Schwarz-Pick lemma at the boundary. This suggests that α might be considered as a sort of dilation coefficient: f expands horocycles by a ratio of α. If σ were an internal point and $E(\sigma, R)$ an infinitesimal euclidean disk actually centered at σ, one then would be tempted to say that α is (the absolute value of) the derivative of f at σ. This is exactly the content of the classical *Julia-Wolff-Carathéodory theorem:*

Theorem 1.3 (Julia-Wolff-Carathéodory) *Let $f\colon \Delta \to \Delta$ be a bounded holomorphic function such that*

$$\liminf_{\zeta \to \sigma} \frac{1 - |f(\zeta)|}{1 - |\zeta|} = \alpha < +\infty$$

for some $\sigma \in \partial\Delta$, and let $\tau \in \partial\Delta$ be the non-tangential limit of f at σ. Then both the incremental ratio $\big(\tau - f(\zeta)\big)/(\sigma - \zeta)$ and the derivative $f'(\zeta)$ have non-tangential limit $\alpha\bar{\sigma}\tau$ at σ.

So condition (1.1) forces the existence of the non-tangential limit of both f and its derivative at σ. This is the result of the work of several people: Julia [Ju2], Wolff [Wo], Carathéodory [C], Landau and Valiron [L-V], Nevanlinna [N] and others. We refer, for example, to [B] and [A1] for proofs, history and applications.

2 Generalizations to Several Variables

It was first remarked by Korányi and Stein [Ko, K-S, St] in extending Fatou's theorem to several complex variables, that the notion of non-tangential limit is not the right one to consider for domains in $\mathbb{C}^n$. In fact, it turns out that two notions are needed, and to introduce them it is useful to investigate the notion of non-tangential limit in the unit disk Δ.

The non-tangential limit can be defined in two equivalent ways. A function $f\colon \Delta \to \mathbb{C}$ is said to have *non-tangential limit* $L \in \mathbb{C}$ at $\sigma \in \partial\Delta$ if $f\bigl(\gamma(t)\bigr) \to L$ as $t \to 1^-$ for every curve $\gamma\colon [0,1) \to \Delta$ such that $\gamma(t)$ converges to σ non-tangentially as $t \to 1^-$. In $\mathbb{C}$, this is equivalent to having that $f(\zeta) \to L$ as $\zeta \to \sigma$ staying inside any *Stolz region* $K(\sigma, M)$ of *vertex* σ and *amplitude* $M > 1$, where

$$K(\sigma, M) = \left\{ \zeta \in \Delta \;\middle|\; \frac{|\sigma - \zeta|}{1 - |\zeta|} < M \right\} ,$$

since Stolz regions are angle-shaped nearby the vertex σ, and the angle is going to π as $M \to +\infty$. These two approaches lead to different notions in several variables.

In the unit ball $B^n \subset \mathbb{C}^n$ the natural generalization of a Stolz region is the *Korányi region* $K(p, M)$ of *vertex* $p \in \partial B^n$ and *amplitude* $M > 1$ given by

$$K(p, M) = \left\{ z \in B^n \;\middle|\; \frac{|1 - \langle z, p\rangle|}{1 - \|z\|} < M \right\} ,$$

where $\|\cdot\|$ denote the euclidean norm and $\langle \cdot\,, \cdot\rangle$ the canonical hermitian product. Then a function $f\colon B^n \to \mathbb{C}$ has K-*limit* (or *admissible limit*) $L \in \mathbb{C}$ at $p \in \partial B^n$, and we write

$$K - \lim_{z \to p} f(z)$$

if $f(z) \to L$ as $z \to p$ staying inside any Korányi region $K(\sigma, M)$. A Korányi region $K(p, M)$ approaches the boundary non-tangentially along the normal direction at p but tangentially along the complex tangential directions at p. Therefore, having K-limit is stronger than having non-tangential limit. However, as first noticed by Korányi and Stein, for holomorphic functions of several complex variables one is often able to prove the existence of K-limits. For instance, the best generalization of Julia's lemma to B^n is the following result (proved by Hervé [H] in terms of non-tangential limits and by Rudin [R] in general):

Theorem 2.1 (Rudin [R]) *Let $f\colon B^n \to B^m$ be a holomorphic map such that*

$$\liminf_{z \to p} \frac{1 - \|f(z)\|}{1 - \|z\|} = \alpha < +\infty ,$$

for some $p \in \partial B^n$. Then f admits K-limit $q \in \partial B^m$ at p, and furthermore for all $z \in B^n$ one has

$$\frac{|1 - \langle f(z), q\rangle|^2}{1 - \|f(z)\|^2} \le \alpha \frac{|1 - \langle z, p\rangle|^2}{1 - \|z\|^2} .$$

To define Korányi regions for more general domains in $\mathbb{C}^n$ than the unit ball, we need to briefly recall the definition of the Kobayashi distance (we refer, e.g., to [A1, JP] and [Ko] for details and much more). We denote by k_Δ the Poincaré distance on the unit disk $\Delta \subset \mathbb{C}$. Given X a complex manifold, the *Lempert function* $\delta_X : X \times X \to \mathbb{R}^+$ of X is defined as

$$\delta_X(z, w) = \inf\{k_\Delta(\zeta, \eta) \mid \exists \phi\colon \Delta \to X \text{ holomorphic, with } \phi(\zeta) = z \text{ and } \phi(\eta) = w\}$$

for all $z, w \in X$. The *Kobayashi pseudodistance* $k_X : X \times X \to \mathbb{R}^+$ of X is then defined as the largest pseudodistance on X bounded above by δ_X. The manifold X is called *(Kobayashi) hyperbolic* if k_X is indeed a distance; X is called *complete hyperbolic* if k_X is a complete distance.

The main property of the Kobayashi (pseudo)distance is that it is contracted by holomorphic maps: if $f : X \to Y$ is a holomorphic map then

$$\forall z, w \in X \qquad k_Y\big(f(z), f(w)\big) \le k_X(z, w) .$$

In particular, the Kobayashi distance is invariant under biholomorphisms.

It is easy to see that the Kobayashi distance of the unit disk coincides with the Poincaré distance. Furthermore, the Kobayashi distance of the unit ball $B^n \subset \mathbb{C}^n$ coincides with the Bergman distance (see, e.g., [A1, Corollary 2.3.6]); and the Kobayashi distance of a bounded convex domain coincides with the Lempert function (see, e.g., [A1, Proposition 2.3.44]). Moreover, the Kobayashi distance of a bounded convex domain D is complete [A1, Proposition 2.3.45], and thus for each $p \in D$ we have that $k_D(p, z) \to +\infty$ if and only if z tends to the boundary ∂D.

Using the Kobayashi intrinsic distance we obtain the natural generalization to complete hyperbolic domains of Korányi regions of the balls.

Let $D \subset\subset \mathbb{C}^n$ be a complete hyperbolic domain and denote by k_D its Kobayashi distance. A *K-region* of *vertex* $x \in \partial D$, *amplitude* $M > 1$, and *pole* $z_0 \in D$ is the set

$$K_{D,z_0}(x, M) = \left\{ z \in D \mid \limsup_{w \to x} [k_D(z, w) - k_D(z_0, w)] + k_D(z_0, z) < \log M \right\}.$$

This definition clearly depends on the pole z_0. However, this dependence is not too relevant since changing the pole corresponds to shifting amplitudes. Moreover, it is elementary to check that in the unit ball K-regions coincide with Korányi regions, $K_{B^n,0}(x, M) = K(x, M)$. Therefore K-regions are a natural generalization of Korányi regions allowing us to generalize the notion of K-limit. A function $f : D \to \mathbb{C}^m$ has *K-limit* L at $x \in \partial D$ if $f(z) \to L$ as $z \to p$ staying inside any K-region of vertex x. The best generalization of Julia's lemma in this setting is then the following, due to Abate:

Theorem 2.2 (Abate [A2]) *Let $D \subset\subset \mathbb{C}^n$ be a complete hyperbolic domain and let $z_0 \in D$. Let $f : D \to \Delta$ be a holomorphic function and let $x \in \partial D$ be such that*

$$\liminf_{z \to x} [k_D(z_0, z) - k_\Delta(0, f(z))] < +\infty .$$

Then f admits K-limit $\tau \in \partial D$ at x.

In order to obtain a complete generalization of the Julia-Wolff-Carathéodory for B^n, Rudin discovered that he needed a different notion of limit, still stronger than non-tangential limit but weaker than K-limit. This notion is closely related to the other characterization of non-tangential limit in one variable we mentioned at the beginning of this section.

A crucial one-variable result relating limits along curves and non-tangential limits is *Lindelöf's theorem.* Given $\sigma \in \partial\Delta$, a *$\sigma$-curve* is a continuous curve $\gamma\colon [0,1) \to \Delta$ such that $\gamma(t) \to \sigma$ as $t \to 1^-$. Then Lindelöf [Li] proved that if a bounded holomorphic function $f\colon \Delta \to \mathbb{C}$ admits limit $L \in \mathbb{C}$ along a given σ-curve then it admits limit L along all non-tangential σ-curves — and thus it has non-tangential limit L at σ.

In generalizing this result to several complex variables, Čirka [Č] realized that for a bounded holomorphic function the existence of the limit along a (suitable) p-curve (where $p \in \partial B^n$) implies not only the existence of the non-tangential limit, but also the existence of the limit along any curve belonging to a larger class of curves, including some tangential ones — but it does not in general imply the existence of the K-limit. To describe the version (due to Rudin [R]) of Čirka's result we shall state in this survey, let us introduce a bit of terminology.

Let $p \in \partial B^n$. As before, a *p-curve* is a continuous curve $\gamma\colon [0,1) \to B^n$ such that $\gamma(t) \to p$ as $t \to 1^-$. A p-curve is *special* if

$$\lim_{t\to 1^-} \frac{\|\gamma(t) - \langle\gamma(t), p\rangle p\|^2}{1 - |\langle\gamma(t), p\rangle|^2} = 0\,; \tag{2.1}$$

and, given $M > 1$, it is *M-restricted* if

$$\frac{|1 - \langle\gamma(t), p\rangle|}{1 - |\langle\gamma(t), p\rangle|} < M$$

for all $t \in [0,1)$. We also say that γ is *restricted* if it is M-restricted for some $M > 1$. In other words, γ is restricted if and only if $t \mapsto \langle\gamma(t), p\rangle$ goes to 1 non-tangentially in Δ.

It is not difficult to see that non-tangential curves are special and restricted; on the other hand, a special restricted curve approaches the boundary non-tangentially along the normal direction, but it can approach the boundary tangentially along complex tangential directions. Furthermore, a special M-restricted p-curve is eventually contained in any $K(p, M')$ with $M' > M$, and conversely a special p-curve eventually contained in $K(p, M)$ is M-restricted. However, $K(p, M)$ can contain p-curves that are restricted but not special: for these curves the limit in (2.1) might be a strictly positive number.

With these definitions in place, we shall say that a function $f\colon B^n \to \mathbb{C}$ has *restricted K-limit* (or *hypoadmissible limit*) $L \in \mathbb{C}$ at $p \in \partial B^n$, and we shall write

$$K' - \lim_{z \to p} f(z) = L ,$$

if $f\big(\gamma(t)\big) \to L$ as $t \to 1^-$ for any special restricted p-curve $\gamma\colon [0,1) \to B^n$. It is clear that the existence of the K-limit implies the existence of the restricted K-limit, that in turns implies the existence of the non-tangential limit; but none of these implications can be reversed (see, e.g., [R] for examples in the ball).

Finally, we say that a function $f\colon B^n \to \mathbb{C}$ is *K-bounded* at $p \in \partial B^n$ if it is bounded in any Korányi region $K(p, M)$, where the bound can depend on $M > 1$. Then Rudin's version of Čirka's generalization of Lindelöf's theorem is the following:

Theorem 2.3 (Rudin [R]) *Let $f\colon B^n \to \mathbb{C}$ be a holomorphic function K-bounded at $p \in \partial B^n$. Assume there is a special restricted p-curve $\gamma^o\colon [0,1) \to B^n$ such that $f\big(\gamma^o(t)\big) \to L \in \mathbb{C}$ as $t \to 1^-$. Then f has restricted K-limit L at p.*

As before, it is possible to generalize this approach to a domain $D \subset \mathbb{C}^n$ different from the ball. A very precise and systematic presentation, providing clear proofs, details and examples, of various aspects of the problem of generalization of the classical Julia-Wolff-Carathéodory theorem to domains in several complex variables, and updated until 2004, can be found in [A6].

For the sake of simplicity we state here only the definitions needed to state Abate's version of Lindelöf's theorem in this setting. Given a point $x \in \partial D$, a *x-curve* is again a continuous curve $\gamma\colon [0,1) \to D$ so that $\lim_{t\to 1^-} \gamma(t) = x$. A *projection device* at $x \in \partial D$ is the data of: a neighbourhood U of x in $\mathbb{C}^n$, a holomorphic embedded disk $\varphi_x\colon \Delta \to D \cap U$, such that $\lim_{\zeta\to 1} \varphi_x(\zeta) = x$, a family $\mathcal{P}$ of x-curves in $D \cap U$, and a device associating to every x-curve $\gamma \in \mathcal{P}$ a 1-curve $\tilde{\gamma}_x$ in Δ, or equivalently a x-curve $\gamma_x = \varphi_x \circ \tilde{\gamma}_x$ in $\varphi_x(\Delta)$. If D is equipped with a projection device at $x \in \partial D$, then a curve $\gamma \in \mathcal{P}$ is *special* if $\lim_{t\to 1^-} k_{D\cap U}(\gamma(t), \gamma_x(t)) = 0$, and it is *restricted* if γ_x is a non-tangential 1-curve in Δ. A function $f\colon D \to C$ has *restricted K-limit $L \in \mathbb{C}$ at x* if $\lim_{t\to 1^-} f(\gamma(t)) = L$ for all special restricted x-curves. A projection device is *good* if: for any $M > 1$ there is a $M' > 1$ so that $\varphi_x(K(1, M)) \subset K_{D\cap U, z_0}(x, M')$, and for any special restricted x-curve γ there exists $M_1 = M_1(\gamma)$ such that $\lim_{t\to 1^-} k_{K_{D\cap U, z_0}(x, M_1)}(\gamma(t), \gamma_x(t)) = 0$. Good projection devices exist, and several examples can be found for example in [A6]. Finally, we say that a function $f\colon D \to \mathbb{C}$ is *K-bounded* at $p \in \partial B^n$ if it is bounded in any K-region $K_{D,z_0}(x, M)$, where the bound can depend on $M > 1$.

With these definitions we can state the generalization of Lindelöf principle given by Abate.

Theorem 2.4 (Abate [A6]) *Let $D \subset \mathbb{C}^n$ be a domain equipped with a good projection device at $x \in \partial D$. Let $f\colon D \to \Delta$ be a holomorphic function K-bounded at x. Assume there is a special restricted x-curve $\gamma^o\colon [0,1) \to D$ such that $f\big(\gamma^o(t)\big) \to L \in \mathbb{C}$ as $t \to 1^-$. Then f has restricted K-limit L at x.*

We can now deal with the generalization of the Julia-Wolff-Carathéodory theorem to several complex variables. With respect to the one-dimensional case there is an obvious difference: instead of only one derivative one has to deal with a whole (Jacobian) matrix of them, and there is no reason they should all behave in the same way. And indeed they do not, as shown in Rudin's version of the Julia-Wolff-Carathéodory theorem for the unit ball:

Theorem 2.5 (Rudin [R]) *Let $f : B^n \to B^m$ be a holomorphic map such that*

$$\liminf_{z \to p} \frac{1 - \|f(z)\|}{1 - \|z\|} = \alpha < +\infty ,$$

for some $p \in \partial B^n$. Then f admits K-limit $q \in \partial B^m$ at p. Furthermore, if we set $f_q(z) = \big(f(z), p\big)q$ and denote by df_z the differential of f at z, we have:

(i) *the function $\big[1 - \big(f(z), q\big)\big]/[1 - (z, p)]$ is K-bounded and has restricted K-limit α at p;*
(ii) *the map $[f(z) - f_q(z)]/[1 - (z, p)]^{1/2}$ is K-bounded and has restricted K-limit O at p;*
(iii) *the function $\big(df_z(p), q\big)$ is K-bounded and has restricted K-limit α at p;*
(iv) *the map $[1 - (z, p)]^{1/2} d(f - f_q)_z(p)$ is K-bounded and has restricted K-limit O at p;*
(v) *if v is any vector orthogonal to p, the function $\big(df_z(v), q\big)/[1 - (z, p)]^{1/2}$ is K-bounded and has restricted K-limit 0 at p;*
(vi) *if v is any vector orthogonal to p, the map $d(f - f_q)_z(v)$ is K-bounded at p.*

In the last twenty years this theorem (as well as Theorems 2.1 and 2.3) has been extended to domains much more general than the unit ball: for instance, strongly pseudoconvex domains [A1, A2, A3], convex domains of finite type [AT], and polydisks [A5] and [AMY], (see also [A6] and references therein).

We end this section with the general version of the Julia-Wolff-Carathódory theorem obtained by Abate in [A6] for a complete hyperbolic domain D in $\mathbb{C}^n$. To formulate it, we need to introduce a couple more definitions. A projection device at $x \in \partial D$ is *geometrical* if there is a holomorphic function $\tilde{p}_x : D \cap U \to \Delta$ such that $\tilde{p}_x \circ \varphi_x = \mathrm{id}_\Delta$ and $\tilde{\gamma}_x = \tilde{p}_x \circ \gamma$ for all $\gamma \in \mathcal{P}$. A geometrical projection device at x is *bounded* if $d(z, \partial D)/|1 - \tilde{p}_x(z)|$ is bounded in $D \cap U$, and $|1 - \tilde{p}_x(z)|/d(z, \partial D)$ is K-bounded in $D \cap U$. The statement is then the following, where κ_D denotes the Kobayashi metric.

Theorem 2.6 (Abate [A6]) *Let $D \subset \mathbb{C}^n$ be a complete hyperbolic domain equipped with a bounded geometrical projection device at $x \in \partial D$. Let $f : D \to \Delta$ be a holomorphic function such that*

$$\liminf_{z \to x} [k_D(z_0, z) - k_\Delta(0, f(z))] = \frac{1}{2} \log \beta < +\infty .$$

Then for every $v \in \mathbb{C}^n$ and every $s \geq 0$ such that $d(z, \partial D)^s \kappa_D(z; v)$ is K-bounded at x the function

$$d(z, \partial D)^{s-1} \frac{\partial f}{\partial v} \tag{2.2}$$

is K-bounded at x. Moreover, if $s > \inf\{s \geq 0 \mid d(z, \partial D)^s \kappa_D(z; v)$ *is K-bounded at* $x\}$, *then* (2.2) *has vanishing K-limit at x.*

Depending on more specific properties of the projection device, it is indeed possible to deduce the existence of restricted K-limits, see [A6, Sect. 5].

Further generalizations of Julia-Wolff-Carathéodory theorem have been obtained in infinite-dimensional Banach and Hilbert spaces, and we refer to [EHRS, ELRS, ERS, F, MM, SW, W11, W12, W13, Z], and references therein.

3 Julia-Wolff-Carathéodory Theorem for Infinitesimal Generators

We conclude this survey focusing on a different kind of generalization in several complex variables: infinitesimal generators of one-parameter semigroups of holomorphic self-maps of B^n.

We consider $\mathrm{Hol}(B^n, B^n)$, the space of holomorphic self-maps of B^n, endowed with the usual compact-open topology. A *one-parameter semigroup* of holomorphic self-maps of B^n is a continuous semigroup homomorphism $\Phi : \mathbb{R}^+ \to \mathrm{Hol}(B^n, B^n)$. In other words, writing φ_t instead of $\Phi(t)$, we have $\varphi_0 = \mathrm{id}_{B^n}$, the map $t \mapsto \varphi_t$ is continuous, and the semigroup property $\varphi_t \circ \varphi_s = \varphi_{t+s}$ holds. An introduction to the theory of one-parameter semigroups of holomorphic maps can be found in [A1, RS2] or [S].

One-parameter semigroups can be seen as the flow of a vector field (see, e.g., [A4]). Given a one-parameter semigroup Φ, it is possible to prove that there exists a holomorphic map $G\colon B^n \to \mathbb{C}^n$, the *infinitesimal generator* of the semigroup, such that

$$\frac{\partial \Phi}{\partial t} = G \circ \Phi\,. \tag{3.1}$$

It should be kept in mind, when reading the literature on this subject, that in some papers (e.g., in [ERS] and [RS1]) there is a change of sign with respect to our definition, due to the fact that the infinitesimal generator is defined there as the solution of the equation

$$\frac{\partial \Phi}{\partial t} + G \circ \Phi = O\,.$$

A Julia's lemma for infinitesimal generators was proved by Elin, Reich and Shoikhet in [ERS] in 2008, assuming that the radial limit of the generator at a point $p \in \partial B^n$ vanishes:

Theorem 3.1 ([ERS, Theorem, p. 403]) *Let* $G\colon B^n \to \mathbb{C}^n$ *be the infinitesimal generator on* B^n *of a one-parameter semigroup* $\Phi = \{\varphi_t\}$, *and let* $p \in \partial B^n$ *be such* *that*

$$\lim_{t\to 1^-} G(tp) = O. \tag{3.2}$$

Then the following assertions are equivalent:

(I) $\alpha = \liminf_{t\to 1^-} \mathrm{Re}\frac{\langle G(tp),p\rangle}{t-1} < +\infty$;

(II) $\beta = 2\sup_{z\in B^n} \mathrm{Re}\left[\frac{\langle G(z),z\rangle}{1-\|z\|^2} - \frac{\langle G(z),p\rangle}{1-\langle z,p\rangle}\right] < +\infty$;

(III) *there exists* $\gamma \in \mathbb{R}$ *such that for all* $z \in B^n$ *we have* $\frac{|1-\langle\varphi_t(z),p\rangle|^2}{1-\|\varphi_t(z)\|^2} \le e^{\gamma t}\frac{|1-\langle z,p\rangle|^2}{1-\|z\|^2}$.

Furthermore, if any of these assertions holds then $\alpha = \beta = \inf\gamma$, *and we have*

$$\lim_{t\to 1^-} \frac{\langle G(tp), p\rangle}{t-1} = \beta\ . \tag{3.3}$$

If (3.2) and any (whence all) of the equivalent conditions (I)–(III) holds, $p \in \partial B^n$ is called a *boundary regular null point* of G with *dilation* $\beta \in \mathbb{R}$.

This result suggested that a Julia-Wolff-Carathéodory theorem could hold for infinitesimal generators along the line of Rudin's Theorem 2.5. A first partial generalization has been achieved by Bracci and Shoikhet in [BS]. In collaboration with Abate, in [AR] we have been able to give a full generalization of Julia-Wolff-Carathéodory theorem for infinitesimal generators, proving the following result.

Theorem 3.2 ([AR]) *Let* $G\colon B^n \to \mathbb{C}^n$ *be an infinitesimal generator on* B^n *of a one-parameter semigroup, and let* $p \in \partial B^n$. *Assume that*

$$\frac{\langle G(z), p\rangle}{\langle z, p\rangle - 1} \quad \text{is } K\text{-bounded at } p \tag{3.4}$$

and

$$\frac{G(z) - \langle G(z), p\rangle p}{(\langle z, p\rangle - 1)^\gamma} \quad \text{is } K\text{-bounded at } p \text{ for some } 0 < \gamma < 1/2. \tag{3.5}$$

Then $p \in \partial B^n$ *is a boundary regular null point for* G. *Furthermore, if* β *is the dilation of* G *at* p *then:*

(i) *the function* $\langle G(z), p\rangle / (\langle z, p\rangle - 1)$ *(is* K*-bounded and) has restricted* K*-limit* β *at* p;
(ii) *if* v *is a vector orthogonal to* p, *the function* $\langle G(z), v\rangle/(\langle z, p\rangle - 1)^\gamma$ *is* K*-bounded and has restricted* K*-limit* 0 *at* p;
(iii) *the function* $\langle dG_z(p), p\rangle$ *is* K*-bounded and has restricted* K*-limit* β *at* p;
(iv) *if* v *is a vector orthogonal to* p, *the function* $(\langle z, p\rangle - 1)^{1-\gamma}\langle dG_z(p), v\rangle$ *is* K*-bounded and has restricted* K*-limit* 0 *at* p;
(v) *if* v *is a vector orthogonal to* p, *the function* $\langle dG_z(v), p\rangle \,/\, (\langle z, p\rangle - 1)^\gamma$ *is* K*-bounded and has restricted* K*-limit* 0 *at* p.

(vi) *if* v_1 *and* v_2 *are vectors orthogonal to* p *the function* $(\langle z, p\rangle - 1)^{1/2-\gamma}$ $\langle dG_z(v_1), v_2\rangle$ *is* K*-bounded at* p.

Sketch of Proof of Theorem 3.2. Statement (i) follows immediately from our hypotheses, thanks to Theorems 2.3 and 3.1. Statement (iii) follows by standard arguments, and (iv) follows from (ii), again by standard arguments.

The main point is the proof of part (ii). By Theorem 2.3, it suffices to compute the limit along a special restricted curve. We use the curve

$$\sigma(t) = tp + e^{-i\theta}\varepsilon(1-t)^{1-\gamma}v$$

which is always restricted, and *it is special if and only if* $\gamma < 1/2$. We then plug (i) and this curve into Theorem 3.1.(II), and we then let $\varepsilon \to 0^+$, using θ to get rid of the real part.

Statement (v) follows from (i), (ii) and by Theorem 3.1 using somewhat delicate arguments involving a curve of the form

$$\gamma(t) = \big(t + ic(1-t)\big)p + \eta(t)v\ ,$$

where $1 - t < |\eta(t)|^2 < 1 - |t + ic(1-t)|^2$, and the argument of $\eta(t)$ is chosen suitably. □

A first difference with respect to Theorem 2.5 is that we have to assume (3.4) and (3.5) as separate hypotheses, whereas they appear as part of Theorem 2.5(i) and (ii). Indeed, when dealing with holomorphic maps, conditions (3.4) and (3.5) are a consequence of the equivalent of condition (I) in Theorem 3.1, but in that setting the proof relies in the fact that there we have holomorphic *self-maps* of the ball. In our context, (3.5) is *not* a consequence of Theorem 3.1(I), as shown in [AR, Example 1.2]; and it also seems that (3.4) is stronger than Theorem 3.1(I).

A second difference is the exponent $\gamma < 1/2$. Bracci and Shoikhet proved Theorem 3.2 with $\gamma = 1/2$ but they couldn't prove the statements about restricted K-limits in cases (ii), (iv) and (v). This is due to an obstruction, which is not just a technical problem, but an inevitable feature of the theory. As mentioned in the sketch of the proof, in showing the existence of restricted K-limits, the curves one would like to use for obtaining the exponent $1/2$ in the statements are *restricted but not special*, in the sense that the limit in (2.1) is a strictly positive (though finite) number. Actually the exponent $1/2$ might not be the right one to consider in the setting of infinitesimal generators, as shown in [AR, Example 1.2].

An exact analogue of Theorem 2.5 with $\gamma = 1/2$ can be recovered assuming a slightly stronger hypothesis on the infinitesimal generator. Under assumptions (3.4) and (3.5) we have

$$\frac{\big\langle G\big(\sigma(t)\big), p\big\rangle}{\langle\sigma(t), p\rangle - 1} = \beta + o(1) \tag{3.6}$$

as $t \to 1^-$ for any special restricted p-curve $\sigma\colon [0,1) \to B^n$. Following ideas introduced in [ESY, EKRS, EJ] in the context of the unit disk, p is said to be a

Hölder boundary null point if there is $\alpha > 0$ such that

$$\frac{\langle G(\sigma(t)), p\rangle}{\langle \sigma(t), p\rangle - 1} = \beta + o\big((1-t)^\alpha\big) \tag{3.7}$$

for any special restricted p-curve $\sigma\colon [0,1) \to B^n$ such that $\langle \sigma(t), p\rangle \equiv t$. Using this notion we obtain the following result.

Theorem 3.3 ([AR]) *Let $G\colon B^n \to \mathbb{C}^n$ be the infinitesimal generator on B^n of a one-parameter semigroup, and let $p \in \partial B^n$. Assume that*

$$\frac{\langle G(z), p\rangle}{\langle z, p\rangle - 1} \quad \textit{and} \quad \frac{G(z) - \langle G(z), p\rangle p}{(\langle z, p\rangle - 1)^{1/2}}$$

are K-bounded at p, and that p is a Hölder boundary null point. Then the statement of Theorem 3.2 holds with $\gamma = 1/2$.

Examples of infinitesimal generators with a Hölder boundary null point and satisfying the hypotheses of Theorem 3.3 are provided in [AR].

In a forthcoming paper in collaboration with Abate, we will deal with the generalization of Theorem 3.2 to strongly convex domains in $\mathbb{C}^n$.

Acknowledgments Partially supported by the FIRB2012 grant "Differential Geometry and Geometric Function Theory", and by the ANR project LAMBDA, ANR-13-BS01-0002.

References

[A1] Abate, M.: Iteration Theory of Holomorphic Maps on Taut Manifolds. Mediterranean Press, Rende (1989)

[A2] Abate, M.: The Lindelöf principle and the angular derivative in strongly convex domains. J. Anal. Math. **54**, 189–228 (1990)

[A3] Abate, M.: Angular derivatives in strongly pseudoconvex domains. Proc. Symp. Pure Math. **52**(Part 2), 23–40 (1991)

[A4] Abate, M.: The infinitesimal generators of semigroups of holomorphic maps. Ann. Mat. Pura Appl. **161**, 167–180 (1992)

[A5] Abate, M.: The Julia-Wolff-Carathéodory theorem in polydisks. J. Anal. Math. **74**, 275–306 (1998)

[A6] Abate, M.: Angular derivatives in several complex variables. In Zaitsev, D., Zampieri, G. (eds.) Real Methods in Complex and CR Geometry. Lecture Notes in Mathematics, vol. 1848, pp. 1–47. Springer, Berlin (2004)

[AR] Abate, M., Raissy, J.: A Julia-Wolff-Carathéodory theorem for infinitesimal generators in the unit ball. Trans. AMS 1–17 (2014)

[AT] Abate, M., Tauraso, R.: The Lindelöf principle and angular derivatives in convex domains of finite type. J. Aust. Math. Soc. **73**, 221–250 (2002)

[AMY] Agler, J., McCarthy, J.E., Young, N.J.: A Carathéodory theorem for the bidisk via Hilbert space methods. Math. Ann. **352**, 581–624 (2012)

[BCD] Bracci, F., Contreras, M.D., Díaz-Madrigal, S.: Pluripotential theory, semigroups and boundary behavior of infinitesimal generators in strongly convex domains. J. Eur. Math. Soc. **12**, 23–53 (2010)

[BS] Bracci, F., Shoikhet, D.: Boundary behavior of infinitesimal generators in the unit ball. Trans. Am. Math. Soc. **366**, 1119–1140 (2014)

[B] Burckel, R.B.: An Introduction to Classical Complex Analysis. Academic Press, New York (1979)

[C] Carathéodory, C.: Über die Winkelderivierten von beschränkten analytischen Funktionen, pp. 39–54. Sitzungsber. Preuss. Akad. Wiss, Berlin (1929)

[Č] Cirka, E.M.: The Lindelöf and Fatou theorems in $\mathbb{C}^n$. Math. USSR-Sb. **21**, 619–641 (1973)

[EHRS] Elin, M., Harris, L.A., Reich, S., Shoikhet, D.: Evolution equations and geometric function theory in J^*- algebras. J. Nonlinear Conv. Anal. **3**, 81–121 (2002)

[EKRS] Elin, M., Khavinson, D., Reich, S., Shoikhet, D.: Linearization models for parabolic dynamical systems via Abel's functional equation. Ann. Acad. Sci. Fen. **35**, 1–34 (2010)

[ELRS] Elin, M., Levenshtein, M., Reich, S., Shoikhet, D.: Some inequalities for the horosphere function and hyperbolically nonexpansive mappings on the Hilbert ball. J. Math. Sci. (N.Y.) **201**(5), 595–613 (2014)

[EJ] Elin, M., Jacobzon, F.: Parabolic type semigroups: asymptotics and order of contact. Preprint. (2013). arxiv:1309.4002

[ERS] Elin, M., Reich, S., Shoikhet, D.: A Julia-Carathéodory theorem for hyperbolically monotone mappings in the Hilbert ball. Israel J. Math. **164**, 397–411 (2008)

[ES] Elin, M., Shoikhet, D.: Linearization models for complex dynamical systems. Topics in univalent functions, functional equations and semigroup theory. Operator Theory: Advances and Applications, vol. 208, xii+265 pp. Linear Operators and Linear Systems Birkhuser Verlag, Basel (2010). ISBN: 978-3-0346-0508-3

[ESY] Elin, M., Shoikhet, D., Yacobzon, F.: Linearization models for parabolic type semigroups. J. Nonlinear Convex Anal. **9**, 205–214 (2008)

[F] Fan, K.: The angular derivative of an operator-valued analytic function. Pac. J. Math. **121**, 67–72 (1986)

[Fa] Fatou, P.: Séries trigonométriques et séries de Taylor. Acta Math. **30**, 335–400 (1906)

[H] Hervé, M.: Quelques propriétés des applications analytiques d'une boule à m dimensions dans elle-même. J. Math. Pures Appl. **42**, 117–147 (1963)

[JP] Jarnicki, M., Pflug, P.: Invariant Distances and Metrics in Complex Analysis. Walter de Gruyter & co., Berlin (1993)

[Ju1] Julia, G.: Mémoire sur l'itération des fonctions rationnelles. J. Math. Pures Appl. **1**, 47–245 (1918)

[Ju2] Julia, G.: Extension nouvelle d'un lemme de Schwarz. Acta Math. **42**, 349–355 (1920)

[Ko] Korányi, A.: Harmonic functions on hermitian hyperbolic spaces. Trans. Am. Math. Soc. **135**, 507–516 (1969)

[K-S] Korányi, A., Stein, E.M.: Fatou's theorem for generalized half-planes. Ann. Scuola Norm. Sup. Pisa **22**, 107–112 (1968)

[L-V] Landau, E., Valiron, G.: A deduction from Schwarz's lemma. J. Lond. Math. Soc. **4**, 162–163 (1929)

[L] Lempert, L.: La métrique de Kobayashi et la représentation des domaines sur la boule. Bull. Soc. Math. France **109**, 427–474 (1981)

[Li] Lindelöf, E.: Sur un principe générale de l'analyse et ses applications à la theorie de la représentation conforme. Acta Soc. Sci. Fennicae **46**, 1–35 (1915)

[MM] Mackey, M., Mellon, P.: Angular derivatives on bounded symmetric domains. Israel J. Math. **138**, 291–315 (2003)

[N] Nevanlinna, R.: Remarques sur le lemme de Schwarz. C.R. Acad. Sci. Paris **188**, 1027–1029 (1929)

[RS1] Reich, S., Shoikhet, D.: Semigroups and generators on convex domains with the hyperbolic metric. Atti Acc. Naz. Lincei Cl. Sc. Fis. Mat. Nat. Rend. Lincei **8**, 231–250 (1997)

[RS2] Reich, S., Shoikhet, D.: Nonlinear Semigroups, Fixed Points, and Geometry of Domains in Banach Spaces. Imperial College Press, London (2005)

[R] Rudin, W.: Function Theory in the Unit Ball of $\mathbb{C}^n$. Springer, Berlin (1980)

[S] Shoikhet, D.: Semigroups in Geometrical Function Theory. Kluwer Academic Publishers, Dordrecht (2001)

[St] Stein, E.M.: The Boundary Behavior of Holomorphic Functions of Several Complex variables. Princeton University Press, Princeton (1972)

[SW] Szałowska, A., Włodarczyk, K.: Angular derivatives of holomorphic maps in infinite dimensions. J. Math. Anal. Appl. **204**, 1–28 (1996)

[W] Wachs, S.: Sur quelques propriétés des transformations pseudo-conformes avec un point frontière invariant. Bull. Soc. Math. Fr. **68**, 177–198 (1940)

[Wl1] Włodarczyk, K.: The Julia-Carathéodory theorem for distance decreasing maps on infinite-dimensional hyperbolic spaces. Atti Accad. Naz. Lincei **4**, 171–179 (1993)

[Wl2] Włodarczyk, K.: Angular limits and derivatives for holomorphic maps of infinite dimensional bounded homogeneous domains. Atti Accad. Naz. Lincei **5**, 43–53 (1994)

[Wl3] Włodarczyk, K.: The existence of angular derivatives of holomorphic maps of Siegel domains in a generalization of C^*-algebras. Atti Accad. Naz. Lincei **5**, 309–328 (1994)

[Wo] Wolff, J.: Sur une généralisation d'un théorème de Schwarz. C.R. Acad. Sci. Paris **183**, 500–502 (1926)

[Z] Zhu, J.M.: Angular derivatives of holomorphic maps in Hilbert spaces. J. Math. (Wuhan) **19**, 304–308 (1999)

A Brief Survey on Local Holomorphic Dynamics in Higher Dimensions

Feng Rong

Abstract We give a brief survey on local holomorphic dynamics in higher dimensions. The main novelty of this note is that we will organize the material by the "level" of local invariants rather than the type of maps.

Keywords Local holomorphic dynamics · Local invariant · Attracting flower · Attracting domain

1 Introduction

Let f be a holomorphic map in $\mathbf{C}^n$ with a fixed point, which we assume to be the origin. The local (discrete) holomorphic dynamics studies the asymptotic behavior of f in a neighborhood of the fixed point under iterations. There are several well-written surveys on this subject, see e.g. [A4, A5, B2]. The aim of this short note is twofold. First, we will present some more recent results in this area which were not covered in previous surveys. Second, we organize the material in a different way than before so as to emphasize the importance of the "third-level" local invariants in future studies.

A quantity will be called a *local invariant* if it only depends on the map f, i.e. invariant under changes of local coordinates. (Écalle [E] gave a detailed study on local invariants of holomorphic maps, although the dynamics associated with these invariants are not clear.) We will divide local invariants into three levels, depending on from which "level" of the Taylor expansion of f at 0 the invariants are defined. Roughly speaking, a *first-level* invariant comes from the linear part of the Taylor expansion; a *second-level* invariant comes from the leading nonlinear part of the Taylor expansion; and a *third-level* invariant comes from higher order nonlinear part of the Taylor expansion. When defining these local invariants, we will always use

F. Rong (✉)
Department of Mathematics, Shanghai Jiao Tong University, 800 Dong Chuan Road, Shanghai 200240, People's Republic of China
e-mail: frong@sjtu.edu.cn

F. Bracci et al. (eds.), *Complex Analysis and Geometry*, Springer Proceedings in Mathematics & Statistics 144, DOI 10.1007/978-4-431-55744-9_22

some suitable local coordinates. However, all these local invariants have been shown to be well-defined, i.e. independent of the choice of (allowable) local coordinates.

There are two typical types of results in local holomorphic dynamics. One type is to give normal forms or even linearizations via conjugations, the other is to describe the attracting set of a given map.

Two maps f and g are said to be (holomorphically) *conjugate* if there exists a biholomorphism φ such that $f \circ \varphi = \varphi \circ g$. Obviously, if f and g are conjugate then their local dynamics are equivalent. For a given map f, the "simplest" g it is conjugate to is called its *normal form*. The best one can hope for is that g is the linear part of f, in which case we say that f is *linearizable*. The well-known Poincaré-Dulac normal form and Brjuno's linearization theorem are typical examples. This type of results are certainly important. However, the majority of the note will be devoted to results on the attracting sets.

A point p in a neighborhood of 0 is in the *attracting set* of f if $f^k(p)$ goes to 0 as k goes to the infinity. Here, of course, f^k stands for the k-th iteration of f. The ultimate goal of the local dynamical study is to give a complete description of the asymptotic behavior of a map in a *full* neighborhood of the fixed point. The first step in achieving this goal is to give a complete study on the attracting set. The well-known Leau-Fatou Flower Theorem is the "model" result. Much of the work on the attracting sets in higher dimensions can be viewed as generalizations of the Leau-Fatou Flower Theorem.

Due to the limit of space, the results surveyed in the note are by no means complete. Our focus will be on results obtained in the past few years. For more detailed information on earlier results and results in one dimension, please see the existing surveys cited above.

The author would like to thank the organizers of the KSCV10 conference, especially Prof. Kang-Tae Kim, for the invitation and hospitality. He also thanks Filippo Bracci and the referee for valuable comments.

2 The First-Level Invariants

2.1 The Multipliers

Let f be a holomorphic map in $\mathbf{C}^n$ with the origin as a fixed point. Write f as

$$f(z) = L(z) + P_2(z) + P_3(z) + \cdots ,$$

where $L(z)$ is the linear part of $f(z)$ and $P_k(z)$ are homogeneous of degree k, $k \geq 2$. Write $L(z) = L \cdot z$, where L is an $n \times n$ matrix. The *multipliers* of f are defined to be the eigenvalues $\{\lambda_j\}_{j=1}^n$ of L.

Denote by $\mathbf{N}$ the set of non-negative integers. For $\alpha = (\alpha_1, \cdots, \alpha_n) \in \mathbf{N}^n$, set $\lambda^\alpha = \prod_{j=1}^n \lambda_j^{\alpha_j}$ and $|\alpha| = \sum_{j=1}^n \alpha_j$. Define

$$\omega(m) = \min_{2\le|\alpha|\le m} \min_{1\le j\le n} |\lambda^\alpha - \lambda_j|, \qquad m \ge 2.$$

We say that the multipliers of f satisfy the *Brjuno condition* if

$$\sum_{i\ge0} \frac{1}{p_i} \log \frac{1}{\omega(p_{i+1})} < \infty,$$

where $\{p_i\}_{i=0}^\infty$ is a sequence of integers with $1 = p_0 < p_1 < \cdots$. The following is the best known linearization result (improving earlier results by Siegel [S]).

Theorem 2.1 (Brjuno, [Br]) *Let f be a holomorphic map in $\mathbf{C}^n$ with the origin as a fixed point. If df_0 is diagonalizable and the multipliers of f satisfy the Brjuno condition, then f is holomorphically linearizable.*

A *resonance* for f is a relation of the form

$$\lambda^\alpha - \lambda_j = 0, \qquad |\alpha| \ge 2,\ 1 \le j \le n.$$

Obviously, Theorem 2.1 is not applicable in the presence of resonances. A map f is said to be *quasi-parabolic*, if df_0 is diagonalizable and $\lambda_j = 1$ for $1 \le j \le m < n$ and $\lambda_j \ne 1$ but $|\lambda_j| = 1$ for $m+1 \le j \le n$. Note that quasi-parabolic maps always have resonances. However, inspired by a partial linearization result by Pöschel [P], the author proved the following linearization result for quasi-parabolic maps.

Theorem 2.2 (Rong, [R1]) *Let f be a holomorphic map in $\mathbf{C}^n$ with the origin as a quasi-parabolic fixed point. Assume that there exists an m-dimensional manifold M of fixed points through 0 such that $df_p = df_0$ for every $p \in M$. If $\{\lambda_j\}_{j=m+1}^n$ satisfy the Brjuno condition, then f is holomorphically linearizable.*

This was later generalized to more general settings by Raissy [Ra].

2.2 Multi-resonance

As we have seen above, the presence of resonances is usually an obstacle for the local dynamical study. However, in a recent development, the presence of resonances has been used in a positive way to study the attracting sets of certain maps.

Assume that df_0 is diagonalizable and that there are resonances among the multipliers $\{\lambda_j\}_{j=1}^n$. If the resonances are generated over $\mathbf{N}$ by a finite number of $\mathbf{Q}$-linearly independent multi-indices, f is said to be *multi-resonant*. In [BZ], Bracci and Zaitsev studied *one-resonant* maps and obtained sufficient conditions for the existence of local attracting basins. This was later generalized to multi-resonant maps by Bracci et al. [BRaZ]. More recently, Raissy and Vivas [RaV] gave a more detailed study on *two-resonant* maps, and Bracci and the author [BR] studied *quasi-parabolic one-resonant* maps.

The basic idea of this line of study is as follows: first by using the multi-resonance, f can be projected into a lower-dimensional map $\hat{f}$, the so-called *parabolic shadow* of f, such that $\hat{f}$ is *tangent to the identity* at the origin; then using the local attracting basin of $\hat{f}$ and some attracting conditions on the "fibers," one can create an attracting basin for f. Recall that a holomorphic map f is said to be tangent to the identity at 0 if $df_0 = I_n$, the identity matrix.

2.3 Diagonalization

Most of the results in local holomorphic dynamics assume the linear part of the maps under study to be diagonalizable. However, the non-diagonalizable case is certainly important and worth studying. For instance, Yoccoz [Y] pointed out that the Brjuno condition is in general not sufficient for holomorphic linearization in the non-diagonalizable case (see also [DG]).

The method of *blow-up* is a very important tool in the study of local holomorphic dynamics. It is particularly so for the study on attracting sets in the non-diagonalizable case. Indeed, Abate [A1] gave an explicit description of how to systematically diagonalize a non-diagonalizable map via blow-ups.

There are very few results in the non-diagonalizable case, see e.g. [A3]. Recently, the author [R5] gave a somewhat systematic study of the non-diagonalizable case in dimension two. In particular, the attracting basin studied in [A3] was recovered as a special case.

3 The Second-Level Invariants

3.1 The Order and Characteristic Directions

Let us first recall the well-known *Leau-Fatou Flower Theorem* from the one-dimensional theory (see e.g. [M]).

Theorem 3.1 (Leau, [L]; Fatou, [F1]) *Let f be a holomorphic map in $\mathbf{C}$ with the origin as a fixed point. Assume that f is tangent to the identity with order ν, i.e. f can be written as*

$$f(z) = z + az^{\nu} + O(z^{\nu+1}), \qquad \nu \geq 2,\ a \neq 0.$$

Then there exist $\nu - 1$ "attracting petals" for f at the origin.

A central theme in the study on attracting sets for holomorphic maps in higher dimensions has been to generalize the Leau-Fatou Flower Theorem. To state the known results so far, let us first make several definitions.

Let f be a holomorphic map in $\mathbf{C}^n$, tangent to the identity at the origin. Write f as

$$f(z) = z + P_2(z) + P_3(z) + \cdots ,$$

where $P_k(z)$, $k \geq 2$, are n-tuples of homogeneous polynomials of degree k. The *order* ν of f is defined as

$$\nu := \min\{k : \; P_k(z) \not\equiv 0\}.$$

Write $z = (z_1, \cdots , z_n)$ and $P_\nu(z) = (P_{\nu,1}(z), \cdots , P_{\nu,n}(z))$ and denote by $[\cdot]$ the canonical projection from $\mathbf{C}^n \backslash \{0\}$ to $\mathbf{P}^{n-1}$. A direction $[v] = [z_1 : \cdots : z_n]$ is called a *characteristic direction* of f if

$$(P_{\nu,1}(z), \cdots , P_{\nu,n}(z)) = \lambda(z_1, \cdots , z_n), \qquad \lambda \in \mathbf{C}.$$

If $\lambda \neq 0$, then $[v]$ is said to be *non-degenerate*, otherwise *degenerate*.

An *attracting petal* of dimension d for f at the origin is an injective holomorphic map $\varphi : \Delta \to \mathbf{C}^n$ satisfying the following properties:
(i) Δ is a simply connected domain in $\mathbf{C}^d$ with $0 \in \partial\Delta$;
(ii) φ is continuous on $\partial\Delta$ and $\varphi(0) = 0$;
(iii) $\varphi(\Delta)$ is invariant under f and $f^k(\varphi(\zeta)) \to 0$ as $k \to \infty$ for any $\zeta \in \Delta$.
Furthermore, if $[\varphi(\zeta)] \to [v] \in \mathbf{P}^{n-1}$ as $\zeta \to 0$, then φ is said to be *tangent to* $[v]$ at 0. If there are $\nu - 1$ attracting petals tangent to $[v]$ at 0, then we say they form an *attracting flower* tangent to $[v]$ at 0. When $d = 1$, an attracting petal is also called a *parabolic curve*. When $1 < d < n$, an attracting petal is also called a *parabolic manifold*. When $d = n$, an attracting petal is a (parabolic) *attracting domain*.

The first main generalization of the Leau-Fatou Flower Theorem to higher dimensions is the following

Theorem 3.2 (Écalle, [E]; Hakim, [H2]) *Let f be a holomorphic map in $\mathbf{C}^n$, tangent to the identity at the origin. If f is of order $\nu < \infty$ and $[v]$ is a non-degenerate characteristic direction of f, then there exists a one-dimensional attracting flower of f tangent to $[v]$ at 0.*

A similar result holds for quasi-parabolic maps, which was proven in dimension two by Bracci and Molino [BMo] and in higher dimensions by the author [R2]. To be more precise, we need some definitions.

Let f be a quasi-parabolic map in $\mathbf{C}^n$ with eigenvalue 1 of multiplicity l and other eigenvalues λ_j, $1 \leq j \leq m = n - l$. Set $\Lambda = \mathrm{Diag}(\lambda_1, \cdots , \lambda_m)$. In a suitable local coordinates $(z, w) \in \mathbf{C}^l \times \mathbf{C}^m$, we can then write f as

$$\begin{cases} z_1 = z + p(z) + r(z, w), \\ w_1 = \Lambda w + q(z) + s(z, w), \end{cases}$$

where p, q, r, s are all of degree greater or equal to two and $r(z, 0) = s(z, 0) = 0$.

We say that f is in *ultra-resonant* form if $\operatorname{ord} p(z) \leq \operatorname{ord} q(z)$, in which case we call $\nu = \operatorname{ord} p(z)$ the *order* of f. Assume that $\nu < \infty$, and let $p_\nu(z)$ be the lowest order term of $p(z)$. A *characteristic direction* of f is of the form $[v] = [z_1 : \cdots : z_l : 0 : \cdots : 0]$ where $[u] = [z_1 : \cdots : z_l]$ is a characteristic direction of $p_\nu(z)$, i.e. $p_\nu(z) = \lambda z$ for some $\lambda \in \mathbf{C}$. And $[v]$ is said to be *non-degenerate* if $\lambda \neq 0$, otherwise *degenerate*.

If f has a characteristic direction $[v]$, then in a suitable local coordinates it can be assumed that $[v] = [1 : 0 : \cdots : 0]$. Write $z = (x, y) \in \mathbf{C} \times \mathbf{C}^{l-1}$. Set $\mu = \min\{k;\ x^k w_j \text{ in } s(z, w)\}$. We say that f is *dynamically separating* in $[v]$ if $\mu \geq \nu - 1$.

Theorem 3.3 (Bracci-Molino, [BMo]; Rong, [R2]) *Let f be a holomorphic map in $\mathbf{C}^n$, with a quasi-parabolic fixed point at the origin. If f is of order $\nu < \infty$, has a non-degenerate characteristic direction $[v]$, and f is dynamically separating in $[v]$, then there exists a one-dimensional attracting flower of f tangent to $[v]$ at 0.*

3.2 The Director and Residual Index

Let f be a holomorphic map in $\mathbf{C}^n$, tangent to the identity at the origin. Assume that f has order $\nu < \infty$ and has a non-degenerate characteristic direction $[v]$. In suitable local coordinates $z = (x, y) \in \mathbf{C} \times \mathbf{C}^{n-1}$, it can be assumed that $[v] = [1 : 0]$. Write f as

$$\begin{cases} x_1 = x + p_\nu(x, y) + O(\nu + 1), \\ y_1 = y + q_\nu(x, y) + O(\nu + 1), \end{cases} \tag{3.1}$$

where $p_\nu(x, y)$ and $q_\nu(x, y)$ are homogeneous of degree ν.

Under the blow-up $y = xu$, the blow-up map $\tilde{f}$ takes the form

$$\begin{cases} x_1 = x + x^\nu p_\nu(1, u) + O(x^{\nu+1}), \\ u_1 = u + x^{\nu-1} r(u) + O(x^\nu), \end{cases} \tag{3.2}$$

where $r(u) = q_\nu(1, u) - p_\nu(1, u)u$. The matrix

$$A := p_\nu^{-1}(1, 0) r'(0)$$

is a local invariant associated with f and its $n - 1$ eigenvalues are called the *directors* of f in the non-degenerate characteristic direction $[v]$.

Theorem 3.4 (Hakim, [H3]) *Let f be a holomorphic map in $\mathbf{C}^n$, tangent to the identity at the origin. Assume that f has order $\nu < \infty$ and has a non-degenerate characteristic direction $[v]$. Let α_i, $1 \leq i \leq n - 1$, be the directors of f in $[v]$. Suppose that there exists $\alpha > 0$ such that $\operatorname{Re}\alpha_j > \alpha$ for $1 \leq j \leq l$ and $\operatorname{Re}\alpha_{l+k} < \alpha$*

for $1 \leq k \leq n-1-l$. *Then there exists an* $(l+1)$*-dimensional attracting flower of* f *tangent to* $[v]$ *at* 0.

Similar results hold for quasi-parabolic maps (see [R3]) and semi-attractive maps (see e.g. [F2, U, H1, Ri, R4]). Recall that a holomorphic map f is said to be *semi-attractive* at 0 if $df_0 = \mathrm{Diag}(I_l, A)$, where I_l is the identity matrix with size $1 \leq l \leq n-1$ and the eigenvalues of A all have modulus less than one.

Theorems 3.2 and 3.4 deal with holomorphic maps tangent to the identity with a non-degenerate characteristic direction, which is a generic condition. It would be desirable to obtain a "full" generalization of the Leau-Fatou Flower Theorem without this generic condition. So far, this has only been achieved in dimension two by Abate [A2].

Theorem 3.5 (Abate, [A2]) *Let* f *be a holomorphic map in* $\mathbf{C}^n$. *Assume that the origin is an isolated fixed point of* f *and* f *is tangent to the identity at* 0. *Then there exists a one-dimensional attracting flower of* f *at* 0.

The main point of the proof of Theorem 3.5 is to show that after a sequence of blow-ups one gets a blow-up map which is generic, i.e. with a non-degenerate characteristic direction, as in Theorem 3.2. For this purpose, Abate introduced a key local invariant, the *residual index*, defined as follows.

Let $\tilde{f}$ be a holomorphic map in $\mathbf{C}^2$, tangent to the identity at the origin. Assume that there is a line S of fixed points of $\tilde{f}$ through 0. In local coordinates $(x, u) \in \mathbf{C} \times \mathbf{C}$, such that S is given by $\{x = 0\}$, we can write $\tilde{f}$ as

$$\begin{cases} x_1 = x + x^{\nu} p(x, u), \\ u_1 = u + x^{\mu} q(x, u), \end{cases}$$

where $p(0, u) \not\equiv 0$ and $q(0, u) \not\equiv 0$, $\nu \geq 2$ and $\mu \geq 1$.

Define a meromorphic function, the *residual function*, $\kappa(u)$ by

$$\kappa(u) := \lim_{x \to 0} \frac{p(x, u)}{x^{\mu-\nu+1} q(x, u)}.$$

If $\mu < \nu - 1$, then $\kappa(u) \equiv 0$. If $\mu > \nu - 1$, then $\kappa(u) \equiv \infty$. If $\mu = \nu - 1$, then $\kappa(u) = p(0, u)/q(0, u)$. The *residual index* $\iota_0(\tilde{f}, S)$ of $\tilde{f}$ at 0 along S is defined as $\mathrm{Res}(\kappa(u); 0)$. Note that if $\tilde{f}$ is the blow-up map of a holomorphic map tangent to the identity in a non-degenerate characteristic direction and S is the exceptional divisor, then the residual index is exactly the reciprocal of the director as defined above.

Although Theorem 3.5 gives a Leau-Fatou Flower Theorem in dimension two, it leaves open two questions: 1. What happens if the origin is non-isolated? 2. Given a *degenerate* characteristic direction, is there always an attracting petal tangent to it?

For results related to the first question, see e.g. [B1, De]. Note that the residual index theorems used in [A2, B1] have been developed systematically by Abate, Bracci and Tovena [ABT] to much more general settings and also to higher dimensions. It would be desirable to find an effective use of such more general index theorems to the study of local holomorphic dynamics.

For the second question, Abate [A2] already showed that the answer is yes if the residual index of the blow-up map at the given direction along the exceptional divisor does not belong to $\mathbf{Q}^+$. This result was later generalized by Molino [Mo] to the case of non-vanishing residual index (under a mild "regularity" condition).

In dimension two, using the residual function $\kappa(u)$ defined above, the characteristic directions can be divided into *three* types (cf. [AT]).

Let f be a holomorphic map in $\mathbf{C}^2$, tangent to the identity at the origin. Assume that f has a characteristic direction, which we assume to be $[1:0]$. Then we can write f as in (3.1), and its blow-up map $\tilde{f}$ as in (3.2). If $r(u) \equiv 0$, then f is said to be *dicritical* at 0.

Assume that f is not dicritical at 0. Then the residual function is given by $\kappa(u) = p_\nu(1,u)/r(u)$. If 0 is a simple pole of $\kappa(u)$, then $[1:0]$ is a *Fuchsian* characteristic direction of f. If 0 is a pole of $\kappa(u)$ of order greater than one, then $[1:0]$ is an *irregular* characteristic direction of f. If $\kappa(u) \equiv 0$ or if 0 is a removable singularity of $\kappa(u)$, then $[1:0]$ is an *apparent* characteristic direction of f.

Theorem 3.6 (Vivas, [V2]) *Let f be a holomorphic map in $\mathbf{C}^2$, tangent to the identity at the origin. Assume that f has an irregular characteristic direction $[v]$. Then there exists an attracting domain of f tangent to $[v]$ at 0.*

Vivas [V2] also gave sufficient conditions (in terms of the residual index) for the existence of attracting domains in Fuchsian characteristic directions, and studied examples of apparent characteristic directions. See also [V1, La] for related results.

3.3 The Non-dicritical Order

Let f be a holomorphic map in $\mathbf{C}^n$, tangent to the identity at the origin. Assume that f has order $\nu < \infty$ and has a non-degenerate characteristic direction $[v]$. Let α_i, $1 \le i \le n-1$, be the directors of f in $[v]$. If $\mathrm{Re}\,\alpha_j > 0$ for $1 \le j \le l$ and $\mathrm{Re}\,\alpha_{l+k} < 0$ for $1 \le k \le m = n-1-l$, then by Theorem 3.4 there exists an $(l+1)$-dimensional attracting flower of f tangent to $[v]$ at 0. In fact, from [H3, ArRa], it follows that $l+1$ is the maximal dimension of attracting flowers in this case. It is then natural to ask what happens when $\mathrm{Re}\,\alpha_{l+k} = 0$ for all $1 \le k \le m$.

In suitable local coordinates $(x, y, z) \in \mathbf{C} \times \mathbf{C}^l \times \mathbf{C}^m$, we can assume that $[v] = [1:0:0]$. Under the blow-up $(y = xu, z = xv)$, the blow-up map $\tilde{f}$ can be written as (after possible scaling and suitable linear transformations)

$$\begin{cases} x_1 = (1 - x^{\nu-1})x + O(x^\nu \|w\|, x^{\nu+1}), \\ u_1 = (I_l - x^{\nu-1}B)u + O(x^{\nu-1}\|w\|^2, x^\nu), \\ v_1 = (I_m - x^{\nu-1}C)v + O(x^{\nu-1}\|w\|^2, x^\nu), \end{cases} \tag{3.3}$$

where $w = (u, v)$, B is an $l \times l$ matrix with eigenvalues α_j, $1 \le j \le l$, and C is an $m \times m$ matrix with eigenvalues α_{l+k}, $1 \le k \le m$.

Rewrite v_1 in (3.3) as

$$v_1 = v + x^{\nu-1} \sum_{|k|=1}^{\nu+1} v^k \gamma_k + O(x^{\nu-1} \|u\| \|w\|, x^\nu),$$

where $k = (k_1, \cdots, k_m) \in \mathbf{N}^m$ is a multi-index, $|k| = k_1 + \cdots + k_m$, and $\gamma_k \in \mathbf{C}^m$. Then the *non-dicritical order* of f in the characteristic direction $[v]$ is defined as

$$\tau := \min\{|k| - 1;\ \gamma_k \neq 0\}.$$

The name "non-dicritical" refers to the fact that if $l = 0$ then f is dicritical at the origin if and only if all γ_k vanish. It will always be assumed that some $\gamma_k \neq 0$, in which case $0 \le \tau \le \nu$ (see e.g. [Bro] for a study in the dicritical case).

Theorem 3.7 (Rong, [R6]) *Let f be a holomorphic map in $\mathbf{C}^2$, tangent to the identity at the origin, with a non-degenerate characteristic direction $[v]$. Let τ be the non-dicritical order of f in $[v]$. If $\tau \ge 1$, then there exists an attracting domain of f tangent to $[v]$ at 0.*

A similar result holds in higher dimensions with extra conditions. When $\tau = 0$, it is possible for f to admit a "spiral domain" at the origin (see [R6] for more details). Note also that in dimension two, if $\tau \ge 1$ then $[v]$ is an irregular characteristic direction of f.

4 The Third-Level Invariants

4.1 Essentially Non-degenerate

From the discussion above, it is clear that one of the main problems in the study of local holomorphic dynamics of maps tangent to the identity is to understand the dynamics in degenerate characteristic directions. So far there are only very few results, and only in dimension two (see e.g. [R7, V2]).

Let f be a holomorphic map in $\mathbf{C}^2$, tangent to the identity at the origin. Assume that f has order $\nu < \infty$ and $[v] = [x : y] = [1 : 0]$ is a degenerate characteristic direction of f. Write f as

$$\begin{cases} x_1 = x + y p_{\nu-1}(x, y) + O(\nu + 1), \\ y_1 = y + y q_{\nu-1}(x, y) + O(\nu + 1), \end{cases}$$

where $p_{\nu-1}(x, y)$ and $q_{\nu-1}(x, y)$ are homogeneous of degree $\nu-1$. We say that $[1:0]$ is a *generically degenerate* characteristic direction of f if $q_{\nu-1}(x, 0) \not\equiv 0$. Note that a generically degenerate characteristic direction is an apparent characteristic direction.

Let G be the group of local changes of coordinates which preserves $[1:0]$ as the degenerate characteristic direction. For each $\phi \in G$, write $(x_\phi, y_\phi) = \phi(x, y)$. Then f is transformed under ϕ as

$$\begin{cases} x_{\phi,1} = x_\phi + P_\phi(x_\phi) + y_\phi O(\nu - 1), \\ y_{\phi,1} = y_\phi + Q_\phi(x_\phi) + y_\phi R_\phi(x_\phi) + y_\phi^2 O(\nu - 2), \end{cases}$$

where $\operatorname{ord} P_\phi \geq \nu + 1$, $\operatorname{ord} Q_\phi \geq \nu + 1$ and $\operatorname{ord} R_\phi = \nu - 1$. The *essential order* of f in $[1:0]$ is defined as

$$\mu := \max_{\phi \in G} \operatorname{ord} P_\phi(x_\phi).$$

We say that $[1:0]$ is an *essentially non-degenerate* characteristic direction of f if $\mu < \infty$.

Remark 4.1 The above definition is slightly different than the original definition given in [R7], where μ is required to be less than the so-called *virtual order*. However, using simple linear transformations of the form $(X = x, Y = y + \alpha x^k)$ for $k \geq 2$, it is easy to check that the virtual order is always greater than μ if $\mu < \infty$.

Now assume that $\mu < \infty$ and rewrite f as

$$\begin{cases} x_1 = x - a x^\mu + P(x) + y O(\nu - 1), \\ y_1 = y - b y x^{\nu-1} + Q(x) + y R(x) + y^2 O(\nu - 2), \end{cases}$$

where $a, b \neq 0$, $\operatorname{ord} Q > \operatorname{ord} P \geq \mu + 1$ and $\operatorname{ord} R \geq \nu$. Then the *director* of f in $[1:0]$ is defined as

$$\alpha := \frac{b}{a^{(\nu-1)/(\mu-1)}}.$$

Theorem 4.1 (Rong, [R7]) *Let f be a holomorphic map in $\mathbf{C}^2$, tangent to the identity at the origin. Assume that f has an essentially non-degenerate characteristic direction $[v]$ and the director of f in $[v]$ is α. If $\operatorname{Re}\alpha > 0$, then there exists an attracting domain of f tangent to $[v]$ at 0.*

Note that the above theorem gives an attracting petal instead of an attracting flower. As first observed by the author [R2], this "symmetry break-down" is indeed expected in degenerate characteristic directions.

4.2 Non-dynamically-Separating

Let f be a holomorphic map in $\mathbf{C}^2$, quasi-parabolic at the origin. Then $[1:0]$ is the only characteristic direction of f. Write f as

$$\begin{cases} z_1 = z + P(z) + wS(z,w), \\ w_1 = \lambda w + Q(z) + wR(z) + w^2 T(z,w), \end{cases}$$

with $|\lambda| = 1$ and $\lambda \neq 1$. Set $p = \text{ord}P$, $q = \text{ord}Q$ and $r = \text{ord}R$. Assume that f is in ultra-resonant form, i.e. $q \geq p$, and non-dynamically-separating, i.e. $r < p - 1$.

Now rewrite f as

$$\begin{cases} z_1 = z - az^p + O(z^{p+1}, w), \\ w_1 = \lambda w - bwz^r + O(z^p, wz^{r+1}, w^2), \end{cases}$$

with $a, b \neq 0$. We call p the *essential order* and r the *generic order*. The *director* of f is then defined as

$$\alpha := \frac{b}{\lambda a^{r/(p-1)}}.$$

Theorem 4.2 (Rong, [R8]) *Let f be a holomorphic map in $\mathbf{C}^2$, quasi-parabolic at the origin. Assume that f is non-dynamically-separating and the director of f is α. If $\text{Re}\,\alpha > 0$, then there exists an attracting domain of f tangent to* $[1:0]$ *at* 0.

Acknowledgments The author is partially supported by the National Natural Science Foundation of China (Grant No. 11371246).

References

[A1] Abate, M.: Diagonalization of non-diagonalizable discrete holomorphic dynamical systems. Am. J. Math. **122**, 757–781 (2000)

[A2] Abate, M.: The residual index and the dynamics of holomorphic maps tangent to the identity. Duke Math. J. **107**, 173–207 (2001)

[A3] Abate, M.: Basins of attraction in quadratic dynamical systems with a Jordan fixed point. Nonlinear Anal. **51**, 271–282 (2002)

[A4] Abate, M.: Discrete local holomorphic dynamics. In: Gentili, G., Guenot, J., Patrizio, G. (eds.) Holomorphic Dynamical Systems. Lecture Notes in Mathematics, vol. 1998. Springer, Berlin (2010)

[A5] Abate, M.: Open problems in local discrete holomorphic dynamics. Anal. Math. Phys. **1**, 261–287 (2011)

[ABT] Abate, M., Bracci, F., Tovena, F.: Index theorems for holomorphic self-maps. Ann. Math. (**159**(2), 819–864 (2004)

[AT] Abate, M., Tovena, F.: Poincaré-Bendixson theorems for meromorphic connections and homogeneous vector fields. J. Diff. Equ. **251**, 2612–2684 (2011)

[ArRa] Arizzi, M., Raissy, J.: On Écalle-Hakim's theorems in holomorphic dynamics. In: Bonifant, A., Lyubich, M., Sutherland, S. (eds.) Frontiers in Complex Dynamics: In Celebration of John Milnor's 80th Birthday. Princeton University Press, Princeton (2014)

[B1] Bracci, F.: The dynamics of holomorphic maps near curves of fixed points. Ann. Sc. Norm. Super. Pisa Cl. Sci. **2**(5), 493–520 (2003)

[B2] Bracci, F.: Local dynamics of holomorphic diffeomorphisms, Boll. Unione Mat. Ital. Sez. B Artic. Ric. Mat. **7**(8), 609–636 (2004)

[BMo] Bracci, F., Molino, L.: The dynamics near quasi-parabolic fixed points of holomorphic diffeomorphisms in $\mathbf{C}^2$. Am. J. Math. **126**, 671–686 (2004)

[BRaZ] Bracci, F., Raissy, J., Zaitsev, D.: Dynamics of multi-resonant biholomorphisms. Int. Math. Res. Not. **20**, 4772–4797 (2013)

[BR] Bracci, F., Rong, F.: Dynamics of quasi-parabolic one-resonant biholomorphisms. J. Geom. Anal. **24**, 1497–1508 (2014)

[BZ] Bracci, F., Zaitsev, D.: Dynamics of one-resonant biholomorphisms. J. Eur. Math. Soc. **15**, 179–200 (2013)

[Br] Brjuno, A.D.: Analytical form of differential equations, I; II. Trans. Moscow Math. Soc. **25**, 131–288 (1971); **26**, 199–239 (1972)

[Bro] Brochero-Martínez, F.E.: Groups of germs of analytic diffeomorphisms in $C^2, 0)$. J. Dyn. Control Syst. **9**, 1–32 (2003)

[De] Degli Innocenti, F.: Holomorphic dynamics near germs of singular curves. Math. Z. **251**, 943–958 (2005)

[DG] DeLatte, D., Gramchev, T.: Biholomorphic maps with linear parts having Jordan blocks: linearization and resonance type phenomena. Math. Phys. Electron. J. **8**, 1–27 (2002)

[E] Écalle, J.: Les fonctions résurgentes, Tome III: L'équation du pont et la classification analytiques des objects locaux. Prépubl. Math. Orsay **85-05**, Univ. de Paris-Sud, Orsay (1985)

[F1] Fatou, P.: Sur les équations fonctionnelles, I; II; III. Bull. Soc. Math. France **47**, 161–271 (1919); **48**, 33–94 (1920); **48**, 208–314 (1920)

[F2] Fatou, P.: Substitutions analytiques et equations fonctionnelles de deux variables. Ann. Sci. Éc. Norm. Supér. **40**, 67–142 (1924)

[H1] Hakim, M.: Attracting domains for semi-attractive transformations of $\mathbf{C}^p$. Publ. Mat. **38**, 479–499 (1994)

[H2] Hakim, M.: Analytic transformations of $(\mathbf{C}^p, 0)$ tangent to the identity. Duke Math. J. **92**, 403–428 (1998)

[H3] Hakim, M.: Transformations tangent to the identity. Stable pieces of manifolds. Prépubl. Math. Orsay **97-30**, Univ. de Paris-Sud, Orsay (1997)

[La] Lapan, S.: Attracting domains of maps tangent to the identity whose only characteristic direction is non-degenerate. Int. J. Math. **24**, 1350083, 27 pp (2013)

[L] Leau, L.: Étude sur les équations fonctionelles àune ou plusieurs variables. Ann. Fac. Sci. Toulouse Math. **11**, E1–E110 (1897)

[M] Milnor, J.: Dynamics in One Complex Variable. Annals of Mathematics Studies, vol. 160. Princeton University Press, Princeton (2006)

[Mo] Molino, L.: The dynamics of maps tangent to the identity and with nonvanishing index. Trans. Am. Math. Soc. **361**, 1597–1623 (2009)

[P] Pöschel, J.: On invariant manifolds of complex analytic mappings near fixed points. Expo. Math. **4**, 97–109 (1986)

[Ra] Raissy, J.: Linearization of holomorphic germs with quasi-Brjuno fixed points. Math. Z. **264**, 881–900 (2010)

[RaV] Raissy, J., Vivas, L.: Dynamics of two-resonant biholomorphisms. Math. Res. Lett. **20**, 757–771 (2013)

[Ri] Rivi, M.: Parabolic manifolds for semi-attractive holomorphic germs. Mich. Math. J. **49**, 211–241 (2001)

[R1] Rong, F.: Linearization of holomorphic germs with quasi-parabolic fixed points. Ergod. Theory Dyn. Syst. **28**, 979–986 (2008)

[R2] Rong, F.: Quasi-parabolic analytic transformations of $\mathbf{C}^n$. J. Math. Anal. Appl. **343**, 99–109 (2008)
[R3] Rong, F.: Quasi-parabolic analytic transformations of $\mathbf{C}^n$. Parabolic manifolds. Ark. Mat. **48**, 361–370 (2010)
[R4] Rong, F.: Parabolic manifolds for semi-attractive analytic transformations of $\mathbf{C}^n$. Trans. Am. Math. Soc. **363**, 5207–5222 (2011)
[R5] Rong, F.: Local dynamics of holomorphic maps in $\mathbf{C}^2$ with a Jordan fixed point. Mich. Math. J. **62**, 843–856 (2013)
[R6] Rong, F.: The non-dicritical order and attracting domains of holomorphic maps tangent to the identity. Int. J. Math. **25**, 1450003, 10 pp (2014)
[R7] Rong, F.: New invariants and attracting domains for holomorphic maps in $\mathbf{C}^2$ tangent to the identity. Publ. Mat. **59**, 235–243 (2015)
[R8] Rong, F.: Attracting domains for quasi-parabolic analytic transformations of $\mathbf{C}^2$. Proc. Roy. Soc. Edinburgh Sect. A (to appear)
[S] Siegel, C.L.: Iteration of analytic functions. Ann. Math. **43**(2), 607–612 (1942)
[U] Ueda, T.: Local structure of analytic transformations of two complex variables, I; II. J. Math. Kyoto Univ. **26**, 233–261 (1986); **31**, 695–711 (1991)
[V1] Vivas, L.: Fatou-Bieberbach domains as basins of attraction of automorphisms tangent to the identity. J. Geom. Anal. **22**, 352–382 (2012)
[V2] Vivas, L.: Degenerate characteristic directions for maps tangent to the identity. Indiana U. Math. J. **61**, 2019–2040 (2012)
[Y] Yoccoz, J.-C.: Théorème de Siegel, nombres de Brjuno et polynômes quadratiques. Astérisque **231**, 3–88 (1995)

L^2-Serre Duality on Singular Complex Spaces and Applications

J. Ruppenthal

Abstract In this survey, we explain a version of topological L^2-Serre duality for singular complex spaces with arbitrary singularities. This duality can be used to deduce various L^2-vanishing theorems for the $\overline{\partial}$-equation on singular spaces. As one application, we prove Hartogs' extension theorem for $(n-1)$-complete spaces. Another application is the characterization of rational singularities. It is shown that complex spaces with rational singularities behave quite tame with respect to some $\overline{\partial}$-equation in the L^2-sense. More precisely: a singular point is rational if and only if the appropriate L^2-$\overline{\partial}$-complex is exact in this point. So, we obtain an L^2-$\overline{\partial}$-resolution of the structure sheaf in rational singular points.

Keywords Cauchy-Riemann equations · L^2-theory · Serre duality · Dolbeault cohomology · Vanishing theorems · Singular complex spaces · Rational singularities

1 Introduction

Classical Serre duality, [S1], can be formulated as follows: Let X be a complex n-dimensional manifold, let $V \to X$ be a complex vector bundle, and let $\mathscr{E}^{0,q}(X, V)$ and $\mathscr{E}_c^{n,q}(X, V^*)$ be the spaces of global smooth $(0,q)$-form with values in V and global smooth compactly supported (n,q)-forms with values in the dual bundle V^*, respectively. Then the following pairing is non-degenerate

$$H^q\big(\mathscr{E}^{0,\bullet}(X,V),\bar\partial\big) \times H^{n-q}\big(\mathscr{E}_c^{n,\bullet}(X,V^*),\bar\partial\big) \to \mathbb{C}, \quad ([\varphi]_{\bar\partial},[\psi]_{\bar\partial}) \mapsto \int_X \varphi \wedge \psi \tag{1}$$

provided that $H^q\big(\mathscr{E}^{0,\bullet}(X,V),\bar\partial\big)$ and $H^{q+1}\big(\mathscr{E}^{0,\bullet}(X,V),\bar\partial\big)$ are Hausdorff topological vector spaces.

J. Ruppenthal (✉)
Department of Mathematics, University of Wuppertal,
Gaußstr. 20, 42119 Wuppertal, Germany
e-mail: ruppenthal@uni-wuppertal.de

F. Bracci et al. (eds.), *Complex Analysis and Geometry*, Springer Proceedings in Mathematics & Statistics 144, DOI 10.1007/978-4-431-55744-9_23

If X is allowed to have singularities, then, traditionally, Serre duality takes a more algebraic and much less explicit form. To explain that more precisely, let $\mathscr{F} := \mathscr{O}(F)$, $\mathscr{F}^* := \mathscr{O}(F^*)$ and let Ω_X^n denote the sheaf of holomorphic n-forms on X. Then we can rephrase (1) via the Dolbeault isomorphism algebraically: There is a non-degenerate topological pairing

$$H^q(X, \mathscr{F}) \times H^{n-q}(X, \mathscr{F}^* \otimes \Omega_X^n) \to \mathbb{C}, \tag{2}$$

realized by the cup-product, provided that $H^q(X, \mathscr{F})$ and $H^{q+1}(X, \mathscr{F})$ are Hausdorff. In this formulation, Serre duality has been generalized to singular complex spaces, see, e.g., Hartshorne [H1, H2] and Conrad [C] for the algebraic setting and Ramis-Ruget [RR] and Andreotti-Kas [AK] for the analytic setting. In fact, if X is of pure dimension n, paracompact and Cohen-Macaulay, then there is again a non-degenerate topological pairing (2) if we replace Ω_X^n by the Grothendieck dualizing sheaf ω_X. If X is not Cohen-Macaulay, then $H^{n-q}(X, \mathscr{F}^* \otimes \Omega_X^n)$ has to be the replaced by the cohomology of a certain complex of $\mathscr{O}_X$-modules, called a dualizing complex.

In this survey, we will explain how L^2-theory for the $\overline{\partial}$-operator can be used to obtain an L^2-version of Serre duality on singular spaces which has an analytic realization completely analogous to (1). More precisely, we will show how (1) generalizes to singular spaces by replacing the Dolbeault cohomology groups of smooth $(0, q)$ and (n, q)-forms, respectively, by L^2-Dolbeault cohomology groups.

2 L^2-Theory for the $\overline{\partial}$-Operator on Singular Spaces

The Cauchy-Riemann operator $\overline{\partial}$ plays a fundamental role in Complex Analysis and Complex Geometry. On complex manifolds, functions— or more generally distributions—are holomorphic if and only if they are in the kernel of the $\overline{\partial}$-operator, and the same holds in a certain sense on normal complex spaces. For forms of arbitrary degree, the importance of the $\overline{\partial}$-operator appears strikingly for example in the notion of $\overline{\partial}$-cohomology which can be used to represent the cohomology of complex manifolds by the Dolbeault isomorphism.

The L^2-theory for the $\overline{\partial}$-operator is of particular importance in Complex Analysis and Geometry and has become indispensable for the subject after the fundamental work of Hörmander on L^2-estimates and existence theorems for the $\overline{\partial}$-operator [H3] and the related work of Andreotti and Vesentini [AV]. Important applications of the L^2-theory are e.g. the Ohsawa-Takegoshi extension theorem [OT], Siu's analyticity of the level sets of Lelong numbers [S2] or the invariance of plurigenera [S3]—just to name some.

The first problem one has to face when studying the $\overline{\partial}$-equation on singular spaces is that it is not clear what kind of differential forms and operators one should consider. Recently, there has been considerable progress by different approaches.

Andersson and Samuelsson developed in [AS] Koppelman integral formulas for the $\overline{\partial}$-equation on arbitrary singular complex spaces which allow for a $\overline{\partial}$-resolution of the structure sheaf in terms of certain fine sheaves of currents, called $\mathscr{A}$-sheaves. These $\mathscr{A}$-sheaves are defined by an iterative procedure of repeated application of singular integral operators, which makes them pretty abstract and hard to understand (and difficult to work with in concrete situations).

A second, more explicit approach is as follows: Consider differential forms which are defined on the regular part of a singular variety and which are square-integrable up to the singular set. This setting seems to be very fruitful and has some history by now (see [PS]).[1] Also in this direction, considerable progress has been made recently. Øvrelid–Vassiliadou and the author obtained in [OV2] and [R3] a pretty complete description of the L^2-cohomology of the $\overline{\partial}$-operator (in the sense of distributions) at isolated singularities.

In this setting, we understand the class of objects with which we deal very well (just L^2-forms), but the disadvantage is a different one. Whereas the $\overline{\partial}$-equation is locally solvable for closed $(0, q)$-forms in the category of $\mathscr{A}$-sheaves by the Koppelman formulas in [AS], there are local obstructions to solving the $\overline{\partial}$-equation in the L^2-sense at singular points (see e.g. [FOV, OV2, R3]). So, there can be no L^2-$\overline{\partial}$-resolution for the structure sheaf in general.

In this survey, we will see that the $\overline{\partial}$-operator in the L^2-sense behaves pretty well on spaces with canonical singularities which play a prominent role in the minimal model program. The underlying idea is that canonical Gorenstein singularities are rational (see e.g. [K], Theorem 11.1), i.e., we expect that the singularities do not contribute to the local cohomology.

Pursuing this idea, it turned out that there is a notion of L^2-$\overline{\partial}$-cohomology for $(0, q)$-forms which can be described completely in terms of a resolution of singularities (see (6) below). A singular point is rational if and only if this certain L^2-$\overline{\partial}$-complex is exact in this point. If the underlying space has rational singularities, particularly on a Gorenstein space with canonical singularities, then we obtain an L^2-$\overline{\partial}$-resolution of the structure sheaf, i.e., a resolution of the structure sheaf in terms of a well-known and easy to handle class of differential forms. One of our main tools is a version of topological L^2-Serre duality for singular complex spaces with arbitrary singularities, which seems to be useful in other contexts, too (Theorem 4.1).

[1] The interest in this setting goes back to the invention of intersection (co-)homology by Goresky and MacPherson which has very tight connections to the L^2-deRham cohomology of the regular part of a singular variety. We refer here to the solution of the Cheeger-Goresky-MacPherson conjecture [CGM] for varieties with isolated singularities by Ohsawa [O] (see [PS] for more details).

3 Two $\overline{\partial}$-Complexes on Singular Complex Spaces

We need to specify what we mean by differential forms and the $\overline{\partial}$-operator in the presence of singularities. Let X be a Hermitian complex space[2] of pure dimension n and $F \to X$ a Hermitian holomorphic line bundle. We denote by $\mathscr{L}^{p,q}(F)$ the sheaf of germs of F-valued (p,q)-forms on the regular part of X which are square-integrable on $K^* = K \setminus \operatorname{Sing} X$ for any compact set K in their domain of definition.[3] Note that $\mathscr{L}^{p,q}(F)$ becomes a Fréchet sheaf with the $L^{2,loc}$-topology on open subsets of X.

Due to the incompleteness of the metric on $X^* = X \setminus \operatorname{Sing} X$, there are different reasonable definitions of the $\overline{\partial}$-operator on $\mathscr{L}^{p,q}(F)$-forms. To be more precise, let $\overline{\partial}_{cpt}$ be the $\overline{\partial}$-operator on smooth forms with support away from the singular set $\operatorname{Sing} X$. Then $\overline{\partial}_{cpt}$ can be considered as a densely defined operator $\mathscr{L}^{p,q}(F) \to \mathscr{L}^{p,q+1}(F)$. One can now consider various closed extensions of this operator. The two most important are the maximal closed extension, i.e., the $\overline{\partial}$-operator in the sense of distributions which we denote by $\overline{\partial}_w$, and the minimal closed extension, i.e., the closure of the graph of $\overline{\partial}_{cpt}$ which we denote by $\overline{\partial}_s$. Let $\mathscr{C}^{p,q}(F)$ be the domain of definition of $\overline{\partial}_w$ which is a subsheaf of $\mathscr{L}^{p,q}(F)$, and $\mathscr{F}^{p,q}(F)$ the domain of definition of $\overline{\partial}_s$ which in turn is a subsheaf of $\mathscr{C}^{p,q}(F)$. We obtain complexes of fine sheaves

$$\mathscr{C}^{p,0}(F) \xrightarrow{\overline{\partial}_w} \mathscr{C}^{p,1}(F) \xrightarrow{\overline{\partial}_w} \mathscr{C}^{p,2}(F) \xrightarrow{\overline{\partial}_w} \dots \tag{3}$$

and

$$\mathscr{F}^{p,0}(F) \xrightarrow{\overline{\partial}_s} \mathscr{F}^{p,1}(F) \xrightarrow{\overline{\partial}_s} \mathscr{F}^{p,2}(F) \xrightarrow{\overline{\partial}_s} \dots \tag{4}$$

We refer to [R4] for more details, but let us explain the $\overline{\partial}_s$-operator more precisely for convenience of the reader. Let f be a germ in $\mathscr{C}^{p,q}(F)$, i.e., an F-valued (p,q)-form on an open set U in X (living on the regular part of U) which is L^2 on compact subsets of U and such that the $\overline{\partial}$ in the sense of distributions, $\overline{\partial}_w f$, is in the same class of forms. Then f is in the domain of the $\overline{\partial}_s$-operator (and we set $\overline{\partial}_s f = \overline{\partial}_w f$) if there exists a sequence of forms $\{f_j\}_j \subset \mathscr{C}^{p,q}(U,F)$ with support away from the singular set, $\operatorname{supp} f_j \cap \operatorname{Sing} X = \emptyset$, such that

$$f_j \to f \ \text{ in } \ \mathscr{L}^{p,q}(U,F),$$
$$\overline{\partial}_w f_j \to \overline{\partial}_w f \ \text{ in } \ \mathscr{L}^{p,q+1}(U,F).$$

---

[2]A Hermitian complex space (X,g) is a reduced complex space X with a metric g on the regular part such that the following holds: If $x \in X$ is an arbitrary point there exists a neighborhood $U = U(x)$ and a biholomorphic embedding of U into a domain G in $\mathbb{C}^N$ and an ordinary smooth Hermitian metric in G whose restriction to U is $g|_U$.

[3]This is what we mean by square-integrable up to the singular set.

This means that the $\overline{\partial}_s$-operator comes with a certain Dirichlet boundary at the singular set of X, which can also be interpreted as a growth condition. We have e.g. the following:

Lemma 3.1 ([R4]) *Bounded forms in the domain of $\overline{\partial}_w$ are in the domain of $\overline{\partial}_s$.*

If F is just the trivial line bundle, then $\mathscr{K}_X := \ker \overline{\partial}_w \subset \mathscr{C}^{n,0}$ is the canonical sheaf of Grauert–Riemenschneider (see [GR]) and $\mathscr{K}_X^s := \ker \overline{\partial}_s \subset \mathscr{F}^{n,0}$ is the sheaf of holomorphic n-forms with Dirichlet boundary condition that was introduced in [R3]. We will see below that $\widehat{\mathscr{O}}_X = \ker \overline{\partial}_s \subset \mathscr{F}^{0,0}$ for the sheaf of weakly holomorphic functions $\widehat{\mathscr{O}}_X$.

It is clear that (3) and (4) are exact in regular points of X. Exactness in singular points is equivalent to the difficult problem of solving $\overline{\partial}$-equations locally in the L^2-sense at singularities, which is not possible in general (see e.g. [FOV, OV1, OV2, R1, R2, R3]). However, it is known that (3) is exact for $p = n$ (see [PS]), and that (4) is exact for $p = n$ if X has only isolated singularities (see [R3]). In these cases, the complexes (3) and (4) are fine resolutions of the canonical sheaves $\mathscr{K}_X$ and $\mathscr{K}_X^s$, respectively.

For an open set $\Omega \subset X$, we denote by $H^{p,q}_{w,loc}(\Omega, F)$ the cohomology of the complex (3), and by $H^{p,q}_{w,cpt}(\Omega, F)$ the cohomology of (3) with compact support. Analogously, let $H^{p,q}_{s,loc}(\Omega, F)$ and $H^{p,q}_{s,cpt}(\Omega, F)$ be the cohomology groups of (4). These L^2-cohomology groups inherit the structure of topological vector spaces, which are locally convex Hausdorff spaces if the corresponding $\overline{\partial}$-operators have closed range.[4]

4 L^2-Serre Duality

We can now formulate the L^2-version of (1) for singular complex spaces:

Theorem 4.1 (Serre duality [R4]) *Let X be a Hermitian complex space of pure dimension n, $F \to X$ a Hermitian holomorphic line bundle, and let $0 \leq p, q \leq n$. If $H^{p,q}_{w,loc}(\Omega, F)$ and $H^{p,q+1}_{w,loc}(\Omega, F)$ are Hausdorff, then the mapping*

$$\mathscr{L}^{p,q}(\Omega, F) \times \mathscr{L}^{n-p,n-q}_{cpt}(\Omega, F^*) \to \mathbb{C} \ , \ (\eta, \omega) \mapsto \int_{\Omega^*} \eta \wedge \omega,$$

induces a non-degenerate pairing of topological vector spaces

$$H^{p,q}_{w,loc}(\Omega, F) \times H^{n-p,n-q}_{s,cpt}(\Omega, F^*) \to \mathbb{C}$$

such that $H^{n-p,n-q}_{s,cpt}(\Omega, F^*)$ is the topological dual of $H^{p,q}_{w,loc}(\Omega, F)$ and vice versa.

---

[4] Note that different Hermitian metrics lead to $\overline{\partial}$-complexes which are equivalent on relatively compact subsets. So, one can put any Hermitian metric on X in many of the results below.

The same statement holds with the indices $\{s, w\}$ in place of $\{w, s\}$. Then there is a non-degenrate pairing

$$H^{p,q}_{s,loc}(\Omega, F) \times H^{n-p,n-q}_{w,cpt}(\Omega, F^*) \to \mathbb{C}.$$

If the topological vector spaces $H^{p,q}_{w/s,loc}(\Omega, F)$, $H^{p,q+1}_{w/s,loc}(\Omega, F)$ are non-Hausdorff, then the statement of Theorem 4.1 holds at least for the separated cohomology groups $\overline{H}_{w/s} = \ker \overline{\partial}_{w/s} / \overline{\operatorname{Im} \overline{\partial}_{w/s}}$.[5] The two main difficulties in the proof of Theorem 4.1 are as follows. First, the $\overline{\partial}$-operators under consideration are just closed densely defined operators in the Fréchet spaces $\mathscr{L}^{p,q}(\Omega, F)$ and the (LF)-spaces $\mathscr{L}^{n-p,n-q}_{cpt}(\Omega, F^*)$. Second, we have to show that the operators $\overline{\partial}_w$ and $\overline{\partial}_s$ are topologically dual, even at singularities. Note that $H^{p,q}_{w/s,loc}(\Omega, F)$ is Hausdorff if and only if $\overline{\partial}_{w/s}$ has closed range in $\mathscr{L}^{p,q}(\Omega, F)$, and to decide whether this is the case is usually as difficult as solving the corresponding $\overline{\partial}$-equation. Using local L^2-$\overline{\partial}$-solution results for singular spaces, one can show at least:

Theorem 4.2 ([R4]) *Let X be a Hermitian complex space of pure dimension n, $F \to X$ a Hermitian holomorphic line bundle, and let $0 \le p, q \le n$. Let $\Omega \subset X$ be a holomorphically convex open subset. Then the topological vector spaces*

$$H^{n,q}_{w,loc}(\Omega, F)\ ,\quad H^{n,q}_{w,cpt}(\Omega, F)\ ,\quad H^{0,n-q}_{s,cpt}(\Omega, F^*)\ ,\quad H^{0,n-q}_{s,loc}(\Omega, F^*)$$

are Hausdorff for all $0 \le q \le n$.

A main point in the proof of Theorem 4.2 is to show that the canonical Fréchet sheaf structure of compact convergence on the coherent analytic canonical sheaf $\mathscr{K}_X$ coincides with the Fréchet sheaf structure of L^2-convergence on compact subsets. This allows then to show also the topological equivalence of Čech cohomology and L^2-cohomology. If X has only isolated singularities, then the Hausdorff property is known also for some cohomology spaces of different degree (see [R4]).

As a direct application of Serre duality, Theorem 4.1, one can deduce:

Theorem 4.3 *Let X be a Hermitian complex space of pure dimension n, $F \to X$ a Hermitian holomorphic line bundle and $\Omega \subset X$ a cohomologically q-complete open subset, $q \ge 1$. Then*

$$H^{n,r}_{w,loc}(\Omega, F) = H^{0,n-r}_{s,cpt}(\Omega, F^*) = 0 \quad \text{for all } r \ge q.$$

Note that Ω is cohomologically q-complete if it is q-complete by the Andreotti-Grauert vanishing theorem [AG]. So, Theorem 4.3 allows to solve the $\overline{\partial}_s$-equation with compact support for $(0, n-q)$-forms on q-complete spaces, which is of particular interest for 1-complete spaces, i.e., Stein spaces.

[5]The notation w/s refers either to the index w or the index s in the whole statement.

5 Hartogs' Extension Theorem

Let us mention some applications. As an interesting consequence of Theorem 4.3, we obtain Hartogs' extension theorem in its most general form. This version of the Hartogs' extension was first obtained by Merker-Porten [MP] and shortly thereafter also by Coltoiu-Ruppenthal [CR]. Merker and Porten gave an involved geometrical proof by using a finite number of parameterized families of holomorphic discs and Morse-theoretical tools for the global topological control of monodromy, but no $\overline{\partial}$-theory. Shortly after that, Coltoiu and Ruppenthal were able to give a short $\overline{\partial}$-theoretical proof by the Ehrenpreis-$\overline{\partial}$-technique (cf. [CR]). This approach involves Hironaka's resolution of singularities which may be considered a very deep theorem. In the present survey, we give a very short proof of the extension theorem by the Ehrenpreis-$\overline{\partial}$-technique without needing a resolution of singularities. We just use the vanishing result $H^{0,1}_{s,cpt}(X) = 0$

Theorem 5.1 *Let X be a connected normal complex space of dimension $n \geq 2$ which is cohomologically $(n-1)$-complete. Furthermore, let D be a domain in X and $K \subset D$ a compact subset such that $D \setminus K$ is connected. Then each holomorphic function $f \in \mathscr{O}(D \setminus K)$ has a unique holomorphic extension to the whole set D.*

Proof Let $f \in \mathscr{O}(D \setminus K)$. Choose a cut-off function $\chi \in C^\infty_{cpt}(D)$ such that χ is identically 1 in a neighborhood of K. Then $g := (1-\chi)f$ is an extension of f, but unfortunately not holomorphic. However, we can fix it by the $\overline{\partial}$-strategy of Ehrenpreis. By Lemma 3.1, g is in the domain of $\overline{\partial}_s$ and $H^{0,1}_{s,cpt}(X) = 0$ by Theorem 4.3. So, there exists a solution h to the $\overline{\partial}_s$-equation with compact support $\overline{\partial}_s h = \overline{\partial}_s g$ and $F := g - h$ is the desired extension of f to the whole of D. That can be seen by use of the identity theorem and the fact that X cannot be compact (because Theorem 4.3 implies also that $H^{0,0}_{s,cpt}(X) = 0$). □

6 Rational Singularities

Another, very interesting application of L^2-Serre duality is the following characterization of rational singularities. Let $\pi : M \to X$ be a resolution of singularities and $\Omega \subset\subset X$ holomorphically convex. Give M any Hermitian metric. Then pullback of L^2-(n,q)-forms under π induces an isomorphism

$$\pi^* : H^{n,q}_{w,cpt}(\Omega) \xrightarrow{\cong} H^{n,q}_{w,cpt}\big(\pi^{-1}(\Omega)\big) \cong H^q_{cpt}\big(\pi^{-1}(\Omega), \mathscr{K}_M\big) \tag{5}$$

for all $0 \leq q \leq n$ by use of Pardon–Stern [PS] and the Takegoshi vanishing theorem [T] (see [R4] for more details). Now we can use the L^2-Serre duality, Theorem 4.1, and classical Serre duality on the smooth manifold $\pi^{-1}(\Omega)$ to deduce that push-forward of forms under π induces another isomorphism

$$\pi_* : H^{n-q}\big(\pi^{-1}(\Omega), \mathscr{O}_M\big) \xrightarrow{\cong} H^{0,n-q}_{s,loc}\big(\Omega\big) \tag{6}$$

for all $0 \le q \le n$ (see [R4], Theorem 1.1). This shows that the obstructions to solving the $\overline{\partial}_s$-equation locally for $(0, q)$-forms can be expressed in terms of a resolution of singularities. For the cohomology sheaves of the complex $(\mathscr{F}^{0,\bullet}, \overline{\partial}_s)$, we see that

$$\Big(\mathscr{H}^q\big(\mathscr{F}^{0,\bullet}, \overline{\partial}_s\big)\Big)_x \cong \big(R^q\pi_*\mathscr{O}_M\big)_x$$

in any point $x \in X$ for all $q \ge 0$. It follows that the functions in the kernel of $\overline{\partial}_s$ are precisely the weakly holomorphic functions, and for $p = 0$ the complex (4) is exact in a point $x \in X$ exactly if x is a rational point:

Theorem 6.1 ([R4], Theorem 1.3) *Let X be a Hermitian complex space. Then the L^2-$\overline{\partial}$-complex*

$$0 \to \mathscr{O}_X \longrightarrow \mathscr{F}^{0,0} \xrightarrow{\overline{\partial}_s} \mathscr{F}^{0,1} \xrightarrow{\overline{\partial}_s} \mathscr{F}^{0,2} \xrightarrow{\overline{\partial}_s} \mathscr{F}^{0,3} \xrightarrow{\overline{\partial}_s} \dots \tag{7}$$

is exact in a point $x \in X$ if and only if x is a rational point.

Hence, if X has only rational singularities, then (7) *is a fine resolution of the structure sheaf $\mathscr{O}_X$.*

Recall that a point $x \in X$ is rational if it is a normal point and $\big(R^q\pi_*\mathscr{O}_M\big)_x = 0$ for all $q \ge 1$. If X has only rational singularities, then Theorem 6.1 yields immediately further finiteness and vanishing results, e.g. if X is q-convex or q-complete.

Let us point out also the following interesting fact. Let X be a Gorenstein space with canonical singularities. By exactness of (7) and exactness of (3) for $p = n$, the non-degenerate L^2-Serre duality pairing

$$H^{0,q}_{s,loc}(\Omega) \times H^{n,n-q}_{w,cpt}(\Omega) \to \mathbb{C}\ , \ ([\eta], [\omega]) \mapsto \int_{\Omega^*} \eta \wedge \omega,$$

is for $0 \le q \le n$ then an explicit realization of Grothendieck duality after Ramis-Ruget [RR],

$$\big(H^q(\Omega, \mathscr{O}_X)\big)^* \cong H^{n-q}_{cpt}(\Omega, \omega_X),$$

given the cohomology groups under consideration are Hausdorff. Here, ω_X denotes the Grothendieck dualizing sheaf which coincides with the Grauert-Riemenschneider canonical sheaf $\mathscr{K}_X$ as X has canonical Gorenstein singularities.

7 $\mathscr{A}$-sheaf duality

We conclude by mentioning another approach to analytic Serre duality on singular complex spaces which is based on the so-called $\mathscr{A}_{0,q}$-sheaves introduced by Andersson and Samuelsson in [AS]. These are certain sheaves of $(0, q)$-currents on singular complex spaces which are smooth on the regular part of the variety and such that the $\overline{\partial}$-complex

$$0 \to \mathscr{O}_X \hookrightarrow \mathscr{A}_{0,0} \xrightarrow{\overline{\partial}} \mathscr{A}_{0,1} \xrightarrow{\overline{\partial}} \mathscr{A}_{0,2} \longrightarrow \dots \tag{8}$$

is a fine resolution of the structure sheaf. The $\mathscr{A}$-sheaves are defined via Koppelman integral formulas on singular complex spaces.

Analogously, in [RSW], we introduced a $\overline{\partial}$-complex of fine sheaves of (n, q)-currents (smooth on the regular part of the variety)

$$0 \to \omega_X \hookrightarrow \mathscr{A}_{n,0} \xrightarrow{\overline{\partial}} \mathscr{A}_{n,1} \xrightarrow{\overline{\partial}} \mathscr{A}_{n,2} \longrightarrow \dots \tag{9}$$

where X is of pure dimension n and ω_X denotes the Grothendieck dualizing sheaf. The complex (9) is exact only under some additional assumptions, e.g. if X is Cohen-Macaulay. We call $(\mathscr{A}_{n,\bullet}, \overline{\partial})$ a dualizing Dolbeault complex for $\mathscr{O}_X$ because we obtain in [RSW] a non-degenerate topological pairing

$$H^q\big(\mathscr{A}_{0,\bullet}(X), \bar{\partial}\big) \times H^{n-q}_{cpt}\big(\mathscr{A}_{n,\bullet}(X), \bar{\partial}\big) \to \mathbb{C}, \quad ([\varphi]_{\bar{\partial}}, [\psi]_{\bar{\partial}}) \mapsto \int_X \varphi \wedge \psi, \tag{10}$$

provided that $H^q(X, \mathscr{O}_X) \cong H^q\big(\mathscr{A}_{0,\bullet}(X), \bar{\partial}\big)$ and $H^{q+1}(X, \mathscr{O}_X) \cong H^{q+1}\big(\mathscr{A}_{0,\bullet}(X), \bar{\partial}\big)$ are Hausdorff topological spaces.

Acknowledgments The author was supported by the Deutsche Forschungsgemeinschaft (DFG, German Research Foundation), grant RU 1474/2 within DFG's Emmy Noether Programme.

References

[AS] Andersson, M., Samuelsson, H.: A Dolbeault-Grothendieck lemma on complex spaces via Koppelman formulas. Invent. Math. **190**(2), 261–297 (2012)

[AG] Andreotti, A., Grauert, H.: Théorème de finitude pour la cohomologie des espaces complexes. Bull. Soc. Math. France **90**, 193–259 (1962)

[AK] Andreotti, A., Kas, A.: Duality on complex spaces. Ann. Scuola Norm. Sup. Pisa **27**(3), 187–263 (1973)

[AV] Andreotti, A., Vesentini, E.: Carleman estimates for the Laplace Beltrami equation on complex manifolds. Publ. Math. Inst. Hautes Etudes Sci. **25**, 81–130 (1965)

[CGM] Cheeger, J., Goresky, M., MacPherson, R.: L^2-Cohomology and Intersection Homology of Singular Algebraic Varieties. Annals of Mathematics Studies, vol. 102, pp. 303–340. Princeton University Press, Princeton (1982)

[CR] Coltoiu, M., Ruppenthal, J.: On Hartogs' extension theorem on $(n-1)$-complete spaces. J. Reine Angew. Math. **637**, 41–47 (2009)

[C] Conrad, B.: Grothendieck Duality and Base Change. Lecture Notes in Mathematics, vol. 1750. Springer, Berlin (2000)

[FOV] Fornæss, J.E., Øvrelid, N., Vassiliadou, S.: Local L^2 results for $\overline{\partial}$: the isolated singularities case. Int. J. Math. **16**(4), 387–418 (2005)

[GR] Grauert, H., Riemenschneider, O.: Verschwindungssätze für analytische Kohomologiegruppen auf komplexen Räumen. Invent. Math. **11**, 263–292 (1970)

[H1] Hartshorne, R.: Algebraic Geometry. Graduate Texts in Mathematics, vol. 52. Springer, New York (1977)
[H2] Hartshorne, R.: Residues and Duality. Lecture Notes in Mathematics, vol. 20. Springer, Berlin (1966)
[H3] Hörmander, L.: L^2-estimates and existence theorems for the $\overline{\partial}$-operator. Acta Math. **113**, 89–152 (1965)
[K] Kollár, J.: Singularities of pairs. Algebraic geometry—Santa Cruz 1995, pp. 221–287 (Proc. Symp. Pure Math. **62**, Part 1, Amer. Math. Soc., Providence, RI) (1997)
[MP] Merker, J., Porten, E.: The Hartogs' extension theorem on $(n-1)$-complete complex spaces. J. Reine Angew. Math. **637**, 23–39 (2009)
[O] Ohsawa, T.: Cheeger-Goresky-MacPherson's conjecture for the varieties with isolated singularities. Math. Z. **206**, 219–224 (1991)
[OT] Ohsawa, T., Takegoshi, K.: On the extension of L^2-holomorphic functions. Math. Z. **195**(2), 197–204 (1987)
[OV1] Øvrelid, N., Vassiliadou, S.: Solving $\overline{\partial}$ on product singularities. Complex Var. Ellipitic Equ. **51**(3), 225–237 (2006)
[OV2] Øvrelid, N., Vassiliadou, S.: L^2-$\overline{\partial}$-cohomology groups of some singular complex spaces. Invent. Math. **192**(2), 413–458 (2013)
[PS] Pardon, W., Stern, M.: L^2-$\overline{\partial}$-cohomology of complex projective varieties. J. Am. Math. Soc. **4**(3), 603–621 (1991)
[RR] Ramis, J.-P., Ruget, G.: Complexe dualisant et théorème de dualité en géométrie analytique complexe. Inst. Hautes Études Sci. Publ. Math. **38**, 77–91 (1970)
[R1] Ruppenthal, J.: About the $\overline{\partial}$-equation at isolated singularities with regular exceptional set. Int. J. Math. **20**(4), 459–489 (2009)
[R2] Ruppenthal, J.: The $\overline{\partial}$-equation on homogeneous varieties with an isolated singularity. Math. Z. **263**, 447–472 (2009)
[R3] Ruppenthal, J.: L^2-theory for the $\overline{\partial}$-operator on compact complex spaces. Duke Math. J. **163**, 2887–2934 (2014)
[R4] Ruppenthal, J.: L^2-Serre duality on singular complex spaces and rational singularities. Preprint 2014 (submitted). arXiv:1401.4563
[RSW] Ruppenthal, J., Samuelsson Kalm, H., Wulcan, E.: Explicit Serre duality on complex spaces. Preprint 2014 (submitted). arXiv:1401.8093
[S1] Serre, J.-P.: Un théorème de dualité. Comm. Math. Helv. **29**, 9–26 (1955)
[S2] Siu, Y.-T.: Analyticity of sets associated to Lelong numbers and the extension of closed positive currents. Invent. Math. **27**, 53–156 (1974)
[S3] Siu, Y.-T.: Invariance of plurigenera. Invent. Math. **134**(3), 661–673 (1998)
[T] Takegoshi, K.: Relative vanishing theorems in analytic spaces. Duke Math. J. **51**(1), 273–279 (1985)

Proper Holomorphic Maps Between Bounded Symmetric Domains

Aeryeong Seo

Abstract In this article, we survey the background and recent development on proper holomorphic maps between bounded symmetric domains.

Keywords Bounded symmetric domain · Proper holomorphic map

A bounded domain Ω is called symmetric if for each $p \in \Omega$, there is a holomorphic automorphism I_p such that I_p^2 is the identity map of Ω which has p as an isolated fixed point. All bounded symmetric domains are homogeneous domains, i.e. the automorphism group acts transitively on the domain. In 1920s, E. Cartan classified all irreducible bounded symmetric domains which consist of 4 classical types and 2 exceptional types [CA35]. The classical types are the following:

1. Type I : $\Omega^I_{m,n} = \{Z \in M(m,n,\mathbb{C}) : I_n - ZZ^* > 0\}$ where $m \geq n = \text{rank}(\Omega^I_{m,n})$
2. Type II : $\Omega^{II}_m = \{Z \in M(m,m,\mathbb{C}) : I_m - ZZ^* > 0,\ Z^t = -Z\}$ $\text{rank}(\Omega^{II}_m) = [\frac{1}{2}m]$
3. Type III : $\Omega^{III}_m = \{Z \in M(m,m,\mathbb{C}) : I_m - ZZ^* > 0,\ Z^t = Z\}$ $\text{rank}(\Omega^{III}_m) = m$
4. Type IV : $\Omega^{IV} = \{z = (z_1,\ldots,z_n) \in \mathbb{C}^n : ||z||^2 < 2,\ ||z||^2 < 1 + \left|\frac{1}{2}\sum z_k^2\right|^2\}, \text{rank}(\Omega^{IV}) = 2$

Note that $\Omega_{m,1}$ is the m-dimensional unit ball and irreducible bounded symmetric domain of rank 1 is the unit ball. Bounded symmetric domains have abundant symmetries, hence there are many structures to make these domains rigid. For domains Ω_1, Ω_2, a map $f : \Omega_1 \to \Omega_2$ is called *proper* if for every compact set $K \subset \Omega_2$, $f^{-1}(K)$ is compact. Equivalently, f is proper if and only if for every sequence $\{a_j\}$ in Ω_1 which converges to $\partial\Omega_1$, $\{f(a_j)\}$ converges to $\partial\Omega_2$. This is the reason why proper holomorphic maps are deeply related to boundary structures of bounded

A. Seo (✉)
The Center for Geometry and its Applications (GAIA), Mathematical Sciences Building, Pohang University of Science and Technology, Pohang,Gyeong-sang-bugdo 790-784, South Korea
e-mail: inno827@postech.ac.kr

F. Bracci et al. (eds.), *Complex Analysis and Geometry*, Springer Proceedings in Mathematics & Statistics 144, DOI 10.1007/978-4-431-55744-9_24

domains. There are several interesting results on proper holomorphic maps between strongly pseudoconvex domains or general weakly pseudoconvex domains. In this article, we will only focus on proper holomorphic maps between bounded symmetric domains of classical type.

1 Proper Holomorphic Maps Between the Unit Balls

The simplest case of classifying proper holomorphic maps is those between discs. Denote the unit disc by $\Delta = \{z \in \mathbb{C} : |z| < 1\}$ and the n-dimensional unit ball by $\mathbb{B}^n = \{z \in \mathbb{C}^n : |z| < 1\}$.

Theorem 1.1 *Let $f : \Delta \to \Delta$ be a proper holomorphic map between unit discs. Then there are finitely many points $\{a_j\}$ in the unit disc, positive integers $\{m_j\}$ and θ such that*

$$f(z) = e^{i\theta} \prod_{j=1}^{m} \left(\frac{a_j - z}{1 - \overline{a}_j z} \right)^{m_j} \tag{1.1}$$

The form of (1.1) is called *Blaschke product*. In higher dimensional ball, proper holomorphic maps are more rigid as H. Alexander proved the following theorem.

Theorem 1.2 (Alexander (1974) [AL74]) *Every proper holomorphic self-maps of the unit ball $\mathbb{B}^n$ where $n \geq 2$ are holomorphic automorphisms.*

We say that proper holomorphic maps $f_1, f_2 : \Omega_1 \to \Omega_2$ are equivalent if and only if $f_1 = g_2 \circ f_2 \circ g_1$ for some $g_i \in \mathrm{Aut}(\Omega_i)$. Hence every proper holomorphic self-maps of the unit ball are equivalent to the identity map. A first result on rigidity of proper holomorphic maps between balls with different dimensions is due to Webster ([W79]). He proved that every proper holomorphic maps from $\mathbb{B}^n$ to $\mathbb{B}^{n+1}$ for $n \geq 3$, which extends C^3 up to the boundary, are equivalent to

$$(z_1, \ldots, z_n) \mapsto (z_1, \ldots, z_n, 0).$$

For $n = 2$, Alexander suggested a proper map $(z, w) \mapsto (z^2, \sqrt{2}zw, w^2)$ and Faran classified sufficiently big subset of proper holomorphic maps as following:

Theorem 1.3 (Faran (1982) [F82]) *Let $f : \mathbb{B}^2 \to \mathbb{B}^3$ be a proper holomorphic map that is C^3 up to the boundary. Then f is equivalent to one of the following:*

$$(z, w) \mapsto (z^3, w^3, \sqrt{3}zw)\,, \qquad (z, w) \mapsto (z, zw, w^2)\,,$$
$$(z, w) \mapsto (z^2, \sqrt{2}zw, w^2)\,, \qquad (z, w) \mapsto (z, w, 0)\,.$$

If the difference of the dimension gets bigger, there are infinitely many inequivalent proper holomorphic maps between the unit balls. For example, proper holomorphic maps $Q_t : \mathbb{B}^n \to \mathbb{B}^{2n}$ with $t \in [0, 1]$ defined by

$$Q_t(z) = \left(z_1, \ldots, z_{n-1}, tz_n, \left(1-t^2\right) z_1 z_n, \left(1-t^2\right) z_2 z_n, \ldots, \left(1-t^2\right) z_n^2\right)$$

are inequivalent for every $t \in [0, 1]$. For the projection $\pi_0 : \mathbb{C}^{2n} \to \mathbb{C}^{2n-1}$ given by $(z_1, \ldots, z_{2n}) \mapsto (z_1, \ldots, z_{n-1}, z_{n+1}, \ldots, z_{2n})$ and $\pi_1 : \mathbb{C}^{2n} \to \mathbb{C}^n$ given by $(z_1, \ldots, z_{2n}) \mapsto (z_1, \ldots, z_n)$, $\pi_0 \circ Q_0 : \mathbb{B}^n \to \mathbb{B}^{2n-1}$ is the *Whitney map* and $\pi_1 \circ Q_1 : \mathbb{B}^n \to \mathbb{B}^n$ is the identity map. The Whitney map and the identity map are homotopic to each other by Q_t. On the other hand, if one considers only monomial proper holomorphic maps, there are 12 number of maps.

Theorem 1.4 (D'Angelo (1988) [DA88]) *Let $f : \mathbb{B}^2 \to \mathbb{B}^4$ be a monomial proper holomorphic map. Then f is equivalent to one of the following:*

$$(z, w, 0, 0),\ (z^2, zw, w, 0),\ (z^2, \sqrt{2}zw, w^2, 0),\ (z^3, \sqrt{3}zw, w^3, 0),$$
$$(z^3, \sqrt{3}z^2w, \sqrt{3}zw^2, w^3),\ (z^3, z^2w, zw, w),\ (z^2, z^2w, zw^2, w),$$
$$(z^2, \sqrt{2}z^2w, \sqrt{2}zw^2, w^2),\ (z^3, \sqrt{3}z^2w, \sqrt{2}zw^2, w^2),\ (z, z^2w, \sqrt{2}zw^2, w^3),$$
$$(z^4, z^3w, \sqrt{3}zw, w^3),\ (z^4, \sqrt{3}z^2w, zw^3, w),\ (z^5, \sqrt{5}z^3w, \sqrt{5}zw^2, w^5),$$
$$(z, \cos(\theta)w, \sin(\theta)zw, \sin(\theta)w^2),\ (z^2, \sqrt{(1+\cos^2(\theta))}zw, \cos(\theta)w^2, \sin(\theta)w).$$

There are many other result finding difference of the dimension to make proper holomorphic maps rigid. For more detail, see [F86, Fo89, Ds09, D07, L12, HU01, Ds88, HA05, HJ06, Hu04] and the references therein.

2 Rigidity of Proper Holomorphic Maps Between Bounded Symmetric Domains

The first work on the rigidity of proper holomorphic maps between bounded symmetric domains of rank greater than or equal to two are given by G. M. Khenkin and R. Novikov in 1982. They proved that every proper holomorphic self-maps of irreducible bounded symmetric domains of rank ≥ 2 are holomorphic automorphisms. In case of proper holomorphic maps between two different bounded symmetric domains, the difference of rank is more crucial than that of dimension because of fine structures of bounded symmetric domains. Especially, orbits of domain's automorphism group at a given point in its compact dual and boundary arc components are explicitly described in [W72].

Definition 2.1 ([W72]) Let V is a complex analytic space and $S \subset V$ is a subset. We call a holomorphic map $f : \Delta \to V$ with $f(\Delta) \subset S$ a *holomorphic arc* in S. We call a finite sequence $\{f_1, \ldots, f_k\}$ of holomorphic arcs in S such that $f_j(\Delta) \bigcap f_{j+1}(\Delta) \neq \emptyset$ for $1 \leq j \leq k-1$ *chain of holomorphic arcs* in S. Then for $v_1, v_2 \in S$, $v_1 \sim v_2$ if and only if there is a chain $\{f_1, \ldots, f_k\}$ such that $v_1 \in f_1(\Delta)$ and $v_2 \in f_k(\Delta)$. We call the equivalence classes of $\sim$ the *holomorphic arc components* of S in V.

Using the property that proper holomorphic maps between bounded symmetric domains which can be extended over the boundary should preserve these boundary components, N. Mok and I. H. Tsai proved that every proper holomorphic maps between bounded symmetric domain should preserve *the maximal characteristic subspaces* (See [Mo92].) which are totally geodesic subspaces in its ambient bounded symmetric domain. With these fine structures, I. H. Tsai proved that proper holomorphic maps between bounded symmetric domains are very rigid in some special case.

Theorem 2.1 (Tsai (1993) [T93]) *Let (Ω_1, g_1) and (Ω_2, g_2) be bounded symmetric domains. Suppose that Ω_1 is irreducible and $rank(\Omega_1) \geq rank(\Omega_2) \geq 2$. Then $rank(\Omega_1) = rank(\Omega_2)$ and any proper holomorphic map $f : \Omega_1 \to \Omega_2$ is a totally geodesic isometric embedding up to normalizing constants.*

Based on I. H. Tsai's result and structures given by Wolf et al. in [W72, Mo92], it is proved that for equidimensional bounded symmetric domains Ω_1, Ω_2 where Ω_1 is irreducible and rank is greater than 1, then every proper holomorphic maps should be biholomorphism [Tu02]. For the specific case, when f is a proper holomorphic map from $\Omega^I_{p,p-1}$ to $\Omega^I_{p,p}$, f should be a totally geodesic isometric embedding with respect to their Bergman metrics [Tu03]. Furthermore, for proper holomorphic maps from $\Omega^I_{r,s}$ to $\Omega^I_{r',s'}$ where $s \geq 2$ and $s \geq r' \geq r$, if $r' \leq 2r - 1$, then the map is equivalent to $Z \mapsto \begin{pmatrix} Z & 0 \\ 0 & 0 \end{pmatrix}$ ([Ng13]).

Bounded symmetric domains can be canonically embedded into some compact manifold, so called, the *compact dual* by *the Borel embedding*. The compact dual of $\Omega^I_{r,s}$ is the Grassmannian of s-dimensional plane in $\mathbb{C}^{r+s}$ and the Borel embedding is given by

$$Z \mapsto [v_1 \wedge \cdots \wedge v_r]$$

where $[v_1 \wedge \cdots \wedge v_r]$ is the r-dimensional plane in $\mathbb{C}^{r+s}$ generated by row vectors $(v_1, \ldots, v_r)$ of $\left(I_r, Z\right)$. In [Mo92, T93], authors considered the subspaces of bounded symmetric domains which are called *the invariantly geodesic subspaces* which are totally geodesic under the action of automorphism group of its compact dual.

Definition 2.2 For $W \in \Omega_{r',s'}$ with $r' \leq r$ and $s' \leq s$, consider the image of the embedding

$$W \hookrightarrow \begin{pmatrix} 0 & 0 \\ 0 & W \end{pmatrix} \in \Omega_{r,s}$$

which is an invariantly geodesic subspace in $\Omega^I_{r,s}$. The totally geodesic subspaces which are equivalent under the action of $\mathrm{Aut}(\Omega^I_{r,s}) = SU(r, s)$ to this subspace are called *(r', s')-subspaces* of $\Omega_{r,s}$.

Let $D_{r,s}$ be a generalized ball in $\mathbb{P}^{r+s-1}$ which is defined by

$$D_{r,s} = \{[z_1, \ldots, z_{r+s}] \in \mathbb{P}^{r+s-1} : |z_1|^2 + \cdots + |z_r|^2 > |z_{r+1}|^2 + \cdots + |z_{r+s}|^2\}.$$

Bounded symmetric domains of type I are more intensively studied by S. C. Ng using maximal invariantly geodesic subspaces. For $X \in M(r, r+s, \mathbb{C})$ of rank r, denote $[X]$ a r-plane in $\mathbb{C}^{r+s}$ which is generated by the row vectors of X. For $\Omega_{r,s}$ and $D_{r,s}$, consider the two surjective maps

$$\phi : \mathbb{P}^{r-1} \times \Omega_{r,s} \to \Omega_{r,s}, \ ([X], Z) \mapsto Z \tag{2.1}$$

$$\psi : \mathbb{P}^{r-1} \times \Omega_{r,s} \to D_{r,s}, \ ([X], Z) \mapsto [X, XZ]. \tag{2.2}$$

For $Z \in \Omega_{r,s}$, denote $Z^\# = \psi(\phi^{-1}(Z)) \subset D_{r,s}$. Similarly for $X \in D_{r,s}$, denote $X^\# = \phi(\psi^{-1}(X)) \subset \Omega_{r,s}$. $Z^\#$ and $X^\#$ are called fibral images of Z and X respectively. Then for $Z \in \Omega_{r,s}$ and $X = [A, B] \in D_{r,s}$ where $A \in M(1, r, \mathbb{C})$ and $B \in M(1, s, \mathbb{C})$,

$$Z^\# = \{[A, AZ] \in D_{r,s} : [A] \in \mathbb{P}^{r-1}\} \cong \mathbb{P}^{r-1} \tag{2.3}$$

$$X^\# = \{Z \in \Omega_{r,s} : AZ = B\} \cong (r-1, s)\text{-subspace} \tag{2.4}$$

This implies that the maximal subspaces, $(r-1, s)$-subspaces are parametrized by the generalized ball $D_{r,s}$ in $\mathbb{P}^{r+s-1}$.

Theorem 2.2 (Ng (2013) [Ng13]) *Let $r \geq r' \geq 2$ and $f : \Omega_{r,s} \to \Omega_{r',s'}$ be a proper holomorphic map. Suppose that f maps $(r-1, s)$-subspaces into $(r'-1, s')$-subspaces and $f(\Omega_{r,s})$ is not contained in a single $(r'-1, s')$-subspace. Then $r = r'$, $s \leq s'$ and f is equivalent to $Z \mapsto \begin{pmatrix} Z | 0 \end{pmatrix}$.*

The condition mapping $(r-1, s)$-subspaces to $(r'-1, s')$-subspaces may be considered strange. However since every proper holomorphic maps should send $(r-1, s-1)$-subspaces to $(r'-1, s'-1)$-subspaces, this condition is a little bit stronger, but reasonable.

3 Other Proper Holomorphic Maps Between Bounded Symmetric Domains of Type I

Now we can ask that what kind of proper holomorphic maps are there between bounded symmetric domains non-equivalent to

$$Z \mapsto \begin{pmatrix} Z & 0 \\ 0 & h(Z) \end{pmatrix}$$

for some holomorphic function $h : \Omega^I_{r,s} \to \mathbb{C}$ satisfying $I_{s-r,s-r} - h(Z)h(Z)^* > 0$ for all $Z \in \Omega^I_{r,s}$? If the difference of the rank gets bigger, there exists many other proper holomorphic maps.

Using the relations between proper holomorphic maps from $\Omega^I_{r,s}$ to $\Omega^I_{r',s'}$ and proper rational maps from $D_{r,s}$ to $D_{r',s'}$ which parametrize $(r-1, s)$-subspaces of $\Omega^I_{r,s}$ and $(r'-1,s')$-subspaces of $\Omega^I_{r',s'}$ [Ng13] and [SE], one can find proper holomorphic maps between $\Omega^I_{r,s}$ and $\Omega^I_{r',s'}$ which maps $(r-1,s)$-subspaces to $(r'-1,s')$-subspaces explicitly. Since finding proper holomorphic maps between bounded symmetric domains of type I is relatively harder than finding proper rational maps between generalized balls, it gives an effective way.

Example 3.1 (Generalized Whitney map) In [SE], there is a proper holomorphic map $f:\Omega^I_{2,2}\to\Omega^I_{3,3}$ given by

$$f\left(\begin{pmatrix} z_1 & z_2 \\ z_3 & z_4 \end{pmatrix}\right)=\begin{pmatrix} z_1^2 & z_1z_2 & z_2 \\ z_1z_3 & z_2z_3 & z_4 \\ z_3 & z_4 & 0 \end{pmatrix}.$$

This map can be generalized to proper holomorphic map from $\Omega^I_{r,s}$ to $\Omega^I_{2r-1,2s-1}$ as

$$f\left(\begin{pmatrix} z_{11} & \cdots & z_{1s} \\ \vdots & \ddots & \vdots \\ z_{r1} & \cdots & z_{rs} \end{pmatrix}\right)=\begin{pmatrix} z_{11}^2 & z_{11}z_{12} & \cdots & z_{11}z_{1s} & z_{12} & \cdots & z_{1s} \\ z_{11}z_{21} & z_{21}z_{12} & \cdots & z_{21}z_{1s} & z_{22} & \cdots & z_{2s} \\ \vdots & \vdots & \ddots & \vdots & \vdots & \ddots & \vdots \\ z_{11}z_{r1} & z_{r1}z_{12} & \cdots & z_{r1}z_{1s} & z_{r2} & \cdots & z_{rs} \\ z_{21} & z_{22} & \cdots & z_{2s} & 0 & \cdots & 0 \\ \vdots & \vdots & \ddots & \vdots & \vdots & \ddots & \vdots \\ z_{r1} & z_{r2} & \cdots & z_{rs} & 0 & \cdots & 0 \end{pmatrix}$$

Under the condition of Theorem 2.1, every proper holomorphic maps can be extended to its compact dual as a meromorphic maps. But this is no longer true if the rank of the target domain is bigger than that of the source domain as we can see via the generalized Whitney maps.

Example 3.2 (Homogeneous map) In [SE], there is a proper holomorphic map given by

$$f\left(\begin{pmatrix} z_1 & z_2 \\ z_3 & z_4 \end{pmatrix}\right)=\begin{pmatrix} z_1^2 & \sqrt{2}z_1z_2 & z_2^2 \\ \sqrt{2}z_1z_3 & z_1z_4+z_2z_3 & \sqrt{2}z_2z_4 \\ z_3^2 & \sqrt{2}z_3z_4 & z_4^2 \end{pmatrix}$$

which is first found by S.C. Ng. This homogeneous map can be also generalized to higher dimensional domain, i.e. from $\Omega^I_{r,s}$ to $\Omega^I_{\frac{1}{2}r(r+1),\frac{1}{2}s(s+1)}$.

Definition 3.1 Let $f,g:\Omega_{r,s}\to\Omega_{r',s'}$ be holomorphic maps. Then we call f and g are *isotropically equivalent* if there are $U\in \mathrm{Isot}_0(\Omega_{r,s})$ and $V\in\mathrm{Isot}_0(\Omega_{r',s'})$ such that $f=V\circ g\circ U$.

In [Ds88], D'Angelo proved that for $f,g:\mathbb{B}^n\to\mathbb{B}^N$ is equivalent if and only if f and g are isotropically equivalent. Extending this result to the bounded symmetric domains, in [SE], the author proved that $f, g : \Omega_{r,s} \to \Omega_{r',s'}$ is equivalent if and only if f, g are isotropically equivalent. Using this property, one can find infinitely many inequivalent proper holomorphic maps between $\Omega_{2,2}$ to $\Omega_{4,4}$.

Example 3.3 Let $f_t : \Omega_{2,2} \to \Omega_{4,4}$ with $t \in [0, 1]$ be maps defined by

$$f_t\left(\begin{pmatrix} z_1 & z_2 \\ z_3 & z_4 \end{pmatrix}\right) = \begin{pmatrix} z_1^2 & \frac{1}{1+t}z_2^2 & \frac{2+t}{1+t}z_1 z_2 & \frac{1}{1+t}z_2 \\ 0 & z_4^2 & (2+t)z_3 z_4 & 0 \\ 0 & \frac{2}{2+t}z_2 z_4 & z_1 z_4 + z_2 z_3 & \frac{1}{2+t}z_4 \\ (1+t)z_3 & 0 & 0 & 0 \end{pmatrix}. \tag{3.1}$$

f_t are inequivalent proper holomorphic maps for all $t \in [0, 1]$.

References

[AL74] Alexander, H.: Holomorphic mappings from the ball and polydisc. Math. Ann. **209**, 249–256 (1974)

[CA35] Cartan, E.: Domaines bornes dans l'espace de n variables complexes. Abh. Math. Sem. Hamburg **11**, 116–162 (1935)

[Ds88] D'Angelo, J.P.: Proper holomorphic maps between balls of different dimensions. Mich. Math. J. **35**(1), 83–90 (1988)

[DA88] D'Angelo, J.P.: Polynomial proper maps between balls. Duke Math. J. **57**(1), 211–219 (1988)

[Ds09] D'Angelo, J.P., Lebl, J.: On the complexity of proper holomorphic mappings between balls. Complex Var. Elliptic Equ. **54**(3–4), 187–204 (2009)

[D07] D'Angelo, J.P., Lebl, J., Peters, H.: Degree estimates for polynomials constant on a hyperplane. Mich. Math. J. **55**(3), 693–713 (2007)

[F82] Faran, J.J.: Maps from the two-ball to the three-ball. Invent. Math. **68**(3), 441–475 (1982)

[F86] Faran, J.J.: The linearity of proper holomorphic maps between balls in the low codimension case. J. Differ. Geom. **24**(1), 15–17 (1986)

[Fo89] Forstneric, F.: Extending proper holomorphic mappings of positive codimension. Invent. Math. **95**(1), 31–61 (1989)

[HA05] Hidetaka, H.: Rational proper holomorphic maps from $\mathbb{B}^n$ into $\mathbb{B}^{2n}$. Math. Ann. **331**(3), 693–711 (2005)

[HU01] Huang, X., Ji, S.: Mapping $\mathbb{B}^n$ into $\mathbb{B}^{2n-1}$. Invent. Math. **145**(2), 219–250 (2001)

[Hu04] Huang, X., Ji, S., Yin, W.: On the third gap for proper holomorphic maps between balls. Math. Ann. **358**(1–2), 115–142 (2014)

[HJ06] Huang, X., Ji, S., Xu, D.: A new gap phenomenon for proper holomorphic mappings from $\mathbb{B}^n$ into $\mathbb{B}^N$. Math. Res. Lett. **13**(4), 515–529 (2006)

[L12] Lebl, J., Peters, H.: Polynomials constant on a hyperplane and CR maps of spheres, Ill. J. Math. **56**(1), 155–175 (2012) 2013

[Mo92] Mok, N., Tsai, I.H.: Rigidity of convex realizations of irreducible bounded symmetric domains of rank $\geq$ 2. J. Reine Angew. Math. **431**, 91–122 (1992)

[Ng13] Ng, S.-C.: Holomorphic double fibration and the mapping problems of classical domains. Int. Math. Res. Not. first published online september 18 (2013) doi:10.1093/imrn/rnt200

[SE] Seo, Aeryeong: New examples of proper holomorphic maps among symmetric domains. Michigan Math. J. **64**(2), 435–448 (2015)

[T93] Tsai, I.-H.: Rigidity of proper holomorphic maps between symmetric domains. J. Differ. Geom. **37**(1), 123–160 (1993)

[Tu02] Tu, Z.-H.: Rigidity of proper holomorphic mappings between nonequidimensional bounded symmetric domains. Math. Z. **240**(1), 13–35 (2002)

[Tu03] Tu, Z.-H.: Rigidity of proper holomorphic mappings between equidimensional bounded symmetric domains. Proc. Am. Math. Soc. **130**(4), 1035–1042 (2002)

[W72] Wolf, J.A.: Fine structure of Hermitian symmetric spaces. In: Symmetric Spaces (Short Courses, Washington Univ., St. Louis, Mo., 1969–1970), pp. 271–357. Pure and App. Math., vol. 8, Dekker, New York (1972)

[W79] Webster, S.M.: The rigidity of C-R hypersurfaces in a sphere. Indiana Univ. Math. J. **28**(3), 405–416 (1979)

Some Dynamical Systems of Extremal Measures

Hajime Tsuji

Abstract We consider the dynamical system of extremal measures on a compact Kähler manifold. And we show that the dynamical system converges to the canonical measure, if we assume the abundance of the canonical bundle.

Keywords Kähler-Einstein metrics · Bergman kernels · Extremal measures

1 Introduction

In complex geometry, there has been introduced several intrinsic (pseudo)volume forms on complex manifolds such as Bergman volume forms, Kähler-Einstein volume forms, Carathéodory volume forms and Kobayashi volume forms (cf. [B, C, Y, C-Y, K]). It is interesting to study the relation between these volume forms.

In this survey article, I would like to review my recent works about the construction of Kähler-Einstein volume forms in terms of the dynamical systems of twisted Bergman volume forms (cf. Sect. 1.3) and twisted estremal measures (cf. Sect. 2). These new constructions of Kähler-Einstein volume forms or more generally canonical measures will be used to study the pluricanonical systems of compact Kähler manifolds in [T9].

Let us briefly review several invariant volume forms which are used in this article.

1.1 Bergman Volume Forms

The Bergman volume forms is the most basic invariant volume form in complex geometry.

H. Tsuji (✉)
Department of Mathematics, Sophia University, 7-1 Kioicho,
Chiyoda-ku 102–8554, Japan
e-mail: h-tsuji@h03.itscom.net

F. Bracci et al. (eds.), *Complex Analysis and Geometry*, Springer Proceedings
in Mathematics & Statistics 144, DOI 10.1007/978-4-431-55744-9_25

Let Ω be a bounded pseudoconvex domain in $\mathbb{C}^n$ and let K_Ω denote the canonical bundle of Ω. Then the space of L^2-canonical forms on Ω:

$$A^2(\Omega, K_\Omega) = \left\{\sigma \in \Gamma(\Omega, \mathcal{O}_\Omega(K_\Omega)) \middle| \int_\Omega |\sigma|^2 < +\infty\right\} \quad (|\sigma|^2 = (\sqrt{-1})^{n^2}\sigma \wedge \overline{\sigma})$$

is a Hilbert space with respect to the inner product:

$$(\sigma, \tau) := (\sqrt{-1})^{n^2}\int_\Omega \sigma \wedge \overline{\tau}.$$

We set

$$K(\Omega)(x) := \sup\{|\sigma|^2(x) | \sigma \in A^2(\Omega, K_\Omega), \| \sigma \| = 1\} (x \in \Omega), \tag{1.1}$$

where $\| \sigma \|$ denotes the L^2-norm of σ. We call $K(\Omega)$ *the Bergman volume form* on Ω. This definition is a little bit different from the usual one. But this definition coincides with the usual definition:

$$K(\Omega) := \sum_i |\sigma_i|^2,$$

where $\{\sigma_i\}$ is a complete orthonomal basis of $A^2(\Omega, K_\Omega)$ (The definition is independent of the choice of the complete orthonormal basis $\{\sigma_i\}$). Then

$$\omega_B := -\mathrm{Ric}\, K(\Omega)$$

is the pull back of the Fubini-Study Kähler form by the embedding

$$\Phi : \Omega \longrightarrow \mathbb{P}^\infty$$

defined by

$$\Phi(x) := [\sigma_1(x) : \sigma_2(x) : \cdots : \sigma_k(x) : \cdots].$$

Hence ω_B is C^∞ and

$$\omega_B > 0$$

holds, i.e., ω_B is a C^∞-Kähler form on Ω.

The Bergman kernel can be generalized to the case of compact complex manifolds as follows. Let X be a compact complex manifold of dimension n and let (L, h_L) be a singular hermitian line bundle on X. Let $\{\sigma_j\}$ be a complete orhonormal basis of $H^0(X, \mathcal{O}_X(K_X + L) \otimes \mathcal{I}(h_L))$ with respect to the L^2-inner product:

$$(\sigma, \tau) := (\sqrt{-1})^{n^2}\int_X \sigma \wedge \overline{\tau} \cdot h_L.$$

And we set

$$K(X, K_X + L, h_L) := \sum_j |\sigma_j|^2. \tag{1.2}$$

Then $K(X, K_X+L, h_L)$ is a semipositive $|L|^2 = L \otimes \bar{L}$-valued semipositive (n, n)-form on X such that $K(X, K_X + L, h_L)^{-1}$ is a singular hermitian metric on $K_X + L$ with semipositive curvature current, unless it is not identically $+\infty$.

As above, in the case of a compact complex manifold, the Bergman kernel is defined in terms of line bundle on X,

1.2 Kähler-Einstein Volume Forms

On the other hand, there exists another invariant volume form on a bounded pseudoconvex domain in $\mathbb{C}^n$.

Theorem 1.1 ([C-Y, M-Y]) *Let Ω be a bounded pseudoconvex domain in $\mathbb{C}^n$. Then there exists a unique complete Kähler form ω_E on Ω such that*

$$-\mathrm{Ric}(\omega_E) = \omega_E$$

holds.

By the definition, we see that *the Kähler-Einstein volume form*:

$$dV_E := \frac{1}{n!}\omega_E^n$$

satisfies

$$-\mathrm{Ric}(dV_E) = \omega_E > 0.$$

In the compact case, we have the following theorem,

Theorem 1.2 ([A, Y]) *Let X be a smooth projective variety with ample K_X. Then there eixsts a unique C^∞-Kähler form ω_E such that $-Ric(\omega_E) = \omega_E$ holds on X.*

1.3 Dynamical System of Bergman Kernels

In [T3], I have discovered the relation between a dynamical system of Bergman volumes forms and the Kähler-Einstein volume form on a smooth projective variety with ample canonical bundle.

The relation is described as follows. Let X be a smooth projective variety with ample canonical bundle. Let A be a sufficiently ample line bundle on X and let h_A be a C^∞-hermitian metric on A. We set

$$K_1 := K(X, K_X + A, h_A)$$

and

$$h_1 := K_1^{-1}.$$

We note that $mK_X + A$ is globally generated for every positive integer m. Then h_1 is a C^∞-hermitian metric on $K_X + A$. For every $m \geqq 2$, we define inductively

$$K_m := K(X, mK_X + A, h_{m-1})$$

and

$$h_m = K_m^{-1}.$$

Then we have the following theorem.

Theorem 1.3 ([T3, S-W])

$$\lim_{m\to\infty} \left((m!)^{-n} h_A \cdot K_m\right)^{\frac{1}{m}} = \frac{1}{(2\pi)^n} dV_E$$

holds in the uniform topology.

1.4 Supercanonical Volume Form

In [T5], I have introduced the canonical volume form on a smooth projective variety with pseudoeffective canonical bundle.

Let X be a smooth projective n-fold such that the canonical bundle K_X is pseudoeffective. Let A be a sufficiently ample line bundle such that for every pseudoeffective singular hermitian line bundle (L, h_L) on X, $\mathscr{O}_X(A + L) \otimes \mathscr{I}(h_L)$ and $\mathscr{O}_X(K_X + A + L) \otimes \mathscr{I}(h_L)$ are globally generated. The existence of such an ample line bundle A follows from Nadel's vanishing theorem ([N, p. 561]).

For every $x \in X$ we set

$$\hat{K}_m^A(x) := \sup\left\{ \mid \sigma(x) \mid^{\frac{2}{m}} \middle| \sigma \in \Gamma(X, \mathscr{O}_X(mK_X + A)), \parallel \sigma \parallel_{\frac{1}{m}} = 1 \right\}, \quad (1.3)$$

where

$$\parallel \sigma \parallel_{\frac{1}{m}} := \left| \int_X h_A^{\frac{1}{m}} \cdot (\sigma \wedge \bar{\sigma})^{\frac{1}{m}} \right|^{\frac{m}{2}}. \quad (1.4)$$

Here $\mid \sigma \mid^{\frac{2}{m}}$ is not a function on X, but the supremum is takan as a section of the real line bundle $\mid A \mid^{\frac{2}{m}} \otimes \mid K_X \mid^2$ in the obvious manner.[1] Then $h_A^{\frac{1}{m}} \cdot \hat{K}_m^A$ is a continuous semipositive (n, n)-form on X. Under the above notations, we have the following theorem.

Theorem 1.4 ([T5]) *We set*

$$\hat{K}_\infty^A := \limsup_{m \to \infty} h_A^{\frac{1}{m}} \cdot \hat{K}_m^A$$

and

$$\hat{h}_{can,A} := \text{the lower envelope of } (\hat{K}_\infty^A)^{-1}.$$

Then $\hat{h}_{can,A}$ is an AZD(cf. Definition 1.1) of K_X. And we define

$$\hat{h}_{can} := \text{the lower envelope of } \inf_A \hat{h}_{can,A},$$

where inf *denotes the pointwise infimum and A runs all the ample line bundles on X. Then $\hat{h}_{can}$ is a well defined AZD(cf. Definition 1.1) on K_X with minimal singularities depending only on X.*

Remark 1.1 As one sees later, $\hat{h}_{can,A} = \hat{h}_{can}$ holds for every sufficiently ample A.

Definition 1.1 Let M be a compact complex manifold and let L be a line bundle on X. A singular hermitian metric h_L is said to be an AZD of L if the followings are satisfied:

(1) $\sqrt{-1}\Theta(h_L) \geqq 0$ holds on M,
(2) $H^0(M, \mathcal{O}_M(mL) \otimes \mathcal{I}(h_L^m)) \simeq H^0(M, \mathcal{O}_M(mL))$ holds for every $m \geqq 0$.

Definition 1.2 Let X be a smooth projective variety with pseudoeffective canonical bundle. Let $\hat{h}_{can}$ be as in Theorem 1.4. We set

$$d\hat{\mu}_{can} := \hat{h}_{can}^{-1}$$

and call it the supercanonical measure.

1.5 Extremal Measures

Now we shall introduce an intrisic volume form on a bounded pseudoconvex domain.

Let Ω be a bounded pseudoconvex domain in $\mathbb{C}^n$. We shall introduce a new invariant volume form on Ω:

---

[1] We have abused the notations $\mid A \mid$, $\mid K_X \mid$ here. These notations are similar to the notations of corresponding linear systems. But we shall use the notation if without fear of confusion.

$$d\mu(\Omega)_{ext}(x) := \left(\sup\left\{dV(x)| - \mathrm{Ric}\, dV \geqq 0, \int_{\Omega} dV = 1\right\}\right)^{*} \quad (x \in \Omega),$$

where dV runs all the upper-semicontinuous semipostiive (n, n)-forms such that dV^{-1} is a singular hermitian metric on K_{Ω} and $-\mathrm{Ric}\, dV = \sqrt{-1}\partial\bar{\partial}\log dV \geqq 0$ in the sense of current and $(\ )^{*}$ denotes the uppersemicontinuous envelope. We call $d\mu(\Omega)_{ext}$ *the extremal measure* of Ω. Then we see that

$$-\mathrm{Ric}\, d\mu(\Omega)_{ext} \geqq 0$$

holds in the sense of current.

This definition can be generalized to the case of compact complex manifold with pseudoeffective canonical bundles.

1.6 Relation Between Extremal Measures and Supercanonical Measures

In the case of a smooth projective variety, the extermal measure conicides with the supercanonical measure.

Theorem 1.5 ([T8]) *Let X be a smoorh projective variety with pseudoeffective canonical bundle. Then $d\hat{\mu}(X)_{can} = d\mu(X)_{ext}$ holds.*

By the logarithmic plurisubharmonic variation property of supercanonical measures ([T5]), we have the following corollary.

Corollary 1.1 ([T5]) *Let $f : X \to S$ be a surjective projective morphism with connected fibers such that X and S are smooth. Suppose that $K_{X/S}$ is relatively pseudoeffective. Let S° be the smoorh locus of f. We define the relative extremal measure $d\mu_{X/S}$ on $f^{-1}(S^{\circ})$ by*

$$d\mu_{X/S}|_{X_s} := d\mu(X_s)_{ext}, s \in S^{\circ}.$$

Then we have

(1) *$d\mu_{X/S}^{-1}$ extends to a singular hermitian metric $h_{X/S,ext}$ of $K_{X/S}$ on the whole X.*
(2) *$\sqrt{-1}\Theta(h_{X/S}) \geqq 0$ holds on X.*

2 Dynamical Systems of Extremal Measures

In this section, we shall consider a dynamical system of extremal measures and prove that its normalized limit exists and is the Kähler-Einstein volume form. This result is similar to the dynamical construction of the Kähler-Einstein volume form in [T3].

2.1 Dynamical System of Extremal Measure

Let X be a smooth projective variety with ample canonical bundle. Let A be a sufficiently ample line bundle on X such that for every pseudoeffective singular hermitian line bundle (F, h_F) on X, $\mathscr{O}_X(A+F)\otimes\mathscr{I}(h_F)$ is globally generated on X. Such a A exists by Nadel's vanishing theroem ([N, p. 561]). Let us fix a C^∞-hermitian metric h_A on A. Let $d\mu(A, h_A)_{ext}$ be the extremal measure associated with (A, h_A), i.e.,

$$d\mu(A, h_A)(x) := \left(\sup\left\{dV(x) | -\mathrm{Ric}\, dV + \sqrt{-1}\Theta(h_A) \geqq 0, \int_\Omega dV = 1\right\}\right)^* \quad (x \in X),$$

where ( )* denotes the uppersemicontinuous envelope. And we set

$$h_1 := d\mu(A, h_A)^{-1} \otimes h_A,$$

And for $m \geqq 2$, inductively we define

$$h_m := d\mu((m-1)K_X + A, h_{m-1})^{-1} \otimes h_{m-1},$$

where

$$d\mu((m-1)K_X + A, h_{m-1})(x) := \left(\sup\left\{dV(x) | -\mathrm{Ric}\, dV + \sqrt{-1}\Theta(h_{m-1}) \geqq 0, \int_\Omega dV = 1\right\}\right)^* \quad (x \in X).$$

Then by definition, h_m is a singular hermitian metric on $mK_X + A$ with semipositive curvature current. We set

$$dv_m := h_m^{-1} = d\mu((m-1)K_X + A, h_{m-1}) \otimes h_{m-1}^{-1}.$$

Then $h_A \cdot dv_m$ is a m-ple volume form on X.

By the assumption that K_X is ample, there exists a C^∞-Kähler form ω_E such that $-\mathrm{Ric}(\omega_E) = \omega_E$ holds on X. Let n denote the dimension of X and we set

$$dV_E := \frac{1}{n!}\omega_E^n.$$

Then we have the following theorem.

Theorem 2.1 ([T8])

$$\lim_{m\to\infty} h_A^{\frac{1}{m}} \cdot ((m!)^{-n} \cdot dv_m)^{\frac{1}{m}} = \frac{1}{(2\pi)^n} dV_E$$

holds in the uniform topology on X.

2.2 Proof of Theorem 2.1

Now we shall explain the proof of Theorem 2.1 according to [T8].

As in Sect. 2.1, we set

$$K_1 := K(X, K_X + A, h_A)$$

and for $m \geqq 2$, inductively we define

$$K_m := K(X, K_X + A, K_{m-1}^{-1}).$$

By induction on m and the definition of A, we see that K_m^{-1} is a C^∞-hermitian metric on $mK_X + A$ for every m.

Now we shall compare dv_m and K_m.

Lemma 2.1 $dv_m \geqq K_m$ *holds on* X *for every* $m \geqq 1$.

Proof By the construction we have that

$$dv_m(x) = \sup\{h^{-1}(x) | \sqrt{-1}\Theta(h) \geqq 0, \int_X h^{-1} \cdot h_{m-1} = 1\}, \tag{2.1}$$

holds, where h runs lower semicontinuous singular hermitian metrics on $mK_X + A$. On the other hand by the extremal property of the Bergman kernel, we have that

$$K_m(x) = \sup\{|\sigma(x)|^2 | \sigma \in \Gamma(X, \mathcal{O}_X(mK_X + A)), \int_X |\sigma|^2 \cdot K_{m-1}^{-1} = 1\} \tag{2.2}$$

holds. Comparing (2.1) and (2.2), noting $h_0 = K_0^{-1} = h_A$, the inequality $dv_1 \geqq K_1$ holds on X. This means that $h_1 \leqq K_1^{-1}$. Hence again by (2.1) and (2.2), we have that

$$dv_2 \geqq K(X, 2K_X + A, h_1) \geqq K(X, 2K_X + A.K_1^{-1}) = K_2$$

holds. Hence $d\nu_2 \geqq K_2$ holds on X. Continuing this process, we have the desired inequality $d\nu_m \geqq K_m$ for every m. □

By Theorem 1.3 and Lemma 2.1, we have the following lower estimate:

Lemma 2.2

$$\liminf_{m\to\infty} h_A^{\frac{1}{m}} \cdot ((m!)^{-n} \cdot d\nu_m)^{\frac{1}{m}} \geqq \frac{1}{(2\pi)^n} dV_E$$

holds.

Next we shall estimate $d\nu_m$ from above.

Let $x_0 \in X$ be an arbitrary point. Then by the Kähler-Einstein condition, there exists a holomorphic normal coordinate $(U, (z_1, \ldots, z_n))$ with center x_0 such that

(1) $g_{i\bar{j}} = \delta_{ij}$,
(2)

$$\det(g_{i\bar{j}}) = \prod_{j=1}^{n} \left(1 - \frac{1}{2}|z_j|^2\right)^{-1} + O(\| z \|^3), \tag{2.3}$$

(3) U is biholomorphic to $B(O, r)$ for some $r > 0$.

We shall identify U with $B(O, r)$ and later we let r tend to 0.

For $m = 1$, there exists a positive constant C_1 such that

$$d\nu_1 \leqq C_1 \cdot h_A^{-1} \cdot dV_E$$

holds on X. Suppose that for some $m \geqq 1$, such that

$$d\nu_m \leqq C_m \cdot h_A^{-1} \cdot (dV_E)^m$$

holds on X. We note that

$$d\nu_m(x_0) \leqq d\nu(U, C_m^{-1} \cdot h_A \cdot (dV_E)^{-m})(x_0)$$

holds, where

$$d\nu(U, C_m^{-1} \cdot h_A \cdot dV_E^{-m})(x) = \sup\{h(x)^{-1} | \sqrt{-1}\Theta(h) \geqq 0, C_m^{-1} \int_X h^{-1} \cdot h_A \cdot dV_E^{-m} = 1\}.$$

Here h runs lowersemicontinuous singular hermitian metrics on $mK_X + A$ on U.

On the other hand, by Demailly's approximation theorem ([D]), we have that

$$h^{-1} = \lim_{\ell\to\infty} h_A^{\frac{1}{\ell}} \sqrt[\ell]{K(U, \ell(mK_X + A)|_U, h^\ell)}$$

holds. This implies that

$$dv_{m+1}(x_0) \leqq$$
$$\limsup_{\ell\to\infty} \sup\{|\sigma|^{\frac{2}{\ell}}(x_0)|\sigma \in \Gamma(U, \mathscr{O}_X(\ell(mK_X + A))), \int_U |\sigma|^{\frac{2}{\ell}} \cdot h_A \cdot dV_E^{-m} = 1\} \tag{2.4}$$

holds.

Let $\mathbf{e}_A$ be a local holomorphic frame of A on U. Let $\sigma_0 \in \Gamma(U, \mathscr{O}_X(\ell(mK_X+A)))$ be a section such that

$$|\sigma_0(x_0)|^2 =$$
$$\sup\{|\sigma|^{\frac{2}{\ell}}(x_0)|\sigma \in \Gamma(U, \mathscr{O}_X(\ell((m+1)K_X + A)), \int_U |\sigma|^{\frac{2}{\ell}} \cdot h_A \cdot dV_E^{-m} = 1\}.$$

We define $f \in \mathscr{O}_X(U)$ by

$$\sigma_0 = f \cdot \mathbf{e}_A^{\otimes \ell} \otimes (2^{\frac{n}{2}} dz_1 \wedge \cdots \wedge dz_n)^{\otimes m\ell}. \tag{2.5}$$

Then

$$C_m^{-1} \int_U |f|^{\frac{2}{\ell}} \cdot h_A(\mathrm{e}_A, \mathrm{e}_A) \cdot dV_E^{-m} \cdot (2^{-n}|dz_1 \wedge \cdots \wedge dz_n|^2)^{(m+1)} \leqq 1$$

holds. Hence we have

$$\int_U |f|^{\frac{2}{\ell}} \cdot (dV_E^{-1} \cdot (2^{-n}|dz_1 \wedge \cdots \wedge dz_n|^2))^m \cdot |dz_1 \wedge \cdots \wedge dz_n|^2 \leqq (\inf_U h_A(\mathrm{e}_A, \mathrm{e}_A))^{-1} \cdot C_m.$$

By the Taylor expansion (2.3), there exist a function $\varepsilon(r)$ of $r > 0$ such that

$$dV_E^{-1} \cdot (2^{-n}|dz_1 \wedge \cdots \wedge dz_n|^2) \geqq \prod_{j=1}^{n} \left(1 - \frac{1}{2}(1 + \varepsilon(r))|z_j|^2\right) \tag{2.6}$$

and

$$\lim_{r \downarrow 0} \varepsilon(r) = 0. \tag{2.7}$$

We set

$$K_\ell(x_0) := \sup\{|F|^{\frac{2}{\ell}}(x_0) \mid F \in \mathscr{O}_X(U),$$
$$\int_U |F|^{\frac{2}{\ell}} \cdot (\prod_{j=1}^{n} \left(1 - 2^{-1}(1 + \varepsilon(r))|z_j|^2\right))^m \cdot 2^{-n}|dz_1 \wedge \cdots \wedge dz_n|^2 = 1\}.$$

We note that

$$\frac{\sqrt{-1}}{2}\int_{\Delta(0,\rho)}\left(1-\frac{1}{2}|t|^2\right)^m dt\wedge d\bar{t}=\frac{2\pi}{m+1}\left(1-\left(1-\frac{1}{2}\rho^2\right)^{m+1}\right)$$

holds, where $\Delta(0,\rho)$ denotes the unit open disk in $\mathbb{C}$ with center 0 and radius ρ. By the symmetry. we have that

$$K_\ell(x_0)\leqq(1+\varepsilon(r))^n\left(\frac{2\pi}{m+1}\right)^n$$

holds. By (2.5) and (2.6),

$$|f(x_0)|^2\leqq K_\ell(x_0)$$

and by (2.4) and (2.5), there exists a positive constant $c<1$ and and a positive function $\varepsilon(r)$ of r such that

$$dv_{m+1}(x_0)\leqq C_m\cdot\left(\frac{m+1}{2\pi}\right)^n(1+\varepsilon(r))^n(1-c^{m+1})\cdot dV_E^{m+1}(x_0)\cdot h_A^{-1}(x_0)$$

holds. This implies that

$$\limsup_{m\to\infty} h_A^{\frac{1}{m}}\cdot((m!)^{-n}dv_m)^{\frac{1}{m}}(x_0)\leqq\frac{1}{(2\pi)^n}(1+\varepsilon(r))^n dV_E(x_0)$$

holds. Letting $r\downarrow 0$, by (2.7), we have that

$$\limsup_{m\to\infty} h_A^{\frac{1}{m}}\cdot((m!)^{-n}dv_m)^{\frac{1}{m}}(x_0)\leqq\frac{1}{(2\pi)^n}dV_E(x_0)$$

holds. Since x_0 is arbitrary we have the following lemma.

Lemma 2.3

$$\limsup_{m\to\infty} h_A^{\frac{1}{m}}\cdot((m!)^{-n}dv_m)^{\frac{1}{m}}\leqq\frac{1}{(2\pi)^n}dV_E$$

holds on X.

By Lemmas 2.2 and 2.3, we conclude that

$$\lim_{m\to\infty} h_A^{\frac{1}{m}}\cdot((m!)^{-n}dv_m)^{\frac{1}{m}}=\frac{1}{(2\pi)^n}dV_E$$

holds. By the proof, the convergence is uniform on X. This completes the proof of Theorem 2.1.

3 Dynamical System of Extremal Measures on Compact Kähler manifolds

In this section we shall consider the dynamical systems of extremal measures on a compact Kähler manifold. In this case the main difference is that we start from a Kähler which is not a Chern form of an ample line bundle.

The motivation to investigate the dynamical systems of extremal measure on compact Kähler manifolds is to prove the deformation invariance plurigenra for compact Kähler manifolds ([T9]).

3.1 Abundance of Canonical Line Bundle

Let X be a compact complex manifold. The Kodaira dimension $\kappa(X)$ of X is defined by

$$\kappa(X) := \limsup_{m\to\infty} \frac{\log \dim H^0(X, \mathcal{O}_X(mK_X))}{\log m}.$$

It is known that if $\kappa(X) \geqq 0$, then for every $m >> 1$, the complete linear system $|m!K_X|$ defines a rational fibration (unique up to birational equivalence)

$$f : X - \cdots \to Y$$

with $\dim Y = \kappa(X)$. This fibration is called an *Iitaka fibration.*

Definition 3.1 Let X be a compact complex manifold with $\kappa(X) \geqq 0$ and let $f : X \to Y$ be an Iitaka fibration(We may and do assume that f is a morphism and Y is smooth by taking a suitable modification).

K_X is said to be abundant, if there exists a $\mathbb{Q}$-line bundle L on Y and a singular hermitian metric h_L on L such that

(1) There exists an effective $\mathbb{Q}$-divisor E on X such that

$$K_X = f^*L + E,$$

(2) Let σ_E be a multivalued holomorphic section of E with divisor E. Then

$$(f^*h) \cdot \frac{1}{|\sigma_E|^2}$$

is an AZD(Definition 1.1) of K_X.

Remark 3.1 This definition is birationally invariant. Hence the abundance of K_X is defined for every compact complex manifold X. Also this definition can be generalized for any line bundle F on a compact complex manifold with $\kappa(F) \geqq 0$.

3.2 Twisted Kähler-Einstein Currents and Canonical Measures

First we shall review the twisted Kähler-Einstein current on the canonical model of a smooth projective variety with nonnegative Kodaira dimension.

Let X be a smooth projective variety with $\kappa(X) \geqq 0$. Let $f : X - \cdot \to Y$ be the Iitaka fibration of X. The following argument is birationally invariant. We may and do assume that f is a morphism and Y is smooth. Let $L_{X/Y}$ be the Hodge $\mathbb{Q}$-line bundle on Y defined by

$$L_{X/Y} := \frac{1}{m!}(f_*\mathcal{O}_X(m!K_{X/Y}))^{**} \in Div(Y) \otimes \mathbb{Q},$$

where m is a sufficiently large positive integer. We define a singular hermitian metric on $L_{X/Y}$ by

$$h^{m!}_{L_{X/Y}}(\sigma, \sigma) = \left(\int_{X/Y} |\sigma|^{\frac{2}{m!}}\right)^{m!}. \tag{3.1}$$

$h_{L_{X/Y}}$ is said to be the *Hodge metric* on $L_{X/Y}$. We may take $f : X \to Y$ so that some positive multiple of the Hodge $\mathbb{Q}$-line bundle $L_{X/Y}$ is locally free.

Theorem 3.1 (cf. [T7, Theorem 1.5] and [S-T, Theorem B.2]) *In the above notations, there exists a unique singular hermitian metric on h_K on $K_Y + L_{X/Y}$ such that*

(1) *h_K is an AZD of $K_Y + L_{X/Y}$,*
(2) *f^*h_K is an AZD of $K_X + D$,*
(3) *h_K is C^∞ on a nonempty Zariski open subset U of Y,*
(4) *$\omega_Y = \sqrt{-1}\,\Theta_{h_K}$ is a Kähler form on U,*
(5) *ω_Y satisfies the twisted Kähler-Einstein equation:*

$$-\mathrm{Ric}_{\omega_Y} + \sqrt{-1}\,\Theta_{h_{L_{X/Y}}} = \omega_Y$$

holds on U, where $h_{L_{X/Y}}$ denotes the Hodge metric defined as (3.1).

The above equation:

$$-\mathrm{Ric}_{\omega_Y} + \sqrt{-1}\,\Theta_{h_{L_{X/Y}}} = \omega_Y \tag{3.2}$$

is similar to the Kähler-Einstein equation:

$$-\mathrm{Ric}_{\omega_Y} = \omega_Y.$$

The correction term $\sqrt{-1}\,\Theta_{h_{L_{X/Y}}}$ reflects the isomorphism :

$$R(X, K_X)^{(a)} = R(Y, K_Y + L_{X/Y})^{(a)}$$

for some positive integer a, where for a graded ring $R := \oplus_{i=0}^{\infty} R_i$ and a positive integer b, we set

$$R^{(b)} := \oplus_{i=0}^{\infty} R_{bi}.$$

We note that even if X is a compact Kähler manifold the Iitaka fibration $f : X \to Y$, the Hodge $\mathbf{Q}$-line bundle $L_{X/Y}$ and the Hodge metric $h_{L_{X/Y}}$ is well defined and the curvature $\sqrt{-1}\Theta(h_{L_{X/Y}})$ is semipositive. Since the equation (3.2) is defined on the projective variety Y even in this case, Theorem 3.1 is valid, even if X is a compact Kähler manifold with nonnegative Kodaira dimension.

Now we shall define the twisted Kähler-Einstein current and the canonical measure.

Definition 3.2 ([S-T, T6, T7]) The current ω_Y on Y constructed in Theorem 3.1 is said to be *the twisted Kähler-Einstein current* of the Iitaka fibration $f : X \longrightarrow Y$. Also $\omega_{X,D} := f^*\omega_Y$ is said to be *the canonica* of X. We define the measure $d\mu_{can}$ of X by

$$d\mu_{can} := \frac{1}{n!} f^* \left(\omega_Y^n \cdot h_{L_{X/Y}}^{-1}\right)$$

and is said to be *the canonical measure* of X, where n denotes dim Y. Here we note that ω_Y^n is a singular volume form on Y and $f^*h_{X/Y}^{-1}$ is considered to be a relative singular volume form on $f : X \to Y$ (cf. (3.1)), hence $f^* \left(\omega_Y^n \cdot h_{L_{X/Y}}^{-1}\right)$ is considered to be a singular volume form on X.

Remark 3.2 In Theorem 3.1, the metric h_K depends only on the canonical ring of X. Hence adding effective exceptional $\mathbb{Q}$-divisors does not affect h_K and ω_Y essentially.

We note that the canonical measure $d\mu_{can}$ depends only on the canonical ring of X, hence it is birationally invariant.

3.3 Dynamical Systems of Extremal Measures on a Compact Kähler Manifold

Let X be a compact Kähler n-manifold with pseudoeffective canonical bundle. Let ω_0 be a C^∞-Kähler form on X. We set

$$d\mu_1 = d\mu(X, \omega_0) := \sup\{dV | -\mathrm{Ric}\, dV + \omega_0 \geqq 0, \int_X dV = 1\}^*,$$

where dV runs semipositive (n, n)-forms such that dV^{-1} is a singular hermitian metric on K_X and $\{\ \}^*$ denotes the uppersemicontinuous envelope. And we set

$$T_1 := -\mathrm{Ric}\, d\mu_1 + \omega_0.$$

For $m \geqq 2$, we define

$$d\mu_m = d\mu(X, T_{m-1}) := \sup\{dV \mid -\operatorname{Ric} dV + T_{m-1} \geqq 0, \int_X dV = 1\}^*,$$

and

$$T_m := -\operatorname{Ric} d\mu_m + T_{m-1}.$$

By the construction we see that the de Rham cohomology class $[T_m]$ of T_m satisfies $[T_m] = 2m\pi c_1(K_X) + [\omega_0]$.

We shall consider the normalized limit of $\{T_m\}_{m=1}^{\infty}$.

Theorem 3.2 ([T9]) *Suppose that K_X is abundant. Then*

$$\lim_{m\to\infty} \frac{1}{m} T_m = -\operatorname{Ric} d\mu_{can}$$

holds, where $d\mu_{can}$ denotes the canonical measure (cf. Definition 3.2) on X.

Remark 3.3 The abundance of K_X is necessary because $\lim_{m\to\infty} m^{-1}T_m$ is a current with minimal singularity (the curvature of an AZD of K_X) and $d\mu_{can}^{-1}$ is an AZD of K_X, if and only if K_X is abundant.

References

[A] Aubin, T.: Equations du type Monge-Ampere sur les varietes kahleriennes compactes. C.R. Acad. Sci. Paris, Ser. A **283**, 119 (1976)

[B] Bergman, S.: The Kernel Function and Conformal Mapping. Americal Mathematical Society, Providence (1970)

[Ber] Berndtsson, B.: Curvature of vector bundles associated to holomorphic fibrations, Ann. Math. **169**, 531-560 (2009)

[B-P] Berndtsson, B., Paun, M.: Bergman kernels and the pseudoeffectivity of relative canonical bundles. Duke Math. J. **145**(2), 341–378 (2008)

[C] Caratheodory, C.: Uber die Abbildungen, die durch Systeme von analytischen Funktionen von mehreren Veranderlichen erzeugt werden. Math. Z. **34**, 758–792 (1932)

[C-Y] Cheng, S.-Y., Yau, S.-T.: On the existence of a complete Kahler-Einstein metric on noncompact complex manifolds and the regularity of Fefferman's equation. Com. Pure Appl. Math. **33**, 507–544 (1980)

[D] Demailly, J.P.: Regularization of closed positive currents and intersection theory. J. Algebraic Geom. **1**(3), 361–409 (1992)

[D-P-S] Demailly, J.P., Peternell, T., Schneider, M.: Pseudo-effective line bundles on compact Kähler manifolds. Int. J. Math. **12**, 689–742 (2001)

[H] Hormander, L.: An Introduction to Complex Analysis in Severa! Variables. North-Holland Amsterdam (1973)

[Ka1] Kawamata, Y.: Kodaira dimension of Algebraic fiber spaces over curves. Invent. Math. **66**, 57–71 (1982)

[K] Kobayashi, S.: Intrinsic distances, measures and geometric function theory. Bull. Am. Math. Soc. **82**(3), 357–416 (1976)

[L] Lelong, P.: Fonctions Plurisousharmoniques et Formes Differentielles Positives, Gordon and Breach (1968)
[M-Y] Mok, N., Yau, S.-T.: Completeness of the Kähler-Einstein metric on bounded domains and characterization of domains of holomorphy by curvature conditions. In: The Mathematical Heritage of Henri Poincare. Proceedings of Symposia in Pure Mathematics, vol. 39, Part I, pp. 41-60 (1983)
[N] Nadel, A.M.: Multiplier ideal sheaves and existence of Kähler-Einstein metrics of positive scalar curvature. Ann. Math. **132**, 549–596 (1990)
[O-T] Ohsawa, T., Takegoshi, K.: L^2-extension of holomorphic functions. Math. Z. **195**, 197–204 (1987)
[O] Ohsawa, T.: On the extension of L^2 holomorphic functions V, effects of generalization. Nagoya Math. J. **161**, 1–21 (2001) (Erratum : Nagoya Math. J. **163**(2001))
[S1] Siu, Y.-T.: Invariance of plurigenera. Invent. Math. **134**, 661–673 (1998)
[S2] Siu, Y.-T.: Extension of twisted pluricanonical sections with plurisubharmonic weight and invariance of semipositively twisted plurigenera for manifolds not necessarily of general type. Collected papers Dedicated to Professor Hans Grauert (2002), pp. 223-277
[S-T] Song, J., Tian, G.: Canonical Measures and Kähler-Ricci Flow, Math. arXiv:0802.2570 (2008)
[S-W] Song, J., Weinkove, B.: Construction of Kähler-Einstein metrics with negative scalar curvature. Math. Ann. **347**, 59–79 (2010)
[T1] Tsuji, H.: Analytic Zariski decomposition. Proc. Japan Acad. **61**, 161–163 (1992)
[T2] Tsuji, H.: Existence and Applications of Analytic Zariski Decompositions. Trends in Math. Analysis and Geometry in Several Complex Variables, pp. 253-272 (1999)
[T3] Tsuji, H.: Dynamical construction of Kähler-Einstein metrics. Nagoya Math. J. **199**, 107–122 (2010)
[T4] Tsuji, H.: Canonical Volume Forms on Compact Kähler Manifolds. arXiv:0707.0111 (2007)
[T5] Tsuji, H.: Canonical singular hermitian metrics on relative canonical bundles. Am. J. Math. **133**(6), 1469–1501 (2011)
[T6] Tsuji, H.: Canonical Measures and Dynamical Systems of Bergman Kernels, Math (2008). arXiv:0805.1829
[T7] Tsuji, H.: Ricci Iterations and Canonical Kähler-Einstein Currents on LC Pairs, Math (2009). arXiv:0903.5445 (2009)
[T8] Tsuji, H.: On the Extremal Measure on a Complex Manifold. to appear in Progress in Mathematics, Birkhäuser (2014)
[T9] Tsuji, H.: Stability of Pluricanonical Systems for Compact Kähler Manifolds (in preparation)
[Y] Yau, S.-T.: On the Ricci curvature of a compact Kahier manifold and the complex Monge-Ampere equations I. Commun. Pure Appi. Math. **31**, 339-411 (1978)

On Representative Domains and Cartan's Theorem

Atsushi Yamamori

Abstract This is a short survey article on Cartan's theorem about automorphisms fixing the origin for certain class of quasi-circular domains and non-hyperbolic circular domains. Some open problems are also given.

Keywords Automorphism · Quasi-circular domain · Non-hyperbolic circular domain · Bergman mapping · Representative domain

1 Introduction

Throughout this article, we always assume that D is a domain in $\mathbb{C}^n$ which contains the origin. The focus of interest of the present article is on a certain class of domains which are so-called *representative domains* (Definition 2.1). For instance, this class of domains includes as a special case the bounded circular domains (e.g. the unit ball, the polydisk, the Thullen domain). The purpose of this article is to survey how this class of domains is useful to derive the linearity of automorphisms fixing the origin.

Let us recall the following classical result due to Cartan.

Theorem 1.1 (**Cartan's Uniqueness Theorem**) *Let D be a bounded (or hyperbolic) domain and $f : D \to D$ a holomorphic mapping such that*

- $f(p) = p$,
- *the Jacobian matrix of f at p is the identity matrix,*

for some $p \in D$. Then f must be the identity mapping.

As a consequence of this theorem, one can prove the linearity of origin-preserving automorphisms of the bounded circular domains (for the proof see Proposition 11.1.2 in [Kran]).

A. Yamamori (✉)
The Center for Geometry and Its Applications, Pohang University of Science and Technology, Pohang 790-784, Republic of Korea
e-mail: ats.yamamori@gmail.com; yamamori@postech.ac.kr

F. Bracci et al. (eds.), *Complex Analysis and Geometry*, Springer Proceedings in Mathematics & Statistics 144, DOI 10.1007/978-4-431-55744-9_26

It is also known by Kaup [Kaup] that every origin-preserving automorphism of the quasi-circular domains (Definition 3.1) is a polynomial mapping. The two theorems due to Cartan and Kaup are apparently analogous and thus the problem below arises naturally:

Problem 1 When does Cartan's theorem remain true for quasi-circular domains?

For the circular case, the Jacobian matrix of the rotation mapping $\rho_\theta : z \mapsto e^{i\theta} z$ is a scalar matrix. This fact is substantial when one applies Cartan's Uniqueness Theorem to prove the linearity. On the other hand, for the quasi-circular cases, the same argument does not work. In fact, the Jacobian matrix of the mapping $(z_1, \ldots, z_n) \mapsto (e^{im_1\theta} z_1, \ldots, e^{im_n\theta} z_n)$ is not a scalar matrix unless $m_1 = \cdots = m_n$. This situation indicates that an essentially different method is required for the study of Problem 1. Moreover, the method presented here gives some non-hyperbolic circular domains for which Cartan's theorem remains true.

The organization of this article is as follows. In Sect. 2, we prepare basic definitions and relevant properties. In Sect. 3, we explain how the notion of representative domain is helpful in the study of automorphisms of certain class of quasi-circular domains and non-hyperbolic circular domains. We conclude this article with some open problems.

This article is only an exposition of some aspects of applications of the representative domains. We do not intend to cover all important results in the subject. We refer the readers to [IK, Lu, Tsuboi, Y2014, Ypre] and references therein for more information about the representative domains and applications.

2 Preliminaries

In this section, we prepare some definitions and basic facts. Let D be a domain in $\mathbb{C}^n$, K_D the Bergman kernel of D and T_D the $n \times n$ matrix defined by

$$T_D(z, w) := \begin{pmatrix} \frac{\partial^2}{\partial \overline{w}_1 \partial z_1} \log K_D(z, w) & \cdots & \frac{\partial^2}{\partial \overline{w}_1 \partial z_n} \log K_D(z, w) \\ \vdots & \ddots & \vdots \\ \frac{\partial^2}{\partial \overline{w}_n \partial z_1} \log K_D(z, w) & \cdots & \frac{\partial^2}{\partial \overline{w}_n \partial z_n} \log K_D(z, w) \end{pmatrix},$$

for $K_D(z, w) \neq 0$. In the following, for simplicity, we use the notation

$$T_{\overline{i}j}(z, w) = \frac{\partial^2}{\partial \overline{w}_i \partial z_j} \log K_D(z, w).$$

The matrix $T_D(z, z)$ is a positive definite hermitian matrix for all $z \in D$. The matrix T_D possesses the following transformation formula under the biholomorphism $\varphi : D \to D'$:

$$T_D(z,w) = {}^t\overline{\mathrm{J}(\varphi,w)}T_{D'}(\varphi(z),\varphi(w))\mathrm{J}(\varphi,z), \quad \text{if } K_D(z,w) \neq 0. \tag{1}$$

Here $\mathrm{J}(\varphi, z)$ is the Jacobian matrix of φ at z. The matrix T_D is well-defined if D is bounded and $z = w$. However T_D may not be well-defined for some $z, w \in D$ in general. For instance, it is known that the Bergman kernel of the domain $\mathbb{E}_p$ defined by

$$\mathbb{E}_p := \left\{(z_1, \dots z_n) \in \mathbb{C}^n : |z_1| + |z_2|^{2/p_2} + \dots + |z_m|^{2/p_n} < 1\right\}$$

has zeros if $p_2 + \dots + p_n > 2$. It is also known that the Bergman kernel of the domain defined by

$$\left\{(z_1, z_2) \in \mathbb{C}^2 : |z_2| < \frac{1}{1+|z_1|}\right\}$$

has a zero at the origin. For details of these examples, see Boas's paper [Boas].

We now define a class of domains, which plays an important role in this article (see also [Lu]).

Definition 2.1 A domain D is called *representative* if T_D satisfies

$$T_D(z, t_0) \equiv T_D(t_0, t_0),$$

for some $t_0 \in D$. The point t_0 is called the center of D.

In the case of the unit ball $\mathbb{B}_2$ in $\mathbb{C}^2$, the Bergman kernel is given by

$$K_{\mathbb{B}_2}(z,w) = \frac{2!}{\pi^2(1 - \langle z, w\rangle)^3}.$$

Using this explicit form, one has

$$T_{\mathbb{B}_2}(z,0) \equiv T_{\mathbb{B}_2}(0,0) \equiv \begin{pmatrix} 3 & 0 \\ 0 & 3 \end{pmatrix}.$$

It follows that the unit ball $\mathbb{B}_2$ is a representative domain with the center at the origin. Since we do not know an explicit form of the Bergman kernel for an arbitrary given domain D, it is impossible to verify $T_D(z,0) \equiv T_D(0,0)$ in the same way.

Instead of an explicit form of the Bergman kernel, one can use the transformation formula (1) of T_D. Since $\mathbb{B}_2$ is circular (i.e. $\rho_\theta : z \mapsto e^{i\theta}z$ is an automorphism), one can obtain the following relation on $T_{\mathbb{B}_2}$:

$$T_{\mathbb{B}_2}(z,0) \equiv T_{\mathbb{B}_2}(\rho_\theta(z),0).$$

It is not difficult to see that this relation gives us our desired conclusion. Since we use the circularity of $\mathbb{B}_2$ and (1) in the argument, the same argument works for any bounded circular domains. Namely we have the following proposition (cf. [IK]).

Proposition 2.1 *Every circular domain is a representative domain with the center at the origin.*

In the next section, we will see an example of representative domain which is not circular.

3 Cartan Theorem Revisited

One of the fundamental theorem on holomorphic automorphism groups is the following theorem due to Cartan.

Theorem 3.1 *Let D be a bounded circular domain and f an automorphism of D with $f(0) = 0$. Then f is a linear mapping.*

As we mentioned in the Introduction, this theorem is proved by using Cartan's Uniqueness Theorem. In that proof, what is the most important part is to show the commutative relation $\rho_\theta \circ f = f \circ \rho_\theta$. However this is not the only way to prove this theorem. Indeed, we will see that the theory of representative domain allows us to have another commutative relation involving origin-preserving automorphisms and the *Bergman mapping*.

Before explaining a connection between this theorem and representative domains, let us pause to consider the following *toy observation*. Let D be a domain in $\mathbb{C}^n$, $\mathrm{Hol}(D, D)$ the set of holomorphic mappings from D to D and S a certain class of holomorphic mappings in $\mathrm{Hol}(D, D)$ such that one wishes to show the linearity for all mappings in S. Further we pose the following assumption on S.

Assumption 1 For each element f of S, there exists a linear mapping L and a holomorphic mapping g^D such that the following diagram is commutative:

$$\begin{array}{ccc} D & \xrightarrow{\;f\;} & D \\ {\scriptstyle \exists g^D}\downarrow & \circlearrowleft & \downarrow{\scriptstyle \exists g^D} \\ \mathbb{C}^n & \xrightarrow{\;\exists L\;} & \mathbb{C}^n. \end{array}$$

Here g^D depends only on D and L depends only on f.

Under this assumption, we readily observe the following:

Observation 1 Let S be a set as in Assumption 1. If D is a domain in $\mathbb{C}^n$ such that g^D is a biholomorphic linear mapping, then all elements of S are linear.

Since Assumption 1 poses a strong condition on S, one cannot expect the existence of such a nice mapping g^D for an arbitrary given S. However if one considers the set of all origin-preserving automorphisms of D as S, then one can show that such a mapping g^D exists. Indeed the following theorem is known (cf. [IK]):

Theorem 3.2 *Suppose that $K_D(z,0) \neq 0$ for any $z \in D$ and let σ_0^D be a holomorphic mapping defined by*

$$\sigma_0^D : D \to \mathbb{C}^n,\ (z_1, \dots, z_n) \mapsto T_D(0,0)^{-1/2} \mathrm{grad}_{\overline{w}} \log \left. \frac{K_D(z,w)}{K_D(0,w)} \right|_{w=0},$$

where we set

$$\mathrm{grad}_{\overline{w}} f(w) := {}^t\left(\frac{\partial f}{\partial \overline{w}_1}(w), \dots, \frac{\partial f}{\partial \overline{w}_n}(w) \right),$$

for anti-holomorphic functions f on D. Then the following commutative diagram holds:

$$\begin{array}{ccc} D & \xrightarrow[\sim]{\varphi} & D \\ {\scriptstyle \sigma_0^D}\downarrow & \circlearrowleft & \downarrow{\scriptstyle \sigma_0^D} \\ \mathbb{C}^n & \xrightarrow{L_\varphi} & \mathbb{C}^n \end{array} .$$

Here φ is an automorphism of D with $\varphi(0) = 0$ and L_φ is a certain unitary transformation.

The mapping σ_0^D is called the *Bergman (representative) mapping* of D. If D is a bounded circular or quasi-circular, then the Bergman kernel K_D satisfies $K_D(z,0) \equiv K(0,0) > 0$. Thus the σ_0^D is globally defined on D for these cases. However, in general, the Bergman mapping is defined only on $U_0^D = \{z \in D : K_D(z,0) \neq 0\}$ (cf. [IK]). From this theorem and Observation 1, one can see that Theorem 3.1 remains true for any domains for which the Bergman mapping is a biholomorphic linear mapping. Moreover one can find that the Bergman mapping σ_0^D is linear if and only if D is a representative domain with the center at the origin. In fact, it follows from the following two facts on the Bergman mapping:

- $\sigma_0^D(0) = 0$,
- $\mathrm{J}(\sigma_0^D, z) = T_D(0,0)^{-1/2} T_D(z,0)$.

This, together with Proposition 2.1, gives us another proof of Theorem 3.1 without Cartan Uniqueness Theorem. Then it is appropriate to ask the question below.

Question 1 Can we find a class of representative domains which are not circular domains?

In many standard texts on Several Complex Variables (cf. [Kran, Chapt. 11], [Nar, Chap. 5]), Theorem 3.1 is proved as a consequence of Cartan's uniqueness theorem. It is absolutely unclear whether or not Cartan's uniqueness theorem holds even for some non-hyperbolic unbounded cases. Thus it is also non-trivial to decide whether or not Theorem 3.1 holds for a given non-hyperbolic unbounded circular domain.

Question 2 Can we find a non-hyperbolic unbounded domain for which Theorem 3.1 holds?

Before discussing the results related to these two questions, let us first introduce some definitions.

Definition 3.1 Let $m_1, \dots m_n$ be positive integers such that $m_1 < m_2 < \cdots < m_n$ and $\gcd(m_1, \dots, m_n) = 1$. A domain in $\mathbb{C}^n$ is called *quasi-circular* if it is invariant under

$$\rho_{m,\theta} : D \to \mathbb{C}^n, (z_1, \dots, z_n) \mapsto (e^{im_1\theta} z_1, \dots, e^{im_n\theta} z_n).$$

The n-tuple $m = (m_1, \cdots, m_n)$ is called the *weight* of D.

The domains given below are examples of quasi-circular domains:

$$D_1 = \{(z_1 + z_2, z_1 z_2) : |z_1|, |z_2| < 1\},$$
$$D_2 = \{(z_1, z_2) : |z_1^3 + z_2^2| < 1\}.$$

The weights of these domains are $(1, 2)$ and $(2, 3)$ respectively. Now we are ready to state our result [Y2014], which gives an answer for Question 1:

Theorem 3.3 *Let $D \subset \mathbb{C}^2$ be a quasi-circular domain whose weight satisfies $2 \leq m_1 < m_2$. Then D is a representative domain with center at the origin. In particular, every origin-preserving automorphism is linear.*

Let us give a sketch of the proof:

Step 1. Applying the transformation formula (1) to $\varphi = \rho_{m,\theta}$, one has a relation of T_D:

$$\begin{cases} T_{\bar{i}i}(z, 0) = T_{ii}(f_{m,\theta}(z), 0), \quad i = 1, 2, \\ T_{\bar{1}2}(z, 0) = e^{i(m_2 - m_1)\theta} T_{12}(\rho_{m,\theta}(z), 0), \\ T_{\bar{2}1}(z, 0) = e^{i(m_1 - m_2)\theta} T_{21}(\rho_{m,\theta}(z), 0). \end{cases}$$

Step 2. Put $T_{\bar{i}j}(z, 0) = \sum_{k_1, k_2 \geq 0} a_{k_1 k_2}^{(i,j)} z_1^{k_1} z_2^{k_2}$. Then, except for $T_{\bar{2}1}$, one can show that the coefficient $a_{k_1 k_2}^{(i,j)} = 0$ for any $k \neq (0, 0)$ without the assumption "$2 \leq m_1 < m_2$".

Step 3. Using the assumption of the theorem, one can conclude that $T_{\bar{2}1}(z, 0) \equiv T_{\bar{2}1}(0, 0)$.

Remark 1 If the weight is $(1, 2)$ then there exists a quasi-circular domain such that the automorphism group contains a non-linear mapping [Z]. Thus, we cannot drop the condition "$2 \leq m_1$" in the theorem.

We note that the above theorem holds for any biholomorphism $f : D \to D'$ between quasi-circular domains with the assumption as in the theorem. It is known by Kaup [Kaup] that if two quasi-circular domain are biholomorphic then there exists a biholomorphic mapping fixing the origin. Therefore we obtain the next corollary (cf. [Rong, Ypre]).

Corollary 3.1 *Let D_1, D_2 be quasi-circular domains as in Theorem 3.3. Then the two domains D_1 and D_2 are biholomorphically equivalent if and only if they are linearly equivalent.*

This result is regarded as an analogue of Braun-Kaup-Upmeier's theorem [BKU] for quasi-circular cases.

Theorem 3.4 (Braun, Kaup, Upmeier) *Let D_1, D_2 be circular domains. Then the two domains D_1 and D_2 are biholomorphically equivalent if and only if they are linearly equivalent.*

Let us now turn to Question 2. For unbounded cases, it is a non-trivial question whether or not the Bergman kernel exists. Moreover it is also unclear that $K_D(0,0) > 0$ and $T_D(0,0)$ is positive definite for a unbounded circular domain D. The two domains given below are examples of circular domains (more precisely Reinhardt domains) for which the Bergman mappings are well-defined:

$$\Omega_1 = \{(z, \zeta) \in \mathbb{C}^n \times \mathbb{C}^m : \|\zeta\|^2 < e^{-s\|z\|^2}\}, \quad s > 0,$$
$$\Omega_2 = \{(z, w) \in \mathbb{C}^2 : \log|w|^2 + |z|^2 + |w|^2 < 1\}.$$

Since the Bergman mappings are well-defined for these two, Cartan's theorem remains true for Ω_i ($i = 1, 2$). Using this fact, the automorphism groups of Ω_1, Ω_2 are computed (cf. [KNY, Kim]).

Theorem 3.5 *The automorphism group of Ω_1 is generated by the following mappings:*

$$\varphi_U : (z, \zeta) \mapsto (Uz, \zeta), \quad U \in U(n),$$
$$\varphi_{U'} : (z, \zeta) \mapsto (z, U'\zeta), \quad U' \in U(m),$$
$$\varphi_v : (z, \zeta) \mapsto (z - v, e^{s\langle z, v\rangle - \frac{s}{2}\|v\|^2}\zeta), \quad v \in \mathbb{C}^n.$$

Theorem 3.6 *The automorphism group of Ω_2 is generated by the following mappings:*

$$\varphi_\theta : (z, w) \mapsto (e^{i\theta}z, w), \quad \theta \in \mathbb{R},$$
$$\varphi_{\theta'} : (z, w) \mapsto (z, e^{i\theta}w), \quad \theta' \in \mathbb{R}.$$

Remark 2 More information about Ω_1, Ω_2 can be found in [HST], [Spri] and [Y2013].

4 Open Problems

We conclude this article with some open problems.

In the previous section, we gave the two non-hyperbolic circular examples Ω_1, Ω_2 with explicit descriptions of the automorphism groups. However, we do not know any non-hyperbolic quasi-circular domain whose automorphism group can be computed explicitly.

Problem 2 Can we find a non-hyperbolic quasi-circular domain with the well-defined Bergman mapping and an explicit description of the automorphism group?

As we have seen in the previous section, Cartan's theorem remains true for two domains Ω_1, Ω_2. Since the Bergman mapping is used for the proof, we do not know whether or not Cartan Uniqueness theorem holds for these domains. Thus it is natural to ask to the following problem.

Problem 3 Can we find a non-hyperbolic circular domain D such that the Bergman mapping is well-defined, but Cartan uniqueness theorem does not hold?

In particular, it might also be interesting to study this problem for two specific domains Ω_1, Ω_2. We expect that there does not exist such a non-hyperbolic circular domain. However, we do not succeed in proving it at the time of writing this article. We hope that these problems will be helpful towards a deep understanding of non-hyperbolic (quasi-)circular domains.

The following proposition asserts that if a holomorphic mapping $f : D \to D$ of a representative domain D such that $f(0) = 0, \mathrm{J}(f, 0) = \mathrm{Id}$ and $f \neq \mathrm{Id}$, then f must not be biholomorphic. Although the proof is essentially contained in the article [Lu], we give the proof of the proposition for the convenience of the reader (see also [Kim, Proposition3.1]).

Proposition 4.1 *Let D be a non-hyperbolic representative domain with the center at the origin such that $K_D(0, 0) > 0$ and $T_D(0, 0)$ is positive definite and $f : D \to D$ an automorphism such that $f(0) = 0$ and $\mathrm{J}(f, 0) = \mathrm{Id}$. Then f must be the identity mapping of D.*

Proof Using the assumption $\mathrm{J}(f, 0) = \mathrm{Id}$ and the transformation formula (1) to $\varphi = f$, we obtain

$$\begin{aligned} T_D(z, 0) &= \overline{{}^t\mathrm{J}(f, 0)}T_D(f(z), 0)\mathrm{J}(f, z), \\ &= \overline{{}^t\mathrm{J}(f, 0)}T_D(0, 0)\mathrm{J}(f, z), \\ &= T_D(0, 0)\mathrm{J}(f, z). \end{aligned} \tag{2}$$

Combining two relations (2) and $T_D(z, 0) \equiv T_D(0, 0)$, we see that

$$T_D(0, 0) = T_D(0, 0)\mathrm{J}(f, z). \tag{3}$$

It follows that $\mathrm{J}(f, z) = \mathrm{Id}$. This, together with $f(0) = 0$, implies that $f = \mathrm{Id}$ as desired. □

The argument presented here entirely relies on the transformation formula of T_D. Thus this argument does not work for a holomorphic mapping $f : D \to D$ which is not an automorphism.

Remark 3 The circularity of a domain D implies that $K_D(z, 0) \equiv K_D(0, 0)$. Thus it is enough to assume that $K_D(0, 0) > 0$ to ensure $U_0^D = D$. We also note that the assumption of T_D is needed to derive the relation $\mathrm{J}(f, z) = \mathrm{Id}$ from (3).

Acknowledgments This article is based on the talk "*On Bergman's representative domains and origin-preserving automorphisms of quasi-circular domains*" given by the author at the KSCV10 Symposium, August, 2014. The author would like to thank the organizers for their kind invitation and hospitality. The author also thanks the anonymous referee for valuable comments on this paper. The research of the author is supported in part by SRC-GaiA (Center for Geometry and its Applications), the Grant 2011-0030044 from The Ministry of Education, The Republic of Korea.

References

[Boas] Boas, H.P.: Lu Qi-Keng's problem. J. Korean Math. Soc. **37**, 253–267 (2000)

[BKU] Braun, R., Kaup, W., Upmeier, H.: On the automorphisms of circular and Reinhardt domains in complex Banach spaces. Manuscripta Math. **25**, 97–133 (1978)

[HST] Harz, T., Shcherbina, N., Tomassini, G.: Wermer type sets and extension of CR functions. Indiana Univ. Math. J. **61**, 431–459 (2012)

[IK] Ishi, H., Kai, C.: The representative domain of a homogeneous bounded domain. Kyushu J. Math. **64**, 35–47 (2010)

[Kaup] Kaup, W.: Über das Randverhalten von holomorphen Automorphismen beschränkter Gebiete. Manuscripta Math. **3**, 257–270 (1970)

[KNY] Kim, H., Ninh, V.T., Yamamori, A.: The automorphism group of a certain unbounded non-hyperbolic domain. J. Math. Anal. Appl. **409**, 637–642 (2014)

[Kim] Kim, H.: The automorphism group of an unbounded domain related to Wermer type sets. J. Math. Anal. Appl. **421**, 1196–1206 (2015)

[Kran] Krantz, S.G.: Function Theory of Several Complex Variables, 2nd edn. Wadsworth & Brooks/Cole Advanced Books & Software, Pacific Grove, California (1992)

[Lu] Lu, Q.-K.: On the representative domain, several complex variables (Hangzhou, 1981). Birkhäuser. Boston, pp. 199–211 (1984)

[Nar] Narasimhan, R.: Several Complex Variables. Chicago lecture in mathematics. The University of Chicago Press, Chicago (1971)

[Rong] Rong, F.: On automorphisms of quasi-circular domains fixing the origin. Bull. Sci. Math., (to appear)

[Spri] Springer, G.: Pseudo-conformal transformations onto circular domains. Duke Math. J. **18**, 411–424 (1951)

[Tsuboi] Tsuboi, T.: Bergman representative domains and minimal domains. Japan. J. Math. **29**, 141–148 (1959)

[Y2013] Yamamori, A.: The Bergman kernel of the Fock-Bargmann-Hartogs domain and the polylogarithm function. Complex Var. Elliptic Equ. **58**(6), 783–793 (2013)

[Y2014] Yamamori, A.: Automorphisms of normal quasi-circular domains. Bull. Sci. Math. **138**, 406–415 (2014)

[Ypre] Yamamori, A.: On the linearity of origin-preserving automorphisms of quasi-circular domains in $\mathbb{C}^n$. J. Math. Anal. Appl. **426**, 612–623 (2015)

[Z] Zapałowski, P.: Proper holomorphic mappings between symmetrized ellipsoids. Arch. Math. (Basel) **97**(4), 373–384 (2011)

On Curvature Estimates of Bounded Domains

Liyou Zhang

Abstract We consider the Bergman curvatures estimate for bounded domains in terms of the squeezing function. As applications, we give the asymptotic boundary behaviors of the curvatures near strictly pseudoconvex boundary points, using a recent result given by Fornaess and Wold.

Keywords Bergman curvature · Squeezing function · Intrinsic derivative

1 Introduction

The holomorphic invariant objects related to the Bergman geometry have been extensively studied during the past decades, for instance, the Bergman canonical invariant introduced by Bergman himself [Ber], various kinds of curvatures of the Bergman metric, including the holomorphic sectional curvature, the Ricci curvature and the scalar curvature. Sometimes, these curvatures are called Bergman curvatures in the literatures (see e.g. [KiY, KrY]). It is well known that for any bounded domain in $\mathbb{C}^n$, the upper bound of the holomorphic sectional curvature is 2 (see [Fuk, Hua, Kob]) and the Ricci curvature is strictly less than $n + 1$ (see [Kob, Noz]).

A natural question is to consider the lower bounds of the above mentioned curvatures. Usually, one can not expect there exist uniform lower bounds of the Bergman curvatures for all bounded domains, like the upper bounds. Recently, Lu [Lu4] proposed a program to investigate the lower bounds of the Bergman curvatures on a bounded domain D in $\mathbb{C}^n$, in terms of the Bergman kernel and metric of the domain D_1 contained in D and the domain D_2 containing D. In particular, if D_1 and D_2 are chosen "good" enough, for example, the Euclidean balls, then both the upper and lower bounds can be established explicitly (see Theorem 3.1 and Theorem 3.2 in Sect. 3). Unfortunately, the lower bounds of the Bergman curvatures in Theorem 3.1 and 3.2 tends to $-\infty$ as the point goes to the boundary of the domain D.

L. Zhang (✉)
School of Mathematical Science, Capital Normal University, Beijing 100048, People's Republic of China
e-mail: zhangly@cnu.edu.cn

F. Bracci et al. (eds.), *Complex Analysis and Geometry*, Springer Proceedings in Mathematics & Statistics 144, DOI 10.1007/978-4-431-55744-9_27

The aim of this present paper is to explore the lower bounds of the holomorphic sectional curvature, the Ricci curvature and the scalar curvature. We show that both the explicit uniform lower and upper bounds of these Bergman curvatures exist on some general classes of bounded pseudoconvex domains named the *holomorphically homogeneous regular* manifolds (HHR) or the *uniformly squeezing* domains (USq), which was introduced independently by Liu-Sun-Yau in [LSY1, Definition 7.1] and by Yeung in [Yeu, Definition 1] (see also Definition 4.1 in Sect. 4). The concept, HHR or USq, has been developed in order for the study of completeness and other geometric properties such as the metric equivalence of the invariant metrics including the Carathéodry metric, the Kobayashi-Royden metric, the Teichmüller metric, the Bergman metric, and the Kähler-Einstein metric.

The examples of HHR or USq domains include bounded homogeneous domains, bounded strongly convex domains, bounded domains which cover a compact Kähler manifold, and the Teichmüler spaces $\mathscr{T}_{g,n}$ of hyperbolic Riemann surfaces of genus g with n punctures (see [Yeu, Proposition 1]). Recently, it has been proved that the bounded convex domains [KiZ] and the strictly pseudoconvex domains with C^2 boundary also admit such HHR/USq property (see [DGZ2, KiZ]).

Note that for domains with HHR/USq property, it was S.-K Yeung who proved firstly that the curvature tensor for either the Kähler-Einstein metric or the Bergman metric, as well as any order of covariant derivatives of the curvature tensor is bounded by a uniform constant [Yeu, Proposition 4]. What we present in this article is to describe how the upper or lower bounds depend on the so called *squeezing function* (see Definition 4.2 in Sect. 4) for a bounded domain in $\mathbb{C}^n$. The concept of squeezing function was introduced by Deng, Guan and the author in [DGZ1], inspired by Yeung's talk in Chinese Academy of Science in 2009. It was used to characterize how a bounded domain looks like the unit ball observed at a point of the given domain. A bounded domain is HHR/USq if and only if its squeezing function admits a positive lower bound. Later, it turns out that the squeezing function is powerful in the characterization of the geometric and analytic properties of bounded domains in $\mathbb{C}^n$. In particular, the squeezing function approaches to 1 when the point tends to a strictly pseudoconvex boundary point. In virtue of the squeezing function, we state the main result as follows.

Theorem 1.1 *Let D be a bounded domain in $\mathbb{C}^n$ and $s_D(z)$ be the squeezing function at the point $z \in D$. Denote by $Sec_D(z,\xi)$, $Ric_D(z,\xi)$ and $Scal_D(z,\xi)$ the holomorphic sectional curvature, the Ricci curvature and the scalar curvature at z in the direction ξ, respectively. Then we have*

$$2 - 2\frac{n+2}{n+1}s_D^{-4n}(z) \le Sec_D(z,\xi) \le 2 - 2\frac{n+2}{n+1}s_D^{4n}(z),$$

$$(n+1) - (n+2)s_D^{-2n}(z) \le Ric_D(z,\xi) \le (n+1) - (n+2)s_D^{2n}(z),$$

$$n(n+1) - n(n+2)s_D^{-2n}(z) \le Scal_D(z) \le n(n+1) - n(n+2)s_D^{2n}(z).$$

Furthermore, if D is HHR/USq, that is, the squeezing function $s_D(z)$ admits a positive lower bound, say $\hat{s}_D$ (depends merely on D), then one can substitute $\hat{s}_D$ into Theorem 1.1 to obtain the uniform bounds for the Bergman curvatures on the domain D.

One important application of Theorem 1.1 is the boundary behaviors of the Bergman curvatures near a strictly pseudoconvex point $p \in \partial D$. As mentioned above, the squeezing function at the moment tends to 1 when the point z tends to p. Therefore, Theorem 1.1 immediately yields

$$\lim_{D\ni z\to p} \mathrm{Sec}_D(z,\xi) = -\frac{2}{n+1}, \quad \lim_{D\ni z\to p} \mathrm{Ric}_D(z,\xi) = -1 \text{ and } \lim_{D\ni z\to p} \mathrm{Scal}_D(z) = -n.$$

The asymptotic behavior of the holomorphic sectional curvature near strictly pseudoconvex boundary points was firstly considered by Klecmbeck for the C^∞ boundary case [Kle]. Later, Kim and Yu reduced the boundary condition to the C^2 case by using the scaling method (see [KiY] or [GKK, Chap. 10]).The boundary limit of the Ricci curvature was indicated in [ChY, pp. 510]. The existence of the boundary behaviors of the above Bergman curvatures was given by Krantz and Yu on some h-extendible pseudoconvex domains (see [KrY, Definition 0]), or semiregular domains as in [DiH]. These include, for example, not only the bounded pseudoconvex domains of finite type in $\mathbb{C}^2$ and convex domains of finite type in $\mathbb{C}^n$, but also bounded strictly pseudoconvex domains (see [KrY, Corollary 2]). Green and Krantz showed that for a sufficiently small C^∞ perturbation $\tilde{D}$ of a strongly pseudoconvex domain D, the boundary behavior of the holomorphic sectional curvature is stable near the boundary of $\tilde{D}$ (see [GK1, Theorem 3] or [GK2, Theorem 1.1]).

The paper is organized as follows. In Sect. 2, we introduce some basic notations and terminology, especially the concept of intrinsic derivatives of sections of the canonical line bundle, induced by the Bergman kernel. In Sect. 3, we recall the minimal function method Lu used in [Lu4] to estimate the lower bounds of the Bergman curvatures. In the last section, we will explore the curvature estimates in term of the squeezing function on a bounded domain in $\mathbb{C}^n$ and obtain the uniform estimates depending only on the squeezing constant. Some open questions are proposed in the final conclusion.

2 Intrinsic Derivatives Induced by the Bergman Kernel

Let Ω be a bounded domain in $\mathbb{C}^n$ with C^2 smooth boundary $b\Omega$. Let $A^2(\Omega)$ be the space of L^2 holomorphic $n-$forms on Ω, i.e. ,

$$A^2(\Omega) := \left\{ F \in H^0(\Omega, \mathscr{O}(\wedge^n T^*_{(1,0)})) \middle| \int_\Omega |F|^2 < \infty \right\},$$

where $\wedge^n T^*_{(1,0)}(\Omega)$ denotes the canonical line bundle over Ω and $|F|^2 := (\sqrt{-1})^{n^2} F \wedge \bar{F}$. It is well known that $A^2(\Omega)$ is a Hilbert space with respect to the inner product

$$(F, G) := (\sqrt{-1})^{n^2} \int_{\Omega} F \wedge \bar{G}, \quad \forall F, G \in A^2(\Omega).$$

Let $\{\Phi_j\}_{j=0}^{\infty}$ be a complete orthonormal basis of $A^2(\Omega)$. The holomorphic (n, n)-form $K_{\Omega}(z, \bar{w}) := \sum_{j=0}^{\infty} \Phi_j(z) \wedge \bar{\Phi}_j(w)$ on $\Omega \times \bar{\Omega}$ is called the Bergman kernel form of Ω. Let $z^1, \cdots, z^n$ be a coordinate system on Ω and $K_{\Omega}(z, \bar{z}) = k(z, \bar{z}) dz^1 \wedge \cdots \wedge dz^n \wedge d\bar{z}^1 \cdots \wedge d\bar{z}^n$. The Kähler metric given by

$$ds^2 = \sum_{\alpha, \beta=1}^{n} \frac{\partial^2 \log k(z, \bar{z})}{\partial z^{\alpha} \partial \bar{z}^{\beta}} dz^{\alpha} \otimes d\bar{z}^{\beta}$$

is called the Bergman metric on Ω and $\omega_B := \sqrt{-1}\partial\bar{\partial} \log k(z, \bar{z})$ is the corresponding Kähler form.

Let $h := k^{-1}(z, \bar{z})$. Then h is a C^{∞} Hermitian metric on the canonical line bundle $\wedge^n T^*_{(1,0)}(\Omega)$ with strictly positive curvature. The Hermitian connection of h is given by $B(z, \bar{z}) := \partial h \cdot h^{-1} = -\partial \log k(z, \bar{z})$.

For a smooth section $F \in \Gamma(\Omega, \wedge^n T^*_{(1,0)})$, $F = f dz^1 \wedge \cdots \wedge dz^n$, the covariant derivative of F associated to the Hermitian connection $B(z, z)$ is defined as

$$D^{(1,0)} F := \left(\partial f(z)/\partial z^{\alpha} + f(z) B_{\alpha}(z, \bar{z}) \right) dz^1 \wedge \cdots \wedge dz^n \otimes dz^{\alpha},$$

where $B_{\alpha}(z, \bar{z}) = -\partial \log k(z, \bar{z})/\partial z^{\alpha}$. This shows

$$D^{(1,0)}: \quad \Gamma(\Omega, \wedge^n T^*_{(1,0)}) \rightarrow \Gamma(\Omega, \wedge^n T^*_{(1,0)} \otimes T^*_{(1,0)}),$$

where $T^*_{(1,0)}(\Omega)$ is the holomorphic cotangent bundle of Ω.

Hereinafter, unless otherwise stated, the Einstein Summation Convention is used.

Let

$$\frac{\delta f(z)}{\delta z^{\alpha}} := \frac{\partial f(z)}{\partial z^{\alpha}} + f(z) B_{\alpha}(z, \bar{z}).$$

$\delta f(z)/\delta z^{\alpha}$ is called the intrinsic derivative by Lu in [Lu2], where Lu considered the estimates of higher order intrinsic derivatives of holomorphic mappings.

It is well known that there exists a reduced connection on $T^*_{(1,0)}(\Omega)$ with respect to the Kähler metric $g_{\alpha\bar{\beta}}(z) = \partial^2 \log k(z, \bar{z})/\partial z_{\alpha} \partial \bar{z}_{\beta}$, that is,

$$\Gamma^{\lambda}_{\alpha\beta}(z) = -g^{\bar{\mu}\lambda}(z) \frac{\partial g_{\alpha\bar{\mu}}(z)}{\partial z^{\beta}},$$

where $(g^{\bar{\beta}\alpha})$ denotes the inverse matrix of $(g_{\alpha\bar{\beta}})$. Therefore, one can define the covariant derivative for a section $F \in \Gamma(\Omega, \wedge^n T^*_{(1,0)} \otimes T^*_{(1,0)}(\Omega))$ in terms of the connections $B(z, \bar{z})$ and $\Gamma^{\lambda}_{\alpha\beta}(z)$. More precisely, let $F(z) = f_{\alpha}(z) e(z) \otimes dz^{\alpha}$, where we denote $dz^1 \wedge \cdots \wedge dz^n$ by $e(z)$ for short, we define (Here we still use $D^{(1,0)}$ if no confusions are caused)

$$D^{(1,0)}F := \left(\partial f_\alpha/\partial z^\beta + f_\alpha B_\beta(z,\bar{z}) - f_\lambda \Gamma^\lambda_{\alpha\beta}(z)\right) e(z) \otimes dz^\alpha \otimes dz^\beta.$$

This implies that

$$D^{(1,0)} \circ D^{(1,0)} : \quad \Gamma(\Omega, \wedge^n T^*_{(1,0)}) \to \Gamma(\Omega, \wedge^n T^*_{(1,0)} \otimes T^*_{(1,0)} \otimes T^*_{(1,0)}).$$

Similarly, for a smooth section $F \in \Gamma(\Omega, \wedge^n T^*_{(0,1)})$, one can define the conjugate covariant derivative $D^{(0,1)}$ by

$$D^{(0,1)}F := \left(\partial f(z)/\partial \bar{z}^\alpha + f(z) B_{\bar{\alpha}}(z,\bar{z})\right) \overline{e(z)} \otimes d\bar{z}^\alpha,$$

and for $F \in \Gamma(\Omega, \wedge^n T^*_{(1,0)} \otimes T^*_{(1,0)})$,

$$D^{(0,1)}F := \left(\partial f_\alpha/\partial \bar{z}^\beta + f_\alpha B_{\bar{\beta}}(z,\bar{z}) - f_\lambda \bar{\Gamma}^\lambda_{\alpha\beta}(z)\right) \overline{e(z)} \otimes d\bar{z}^\alpha \otimes d\bar{z}^\beta,$$

where $B_{\bar{\alpha}}(z,\bar{z}) = -\partial \log k(z,\bar{z})/\partial \bar{z}^\alpha$.

In general, given a smooth section $F \in \Gamma(\Omega, \wedge^n T^*_{(1,0)} \otimes \wedge^n T^*_{(0,1)} \otimes T^{*\otimes p}_{(1,0)} \otimes T^{*\otimes q}_{(0,1)})$, one has

$$D^{(1,0)}F \in H^0(\Omega, \wedge^n T^*_{(1,0)} \otimes \wedge^n T^*_{(0,1)} \otimes T^{*\otimes(p+1)}_{(1,0)} \otimes T^{*\otimes q}_{(0,1)}),$$
$$D^{(0,1)}F \in H^0(\Omega, \wedge^n T^*_{(1,0)} \otimes \wedge^n T^*_{(0,1)} \otimes T^{*\otimes p}_{(1,0)} \otimes T^{*\otimes(q+1)}_{(0,1)}).$$

For example, for the Bergman kernel form on the diagonal $K(z,\bar{z}) = k(z,\bar{z})e(z) \wedge \overline{e(z)}$, both $D^{(1,0)}K(z,\bar{z})$ and $D^{(0,1)}K(z,\bar{z})$ vanish, while neither $D^{(1,0)}K(z,\bar{w})$ nor $D^{(0,1)}K(z,\bar{w})$ does.

In the following theorem, we will see the relations between the Bergman curvature tensors and the above mentioned covariant derivatives (see [Lu3]).

Due to the well known reproducing property of the Bergman kernel, for any L^2 holomorphic n−form $F \in A^2(\Omega)$, in local coordinate system, we have

$$f(z) = \int_\Omega f(w) k(z,\bar{w}) dV_w, \tag{2.1}$$

where dV_w is the Lebesgue measure on Ω.

Taking the covariant derivative $D^{(1,0)}$ on both sides of (2.1), we obtain by a direct calculation that

$$\frac{\delta f(z)}{\delta z^\alpha} = \int_\Omega f(w) \frac{\delta k_\Omega(z,\bar{w})}{\delta z^\alpha} dV_w. \tag{2.2}$$

Fix $t \in \Omega$ and let $f(w) = \frac{\delta k(w,u)}{\delta \bar{u}^\beta}\Big|_{u=t}$, it is easy to check that $f(w) \in A^2(\Omega)$. By (2.2) we have

$$\frac{\delta^2 k(z, \bar{t})}{\delta z^\alpha \delta \bar{t}^\beta} = \int_\Omega \frac{\delta k(w, \bar{u})}{\delta \bar{u}^\beta}\Big|_{u=t} \frac{\delta k(z, \bar{w})}{\delta z^\alpha} dV_w.$$

Inductively, one has the following higher order derivatives

$$\frac{\delta^{p+q} k(z, \bar{t})}{\delta z^{\alpha_1} \cdots \delta z^{\alpha_p} \delta \bar{t}^{\beta_1} \cdots \delta \bar{t}^{\beta_q}} = \int_\Omega \frac{\delta^q k(w, \bar{u})}{\delta \bar{u}^{\beta_1} \cdots \delta \bar{u}^{\beta_q}}\Big|_{u=t} \frac{\delta^p k(z, \bar{w})}{\delta z^{\alpha_1} \cdots \delta z^{\alpha_p}} dV_w.$$

Let

$$H_{\alpha_1\cdots\alpha_p\bar{\beta}_1\cdots\bar{\beta}_q}(z, \bar{z}) := \left(\frac{\delta^{p+q} k(z, \bar{t})}{\delta z^{\alpha_1} \cdots \delta z^{\alpha_p} \delta \bar{t}^{\beta_1} \cdots \delta \bar{t}^{\beta_q}} \right)_{t=z},$$

we have the following theorem due to Q.-K. Lu.

Theorem 2.1 ([Lu3, Theorem 1.1]) *Let Ω be a bounded domain in $\mathbb{C}^n$. For any point $z \in \Omega$, we have*

1. *If $p \neq q$, then $H_{\alpha_1\cdots\alpha_p\bar{\beta}_1\cdots\bar{\beta}_q}(z, \bar{z}) = 0$.*
2. *If $p = q$, then $H_{\alpha_1\cdots\alpha_p\bar{\beta}_1\cdots\bar{\beta}_p}(z, \bar{z})$ is positively definite, i.e., for any non-zero vector ζ, one has $H_{\alpha_1\cdots\alpha_p\bar{\beta}_1\cdots\bar{\beta}_p}(z, \bar{z})\zeta^{\alpha_1\cdots\alpha_p}\bar{\zeta}^{\beta_1\cdots\beta_p} > 0$.*
3. *The following recurrence relation holds:*

$$H_{\alpha_1\cdots\alpha_p\bar{\beta}_1\cdots\bar{\beta}_p}(z, \bar{z}) = \sum_{\lambda=1}^{p} H_{\alpha_1\cdots\alpha_{p-1}\bar{\beta}_1\cdots\hat{\bar{\beta}}_\lambda\cdots\bar{\beta}_p} T_{\alpha_p\bar{\beta}_\lambda} - \sum_{\lambda<l} H_{\alpha_1\cdots\alpha_{p-1}\bar{\beta}_1\cdots\bar{\beta}_{\lambda-1}\bar{\gamma}\bar{\beta}_{\lambda+1}\cdots\bar{\beta}_p} R^{\bar{\gamma}}_{\bar{\beta}_\lambda\bar{\beta}_l\bar{\alpha}_p},$$

where $T_{\alpha\bar{\beta}}$ and $R^\lambda_{\alpha\beta\bar{\mu}}$ are the metric tensor and the curvature tensor with respect to the Bergman metric.

Example 2.1 1. For $p = 1$, it's easy to check $H_{\alpha\bar{\beta}}(z, \bar{z}) = k(z, \bar{z})T_{\alpha\bar{\beta}}$.
2. For $p = 2$, the above recurrence formula gives

$$H_{\alpha_1\alpha_2\bar{\beta}_1\bar{\beta}_2}(z, \bar{z}) = k(z, \bar{z}) \left(T_{\alpha_1\bar{\beta}_1} T_{\alpha_2\bar{\beta}_2} + T_{\alpha_2\bar{\beta}_1} T_{\alpha_1\bar{\beta}_2} - R_{\bar{\beta}_1\alpha_1\alpha_2\bar{\beta}_2} \right).$$

3 Lu's Estimates of the Bergman Curvatures

In this section, we recall the lower bound estimates for the Bergman curvatures. First of all, let us briefly recall some basis definitions and notations we will use in the text.

Throughout this section, we focus our attentions on a bounded domain D in $\mathbb{C}^n$ with the Bergman metric $ds_D^2 = T_{\alpha\bar{\beta}}(z, \bar{z})dz^\alpha d\bar{z}^\beta$, where $T_{\alpha\bar{\beta}}(z, \bar{z}) = \partial^2 \log k(z, \bar{z})/$

$\partial z^\alpha \partial \bar{z}^\beta$ is the Bergman metric tensor and $k(z, \bar{z})$ denotes the Bergman kernel. Let $(T^{\bar{\beta}\alpha})$ be the inverse matrix of $T = (T_{\alpha\bar{\beta}})$. The holomorphic curvature tensor is given by

$$R_{\bar{\lambda}\alpha\beta\bar{\mu}} = -\frac{\partial^2 T_{\alpha\bar{\lambda}}}{\partial z^\beta \partial \bar{z}^\mu} + T^{\bar{\nu}\sigma} \frac{\partial T_{\alpha\bar{\nu}}}{\partial z^\beta} \frac{\partial T_{\sigma\bar{\lambda}}}{\partial \bar{z}^\mu},$$

and the holomorphic sectional curvature at z with a complex tangent direction ξ is defined by $\mathrm{Sec}_D(z, \xi) = R_{\bar{\lambda}\alpha\beta\bar{\mu}} \xi^\alpha \bar{\xi}^\lambda \xi^\beta \bar{\xi}^\mu / (T_{\sigma\bar{\nu}} \xi^\sigma \bar{\xi}^\nu)^2$.

The complex Ricci curvature tensor is defined by $R_{\alpha\bar{\lambda}} = T^{\bar{\mu}\beta} R_{\bar{\mu}\beta\alpha\bar{\lambda}}$, which is equivalent to $R_{\alpha\bar{\lambda}} = -\partial^2 \log \det T / \partial z^\alpha \partial \bar{z}^\lambda$. The corresponding Ricci curvature at z with direction ξ is $\mathrm{Ric}_D(z, \xi) = R_{\alpha\bar{\lambda}} \xi^\alpha \bar{\xi}^\lambda / T_{\sigma\bar{\nu}} \xi^\sigma \bar{\xi}^\nu$ and the Scalar curvature is $\mathrm{Scal}_D(z) = T^{\bar{\lambda}\alpha} R_{\alpha\bar{\lambda}}$.

Remark 3.1 Due to Theorem 2.1, we know that $H_{\alpha_1\alpha_2\bar{\beta}_1\bar{\beta}_2}(z, \bar{z})$ in Example 2.1(2) is positive definite. Therefore it follows that $\mathrm{Sec}_D(z, \xi) < 2$ and $\mathrm{Ric}_D(z, \xi) < n+1$ hold for any bounded domain D in $\mathbb{C}^n$.

3.1 Lower and Upper Bounds of the Bergman Curvatures

For any fixed $t \in D$, Lu considered the subspace of $A^2(D)$, say $E_t^N(D)$, consisting of the elements with the following property:

$$\begin{aligned}
&f(t) = 0,\\
&\left.\frac{\delta f(z)}{\delta z^{\alpha_1}}\right|_{z=t} \xi^{\alpha_1} = 0,\\
&\cdots\cdots\cdots\\
&\left.\frac{\delta^N f(z)}{\delta z^{\alpha_1} \cdots \delta z^{\alpha_l}}\right|_{z=t} \xi^{\alpha_1 \cdots \alpha_l} = \delta_N^l, \quad l = 1, \cdots, N,
\end{aligned}$$

where δ_k^j is the Kronecker symbol, $\xi^{\alpha_1}, \cdots, \xi^{\alpha_1 \cdots \alpha_{N-1}}$ are arbitrary complex numbers and $\xi^{\alpha_1 \cdots \alpha_N}$ are some given constants.

A minimal function of $E_t^N(D)$ is the element with the minimal L^2 norm in $E_t^N(D)$. The minimum problem in Bergman geometry goes back to S. Bergman in 1940s, which was used to obtain the extremal function subject to some constraints (see [Ber, Chap. 2]).

In what follows, we'd like to get the minimal function of $E_t^N(D)$. More precisely, for $f \in E_t^N(D)$, $f(z) = \sum a_k \varphi_k(z)$, where $\{\varphi_k\}_{k=0}^\infty$ is a complete orthonormal basis of $A^2(D)$, we will choose the coefficients a_j to minimize the L^2 norm $\|f\|^2 = \sum |a_k|^2$ under the following conditions:

$$\frac{\delta^l f(z)}{\delta z^{\alpha_1}\dots\delta z^{\alpha_l}}\bigg|_{z=t}\xi^{\alpha_1\cdots\alpha_l}=\sum a_k c_{lk}=\delta_N^l,\quad l=0,1,\cdots,N,$$

where

$$c_{lk}=\frac{\delta^l \varphi_k(z)}{\delta z^{\alpha_1}\dots\delta z^{\alpha_l}}\bigg|_{z=t}\xi^{\alpha_1\cdots\alpha_l}.$$

The method of Lagrange multipliers allows us to minimize functions with the constraint. Let

$$\mathscr{L}:=\sum_{k=0}^{\infty}a_k\bar{a}_k-\sum_{l=1}^{N}\lambda_l\left(\sum_{k=0}^{\infty}a_k c_{kl}-\delta_N^l\right)-\sum_{l=1}^{N}\bar{\lambda}_l\left(\sum_{k=0}^{\infty}\bar{a}_k\bar{c}_{kl}-\delta_N^l\right),$$

where $\lambda_0,\lambda_1,\cdots,\lambda_N$ are dummy variables called *Lagrange multipliers*.

Let $\partial\mathscr{L}/\partial a_k=0$, $\partial\mathscr{L}/\partial\bar{a}_k=0$. We have $a_k=\bar{\lambda}_l\bar{c}_{lk}$ and the constraint becomes

$$\sum_{l=1}^{N}\bar{\lambda}_l\sum_{k=0}^{\infty}\bar{c}_{lk}c_{mk}=\delta_N^m. \tag{3.1}$$

Let $A_{lm}:=\sum\bar{c}_{lk}c_{mk}(l,m=0,1,\cdots,N)$. From the definition of c_{lk} and Theorem 2.1, we see that $A=(A_{lm})$ is a diagonal matrix of rank $N+1$, and all the eigenvalues are positive. The equality (3.1) becomes $(\bar{\lambda}_0,\bar{\lambda}_1,\cdots,\bar{\lambda}_N)A=(0,\cdots,0,1)$, which yields $\lambda_0=\cdots=\lambda_{N-1}=0$ and

$$\lambda_N^{-1}=A_{NN}=\bar{c}_{Nk}c_{Nk}=\frac{\delta^{2N}k(z,\bar{w})}{\delta z^{\alpha_1}\dots\delta z^{\alpha_N}\delta\bar{w}^{\beta_1}\dots\delta\bar{w}^{\beta_N}}\bigg|_{z=w=t}\xi^{\alpha_1\cdots\alpha_N}\bar{\xi}^{\beta_1\cdots\beta_N}.$$

The coefficients a_k now is

$$a_k=\bar{\lambda}_N\bar{c}_{Nk}=A_{NN}^{-1}\frac{\delta^N\bar{\varphi}_k(z)}{\delta\bar{z}^{\alpha_1}\dots\delta\bar{z}^{\alpha_N}}\bigg|_{z=t}\bar{\xi}^{\alpha_1\cdots\alpha_N}$$

and the minimal function of $E_t^N(D)$, say f_D, is

$$f_D(z)=\sum_{k=0}^{\infty}a_k\varphi_k(z)=A_{NN}^{-1}\frac{\delta^N k(z,t)}{\delta\bar{t}^{\alpha_1}\dots\delta\bar{t}^{\alpha_N}}\bar{\xi}^{\alpha_1\cdots\alpha_N}. \tag{3.2}$$

The minimal L^2 norm is

$$\|f_D\|^2=\sum_{k=1}^{\infty}|a_k|^2=A_{NN}^{-1}=H^{-1}_{\alpha_1\cdots\alpha_N\bar{\beta}_1\cdots\bar{\beta}_N}(t,\bar{t})\xi^{\alpha_1\cdots\alpha_N}\bar{\xi}^{\beta_1\cdots\beta_N}. \tag{3.3}$$

Let $D_1 \subset D$ be two bounded domains and fix $t \in D_1$. It is easy to see that $E_t^N(D) \subset E_t^N(D_1)$. Denote by f_D and f_{D_1} the minimal functions of $E_t^N(D)$ and $E_t^N(D_1)$, respectively, then we have $\| f_{D_1} \|^2 \leq \| f_D \|^2$, or equivalently,

$$H^D_{\alpha_1\cdots\alpha_N\bar{\beta}_1\cdots\bar{\beta}_N}(t,\bar{t})\xi^{\alpha_1\cdots\alpha_N}\bar{\xi}^{\beta_1\cdots\beta_N} \leq H^{D_1}_{\alpha_1\cdots\alpha_N\bar{\beta}_1\cdots\bar{\beta}_N}(t,\bar{t})\xi^{\alpha_1\cdots\alpha_N}\bar{\xi}^{\beta_1\cdots\beta_N}$$

(see [Lu4, Theorem 2]). In particular, we have the following three inequalities in the cases of $N = 0, 1, 2$.

(1) For $N = 0$, we have $k_D(t,\bar{t}) \leq k_{D_1}(t,\bar{t})$, which is the well known decreasing property of the Bergman kernel.
(2) For $N = 1$, we have $k_D(t,\bar{t})T^D_{\alpha\bar{\beta}}(t,\bar{t}) \leq k_{D_1}(t,\bar{t})T^{D_1}_{\alpha\bar{\beta}}(t,\bar{t})$ due to Example 2.1(1), which can also be regarded as certain decreasing property (see also [JaP, Remark 6.2.7].)
(3) For $N = 2$, we have $k_D(T^D_{\alpha_1\bar{\beta}_1}T^D_{\alpha_2\bar{\beta}_2} + T^D_{\alpha_2\bar{\beta}_1}T^D_{\alpha_1\bar{\beta}_2} - R^D_{\bar{\beta}_1\alpha_1\alpha_2\bar{\beta}_2}) \leq k_{D_1}(T^{D_1}_{\alpha_1\bar{\beta}_1}T^{D_1}_{\alpha_2\bar{\beta}_2} + T^{D_1}_{\alpha_2\bar{\beta}_1}T^{D_1}_{\alpha_1\bar{\beta}_2} - R^{D_1}_{\bar{\beta}_1\alpha_1\alpha_2\bar{\beta}_2})$.

Divided by $(T^D_{\alpha\bar{\beta}}\xi^\alpha\bar{\xi}^\beta)^2$ on both sides of the case (3), then we have

$$\mathrm{Sec}_D(t,\xi) \geq 2 - \frac{k_{D_1}}{k_D}\left(2 - \mathrm{Sec}_{D_1}(t,\xi)\right)\left(\frac{T^{D_1}_{\lambda\bar{\mu}}\xi^\lambda\bar{\xi}^\mu}{T^D_{\alpha\bar{\beta}}\xi^\alpha\bar{\xi}^\beta}\right)^2. \tag{3.4}$$

Furthermore, if we consider three bounded domains $D_1 \subset D \subset D_2$, by the decreasing property of the case (1) and (2), we have $k_{D_2} \leq k_D \leq k_{D_1}$, $k_{D_2}T^{D_2}_{\alpha\bar{\beta}} \leq k_D T^D_{\alpha\bar{\beta}} \leq k_{D_1}T^{D_1}_{\alpha\bar{\beta}}$. Substitute this relations into (3.4), we have

$$\mathrm{Sec}_D(t,\xi) \geq 2 - \left(2 - \mathrm{Sec}_{D_1}(t,\xi)\right)\left(\frac{k_{D_1}T^{D_1}_{\lambda\bar{\mu}}\xi^\lambda\bar{\xi}^\mu}{k_{D_2}T^{D_2}_{\alpha\bar{\beta}}\xi^\alpha\bar{\xi}^\beta}\right)^2. \tag{3.5}$$

Similar arguments give the upper bound estimate,

$$\mathrm{Sec}_D(t,\xi) \leq 2 - \left(2 - \mathrm{Sec}_{D_2}(t,\xi)\right)\left(\frac{k_{D_2}T^{D_2}_{\alpha\bar{\beta}}\xi^\alpha\bar{\xi}^\beta}{k_{D_1}T^{D_1}_{\lambda\bar{\mu}}\xi^\lambda\bar{\xi}^\mu}\right)^2. \tag{3.6}$$

In general, the Bergman kernel and the Bergman metric of D_1 and D_2 are not easy to compute. However, if we choose D_1 and D_1 good enough, for instance, D_1 and D_2 are the Euclidean balls, with the same center t and radius $r(t)$, $R(t)$ respectively, the following estimates were obtained by Lu. Note that at the moment the holomorphic sectional curvature of $D_i (i = 1, 2)$ is $-2/(n + 1)$.

Theorem 3.1 ([Lu4, Corollary 5] *Under the above hypothesis, we have*

$$2-2\frac{n+2}{n+1}\frac{R^{4n}}{r^{4n}} \leq Sec_D(t,\xi) \leq 2-2\frac{n+2}{n+1}\frac{r^{4n}}{R^{4n}}. \tag{3.7}$$

Follows this line, we state the result without proof about the estimates for the Ricci curvature and the scalar curvature.

Theorem 3.2 ([Lu4, Corollary 6]) *Under the above hypothesis, we have*

$$(n+1)-(n+2)\frac{R^{2n}}{r^{2n}} \leq Ric_D(t,\xi) \leq (n+1)-(n+2)\frac{r^{2n}}{R^{2n}}. \tag{3.8}$$

Theorem 3.3 *Under the above hypothesis, we have*

$$n(n+1)-n(n+2)\frac{R^{2n}}{r^{2n}} \leq Scal_D(t) \leq n(n+1)-n(n+2)\frac{r^{2n}}{R^{2n}}. \tag{3.9}$$

Now one natural question occurs that the lower bound tends to $-\infty$ as the point t tends to ∂D, or equivalently $r(t) \to 0$. In the next section, we will see that there exist some general classes of bounded domains with squeezing property whose Bergman curvatures admit uniform bounds.

4 Uniform Estimates of the Bergman Curvatures

In this section, we will give uniform estimates for the Bergman curvatures on the holomorphically homogeneous regular manifolds or the uniform squeezing domains.

Definition 4.1 [[LSY1, Definition 7.2], [Yeu, Definition 1]] A complex manifold X of dimension n is called *holomorphic homogeneous regular* (HHR) or equivalently, *uniformly squeezing* (USq) if $\exists\, r < R$ such that $\forall p \in X$, there is a holomorphic map $f_p : X \to \mathbb{C}^n$ which satisfies

1. $f_p(p) = 0$;
2. $f_p : X \to f_p(X)$ is bi-holomorphic;
3. $B^n(0,r) \subset f_p(X) \subset B^n(0,R)$, where $B^n(0,r)$ and $B^n(0,R)$ are Euclidean balls with center 0 in $\mathbb{C}^n$.

To tell a bounded domain is HHR or USq, i.e., to determine the universal constants r and R is not trivial. In [DGZ1], the authors introduced the concept of *squeezing function* in order to study geometric and analytic properties of the HHR/USq manifolds.

Definition 4.2 [[DGZ1, Definition 1.1]] Let D be a bounded domain in $\mathbb{C}^n$. For $z \in D$ and an (open) holomorphic embedding $f : D \to B^n$ with $f(z) = 0$, we define $s_D(z, f) = \sup\{r | B^n(0, r) \subset f(D)\}$ and $s_D(z) = \sup_f\{s_D(z, f)\}$, where the supremum is taken over all holomorphic embeddings $f : D \to B^n$ with $f(z) = 0$. Here B^n is the unit ball in $\mathbb{C}^n$, and $B^n(0, r)$ is the ball in $\mathbb{C}^n$ with center 0 and radius r. We call s_D the *squeezing function* of D.

For any point $z \in D$, we now consider the extremal holomorphic embedding $f : D \to B^n$ with $f(z) = 0$. Since the Bergman curvatures are invariant under biholomorphic mappings, taking the holomorphic sectional curvature for example, $\mathrm{Sec}_D(z, \xi) = \mathrm{Sec}_{f(D)}(0, df \cdot \xi)$, therefore we have the following analogue of Theorem 3.1, 3.2 and 3.3. Note that at the moment the radius of D_1 and D_2 are $s_D(z)$ and 1, respectively.

Theorem 4.1 *Let $D \subset \mathbb{C}^n$ be a bounded domain. For $\forall z \in D$ and $\xi \in \mathbb{C}^n \setminus \{0\}$, we have*

$$2 - 2\frac{n+2}{n+1}s_D^{-4n}(z) \le Sec_D(z, \xi) \le 2 - 2\frac{n+2}{n+1}s_D^{4n}(z),$$

$$(n+1) - (n+2)s_D^{-2n}(z) \le Ric_D(z, \xi) \le (n+1) - (n+2)s_D^{2n}(z),$$

$$n(n+1) - n(n+2)s_D^{-2n}(z) \le Scal_D(z) \le n(n+1) - n(n+2)s_D^{2n}(z).$$

Furthermore, if we define the *squeezing constant* $\hat{s}_D$ on the domain D as $\hat{s}_D := \inf_{z \in D} s_D(z)$, then D is HHR/USq if $\hat{s}_D > 0$. For HHR/USq domains, we have the following estimates.

Corollary 4.1 *Let D be a HHR/USq domain in $\mathbb{C}^n$. We have*

$$2 - 2\frac{n+2}{n+1}\hat{s}_D^{-4n} \le Sec_D(z, \xi) \le 2 - 2\frac{n+2}{n+1}\hat{s}_D^{4n},$$

$$(n+1) - (n+2)\hat{s}_D^{-2n} \le Ric_D(z, \xi) \le (n+1) - (n+2)\hat{s}_D^{2n},$$

$$n(n+1) - n(n+2)\hat{s}_D^{-2n} \le Scal_D(z) \le n(n+1) - n(n+2)\hat{s}_D^{2n}.$$

Remark 4.1 Note that if $s_D^{4n}(z) > (n+1)/(n+2)$, then all the Bergman curvatures of D are negative. This means that if the domain looks "close" enough to the unit ball observed at z, then the curvatures are negative at this point.

One important application of Theorem 4.1 is boundary behaviors of the Bergman curvatures near strictly pseudoconvex boundary points. First, let us recall the asymptotic behavior of the squeezing function.

Theorem 4.2 ([DGZ2, KiZ]) *Let D be a C^2-smooth bounded strictly pseudoconvex domain in $\mathbb{C}^n$. Then* $\lim_{D \ni z \to p} s_D(z) = 1$ *holds for any $p \in \partial D$.*

Remark 4.2 We should point out that what we proved actually is a little more general than Theorem 4.2. That is, for any bounded domain $D \subset \mathbb{C}^n$, if $p \in \partial D$ is global strongly convex or extremely spherical (see [DGZ2] or [KiZ] for the definition), then $\lim_{D \ni z \to p} s_D(z) = 1$.

The proof of Theorem 4.2 is based on the following remarkable theorem recently given by Diederich, Fornaess and Wold, which asserts that any strictly pseudoconvex boundary point can be exposed to be global strongly convex or extremely spherical.

Theorem 4.3 ([DFW, Theorem 1.1]) *Let $\Omega \subset \mathbb{C}^n$ be a bounded domain which is locally convexifiable and has finite type $2k$ near a point $p \in \partial\Omega$. Assume further that $\partial\Omega$ is C^∞-smooth near p, and that $\overline{\Omega}$ has a Stein neighborhood basis. Then there exists a holomorphic embedding $f : \overline{\Omega} \to \overline{B}^n_k$, where $B^n_k = \{z \in \mathbb{C}^2 : |z_n|^2 + |z'|^{2k} < 1\}$, such that $f(p) = (0, \cdots, 0, 1)$ and $\{z \in \overline{\Omega} : f(z) \in \partial B^n_k\} = \{p\}$.*

In particularly, if $\partial\Omega$ is strongly pseudoconvex near p, i.e. $k = 1$, it is enough to assume that $\partial\Omega$ is C^2-smooth near p.

Combining Theorem 4.1 and Theorem 4.2, one can immediately obtain

Corollary 4.2 *Let D be a bounded p.s.c. domain and $p \in \partial D$ be C^2 strictly pseudoconvex. One has* $\lim_{D \ni z \to p} Sec_D(z, \xi) = -2/(n+1)$, $\lim_{D \ni z \to p} Ric_D(z, \xi) = -1$, $\lim_{D \ni z \to p} Scal_D(z) = -n$.

Very recently, Fornaess and Wold gave an improvement on the estimate of the squeezing function when a bounded strictly pseudoconvex domain has $C^k (k \geq 3)$ boundary.

Theorem 4.4 ([FoW, Theorem 1.1]) *Let $\Omega = \{z \in \mathbb{C}^n : \rho(z) < 0\} \subset \mathbb{C}^n$ be a a strictly pseudoconvex domain with a defining function ρ of class C^k for $k \geq 3$. Then there exists a constant $C > 0$ such that the squeezing function $s_\Omega(z)$ for Ω satisfies the estimate $s_\Omega(z) \geq 1 - C \cdot \sqrt{|\rho(z)|}$.*

If $k \geq 4$, then there exists a fixed constant C' such that $s_\Omega(z) \geq 1 - C' \cdot |\rho(z)|$ for all $z \in \Omega$.

By Theorem 4.1 and Theorem 4.4, we have the following asymptotic behaviors of the Bergman curvatures, which is an improvement of Corollary 4.2.

Corollary 4.3 *Let $D = \{z \in \mathbb{C}^n : \rho(z) < 0\}$ be a strictly pseudoconvex domain with a defining function ρ of class C^k for $k \geq 3$. Then for any point z near the boundary ∂D and a nonvanishing direction ξ, there exists a constant $C > 0$ such that*

$$Sec_D(z,\xi) = -\frac{2}{n+1} + O(\sqrt{|\rho(z)|}), \quad Ric_D(z,\xi) = -1 + O(\sqrt{|\rho(z)|}),$$
$$Scal_D(z) = -n + O(\sqrt{|\rho(z)|}).$$

If $k \geq 4$, there exists a constant $C' > 0$ such that

$$Sec_D(z,\xi) = -\frac{2}{n+1} + O(|\rho(z)|), \quad Ric_D(z,\xi) = -1 + O(|\rho(z)|),$$
$$Scal_D(z) = -n + O(|\rho(z)|),$$

where $O(|\rho(z)|)$ denotes a quantity dominated by $C|\rho|$ with the constant C depending only on the dimension n.

A natural question now may be asked: Can we expect higher order asymptotic behaviors of the Bergman curvatures when the boundary has more higher regularity? The following theorem shows that this is not true in general.

Theorem 4.5 ([JoS, Theorem 1.1]) *Let Ω be a bounded strictly pseudoconvex domain in $\mathbb{C}^n$ with C^∞-smooth boundary and ρ be a defining function of class C^k for $k \geq 4$ such that $g_{i\bar{j}} = -\partial^2 \log(-\rho)/\partial z_i \partial \bar{z}_j$ is a complete Kähler metric on Ω. For $z \in \Omega$ and $\xi \in \mathbb{C}^n \setminus \{0\}$, we have*

1. *For $n \geq 3$, if $Sec_\Omega(z,\xi) = -2/(n+1) + O(\rho^2)$, then $\partial\Omega$ is locally spherical;*
2. *For $n = 2$, if $k \geq 5$ and $Sec_\Omega(z,\xi) = -2/3 + O(\rho^3)$, then $\partial\Omega$ is locally spherical.*

It has been proved by Nemirovskii and Shafikov that a strictly pseudoconvex domain with spherical boundary is universally covered by the unit ball [NS1, Theorem A.2] and if a strictly pseudoconvex domain with real analytic boundary is covered by the unit ball, then its boundary is spherical [NS2, theorem 1.2]. However, the boundary of a strictly pseudoconvex domain is non-spherical in general.

5 Concluding Remarks and Open Questions

In the final section, we present some remarks and open questions.

We have already talked about the Bergman curvature estimates for bounded domains in $\mathbb{C}^n$ by using the squeezing function in the previous sections. Actually, the study of squeezing functions has its own interest since in general it is difficult to have the explicit expression of the squeezing function on a given bounded domain, except bounded symmetric domains on which the squeezing functions have already been calculated (see [Kub, Theorem 1]). For us, the only known example now is the punctured ball $B^n \setminus \{0\}$, of which the squeezing function is $s_{B^n\setminus\{0\}}(z) = \|z\|$, where $\| \bullet \|$ denotes the Euclidean norm (see [DGZ1, Corollary 7.3]). In general, N. Shcherbina posed the following interesting question: Whether the squeezing functions are always plurisubharmonic without knowing the explicit formulae? At the time of writing this paper, we do not know the answer yet.

From the squeezing function of the punctured unit ball, we know that $B^n \setminus \{0\}$ is not HHR/USq since $s_{B^n\setminus\{0\}}(z) = \|z\|$ has no positive lower bound on $B^n \setminus \{0\}$. Yet it is well known that $B^n \setminus \{0\}$ is not a domain of holomorphy for $n > 1$. Hence it is natural to ask whether all smooth bounded pseudoconvex domains in $\mathbb{C}^n (n > 1)$ admit HHR/USq property? One counterexample is the smooth pseudoconvex domain $\Omega \Subset \mathbb{C}^3$, constructed by Diederich and Fornaess [DiF], on which the Bergman metric and the Kobayashi metric are not equivalent. Consequently, Ω is not HHR or USq. However, we do not know for instance whether the bounded pseudoconvex domains of finite type in $\mathbb{C}^2$ are HHR/USq.

One amazing property of the squeezing function is the asymptotic behavior near the strictly pseudoconvex boundary points. Thanks to J.E. Fornaess who proposed the following interesting question at the special workshop of several complex variables held in Chinese Academy of Science in 2014: If the boundary limit of the squeezing function is 1, does this imply the domain is strictly pseudoconvex? The answer is still unknown.

Acknowledgments The author is grateful to the organizers of the 10th Korean Conference on Several Complex Variables, especially Prof. K.-T. Kim and Prof. N. Shcherbina, for their kind invitation. He would also like to thank Prof. Q.-K. Lu for many invaluable communications on this topic. Project partially supported by NSFC (No. 11371025, 11371257).

References

[Ber] Bergman, S.: The kernel function and conformal mapping. American Mathematical Society, Providence, Rhode Island (1970)
[ChY] Cheng, S.-Y., Yau, S.-T.: On the existence of a complete Kähler metric on non-compact complex manifolds and the regularity of Feffermans equation. Comm. Pure Appl. Math. **33**, 507–544 (1980)
[DiF] Diederich, K., Fornss, J.E.: Comparison of the Bergman and the Kobayashi metric. Math. Ann. **254**, 257–262 (1980)
[DFW] Diederich, K., Fornaess, J.E., Wold, E.F.: Exposing points on the boundary of a strictly pseudoconvex or a locally convexifiable domain of finite 1-type, J. Geom. Anal. **24**, 2124–2134. doi:10.1007/s12220-013-9410-0
[DiH] Diederich, K., Herbort, G.: Pseudoconvex domains of semiregular type. In: Contributions to Complex Analysis and Analytic Geometry, Aspects of Mathematics, vol. E26, pp. 127–161 (1994)
[DGZ1] Deng, F., Guan, Q., Zhang, L.: On some properties of squeezing functions of bounded domains. Pac. J. Math. **257**(2), 319–342 (2012)
[DGZ2] Deng, F., Guan, Q., Zhang, L.: Properties of squeezing functions and global transformations of bounded domains. arXiv:1302.5307 [math.CV] (Trans. AMS)
[FoW] Fornaess, J.E., Wold, E.F.: An estimate for the squeezing function and estimates of invariant metrics. In: Proceedings Volume of The KSCV10. arxiv:1411.3846v1 [math.CV]
[Fuk] Fuks, B.A.: Über geodätische Manifaltigkeiten einer invariant Geometrie. Mat. Sb. **2**, 369–394 (1937)
[GK1] Green, R., Krantz, S.: The stability of the Bergman kernel and the the geometry of the Bergman kernel. Bull. AMS **4**, 111–115 (1981)

[GK2] Green, R., Krantz, S.: Deformation of complex structures, estimates for $\bar{\partial}-$ equation, stability of the Bergman kernel. Adv. Math. **43**, 1–86 (1983)

[GKK] Green, R., Kim, K.-T., Krantz, S.: The geometry of complex domains. Birkhauser, Boston (2011)

[Hua] Hua, L.-K.: The estimation of the Riemann curvature in several complex variables. Acta Math. Sin. **4**, 143–170 (1954). in Chinese

[JaP] Jarnicki, M., Pflug, P.: Invariant distances and metrics in complex analysis. De Gruyter Expositions in Mathematics, vol. 9 (1993)

[JoS] Joo, J.-C., Seo, A.: Higher order asymptotic behavior of certain Kähler metrics and uniformization for strongly pseudoconvex domains. J. Korea Math. Soc. **52**, 1–21 (2015)

[KiY] Kim, K.-T., Yu, J.: Boundary behavior of the Bergman curvature in strictly pseudoconvex polyhedral domains. Pac. J. Math. **176**(1), 141–163 (1996)

[KrY] Krantz, S., Yu, J.: On the Bergman invariant and curvatures of the Bergman metric. Ill. J. Math. 40(2), 226-244 (1996)

[KiZ] Kim, K.-T., Zhang, L.: On the uniform squeezing property and the squeezing function. arXiv:1306.2390 [math.CV]

[Kle] Klembeck, P.: Kähler metrics of negative curvature, the Bergmann metric near the boundary, and the Kobayashi metric on smooth bounded strictly pseudoconvex sets. Indiana Univ. Math. J. **27**(2), 275–282 (1978)

[Kob] Koboyashi, S.: Geometry of bounded domains. Trans. Amer. Math. Soc. **93**, 267–290 (1959)

[Kub] Kubota, Y.: A note on holomorphic imbeddings of the classical Cartan domains into the unit ball. Proc. Amer. Math. Soc. **85**(1), 65–68 (1982)

[LSY1] Liu, K.-F., Sun, X.-F., Yau, S.-T.: Canonical metrics on the moduli space of Riemann surfaces. I. J. Differ. Geom. **68**(3), 571–637 (2004)

[Lu1] Lu, Q.-K.: On Kähler manifolds with constant curvature. Acta Math. Sin. **16**, 269–281 (1966)

[Lu2] Lu, Q.-K.: The estimation of the intrinsic derivatives of the analytic mapping of bounded domains. Sci. Sin. Spec. Ser. **II**, 1–17 (1979)

[Lu3] Lu, Q.-K.: Holomorphic invariant forms of a bounded domain. Sci. China Ser. A **51**, 1945–1964 (2008)

[Lu4] Lu, Q.-K.: On the lower bounds of the curvatures in a bounded domain. Sci. China Ser. A **58**, 1–10 (2015)

[NS1] Nemirovskii, S., Shafikov, R.: Uniformization of strictly pseudoconvex domains. I. (Russian) Izv. Ross. Akad. Nauk Ser. Mat. **69**(6), 115–130 (2005) (translation in Izv. Math. **69**(6), 1189–1202 (2005))

[NS2] Nemirovskii, S., Shafikov, R.: Uniformization of strictly pseudoconvex domains. II. (Russian) Izv. Ross. Akad. Nauk Ser. Mat. **69**(6), 131–138 (2005) (translation in Izv. Math. **69**(6), 1203–1210 (2005))

[Noz] Nozarjan, E.: Estimates of Ricci curvature. Nauk. Arm. SSR Ser. Mat. **8**, 418–423 (1973)

[Yeu] Yeung, S.-K.: Geometry of domains with the uniform squeezing property. Adv. Math. **221**(2), 547–569 (2009)

Some Problems

John-Erik Fornaess and Kang-Tae Kim

Abstract We pose some problems for the future research in complex analysis and geometry.

Keywords Complex analysis · Dynamics · Holomorphic map · Invariant metric · Curvature · Automorphisms · Pseudoconvexity

Large part of this problem set is from a lecture by Fornaess at the Center for Geometry and its Applications of POSTECH in August 2014 after the KSCV10 Symposium in Gyeong-Ju. Then the authors agreed to compose this problem set with the addition by the second named author.

1 Worm

Problem 1 Let Ω be the worm domain. Show that every strongly pseudoconvex boundary point of Ω can be exposed.

A boundary point $p \in \Omega$ is called *exposed* if there is an open neighborhood U of p and a 1-1 holomorphic mapping $f = (f_1, f_2)\colon U \cup \Omega \to \mathbb{C}^2$ [Note: the worm domain introduced in [DF] is in $\mathbb{C}^2$.] such that Re $f_1(p) = 0$ and Re $f_1(z) < 0$ for every $z \in \Omega$.

This problem was solved in [DFW] for domains which are strongly pseudoconvex.

See [DF] for definition of the worm . The worm is strongly pseudoconvex except on an annulus. It has no Stein neighborhood basis. But you can still solve the equation $\bar{\partial} u = f$ on the closure, so that u is C^∞ if f is C^∞, see Kohn [Kohn].

J.-E. Fornaess
Department of Mathematical Sciences, NTNU, 7491 Trondheim, Norway
e-mail: johnefo@math.ntnu.no

K.-T. Kim (✉)
Center for Geometry and Its Applications and Department of Mathematics,
POSTECH, Pohang 790-784, The Republic of Korea
e-mail: kimkt@postech.ac.kr

F. Bracci et al. (eds.), *Complex Analysis and Geometry*, Springer Proceedings in Mathematics & Statistics 144, DOI 10.1007/978-4-431-55744-9_28

2 Nirenberg Problem

Problem 2 (*Nirenberg*) Let U be a strongly pseudoconvex bounded domain in $\mathbb{C}^n$ with smooth boundary. Suppose that γ is a smooth curve in the boundary which is transverse to the complex tangent space at each point. Can there exist a continuous function from $\overline{U}$, holomorhic on U, such that it vanishes identically on γ but does not have any zero inside U?

Note that the curve γ is not assumed to be real analytic. If it is real analytic, it extends as a complex curve to the inside. Then the holomorphic function must be zero there.

3 The $\bar{\partial}$-problem

Problem 3 Solve for u in $\bar{\partial}u = f$ with sup-norm estimates on bounded convex domains in $\mathbb{C}^2$ with C^1 boundary.

If the boundary is real analytic, one can solve $\bar{\partial}$ with supnorm estimates. The difficulty is that there might be infinitely flat points.

4 Bergman Space

Problem 4 Let U be an unbounded (smooth strongly) pseudoconvex domain in $\mathbb{C}^2$. Let $A^2(U)$ be the set of L^2 holomorphic functions on U. If $A^2(U) \neq \{0\}$, is $A^2(U)$ infinite dimensional?

There is an example by Wiegerinck (1984), Math. Z., a Reinhardt domain in $\mathbb{C}^2$ with a nontrivial finite dimensional A^2. But this domain is not pseudoconvex. M. Englis showed that the Bergman space is either trivial or infinite dimensional if the domain is pseudoconvex Reinhardt.

5 Polynomial Convexity

Problem 5 Let X be a complex hypersurface in $\mathbb{C}^3$ with an isolated normal singularity at 0. Suppose that $K \subset X$ 0 is compact. Suppose that 0 is in the polynomially convex hull of K. Let $K \subset F \subset X$ be contained in the relative interior. Is 0 in the relative interior of the polynomial hull of F?

This problem originated in some questions about the Levi problem in complex spaces [Fornas].

6 Complex Dynamics/Real Dynamics

Problem 6 Let $P(z)$ be a complex polynomial in $\mathbb{C}$. For $z \in \mathbb{C}$ let $P^n(z) = z_n = x_n + iy_n$. We call the sequence $\{x_n\}$ the *real part of the orbit*. Suppose that one only knows the real orbits. How much can one say about the complex dynamics of P? For example, how can one detect the degree of P?

See the paper by Fornæss and Peters [Fornaess]. Almost nothing is done on this kind of problems. In general, one can iterate a map $F: \mathbb{R}^k \to \mathbb{R}^k$ and one can see only the parts of the orbits, $(x_1^n, \ldots, x_\ell^n)$ where $\ell < k$. What can one then say about the dynamics?

7 Fatou-Bieberbach Domains

7.1 Definition of Fatou-Bieberbach (FB) Domains

They are domains in $\mathbb{C}^2$ which are biholomorphic to $\mathbb{C}^2$ while being proper subsets (Same in $\mathbb{C}^n$, $n > 2$.).

7.2 Standard Construction of Fatou-Bieberbach Domains

See Rosay and Rudin [Ros]. Take a biholomorphic self map of $\mathbb{C}^2$; for example a Henon map H with $0 < a < b < 1$. Then there is a small ball B so that, if $(z, w) \in B$ then, $a\|(z, w)\| \leq \|H(z, w)\| \leq b\|(z, w)\|$, for some constant $c < 1$. This means that, on B, $\|H^n(z, w)\| \to 0$ as $n \to \infty$. We say that B is contained in the basin of attraction of 0. Denote by Ω the set of all points (z, w) such that $\|H^n(z, w)\| \to 0$. This is an open set and it is biholomorphic to $\mathbb{C}^2$. One also sees that, if we start with $(z, w) = (100, 0)$ then, $H^n(z, w) \to \infty$. Hence this point is not in Ω. Hence Ω is a Fatou-Bieberbach domain. Since a Fatou-Bieberbach domain is biholomorphic to $\mathbb{C}^2$, it contains a smaller FB domain. In fact we can find a sequence of FB domains $\Omega_1 \supset \Omega_2 \supset \cdots$. One can then ask if it is possible to find such a sequence such that $\bigcap \Omega_n = \emptyset$. If this is true, then an old conjecture by Michael is true: *All characters on a Frechet algebra are continuous.* (See Dixon and Esterle [DE] for precise statements).

One can have Fatou-Bieberbach domains V with a boundary which is C^∞, Stensønes [Sten].

Problem 7 The boundary of V is a union of Riemann surfaces. Are they all biholomorphic copies of $\mathbb{C}$? Does there exist an FB domain which has real analytic boundary?

Problem 8 (*Harz et al.* [Harz]) Does there exist an FB domain which is contained in a proper strongly pseudoconvex subdomain in $\mathbb{C}^2$?

One can perturb the construction of FB domains: Pick two numbers $0 < a < b < 1$. Let F_n be a sequence of biholomorphisms of $\mathbb{C}^2$ such that, if $\|(z, w)\| < 1$ then, $a\|(z, w)\| \leq \|F_n(z, w)\| \leq b\|(z, w)\|$. Then one can consider the iterates $F^{(n)} = F_n \circ \cdots \circ F_1$. Let $\Omega = \{(z, w)\colon F^{(n)}(z, w) \to 0\}$. We call this a uniformly random basin (or non-autonomous basin), see [AAF] for a recent survey.

Problem 9 Are uniformly random basins FB domains?

There are partial results if a and b are close together. (See Peters and Smit [Pet] on Arxiv recently).

Consider the projective compactifiaction of a ball $B_R = \{z \in \mathbb{C}^n\colon \|z\| < R\}$. Identify the boundary of B_R with the plane at infinity. This gives topology to the closure. The boundary and the inside both have complex structure. But they match poorly. Note that if we take a limit as $R \to \infty$, we get the usual $\mathbb{P}^n$.

Now consider a random basin Ω. This Ω is an increasing union of balls (rather, domains biholomorphic to a ball) $B_n = (F^{(n)})^{-1}(B)$. Use this compactification. And try to pass to the limit.

Problem 10 Does such a limit exist and does it give a complex structure on a compactification of the random basin?

Remark 1 If the random basin is not biholomorphic to $\mathbb{C}^n$, then this gives a new complex structure to $\mathbb{P}^n$. If this can be done when $n = 3$ this might provide a complex structure to the real 6-dimensional sphere S^6 as observed by Siu. In fact, if there is a complex structure on S^6, and we blow up a point then we get a complex manifold which is diffeomorphic to P^3 but not biholomorphic to P^3. So if we get a new complex structure on P^3 so that we can blow down a copy of P^2 we obtain a complex structure on S^6.

Remark 2 An alternative to the uniformly random basins are obtained by removing the condition of the lower bound a. In such a case one can obtain random basins which are not boholomorphic to $\mathbb{C}^n$. (Fornaess: Short $\mathbb{C}^n$, [Fornae]).

If one can use the above compactification in the case of short $\mathbb{C}^3$, then one surely gets a nonstandard $\mathbb{P}^3$.

Problem 11 Does the above compactification work for short $\mathbb{C}^3$? Harz et al. [Harz] has introduced the concept "core of a domain". There is also a notion of core for short $\mathbb{C}^2$.

Problem 12 Describe the core of a short $\mathbb{C}^2$.

Problem 13 Let Ω be a random basin. Show that there exists a proper holomorphic map from $\mathbb{C}$ into Ω [There exist many non-constant holomorphic maps from $\mathbb{C}$ into Ω, see [AAF].].

This problem can be considered to be the next step in the process of showing that uniformly random basins are biholomorphic to $\mathbb{C}^n$. Random basins occur as stable manifolds of hyperbolic maps on complex manifolds. One has similar questions.

8 Convexity

Problem 14 Let Ω be an unbounded convex domain in $\mathbb{C}^n$. When will there exist a biholomorphic mapping-into $\psi : \Omega \to \mathbb{C}^n$ so that the image $\psi(\Omega)$ is bounded convex?

If such ψ exists, Ω has to be Kobayashi hyperbolic. Every convex Kobayashi hyperbolic domain can be mapped biholomorphically onto a bounded pseudoconvex domain in $\mathbb{C}^n$. But the image may not in general be convex. Thus finding some analytic/geometric conditions on the boundary for such mapping ψ to exist is the question here.

Problem 15 (*Gindikin*) Let Ω be a bounded homogeneous domain in $\mathbb{C}^n$. If there exists a biholomorphism-into $\psi : \Omega \to \mathbb{C}^n$ such that $\psi(\Omega)$ is bounded convex, then show that Ω is a bounded symmetric domain.

A theorem by Vinberg, Pyatetsky-Shapiro and Gindikin says that every bounded homogeneous domain is biholomorphic to a Siegel domain (of the second kind). Since all Siegel domains are tubes over a convex cone, they are convex but unbounded. Then the Harish-Chandra realization (via linear fractional transformation) of such domains is convex if and only if the domain is symmetric, [Kai]. There is another special mapping, called the Bergman representative map, that can turn these domains into bounded domains. Ishi and Kai [Ishi] showed that the image under this map of a bounded homogeneous domain is convex if and only if the domain is symmetric.

9 Unbounded Domains

Problem 16 Which unbounded domains can be biholomorphic to a bounded domain?

If one considers for instance the domain

$$W = \{(z, w) \in \mathbb{C}^2 : |w| < e^{-|z|^2}\},$$

then a direct computation shows that this has a finite volume. Any monomial $z^k w^\ell$ is L^2. Since the domain is Reinhardt, these generate the whole $A^2(W)$ the space of square integrable holomorphic functions of W. Its Bergman metric is positive-definite and complete, [AGK]. But the domain contains the complex line defined by $w = 0$. Hence it is not Kobayashi hyperbolic and cannot be realized as a bounded domain.

The Kohn-Nirenberg domain

$$\Omega_{KN} = \{(z, w) \in \mathbb{C}^2 : \operatorname{Re} w + |zw|^2 + |z|^8 + \frac{15}{7}|z|^2 \operatorname{Re} z^6 < 0\}$$

is still not known whether it is biholomorphic to a bounded domain in $\mathbb{C}^2$. The following problem may also be considered.

Problem 17 Which unbouned domains admit positive-definite (and complete) Bergman metric?

The domains W, Ω_{KN} as well as several others are shown to admit positive-definite and complete Bergman metric. See [AGK, CKOh, Herb], e.g. On the other hand Short $\mathbb{C}^k$ does not; A. Seo showed recently, in a private discussion, that there are no nonzero L^2 holomorphic functions on Short $\mathbb{C}^k$.

10 Automorphism Groups

Even when the automorphism group of a bounded domain is non-compact, the understanding of general cases is rather poor.

In the paper of Griffiths [Griffi], a complex two dimensional bounded domain has been constructed as the universal covering space of a Zariski open set. This domain is a disc fibration over the unit open disc and has a noncompact automorphism group.

Problem 18 Is the automorphism group of this domain discrete?

There had been some claims but we are not aware of any written proof.

The Teichmüler space of a compact Riemann surface of genus $g > 1$ is biholomorphic to a bounded domain in $\mathbb{C}^{3g-3}$. The embedding by L. Bers is also a bounded domain. This domain has been shown to be non-convex [Kim].

Problem 19 Can it be re-embedded biholomorphically to be a bounded convex domain?

Of course there is this old guiding question: *Which bounded domains admit non-compact automorphism group?*

For the pseudoconvex bounded domains, there are a few results. Along the line of thoughts, there is this old problem:

Problem 20 (*Greene-Krantz conjecture*) Let Ω be a bouned pseudoconvex domain with smooth boundary. If a boundary point is not of finite type in the sense of D'Angelo, then show that there does not exist any automorphism orbit accumulating at this boundary point.

A weaker problem may be:

Problem 21 Let Ω be a bounded pseudoconvex domain in $\mathbb{C}^2$ with $\mathscr{C}^\infty$ smooth boundary. Then show that no automorphism orbit can accumulate at a boundary point of infinite D'Angelo type if every other neighboring boundary points are of finite type.

It is shown in [Byun] that if boundary is of finite type, but one boundary point has the type strictly larger than the neighboring boundary points, then it cannot be an orbit accumulation point.

11 CR Manifolds and CR Vector Fields

Problem 22 Classify the germs of CR manifolds admitting a parabolic orbit.

The case of CR manifolds of CR codimension 1 that admit CR contractions is understood. See [Yoccoz].

12 Semicontinuity of Automorphism Group

Problem 23 In case the bounded domains converge in the sense of normal convergence (or equivalently, in the sense of Caratheodory kernel convergence) to another bounded domain, show that the automorphism groups show the upper semicontinuity phenomenon, i.e., show: if $\Omega_j \to \Omega_0$ as $j \to \infty$ in the sense described above, show that there exists $N > 0$ such that Aut (Ω_j) is a Lie subgroup of Aut (Ω_0) for every $j > N$.

See [Gree] for the developments concerning this problem up to the early 1980s. Another recent progress is presented [Greene].

13 The Scaling Methods

In the late 1970s, S. Pinchuk came up with the scaling method in complex analysis. See [Pin]. Another scaling method was presented by Frankel [Fran] in the mid 1980s. But there are still some questions left.

Problem 24 Show the "forward convergence" of the Pinchuk scaling sequence for the pseudoconvex domains of finite type when the dimension is 3 or higher.

Problem 25 On which domains, other than convex hyperbolic domains, does the Frankel's scaling sequence converge?

14 Curvature

Problem 26 McNeal proved in [McNeal] that the holomorphic sectional curvature of the Bergman metric for a bounded domain in $\mathbb{C}^2$ with finite type boundary has to be bounded. Can one prove it without using the results on the $\bar{\partial}$-Neumann problem?

There is a counterexample in $\mathbb{C}^3$, see [Herbo]; it is a domain that is pseudoconvex, Reinhardt, of finite type boundary, and defined by a polynomial defining function, but not semiregular (or, h-extendable).

Acknowledgments Research of the second named author is supported in part by the grant 2011-0030044 (The SRC-GAIA) of the NRF of Korea.

References

[AAF] Abbondandolo, A., Arosio, L., Fornaess, J.E., Majer, P., Peters, H., Raissy, J., Vivas, L.: A survey on non-autonomous basins in several complex variables, arxiv:1311.3835

[AGK] Ahn, T., Gaussier, H., Kim, K.-T.: Positivity and completeness of invariant metrics. J. Geom. Anal. doi:10.1007/s12220-015-9587-5 (2015)

[Byun] Byun, J.: On the boundary accumulation points for the holomorphic automorphism groups. Mich. Math. J. **51**(2), 379–386 (2003)

[CKOh] Chen, B.-Y., Kamimoto, J., Ohsawa, T.: Behavior of the Bergman kernel at infinity. Math. Z. **248**, 695–798 (2004)

[DF] Diederich, K., Fornaess, J.E.: Pseudoconvex domains; an example with nontrivial Nebenhülle. Math. Ann. **225**(3), 275–292 (1977)

[DFW] Diederich, K., Fornaess, J.E.,Wold, E.F.: Exposing points on the boundary of a strictly pseudoconvex or a locally convexifiable domain of finite 1-type. J. Geom. Anal. **24**(4), 2124–2134 (2014)

[DE] Dixon, P.G., Esterle, J.: Michaels problem and the Poincar-Fatou-Bieberbach phenomenon. Bull. Am. Math. Soc. (N.S.) **15**(2), 127–187 (1986)

[Forna] Fornaess, J.E.: The Julia set of Henon maps. Math. Ann. **334**(2), 457–464 (2006)

[Fornae] Fornaess, J.E.: Short $\mathbb{C}^k$. In: Complex Analysis in Several Variables, Memorial Conference of Kiyoshi Okas Centennial Birthday, pp. 95–108, Adv. Stud. Pure Math., vol. 42, Math. Soc. Japan, Tokyo (2004)

[Fornas] Fornaess, J.E.: The Levi problem in Stein spaces. Math. Scand. **45**(1), 55–69 (1979)

[Fornaess] Fornaess, J.E. Peters, H.: Complex dynamics with focus on the real parts, arXiv:1310.4673

[Fran] Frankel, S.: Complex geometry of convex domains that cover varieties, Acta Math. **163**(1–2), 109–149 (1989)

[Greene] Greene, R.E., Kim, K.-T.: Stably-interior points and the semicontinuit of the automorphism group. Math. Z. **277**(3–4), 909–916 (2014)

[Gree] Greene, R.E., Krantz, S.G.: The automorphism group of strongly pseudoconvex domains. Math. Ann. **261**, 425–446 (1982)

[Griffi] Griffiths, P.: Complex-analytic properties of certain Zariski open sets on algebraic varieties. Ann. Math. **94**(2), 21–51 (1971)

[Harz] Harz, T., Shcherbina, N., Tomassini, G.: On defining functions for unbounded pseudoconvex domains, arxiv:1405.2250

[Herb] Herbort, G.: Invariant metrics and peak functions on pseudoconvex domains of homogeneous diagonal type. Math. Z. **209**, 223–243 (1992)

[Herbo] Herbort, G.: An example of a pseudoconvex domain whose holomorphic sectional curvature of the Bergman metric is unbounded. Ann. Polon. Math. **92**(1), 29–39 (2007)

[Ishi] Ishi, H., Kai, C.: The representative domain of a homogeneous bounded domain. Kyushu J. Math. **64**(1), 35–47 (2010)

[Kai] Kai, C.: A characterization of symmetric Siegel domains by convexity of Cayley transform images. Tohoku Math. J. (2) **59**(1), 101–118 (2007)

[Kim] Kim, K.-T.: On the automorphism groups of convex domains in $\mathbb{C}^n$. Adv. Geom. **4**(1), 33–40 (2004)

[Yoccoz] Kim, K.-T., Yoccoz, J.-C.: CR manifolds admitting a CR contraction. J. Geom. Anal. **21**(2), 476–493 (2011)

[Kyu] Kim, K.-T., Yu, J.: Boundary behavior of the Bergman curvature in strictly pseudoconvex polyhedral domains. Pac. J. Math. **176**(1), 141–163 (1996)

[Kohn] Kohn, J.J.: Global regularity for $\bar{\partial}$ on weakly pseudo-convex manifolds. Trans. Am. Math. Soc. **181**, 273–292 (See also Math Review of article) (1973)

[McNeal] McNeal, J.: Holomorphic sectional curvature of some pseudoconvex domains. Proc. Am. Math. Soc. **107**(1), 113–117 (1989)

[Pet] Peters, H., Smit, I.M.: Adaptive trains for attracting sequences of holomorphic automorphisms, arxiv:1408.0498

[Pin] Pinchuk, S.: The scaling method and holomorphic mappings. In: Several Complex Variables and Complex Geometry, Part 1 (Santa Cruz, CA, 1989), pp. 151–161, Proc. Sympos. Pure Math. vol. 52, Part 1, Am. Math. Soc., Providence, RI (1991)

[Ros] Rosay, J.-P., Rudin, W.: Holomorphic maps from $\mathbb{C}^n$ to $\mathbb{C}^n$. Trans. Am. Math. Soc. **310**(1), 47–86 (1988)

[Sten] Stensønes, B.: Fatou-Bieberbach domains with C-smooth boundary. Ann. Math. (2) **145**(2), 365–377 (1997)

MIX
Papier aus verantwortungsvollen Quellen
Paper from responsible sources
FSC® C105338

If you have any concerns about our products,
you can contact us on
ProductSafety@springernature.com

In case Publisher is established outside the EU,
the EU authorized representative is:
Springer Nature Customer Service Center GmbH
Europaplatz 3, 69115 Heidelberg, Germany

Printed by Libri Plureos GmbH
in Hamburg, Germany